# Computational Linear and Commutative Algebra

Martin Kreuzer · Lorenzo Robbiano

# Computational Linear and Commutative Algebra

 Springer

Martin Kreuzer
Fakultät für Informatik und Mathematik
Universität Passau
Passau, Germany

Lorenzo Robbiano
Department of Mathematics
University of Genoa
Genoa, Italy

ISBN 978-3-319-82865-7          ISBN 978-3-319-43601-2 (eBook)
DOI 10.1007/978-3-319-43601-2

Mathematics Subject Classification (2010): 13Pxx, 15A27, 15A18, 14Qxx, 13F20

Springer Cham Heidelberg New York Dordrecht London
© Springer International Publishing Switzerland 2016
Softcover reprint of the hardcover 1st edition 2016

Printed on acid-free paper

Springer is part of Springer Science+Business Media (www.springer.com)

*This book is dedicated to the memory of our friend Tony Geramita.*

# Preface

*Books are like imprisoned souls
till someone takes them down from a shelf
and frees them.*
(Samuel Butler)

Dear Readers, a few of you may still remember that eleven years ago, in our previous book, we wrote: *Alas, we have to inform you that this is absolutely and definitively the second and last volume of the trilogy.* Its self-contradictory nature was only meant to be a joke, but imagine how puzzled we were some years later when we received the following somewhat remarkable letter, which appears to have been found tightly corked up in a jug floating on the Mare Tenebrarum.

"Do you know, my dear friends," says the writer, "that not long ago, during my explorations in the wilderness of India, I found an old library. While wandering in its meanders, my attention was drawn to a small manuscript lying upon a high shelf. Its title was 'Computational Linear and Commutative Algebra', and you were its authors! With the utmost care I picked up the manuscript, and with the permission of the librarian, copied it. I append the copy for your perusal."

The date of this letter, we confess, surprised us even more particularly than its contents; for it seemed to have been written in the year 2048, and the writer, a certain E.A.P., remained elusive and enigmatic. Slowly, but clearly, the truth began to shine in front of our eyes: we had to write this third volume. In the following years, we painstakingly transcribed the manuscript and unearthed its secrets. Innumerable struggles, errors, fallacies, and jokes later we are proud to present the result, and hope for your benevolent reception.

While carrying out this demanding job, we were supported and sustained by our wonderful families: our wives Bettina and Gabriella, our children and grandchildren. They cheered us up whenever we were getting too serious about our jokes, and they brought us down to earth whenever our mathematics became too light-headed. Further inspiration was provided by the great achievements of our favorite soccer teams, Bayern München and Juventus Turin.

Special thanks go to many people who helped us in various ways. John Abbott helped us to make the English used in the introductions *commendable* while we take the responsibility for the *derogatory* English in the remaining parts of the book. Anna Bigatti and Elisa Palezzato aided the creation of the CoCoA code for the examples, and Le Ngoc Long supported the proofreading. We are also grateful to them for useful discussions. Last but not least, it our pleasure to thank the Springer team which assisted us in a highly professional and efficient way.

Passau                                                                                       Martin Kreuzer
Genova                                                                                 Lorenzo Robbiano
2016

# Contents

# Introduction

*There has been no exhaustive book on this subject in this country,*
*and [this book] fills a much-needed gap.*
(from an 1895 book review)

## Why Was This Book Written?

Six years ago the authors were invited to give a short series of lectures on applications of Linear Algebra to Polynomial System Solving at the Indian Institute of Science in Bangalore. After being asked to produce lecture notes, we discovered a "much-needed gap" and decided to fill it by writing a short overview article. While working on this endeavour, we experienced numerous eureka moments and discovered traumatic gaps in our presumed knowledge of Linear Algebra. What is the big kernel of an endomorphism? How do you define its eigenspaces if its characteristic polynomial has no zeros, i.e., if there are no eigenvalues? What is the kernel of an ideal?

In the process of answering these natural questions, our manuscript soon exceeded the size acceptable to a journal. So, we planned to go for a new publication form, the *SpringerBriefs* instead. The first tentative title was "**C**omputational **C**ommutative **A**lgebra and **C**omputational **L**inear **A**lgebra," until we realized that its natural acronym had already been trademarked. And while we were worrying for two years about important questions such as whether a vector space is a cyclic module over a commuting family if and only if the dual family is commendable, the *Brief* continued to grow. After breaching the 150-page-mark, we had to abandon all pretence, and admit that we were really working on the third volume of the trilogy.

Fortunately, our troubles with Linear Algebra were soon over, and we considered putting all applications into a last, final chapter. But even the best-laid plans of mice and men go oft awry, which in our case meant that both the discoveries of disconcerting knowledge gaps and the subsequent eureka moments continued unabated. Linear Algebra turned out to have numerous, deep and wonderful applications to the algebra and geometry of zero-dimensional commutative rings, to computing primary and maximal components of polynomial ideals, and finally to solving polynomial systems. Once again, instead of ending in a few short last sections, the book doubled in size.

All told, what have we done? We tried to understand the interconnections between two topics that now make up the title: Computational Linear and Commutative Algebra. We extended a part of Linear Algebra, which had hitherto been primarily used by numerical analysts, to a theory which works for finite dimensional vector spaces over an arbitrary base field. We put many known and unknown algorithms into a general framework for computing with zero-dimensional polynomial ideals and zero-dimensional affine algebras. Did we succeed? Will that much-needed 3 cm gap on your bookshelf be filled forever? You be the Judge!

## What Is This Book About?

*The secret of good writing*
*is to say an old thing in a new way,*
*or to say a new thing in an old way.*
(Richard H. Davis)

Our previous two volumes contained 43 sections in 6 chapters, supported by 349 exercises, 99 tutorials, and 9 appendices. This book contains 45 sections in 6 chapters, supported by 0 exercises, 0 tutorials, and 0 appendices. *What happened?*

As the above brief history of this book indicates already, it has a very different flavour: rather than being the outgrowth of numerous lecture courses, it is an account of our quest to understand some connections between Linear Algebra and Commutative Algebra, a quest which outgrew our original intentions.

Let us have a quick low-voltage ride through the contents, following the current of ideas without being shorted by the surplus of their power. The first three chapters of this book develop what we sometimes, modestly, call "Linear Algebra 1.5" to emphasize its natural connection to the material taught in first year courses, and sometimes, less modestly, the "Linear Algebra of the Third Millennium" to point out its ubiquity, wide applicability, and novel point of view. The initial insight is that one does not need eigenvalues to define the theory of eigenspaces and generalized eigenspaces. In fact, over an arbitrary base field, an endomorphism of a finite dimensional vector space does in general not have eigenvalues. But the role of these numbers can be played by the irreducible factors of the minimal polynomial, for which we therefore coined the new term *eigenfactors*.

Another key concept is the property of an endomorphism to be *commendable*. In spite of this notion having been useful previously for numerical analysts under the derogatory name "non-derogatory endomorphism", its power and beauty seem hitherto to have gone largely unnoticed in algebraic communities. In the last part of the first chapter we aim to change this.

Then, in Chaps. 2 and 3, we extend this theory to commuting families, i.e., to families of endomorphisms which commute pairwise. A commuting family is a finitely generated, commutative algebra over the base field. As such, it has ideals, and one of the main ideas is to form the kernel of an ideal, i.e., the intersection of the kernels of all endomorphisms in the ideal. Then, as a generalization of the well-known result for two commuting matrices, a commuting family has joint eigenspaces, which are the kernels of its maximal ideals, and joint generalized eigenspaces, which are the kernels of the primary components of its zero ideal. Thus we arrive at a deep and intricate connection between the decomposition of a vector space into the direct sum of the joint generalized eigenspaces of a commuting family and the primary decomposition of the zero ideal of the family.

In the third chapter, the central topic is to transfer the concept of commendability to a family. It turns out that this does not imply that any of the endomorphisms in the family has to be commendable, but is strong enough for the main theorem of the chapter: a family is commendable if and only if the vector space is a cyclic module with respect to the dual family. As abstract and ivory-towerly this may seem, it is the heart of some of the most powerful algorithms later on.

Chapters 4, 5, and 6 comprise the Computational Commutative Algebra side of the coin. In the fourth chapter a zero-dimensional affine algebra $R$ over a field $K$ is identified with a commuting family via its multiplication family $\mathscr{F}$. This identification brings the extensive linear algebra preparations to fruition, and surprising connections between the two fields appear: the generalized eigenspaces of $\mathscr{F}$ are the local factors of $R$, the joint eigenvectors of $\mathscr{F}$ are the separators of $R$, there is a commendable endomorphism in $\mathscr{F}$ if and only if $R$ is curvilinear, and the family $\mathscr{F}$ is commendable if and only if $R$ is a Gorenstein ring.

In the fifth chapter we commence the algorithmic applications by gathering together methods for computing the primary and maximal components of a zero-dimensional polynomial ideal. We offer a wide variety of tools, ranging from generically extended linear forms, idempotents, factorization over extension fields, and radical ideal computation to highly specialized instruments for harnessing the power of the Frobenius homomorphism in finite characteristic.

The final chapter is no less ambitious. In our toolbox for solving polynomial systems, we use one-dimensional joint eigenspaces, eigenvalues and eigenvectors, and we even resort to cloning ideals and fields! As long as we are content to determine the $K$-rational solutions of the system, or to deal with systems defined over a finite field $K$, we find reasonably efficient algorithms. But when we try to solve polynomial systems over the rationals, i.e., to describe the solutions in suitable number fields, even the most ingenious exact methods quickly reach their limits.

Altogether, writing this book allowed us to delve into a labyrinth of ideas and techniques combining linear algebra, commutative algebra and computer algebra.

Successfully mastering this material is a *tour de force*, and when you emerge you will be blessed with a quantum leap of knowledge. For us, it also meant achieving full awareness and understanding of our limited ability to comprehend mathematics, life, the universe, and everything. The contents of this book, and a substantial loss of free time, are mainly due to the fact the we did *not* heed the following advice.

> *Write what you know.*
> *That should leave you with a lot of free time.*
> (Howard Nemerov)

## Who Invented This Theory?

> *There is an old saying about those who forget history.*
> *I don't remember it, but it's good.*

Let us not forget the history of this book. A large dose of inspiration came from a few good sources. In spite of us checking dozens of more recent books on Linear Algebra, the single most useful foundation for much of the material in the first chapters was the book by K. Hoffman and R. Kunze which appeared in 1971 (see [9]). The applications of these linear algebra methods to commutative algebra and computer algebra benefited greatly from the wonderful overview article [3] by D. Cox. The original impetus for this book, now underlying some parts of Chap. 6, came from our attempts to transfer the numerical methods in H. Stetter's book [28] to the language of Computational Commutative Algebra.

Besides these main sources, we scoured an immense number of research papers before coming to the generalizations and the new ideas presented here. They are too numerous to be mentioned individually in this introduction, but at the beginning of the relevant sections we cite the most important previous works.

What is the history of studying connections between Linear Algebra and Computational Commutative Algebra? Are there any lessons to be learned from it? After D. Lazard pointed out a relation between eigenvalues and solutions of polynomial systems in [18], not much happened until 1988 when numerical analysts W. Auzinger and H. Stetter constructed a working algorithm to solve polynomial systems via Linear Algebra techniques (see [1]). A few years later H.M. Möller and R. Tenberg contributed some important algorithmic improvements (see [22]).

How did algebraists react to these exciting developments? Very mildly indeed. Besides a useful characterization of commuting matrices by B. Mourrain (see [24]) and the aforementioned overview article by D. Cox, we found little evidence of efforts to bridge the gap between the subjects and to develop a solid algebraic theory. For instance, few authors tried to forego the convenience of an algebraically closed ground field and to work over an arbitrary base field, although this is indispensable for performing symbolic computations.

# What Is This Book *Not* About?

*Judicious omissions create a great _____.*

Just like the previous two volumes, this book is *not* about soccer, chess, gardening, photography, twisty puzzles, and our other favourite pastimes. Even within mathematics, it has gaping holes. For instance, although Linear Algebra is part of its title, you will search in vain for the fields of real and complex numbers. There is a twofold reason for this. Firstly, the Linear Algebra treated here is the theory of finite dimensional vector spaces over an arbitrary field, with the real and complex numbers just being special examples. Secondly, the title of the book also contains the word "computational", and the fields of real or complex numbers are not computable. In our exact world, the real numbers are surreal, and the complex numbers are too complex. For similar reasons you will not find anything about Numerical Linear Algebra here. Such numerical computations involve the set of floating point numbers, and this set is not a field. However, we note that this was the context in which many of the results about commendable matrices and commuting families were first discovered.

Another omission is complexity theory. Since we are in finite dimensional vector spaces, all algorithms have polynomial complexity. But for us what really counts is their *practical complexity*: how long do I have to wait for the output? Unfortunately, for this, even a constant time algorithm can take a lifetime if the constant is too big. The actual running time of the algorithms depends to a large degree on the skill of the implementators which we praise highly, but cannot explain mathematically.

The final gap appears at the end of the last chapter. Our quest to solve polynomial systems using the Computational Linear and Commutative Algebra methods presented here grinds to a halt for *big* polynomial systems over the field of rational numbers. To continue further, we would have to leave the world of exact computation and enter the *world of approximation*; but this world is beyond the reach of the rockets fuelled by the material in this book. The reader who is interested in the approximate realm is invited to look at the books [4, 25], and [28], where many bridges between symbolic and numeric methods are built. Our obsession in this book is to be exact, even if this means that we have to cope with a limited world.

*You're on Earth.*
*There's no cure for that.*
(Samuel Beckett)

## How Was This Book Written?

> *I'm writing a book.*
> *I've got the page numbers done.*
> (Steven Wright)

This book was written in our favourite style, a blend of Italian imagination and German rigour. We filled it with many jokes and quotations to keep you entertained even when the going gets tough. In our opinion, with which we fully agree, an ideal section should contain at least one theorem and one joke. And they should not be the same.

One of the highlights of this book is the inclusion of very many examples. We are convinced that a good grasp of the mathematical ideas we present has to be underpinned by meaningful, non-trivial examples. Almost all of them were computed with CoCoA (see [2]), our favourite computer algebra system. The source files of these examples are freely available on *SpringerLink* as Electronic Supplementary Material (ESM) linked to the corresponding chapters. Even some developments of the theory were only possible because of the inspiration derived from computing appropriate examples with CoCoA. *Thank you, CoCoA!*

One of the main challenges every author faces, besides getting the page numbers done, is to choose the best terminology and notation. Fortunately, many of these difficulties have already been resolved, for better or for worse, in our previous volumes [15] and [16]. So, unless explicitly stated otherwise, we follow the notation and definitions introduced there. There is one minor difference: the ideal generated by a set of elements $\{f_1, \ldots, f_s\}$ is now denoted by $\langle f_1, \ldots, f_s \rangle$, i.e., the parentheses have evolved into angle brackets. Every now and then, we take the liberty to use slightly imprecise notation such as $\mathrm{Ker}(A)$ for the null space of a matrix, or to denote the residue class of an indeterminate $x_i$ in a ring $K[x_1, \ldots, x_n]/I$ by $x_i$ rather than $\bar{x}_i$. We hope that no confusion arises by this occasional lack of German rigour.

Besides the substantial gaps in nomenclature, which we filled generously with our own creations, there is one important point where we decided to ignore the majority and break with tradition. In the literature, an endomorphism of a finite dimensional vector space is called *non-derogatory* if its characteristic and minimal polynomials coincide. This property can be defined using several equivalent conditions, and plays an important role throughout this book.

Frequently, when we lectured about this topic, our listeners would get confused by calling such a good property "non-bad". Would you call a rich person "non-poor"? Would you call an exciting game "non-boring"? Would you call a beautiful picture "non-ugly"? We wouldn't, and hence we introduced the notion of a *commendable* endomorphism. Of course, if an endomorphism is not commendable, it is still derogatory.

Furthermore, we have tried to keep this book as self-contained as possible. Clearly, we require that the reader has purchased and read the previous two volumes, as we do not hesitate to cite from them. Apart from that, we quote a few results from other theories not covered here if their inclusion would have led us too far astray, but in general we believe that the book can be read from cover to cover without missing anything important if the only outside sources are [15] and [16].

Although we tried to write and proofread this book as carefully as we could, we suspect that we overlocked many errors and typos. This suspicion is fuelled by the fact that, whenever we entered a period of intense work on the book, we had to start by changing Sect. 1.1 substantially. Even though one of the early readers of the first volume noted that the book was remarkably error-free and that he found the first misprint after more than 100 pages, we later discovered that in truth Definition 1.1.1, the definition of a ring, contained an error. This proved to us that, no matter how many corrections you do, there are still infinitely many errors left.

> *One man is as good as another*
> *until he has written a book.*
> (Benjamin Jowett)

## And What Is This Book Good for?

> *It is impossible for a man to learn*
> *what he thinks he already knows.*
> (Epictetus, 55–135 A.D.)

Like its two siblings, this book is good for learning, teaching, reading, and most of all, enjoying the topic at hand. Since much of the material presented here is not available elsewhere, the main application of this book is to learn something new. When you want to learn something, it is better not to know everything already. Certainly, we are not young enough to know everything. It is our continuing mission to boldly go where few men have gone before. Come on, join the joyride!

The second application is teaching. We tried to write this book in a way that a curious undergraduate student should not have any serious difficulties understanding it. Thus it can be used as a textbook for advanced undergraduate and for graduate courses in Computational Linear and Commutative Algebra.

The third application is as a research monograph. A couple of months in the laboratory can frequently save a couple of hours in the library. But reading this book for a couple of hours may provide you with research ideas for a couple of years. And if not, there is much pleasure to be gained from useless knowledge.

But are there any *true* applications of this theory? These days, even the purest and most abstract mathematics is in danger of being applied. The technique of solving real world problems by modelling them using polynomial equations has led to unexpected successes and crushing defeats alike. Can the contents of this book be applied to the real world? Even if all is well and good in practice, how does it work in theory?

The polynomial systems considered in this book are all of very modest size. Most algorithms have a running time which is polynomial in the dimension of the underlying vector spaces. In typical real world applications, this dimension will be forbiddingly large. These are the conclusions on which we base our facts, and the upshot is that the road to successful applications is always under construction. On the other hand, how would you classify an algorithm for checking whether a zero-dimensional affine algebra is a Gorenstein ring? Is this *pure* mathematics?

> *Any mathematics is applied mathematics*
> *when it is being applied to something;*
> *when it is not, it is pure mathematics.*
> (Robert D. Richtmyer)

## Some Final Words of Wisdom

> *Mellonta Tauta*
> [Those things that are to be]
> (Sophocles (441 B.C.), Edgar A. Poe (1849), Pundita (2848))

What lies ahead of us? In particular, what comes after a trilogy? A tetralogy? According to Wikipedia, in the Greek theater a tetralogy was a group of three tragedies followed by a satyr play. Including the present book, we have already written three pieces of drama filled with satirical jokes, so we can definitely declare that our job is over. Or is it?

> *There are always so many urgent things,*
> *there's never time for the important ones.*
> (Massimo Caboara)

# Chapter 1
# Endomorphisms

*"Obvious" is the most dangerous word in mathematics.*

(Eric Temple Bell)

The setting of this chapter is a very well-known one: let $K$ be a field, let $V$ be a finite dimensional $K$-vector space, and let $\varphi$ be an endomorphism of $V$, i.e., a $K$-linear map $\varphi : V \longrightarrow V$. Is there anything new or interesting to write about? Clearly, we think so. The main goal of this chapter is to introduce some well-known and some not so well-known topics in Linear Algebra and to furnish the basis for the later chapters in a unified and general style.

An essential foundation stone is a general method of developing the theory of eigenvalues and eigenvectors. Since we are working over an arbitrary base field $K$, the minimal polynomial of $\varphi$ may not split into linear factors. It may even have no linear factor at all. How do we develop a theory of eigenspaces if there are no eigenvalues? As we shall see, the correct new idea is to consider the *eigenfactors* $p_i(z)$ of $\varphi$, i.e., the irreducible factors of the minimal polynomial of $\varphi$, to substitute $\varphi$ into these eigenfactors, and to use the kernels of the resulting endomorphisms $p_i(\varphi)$ as the eigenspaces of $\varphi$. Isn't this merely the obvious generalization of the classical definition? (You should not worry if you have never before encountered the term "eigenfactor": we have just coined it.)

After this idea has been conceived, the further development of the theory proceeds naturally: the generalized eigenspaces are defined as the big kernels of the endomorphisms $p_i(\varphi)$, and the vector space $V$ is the direct sum of the generalized eigenspaces. (If you have not come across big kernels either, you should start reading Sect. 1.1.A.)

There is still one very important notion which will prove beneficial throughout this book, namely the notion of a *commendable* endomorphism. We say that $\varphi$

**Electronic supplementary material** The online version of this chapter (DOI:10.1007/978-3-319-43601-2_1) contains supplementary material, which is available to authorized users.

M. Kreuzer and L. Robbiano, *Computational Linear and Commutative Algebra*, DOI 10.1007/978-3-319-43601-2_1 is commendable if its eigenspaces have the minimal possible dimensions, namely $\deg(p_i(z))$. Equivalently, we may require that the minimal and the characteristic polynomial of $\varphi$ agree, or that $V$ is a cyclic $K[\varphi]$-module. Several further characterizations will be proven in the next chapters. Although the deeper reasons are anything but obvious, commendable endomorphisms are quite ubiquitous. (In the likely event that you have never heard of commendable endomorphisms, please note that this name is yet another of our creations. It generalizes the concept of a non-derogatory endomorphism.)

> *Doesn't matter if the water is cold or warm,*
> *if you're going to have to wade through it anyway.*
> (Teilhard de Chardin)

Overall, this is a warm up chapter. Before plunging into the topic of commuting families of endomorphisms, we want to get you accustomed to this type of Linear Algebra, and we hope that the water is not too cold. Let us also mention a couple of theories that you may have expected to see here, but which are conspicuously absent. We use neither the theory of elementary divisors of the $K[z]$-module structure on $V$ given by $\varphi$, nor the Jordan decomposition of endomorphisms. In both cases the reason is that these theories do not generalize well to families of commuting endomorphisms.

Before we begin in earnest, let us fix some notation that we are going to use throughout the chapter. Unless explicitly specified otherwise, we work over an arbitrary field $K$. We let $V$ be a finitely generated $K$-vector space of dimension $d = \dim_K(V)$, and we let $\mathrm{End}_K(V)$ be the set of all $K$-endomorphisms of $V$, i.e. the set of all $K$-linear maps from $V$ to $V$. The main object of study in the following is a fixed endomorphism $\varphi \in \mathrm{End}_K(V)$. After choosing a $K$-basis of $V$, we can represent $\varphi$ by a matrix. So, if you wish, you can also rewrite everything in the language of matrices. And no, matrices are not what mathematicians sleep on.

## 1.1  Generalized Eigenspaces

> *It is my experience that proofs involving matrices*
> *can be shortened by 50 % if one throws the matrices out.*
> (Emil Artin)

To start our journey into the realm of Linear Algebra, and in particular into the land of endomorphisms, we review and extend the classical notions of eigenspaces and generalized eigenspaces. To keep the proofs short, we concentrate on a single endomorphism $\varphi$ of a finite dimensional vector space $V$ rather than the many matrices representing it.

Our first step is to recall what the big kernel and the small image of $\varphi$ are, namely the kernel and the image of a high enough power of $\varphi$. Why do we need them?

Firstly, the vector space is their direct sum, and secondly, generalized eigenspaces are big kernels. This is the short answer. If you are looking for a longer explanation, including an algorithm for computing the big kernel and the small image from a matrix representing $\varphi$, look at Sect. 1.1.A.

The second step is to recall the minimal polynomial of $\varphi$ and to provide some algorithms for computing it. The irreducible factors of $\mu_\varphi(z)$ generalize the classical notion of an eigenvalue and are one of the central objects of study in this book. We call them the *eigenfactors* of $\varphi$.

The third step is to substitute the endomorphism $\varphi$ into one of its eigenfactors $p(z)$ and to study the kernels of the powers of the resulting map $p(\varphi)$. The kernel of $p(\varphi)$ is the eigenspace of $\varphi$ associated to $p(z)$ and generalizes the usual eigenspace associated to an eigenvalue. Passing to the big kernel of $p(\varphi)$, we get the generalized eigenspaces of $\varphi$ which generalizes the usual concept of a generalized eigenspace.

All generalizations are dangerous, including this one, but, as the last step taken in this section, we succeed in proving the Generalized Eigenspace Decomposition 1.1.17. It provides a direct sum decomposition of $V$ into the big kernels of the maps $p(\varphi)$ and constitutes the origin of numerous later decomposition theorems.

> *The guy who invented the first wheel was an idiot.*
> *The guy who invented the other three, he was a genius.*
> (Sid Caesar)

Notwithstanding some novel nomenclature, the material in this section is well-known and contained in many standard books on Linear Algebra such as [9]. However, it is at the heart of many reinterpretations and extensions in the following sections and chapters which involve true inventive genius.

## 1.1.A Big Kernels and Small Images

> *Experience enables you to recognize an error*
> *when you make it again.*
> (Franklin P. Jones)

One of the first results about an endomorphism $\varphi$ that every students learns is that the dimension of the vector space $V$ is the sum of the dimensions of the kernel and the image of $\varphi$. One of the first errors that many students make is thinking that $V$ is the direct sum of the kernel and the image of $\varphi$. But wait! Let us learn from this error. If we replace $\varphi$ by higher and higher powers, the kernel becomes larger and larger, until it stabilizes, and the image become smaller and smaller, until it stabilizes, too. In this way we get the *big kernel* and the *small image* of $\varphi$, and amazingly, the vector space is their direct sum.

This amusing little observation lies at the core of the most eminent theorem of this section: the Generalized Eigenspace Decomposition Theorem. So, when you make a mistake, never say "oops." Always say, "Ah, interesting." After all this palaver, it is time to start with Definition I of Sect. I of Chap. I.

*I, for one, like Roman numerals.*

In the following we let $K$ be a field, let $V$ be a finite dimensional $K$-vector space, let $d = \dim_K(V)$, and let $\varphi \in \mathrm{End}_K(V)$. The next definition provides names for the kernels and images of high powers of $\varphi$.

**Definition 1.1.1** Let $\mathrm{Ker}(\varphi)$ and $\mathrm{Im}(\varphi)$ denote the kernel and the image of the $K$-linear map $\varphi$, respectively.

(a) The $K$-vector subspace $\mathrm{BigKer}(\varphi) = \bigcup_{i \geq 1} \mathrm{Ker}(\varphi^i)$ of $V$ is called the **big kernel** of $\varphi$.

(b) The $K$-vector subspace $\mathrm{SmIm}(\varphi) = \bigcap_{i \geq 1} \mathrm{Im}(\varphi^i)$ of $V$ is called the **small image** of $\varphi$.

Let us collect some basic properties of the big kernel and the small image of a linear map.

**Proposition 1.1.2 (Fitting's Lemma)**
*Consider the chains of $K$-vector subspaces*

$$\mathrm{Ker}(\varphi) \subseteq \mathrm{Ker}(\varphi^2) \subseteq \cdots \quad and \quad \mathrm{Im}(\varphi) \supseteq \mathrm{Im}(\varphi^2) \supseteq \cdots$$

(a) *There exists a smallest number $m \geq 1$ such that $\mathrm{Ker}(\varphi^m) = \mathrm{Ker}(\varphi^t)$ for all $t \geq m$. It is equal to the smallest number $m \geq 1$ such that $\mathrm{Im}(\varphi^m) = \mathrm{Im}(\varphi^t)$ for all $t \geq m$. In particular, we have strict inclusions*

$$\mathrm{Ker}(\varphi) \subset \mathrm{Ker}(\varphi^2) \subset \cdots \subset \mathrm{Ker}(\varphi^m)$$

(b) *The number $m$ satisfies $\mathrm{BigKer}(\varphi) = \mathrm{Ker}(\varphi^m)$ and $\mathrm{SmIm}(\varphi) = \mathrm{Im}(\varphi^m)$.*
(c) *We have $V = \mathrm{BigKer}(\varphi) \oplus \mathrm{SmIm}(\varphi)$.*

*Proof* To prove (a), we note that the chain $\mathrm{Ker}(\varphi) \subseteq \mathrm{Ker}(\varphi^2) \subseteq \cdots$ has to become stationary, since $V$ is finite dimensional. Suppose we have $\mathrm{Ker}(\varphi^i) = \mathrm{Ker}(\varphi^{i+1})$ for some $i \geq 1$. For every $v \in \mathrm{Ker}(\varphi^{i+2})$ we have $0 = \varphi^{i+2}(v) = \varphi^{i+1}(\varphi(v))$, and therefore $\varphi^{i+1}(v) = \varphi^i(\varphi(v)) = 0$. This shows that $\mathrm{Ker}(\varphi^{i+1}) = \mathrm{Ker}(\varphi^{i+2})$, so that the chain is stationary from $i$ onwards. Hence the first part of (a) holds, and the second part is an immediate consequence of $\dim_K(V) = \dim_K(\mathrm{Ker}(\varphi^i)) + \dim_K(\mathrm{Im}(\varphi^i))$.

Claim (b) follows from (a) and Definition 1.1.1. Finally we prove (c). Since the dimensions of $\mathrm{BigKer}(\varphi)$ and $\mathrm{SmIm}(\varphi)$ add to the dimension of $V$, it suffices to prove $\mathrm{BigKer}(\varphi) \cap \mathrm{SmIm}(\varphi) = \langle 0 \rangle$. Let $m$ be the number defined in (a), and let $v \in \mathrm{Ker}(\varphi^m) \cap \mathrm{Im}(\varphi^m)$. Then there exists a vector $w \in V$ such that $v = \varphi^m(w)$ and $0 = \varphi^m(v)$. But then the equalities $0 = \varphi^m(v) = \varphi^{2m}(w)$ and $\mathrm{Ker}(\varphi^m) = \mathrm{Ker}(\varphi^{2m})$ imply $v = \varphi^m(w) = 0$.                                                                 $\square$

From this proposition we immediately get the following algorithm to compute the big kernel and the small image of an endomorphism.

**Algorithm 1.1.3 (Computing Big Kernels and Small Images)**
*Let $\varphi \in \mathrm{End}_K(V)$.*

(a) *The following instructions define an algorithm which computes* $\mathrm{BigKer}(\varphi)$.

    (1) *Compute* $\mathrm{Ker}(\varphi), \mathrm{Ker}(\varphi^2), \ldots$ *until we have* $\mathrm{Ker}(\varphi^t) = \mathrm{Ker}(\varphi^{t-1})$ *for some* $t \geq 2$.

    (2) *Return* $\mathrm{Ker}(\varphi^t)$ *and stop.*

(a) *The following instructions define an algorithm which computes* $\mathrm{SmIm}(\varphi)$.

    (1) *Compute* $\mathrm{Im}(\varphi), \mathrm{Im}(\varphi^2), \ldots$ *until we have* $\mathrm{Im}(\varphi^t) = \mathrm{Im}(\varphi^{t-1})$ *for some* $t \geq 2$.

    (2) *Return* $\mathrm{Im}(\varphi^t)$ *and stop.*

Let us calculate a big kernel and a small image using this algorithm.

**Example 1.1.4** Let $K = \mathbb{Q}$ and let $\varphi$ be the endomorphism of $V = \mathbb{Q}^8$ represented by the matrix

$$\begin{pmatrix} 0 & 0 & 0 & 0 & 0 & 0 & 0 & 0 \\ 0 & 0 & 0 & 0 & 0 & 0 & 0 & 0 \\ 0 & 0 & 0 & 0 & 1 & 0 & 0 & 1 \\ 1 & 0 & 0 & 0 & 0 & 0 & 0 & 0 \\ 0 & 1 & 0 & 0 & 0 & 0 & 0 & 0 \\ 0 & 0 & 1 & 0 & 0 & 0 & 0 & 0 \\ 0 & 0 & 0 & 1 & 0 & 1 & 0 & 0 \\ 0 & 0 & 0 & 0 & 0 & 0 & 1 & 0 \end{pmatrix}$$

with respect to the canonical basis of $V$. A $K$-basis of $\mathrm{Ker}(\varphi)$ is $(v_1, v_2)$, where $v_1 = (0, 0, 0, 1, 0, -1, 0, 0) = e_4 - e_6$ and $v_2 = (0, 0, 0, 0, 1, 0, 0, -1) = e_5 - e_8$. Then a $K$-basis of $\mathrm{Ker}(\varphi^2)$ is $(v_1, v_2, v_3, v_4)$, where $v_3 = (1, 0, -1, 0, 0, 0, 0, 0) = e_1 - e_3$ and $v_4 = (0, 1, 0, 0, 0, 0, -1, 0) = e_2 - e_7$. Moreover, we have $\mathrm{Ker}(\varphi^2) = \mathrm{Ker}(\varphi^3)$. Thus we conclude that $\mathrm{BigKer}(\varphi) = \mathrm{Ker}(\varphi^2)$. Using Proposition 1.1.2.b we obtain $\mathrm{SmIm}(\varphi) = \mathrm{Im}(\varphi^2)$. It is easy to calculate that $(e_1, e_2, e_4, e_8)$ is a $K$-basis of $\mathrm{Im}(\varphi^2)$.

Given an endomorphism $\varphi$, the following type of subspace plays a useful role in this section.

**Definition 1.1.5** A $K$-vector subspace $U$ of $V$ is called $\varphi$-**invariant** or an **invariant subspace for** $\varphi$ if $\varphi(U) \subseteq U$.

Commuting endomorphisms give rise to the following invariant subspaces.

**Proposition 1.1.6** *Let $\varphi$, $\psi$ be $K$-endomorphisms of $V$ such that $\varphi \circ \psi = \psi \circ \varphi$.*

(a) *The kernel and the image of one endomorphism are invariant subspaces for the other.*

(b) *The big kernel and the small image of one endomorphism are invariant subspaces for the other.*

*Proof* To show (a), we let $v \in \text{Ker}(\psi)$. Then $\psi(\varphi(v)) = \varphi(\psi(v)) = \varphi(0) = 0$ implies $\varphi(v) \in \text{Ker}(\psi)$. Now let $v, w \in V$ be such that $w = \psi(v) \in \text{Im}(\psi)$. Then $\varphi(w) = \varphi(\psi(v)) = \psi(\varphi(v)) \in \text{Im}(\psi)$ yields the claim.

Next we prove (b). By Proposition 1.1.2.a, we have $\text{BigKer}(\psi) = \text{Ker}(\psi^m)$ and $\text{SmIm}(\psi) = \text{Im}(\psi^m)$ for some $m \geq 1$. Since $\psi^m$ commutes with $\varphi^m$, we can apply (a) and obtain the claim.                                                                    □

## 1.1.B Minimal Polynomials and Eigenspaces

> *I was asked to name all the presidents.*
> *I thought they already had names.*
> (Demitri Martin)

In the following we introduce some of the most important new names and definitions of the entire book. The minimal polynomial is a well-known invariant of an endomorphism $\varphi$. The computation of minimal polynomials underlies many of the later algorithms. Therefore we first recall some methods for computing minimal polynomials.

Next we consider polynomials which are even more ubiquitous than the minimal polynomial $\mu_\varphi(z)$ of $\varphi$, namely its monic irreducible factors. Strangely, they did not have their own name before! Borrowing a little from the German language, we named them *eigenfactors*. ("eigen" means own, peculiar, or intrinsic.) Our choice was motivated by the fact that in the special case of a linear eigenfactor $z - \lambda$, the element $\lambda \in K$ is called an *eigenvalue* of $\varphi$. If you view Linear Algebra as the theory of vector spaces over an arbitrary field $K$, as we do, the polynomial $\mu_\varphi(z)$ may not have any zeros in $K$. Hence eigenfactors are the appropriate generalization of eigenvalues.

Given an eigenfactor $p(z)$ of $\varphi$, the next logical step is to define the corresponding eigenspace of $\varphi$ as the kernel of $p(\varphi)$, generalizing the classical definition again. Since it has already a name, we keep the term eigenspace, albeit it is expensive to keep your own space. While in the business of generalizing definitions, we also generalize generalized eigenspaces and define them for any eigenfactor $p(z)$ as the big kernel of $p(\varphi)$. Are you getting this scheming naming scheme? Then you can look forward to the next chapter where we generalize these very generalized eigenspaces even more.

When all is named and done, we proceed to prove the first highlight of this theory. The Generalized Eigenspace Decomposition provides a direct sum decomposition of the vector space into the generalized eigenspaces corresponding to the eigenfactors of $\varphi$. For every eigenfactor $p(z)$, the action of the map $p(\varphi)$ on the various generalized eigenspaces is described in detail. The importance of this theorem cannot be understated. It is related to almost everything in this book except elephants.

*Anything that is unrelated to elephants is irrelephant.*

As before, let $K$ be a field, let $V$ be a finite dimensional $K$-vector space, let $d = \dim_K(V)$, and let $\varphi$ be an endomorphism of $V$. By $\mathrm{End}_K(V)$ we denote the $K$-vector space of all endomorphisms of $V$. Since it is finite-dimensional, the kernel of the substitution homomorphism $\varepsilon : K[z] \longrightarrow K[\varphi]$ given by $z \mapsto \varphi$ is a non-zero polynomial ideal.

**Definition 1.1.7**  The **minimal polynomial** of $\varphi$ is the monic generator of the ideal $\mathrm{Ker}(\varepsilon)$, i.e. the monic polynomial of smallest degree in this ideal. It is denoted by $\mu_\varphi(z)$.

Given a basis of the vector space $V$ and the matrix representing $\varphi$ with respect to this basis, we can compute the minimal polynomial of $\varphi$ as the following algorithm shows. It is a special case of the Buchberger-Möller Algorithm for Matrices (see Algorithm 2.1.4).

---

**Algorithm 1.1.8 (Minimal Polynomials via Reduction)**
*Let $\varphi \in \mathrm{End}_K(V)$, and let $A \in \mathrm{Mat}_d(K)$ be a matrix representing $\varphi$ with respect to a $K$-basis of $V$. Then the following algorithm computes the minimal polynomial $\mu_\varphi(z)$ in $K[z]$.*

(1) *Let $L = \{I_d\}$.*
(2) *For $i = 1, 2, \ldots$, compute the matrix $A^i$ and check whether it is $K$-linearly dependent on the matrices in $L$. If this is not the case, append $A^i$ to $L$ and continue with the next number $i$.*
(3) *If there exist $c_0, \ldots, c_{i-1} \in K$ such that $A^i = c_{i-1}A^{i-1} + \cdots + c_1 A + c_0 I_d$ then return the polynomial $\mu_\varphi(z) = z^i - c_{i-1}z^{i-1} - \cdots - c_0$ and stop.*

---

*Proof* The finiteness of this algorithm follows from the fact that $\mathrm{Mat}_d(K)$ is a finite dimensional $K$-vector space. When this algorithm stops, it returns the monic polynomial in $K[z]$ of smallest degree in the kernel of the substitution homomorphism $z \mapsto A$. Since this polynomial generates that kernel, it is the minimal polynomial of $\varphi$.                                                                           $\square$

Note that the result of this algorithm does not depend on the choice of the basis of $V$ for constructing a matrix representing $\varphi$. If the endomorphism whose minimal polynomial we want to compute is given as a polynomial expression in another endomorphism whose minimal polynomial we already know, we can proceed as follows.

---

**Algorithm 1.1.9 (Minimal Polynomials via Elimination)**
*Let $\varphi \in \mathrm{End}_K(V)$, let $f \in K[z]$, and let $\psi = f(\varphi)$. Given the minimal polynomial of $\varphi$, the following algorithm computes the minimal polynomial of $\psi$.*

(1) *Form the polynomial ring $K[u, z]$ and the ideal $I_\psi = \langle \mu_\varphi(u), z - f(u) \rangle$ in $K[u, z]$.*
(2) *Compute the elimination ideal $J_\psi = I_\psi \cap K[z]$.*
(3) *Return the monic generator of $J_\psi$.*

---

*Proof* By [15], Proposition 3.6.2, this algorithm computes the monic generator of the kernel of the $K$-algebra homomorphism $\varepsilon : K[z] \longrightarrow K[u]/\langle \mu_\varphi(u) \rangle$ which maps $z$ to the residue class of $f(u)$. Since the $K$-algebra on the right-hand side is isomorphic to $K[\varphi]$, the kernel of $\varepsilon$ equals the kernel of the $K$-algebra homomorphism $\varepsilon' : K[z] \longrightarrow \mathrm{End}_K(V)$ defined by $z \mapsto \psi$, and the claim follows.                          $\square$

Notice that in this algorithm we have $\deg(\mu_\psi(z)) \leq \deg(\mu_\varphi(z))$, because $K[\psi]$ is a subalgebra of $K[\varphi]$. The following example shows the preceding two algorithms in action.

**Example 1.1.10** Let $K = \mathbb{Q}$, and let $\varphi$ be the endomorphism of $V = K^4$ defined by the matrix

$$A = \begin{pmatrix} 0 & 0 & 0 & -1 \\ 1 & 0 & 0 & 0 \\ 0 & 1 & 0 & 1 \\ 0 & 0 & 1 & -1 \end{pmatrix}$$

with respect to the canonical basis.

First we use Algorithm 1.1.8 to compute the minimal polynomial of $\varphi$ via reduction. During the first three iterations of the algorithm it turns out that $\{I_4, A, A^2, A^3\}$ is $K$-linearly independent. Then, in the fourth iteration, we discover the relation $A^4 = -A^3 + A^2 - I_4$, and therefore we get $\mu_\varphi(z) = z^4 + z^3 - z^2 + 1$.

Now let us consider the endomorphism $\psi = \varphi^2 - \varphi$ of $V$. To compute its minimal polynomial using Algorithm 1.1.9, we form the ideal $I_\psi = \langle u^4 + u^3 - u^2 + 1, z - (u^2 - u) \rangle$ in $K[u, z]$ and compute the monic generator of $J_\psi = I_\psi \cap K[z]$. The result is $\mu_\psi(z) = z^4 - 4z^3 + z^2 + 5z + 2 = (z - 2)(z^3 - 2z^2 - 3z - 1)$.

Next we consider the endomorphism $\psi = 2\varphi - 3\,\mathrm{id}_V$ of $V$. Its minimal polynomial is $\mu_\psi(z) = z^4 + 14z^3 + 68z^2 + 138z + 115$. Notice that we he equality $\mu_\psi(z) = 2^4 \mu_\varphi(\frac{1}{2}(z + 3))$. It is a special case of a transformation formula which we will prove later on (see Proposition 1.2.4.b).

Next we prove a lemma about the behaviour of the minimal polynomial under restriction to a direct sum of invariant subspaces.

**Lemma 1.1.11** *Let $U', U''$ be $\varphi$-invariant vector subspaces of $V$ such that $V = U' \oplus U''$, and let $\varphi'$ and $\varphi''$ be the restrictions of $\varphi$ to $U'$ and $U''$, respectively. Then we have $\mu_\varphi(z) = \mathrm{lcm}(\mu_{\varphi'}(z), \mu_{\varphi''}(z))$.*

*Proof* Since the polynomial $\mu_\varphi(z)$ vanishes when we substitute $z \mapsto \varphi$, it also vanishes when we substitute $z \mapsto \varphi'$ or $z \mapsto \varphi''$. Therefore it is a multiple of both $\mu_{\varphi'}(z)$ and $\mu_{\varphi''}(z)$, i.e. the minimal polynomial $\mu_\varphi(z)$ is a multiple of $\mathrm{lcm}(\mu_{\varphi'}(z), \mu_{\varphi''}(z))$.

Now let $f(z) = \mathrm{lcm}(\mu_{\varphi'}(z), \mu_{\varphi''}(z)) \in K[z]$. If we restrict $f(\varphi)$ to $U'$, we get $f(\varphi') = 0$. Similarly, we have $f(\varphi'') = 0$. Since we have $V = U' \oplus U''$, it follows that $f(\varphi) = 0$, i.e. the polynomial $f(z)$ is a multiple of $\mu_\varphi(z)$. $\qquad\square$

The following easy example shows that the behaviour of the minimal polynomial in the setting of this lemma differs from the behaviour of the characteristic polynomial (see Sect. 1.2) for which we have $\chi_\varphi(z) = \chi_{\varphi'}(z) \cdot \chi_{\varphi''}(z)$.

**Example 1.1.12** For $V = K^d$, we have $V = K e_1 \oplus \cdots \oplus K e_d$ for the canonical basis $(e_1, \ldots, e_d)$. Clearly, each direct summand $K e_i$ is invariant for the identity endomorphism $\varphi = \mathrm{id}_V$. For the restrictions $\varphi_i = \varphi|_{K e_i}$ we have $\mu_{\varphi_i}(z) = z - 1$. In agreement with the lemma, we have $\mu_\varphi(z) = z - 1 = \mathrm{lcm}(\mu_{\varphi_1}(z), \ldots, \mu_{\varphi_d}(z))$. On the other hand, the characteristic polynomial of $\varphi$ (see Definition 1.2.1) satisfies $\chi_\varphi(z) = (z - 1)^d = \prod_{i=1}^{d} \chi_{\varphi_i}(z)$.

The irreducible factors of the minimal polynomial play a central role in this book. Therefore they deserve a name.

**Definition 1.1.13** Let $\mu_\varphi(z) = p_1(z)^{m_1} \cdots p_s(z)^{m_s}$ be the decomposition of the minimal polynomial of $\varphi$ into pairwise distinct, monic, irreducible factors, where $m_1, \ldots, m_s$ are positive integers.

(a) The polynomials $p_1(z), \ldots, p_s(z) \in K[z]$ are called the **eigenfactors** of $\varphi$.
(b) If an eigenfactor $p_i(z)$ of $\varphi$ is of the form $p_i(z) = z - \lambda$ with $\lambda \in K$ then $\lambda$ is called an **eigenvalue** of $\varphi$.

Notice that the eigenvalues of $\varphi$ can also be described as the zeros of $\mu_\varphi(z)$ in $K$. If we fix a $K$-basis of $V$ and denote the matrix representing $\varphi$ with respect to this basis by $A$, we define eigenfactors of the matrix $A$ analogously.

Clearly, the concept of an eigenfactor of an endomorphism generalizes the concept of an eigenvalue. We observe that if $p(z)$ is an eigenfactor of $\varphi$, then $p(\varphi)$ either is zero or divides $\mu_\varphi(\varphi) = 0$. Hence it is not invertible, and so its kernel contains non-zero vectors. Consequently, we may define more general versions of the notions of eigenspaces and generalized eigenspaces as follows.

**Definition 1.1.14**  Let $\varphi \in \mathrm{End}_K(V)$, and let $\mu_\varphi(z) = p_1(z)^{m_1} \cdots p_s(z)^{m_s}$, where $p_1(z), \ldots, p_s(z) \in K[z]$ are the eigenfactors of $\varphi$.

(a) For $i = 1, \ldots, s$, the $K$-vector subspace $\mathrm{Eig}(\varphi, p_i(z)) = \mathrm{Ker}(p_i(\varphi))$ of $V$ is called the **eigenspace** of $\varphi$ associated to $p_i(z)$. The non-zero elements of $\mathrm{Eig}(\varphi, p_i(z))$ are called $p_i(z)$-**eigenvectors**, or simply **eigenvectors** of $\varphi$.

(b) If $p_i(z)$ is of the form $p_i(z) = z - \lambda_i$ with $\lambda_i \in K$, we also write $\mathrm{Eig}(\varphi, \lambda_i)$ instead of $\mathrm{Eig}(\varphi, z - \lambda_i)$ and call it the **eigenspace** of $\varphi$ associated to the eigenvalue $\lambda_i$. Its non-zero elements are also called $\lambda_i$-**eigenvectors** of $\varphi$.

(c) For $i = 1, \ldots, s$, the $K$-vector subspace $\mathrm{Gen}(\varphi, p_i(z)) = \mathrm{BigKer}(p_i(\varphi))$ is called the **generalized eigenspace** of $\varphi$ associated to $p_i(z)$.

(d) If $p_i(z)$ is of the form $p_i(z) = z - \lambda_i$ with $\lambda_i \in K$, we also write $\mathrm{Gen}(\varphi, \lambda_i)$ instead of $\mathrm{Gen}(\varphi, z - \lambda_i)$ and call it the **generalized eigenspace** of $\varphi$ associated to $\lambda_i$.

If we fix a $K$-basis of $V$ and denote the matrix representing $\varphi$ with respect to this basis by $A$, we define the eigenspaces $\mathrm{Eig}(A, p_i(z))$ and the generalized eigenspaces $\mathrm{Gen}(A, p_i(z))$ of $A$ analogously. For the zero map $\varphi = 0$, we note that we have only one eigenfactor $z$ and that $\mathrm{Eig}(\varphi, p_1(z)) = \mathrm{Gen}(\varphi, p_1(z)) = V$.

**Remark 1.1.15**  The indeterminate $z$ is an eigenfactor of $\varphi$ if and only if we have $\mathrm{Ker}(\varphi) = \mathrm{Eig}(\varphi, z) \neq \{0\}$. This is equivalent to $\varphi$ being not invertible.

Notice that, if $\varphi$ is invertible, then we have $\varphi^{-1} \in K[\varphi]$. To prove this, we write $\mu_\varphi(z) = z^m + c_1 z^{m-1} + \cdots + c_m$ with $c_1, \ldots, c_m \in K$ and observe that $c_m \neq 0$ implies $\varphi^{-1} = -\frac{1}{c_m}(\varphi^{m-1} + c_1 \varphi^{m-2} + \cdots + c_{m-1})$.

The following lemma is the key to proving the main theorem of this section. It says that coprime factors of $\mu_\varphi(z)$ lead to a direct sum decomposition of $V$.

**Lemma 1.1.16**  *Let $p(z) \in K[z]$ be a non-zero multiple of $\mu_\varphi(z)$. Suppose that we have a decomposition $p(z) = q_1(z)q_2(z)$ with coprime polynomials $q_1(z), q_2(z)$ in $K[z]$. Then the two $K$-vector subspaces $W_1 = \mathrm{Ker}(q_1(\varphi))$ and $W_2 = \mathrm{Ker}(q_2(\varphi))$ have the following properties.*

(a) *For every $f(z) \in K[z]$, both $W_1$ and $W_2$ are $f(\varphi)$-invariant.*

(b) *The restriction maps $q_i(\varphi)|_{W_j} : W_j \to W_j$ are isomorphisms for $i, j \in \{1, 2\}$ with $i \neq j$.*

(c) *We have the direct sum decomposition $V = W_1 \oplus W_2$.*

(d) *Suppose now that $p(z) = \mu_\varphi(z)$ is the minimal polynomial of $\varphi$. Then the minimal polynomial of the restriction of $\varphi$ to $W_i$ is $q_i(z)$ for $i = 1, 2$.*

*Proof*  First we prove (a). Clearly, the map $f(\varphi)$ commutes with both $q_1(\varphi)$ and $q_2(\varphi)$. Hence Proposition 1.1.6.a shows that $W_1$ and $W_2$ are $f(\varphi)$-invariant. For the proof of (b) we use the coprimality of $q_1(z)$ and $q_2(z)$ to find two polynomials $f_1(z), f_2(z)$ in $K[z]$ such that $f_1(z)q_1(z) + f_2(z)q_2(z) = 1$. Hence we have

$f_1(\varphi)q_1(\varphi) + f_2(\varphi)q_2(\varphi) = \mathrm{id}_V$. For every vector $v \in V$, this yields

$$f_1(\varphi)q_1(\varphi)(v) + f_2(\varphi)q_2(\varphi)(v) = v$$

In particular, this formula shows that $W_1 \cap W_2 = \langle 0 \rangle$. Hence the restriction maps $q_i(\varphi)_{|W_j} : W_j \to W_j$ are injective, and therefore also surjective.

To prove (c), we write $f_1(z)q_1(z) + f_2(z)q_2(z) = 1$ as above, let $v \in V$, and set $w_1 = f_2(\varphi)q_2(\varphi)(v)$ and $w_2 = f_1(\varphi)q_1(\varphi)(v)$. Using the equality $q_1(\varphi)(w_1) = f_2(\varphi)p(v) = 0$, we conclude that $w_1 \in W_1$. Similarly, the equality $q_2(\varphi)(w_2) = f_1(\varphi)p(v) = 0$ implies $w_2 \in W_2$. By combining these conclusions and the above formula, we see that $V = W_1 + W_2$. This sum is direct since we have shown before that $W_1 \cap W_2 = \langle 0 \rangle$.

Finally we show (d). For $i = 1, 2$, let $q'_i(z)$ be the minimal polynomial of the restriction of $\varphi$ to $W_i$. By the definition of $W_i$, the restriction of $q_i(\varphi)$ to $W_i$ is zero. Hence we see that $q'_i(z)$ divides $q_i(z)$. Consequently, the hypothesis shows that also $q'_1(z)$ and $q'_2(z)$ are coprime. From (c) and Lemma 1.1.11 we deduce that $q_1(z)q_2(z) = \mu_\varphi(z) = q'_1(z)q'_2(z)$, and thus $q'_i(z) = q_i(z)$ for $i = 1, 2$, as was to be shown. $\qquad\square$

Using a straightforward induction argument, this lemma implies the following theorem.

**Theorem 1.1.17 (Generalized Eigenspace Decomposition)**
*Let $K$ be a field, let $V$ be a finite dimensional $K$-vector space, let $\varphi \in \mathrm{End}_K(V)$, and let $\mu_\varphi(z) = p_1(z)^{m_1} \cdots p_s(z)^{m_s}$, where $p_1(z), \ldots, p_s(z) \in K[z]$ are the eigenfactors of $\varphi$.*

(a) *The vector space $V$ is the direct sum of the generalized eigenspaces of $\varphi$, i.e. we have*

$$V = \mathrm{Gen}(\varphi, p_1(z)) \oplus \cdots \oplus \mathrm{Gen}(\varphi, p_s(z))$$

*and the generalized eigenspaces are $\varphi$-invariant.*
(b) *For $i = 1, \ldots, s$, the restriction of the map $p_i(\varphi)$ to $\mathrm{Gen}(\varphi, p_i(z))$ is nilpotent and its index of nilpotency is $m_i$.*
(c) *For $i, j \in \{1, \ldots, s\}$ with $i \neq j$, the restriction of the map $p_i(\varphi)$ to $\mathrm{Gen}(\varphi, p_j(z))$ is an isomorphism.*
(d) *For $i = 1, \ldots, s$, we have*

$$\mathrm{Gen}(\varphi, p_i(z)) = \mathrm{BigKer}(p_i(\varphi)) = \mathrm{Ker}(p_i(\varphi)^{m_i})$$

*In particular, if $m_i = 1$ then $\mathrm{Gen}(\varphi, p_i(z)) = \mathrm{Ker}(p_i(\varphi)) = \mathrm{Eig}(\varphi, p_i(z))$.*

*Proof* For $i = 1, \ldots, s$, we put $G_i = \mathrm{Ker}(\varphi, p_i(z)^{m_i})$.

To prove (a), we proceed by induction on $s$. The induction step follows by applying parts (a) and (c) of the lemma with $q_1(z) = p_1(z)^{m_1} \cdots p_{s-1}(z)^{m_{s-1}}$ and $q_2(z) = p_s(z)^{m_s}$. We get $V = G_1 \oplus \cdots \oplus G_s$. Using part (d) of the lemma at each step of the induction shows that the restriction of $\varphi$ to $G_i$ has the minimal polynomial $p_i(z)^{m_i}$. This yields $\text{Gen}(\varphi, p_i(z)) = G_i$ for $i = 1, \ldots, s$ and proves claims (a), (b), and (d). Using part (b) of the lemma at each step of the induction proves claim (c).                                                    □

In the literature, this theorem is also known as the **Primary Decomposition Theorem**, and the generalized eigenspaces $\text{Gen}(\varphi, p_i(z))$ are sometimes called the $\varphi$-**primary components** of $V$.

Let us have a look at an example.

**Example 1.1.18**  Let $K = \mathbb{Q}$, let $V = K^6$, and let $\varphi \in \text{End}_K(V)$ be defined by

$$A = \begin{pmatrix} -12 & -16 & 3 & -4 & -8 & -20 \\ 78 & 104 & 18 & 26 & 52 & 130 \\ 0 & 0 & 0 & 0 & 0 & 0 \\ -237 & -316 & -72 & -79 & -158 & -395 \\ 78 & 104 & 18 & 26 & 52 & 130 \\ 36 & 48 & -9 & 12 & 24 & 60 \end{pmatrix}$$

with respect to the canonical basis. Then we have $\mu_\varphi(z) = z^3 - 125z^2 = (z - 125)z^2$. We have $\text{Ker}(\varphi - 125\,\text{id}_V) = \langle v_1 \rangle$, where $v_1 = (4, -26, 0, 79, -26, -12)$. A $K$-basis of $\text{Ker}(\varphi^2)$ is $(v_2, v_3, v_4, v_5, v_6)$, where

$$v_2 = (4, -3, 0, 0, 0, 0), \qquad v_3 = (1, 0, 0, -3, 0, 0), \qquad v_4 = e_3$$
$$v_5 = (2, 0, 0, 0, -3, 0), \qquad v_6 = (5, 0, 0, 0, 0, -3)$$

According to the theorem, the tuple $(v_1, v_2, v_3, v_4, v_5, v_6)$ is a $K$-basis of $V$.

## 1.2  Minimal and Characteristic Polynomials

*I'd like to buy a new boomerang,*
*but I don't know how to throw the old one away.*

In the study of an endomorphism $\varphi$ of a finite dimensional vector space $V$, the characteristic polynomial $\chi_\varphi(z)$ is a classical tool. Eigenvalues are traditionally identified as its zeros, and the ubiquitous Cayley-Hamilton Theorem provides a useful property of $\chi_\varphi(z)$. While working on this book, we realized more and more that the minimal polynomial $\mu_\varphi(z)$ of $\varphi$ is an even more useful tool for our purposes, since it generalizes well to commuting families of endomorphisms. However, it is not so easy to throw the old tool away, and the present section is witness to this phenomenon.

The main observation is that the minimal and the characteristic polynomial of $\varphi$ have much in common. They are both invariant under extensions of the base field, they transform similarly when one replaces $\varphi$ by a linear combination $a\varphi + b\,\text{id}_V$, and they have the same irreducible factors. The latter property is the content of the main theorem of this section. Together with the Cayley-Hamilton Theorem which says that $\chi_\varphi(z)$ is a multiple of $\mu_\varphi(z)$, it identifies the main difference in information content of the two polynomials: the multiplicities of the eigenfactors in $\mu_\varphi(z)$ are in general lower.

In spite of these similarities, we believe that the minimal polynomial is more characteristic for the endomorphism than the characteristic polynomial. For instance, the $K$-algebra generated by $\varphi$ satisfies $K[\varphi] \cong K[z]/\langle\mu_\varphi(z)\rangle$, and this ring contains the full information about the powers of $\varphi$. Don't you agree?

*If I agreed with you we'd both be wrong.*

Let $K$ be a field, let $V$ be a $d$-dimensional $K$-vector space, where $d > 0$, let $B$ be a $K$-basis of $V$, and let $\varphi \in \mathrm{End}_K(V)$. We denote the identity matrix of size $d$ by $I_d$ and let $M_B(\varphi) \in \mathrm{Mat}_d(K)$ be the matrix representing $\varphi$ with respect to the basis $B$. Recall that the polynomial $\det(zI_d - M_B(\varphi)) \in K[z]$ is independent of the choice of the basis $B$. It is denoted by $\det(z\,\mathrm{id}_V - \varphi)$.

**Definition 1.2.1**   The polynomial $\chi_\varphi(z) = \det(z\,\mathrm{id}_V - \varphi)$ is called the **characteristic polynomial** of $\varphi$.

Clearly, the degree of the characteristic polynomial is $d$.

**Remark 1.2.2**   Notice that the constant coefficient of $\chi_\varphi(z)$ is $\pm\det(\varphi)$. Therefore $z$ is a factor of $\chi_\varphi(z)$ if and only if $\varphi$ is not invertible.

In the next proposition we examine the behaviour of the minimal and the characteristic polynomial of an endomorphism with respect to field extensions.

**Proposition 1.2.3 (Extending Minimal and Characteristic Polynomials)**
*Let $L \supseteq K$ be a field extension, let $V_L = V \otimes_K L$, and let the $L$-vector space homomorphism $\varphi_L = \varphi \otimes_K L : V_L \longrightarrow V_L$ be the extension of $\varphi$.*

(a) *In $L[z]$ we have $\chi_{\varphi_L}(z) = \chi_\varphi(z)$.*
(b) *In $L[z]$ we have $\mu_{\varphi_L}(z) = \mu_\varphi(z)$.*

*Proof* Both claims follow from the fact that if $A \in \mathrm{Mat}_r(K)$ is a matrix representing $\varphi$ then $A$ also represents $\varphi_L$. On one hand, this implies the equality $\chi_\varphi(z) = \det(zI_r - A) = \chi_{\varphi_L}(z)$. On the other hand, the polynomial $\mu_\varphi(z)$ is the monic polynomial of smallest degree in $K[z]$ such that $\mu_\varphi(A) = 0$. Let $m = \deg(\mu_\varphi(z))$ and let $a_0, \ldots, a_{m-1} \in K$ be such that $\mu_\varphi(z) = a_0 + a_1 z + \cdots + a_{m-1}z^{m-1} + z^m$. Then the condition $a_0 I_d + a_1 A + \cdots + a_{m-1}A^{m-1} + A^m = 0$ means that $(a_0, \ldots, a_{m-1}) \in K^m$ is a solution of a certain linear system. Since the coefficients of this linear system are in $K$, it has a solution in $L^m$ if and only if it has one in $K^m$. Therefore we see that $\mu_\varphi(z)$ is also the monic polynomial of smallest degree in the kernel of the substitution homomorphism $\varepsilon_L : L[z] \longrightarrow \mathrm{Mat}_d(L)$ given by $z \mapsto A$, i.e. the minimal polynomial of $\varphi_L$.                                                                        $\square$

The characteristic polynomial, the minimal polynomial, and the eigenfactors of a linear function of an endomorphism can be computed as follows.

**Proposition 1.2.4 (Transforming Minimal and Characteristic Polynomials)**
*Let $\varphi \in \mathrm{End}_K(V)$, let $a, b \in K$ with $a \neq 0$, and let $\psi = a\varphi + b\,\mathrm{id}_V$.*

(a) *We have $\chi_\psi(z) = a^d \chi_\varphi(\frac{1}{a}(z - b))$ where $d = \dim_K(V)$.*
(b) *We have $\mu_\psi(z) = a^{\deg(\mu_\varphi(z))} \mu_\varphi(\frac{1}{a}(z - b))$.*
(c) *The polynomials $\chi_\varphi(z)$ and $\chi_\psi(z)$ (respectively, the polynomials $\mu_\varphi(z)$ and $\mu_\psi(z)$) have the same number of irreducible factors and corresponding irreducible factors have the same multiplicities.*
(d) *The eigenfactors of $\psi$ are $a^{\deg(p_1(z))} p_1(\frac{1}{a}(z - b)), \ldots, a^{\deg(p_s(z))} p_s(\frac{1}{a}(z - b))$.*

*Proof* To prove (a) it suffices to calculate

$$a^d \chi_\varphi\left(\tfrac{1}{a}(z - b)\right) = a^d \det\left(\tfrac{1}{a}(z - b)\,\mathrm{id}_V - \varphi\right) = a^d \det\left(\tfrac{1}{a}z\,\mathrm{id}_V - \left(\varphi + \tfrac{b}{a}\,\mathrm{id}_V\right)\right)$$
$$= \det\left(z\,\mathrm{id}_V - (a\varphi + b)\right) = \det(z\,\mathrm{id}_V - \psi) = \chi_\psi(z)$$

To prove (b) we apply Algorithm 1.1.9. To compute the minimal polynomial of $\psi$, we form the ideal $I_\psi = \langle \mu_\varphi(u), z - (au + b)\rangle$ in the polynomial ring $K[u, z]$. Clearly, we have $I_\psi = \langle \mu_\varphi(\frac{1}{a}(z - b)), u - \frac{1}{a}(z - b)\rangle$. Therefore we get the elimination ideal $I_\psi \cap K[z] = \langle \mu_\varphi(\frac{1}{a}(z - b))\rangle$. The monic generator of this ideal is the polynomial $a^{\deg(\mu_\varphi(z))} \mu_\varphi(\frac{1}{a}(z - b))$, and this is the minimal polynomial of $\psi$ by the algorithm.

Claim (c) follows immediately from (a) and (b). To prove (d), we consider the factorization $\mu_\varphi(z) = p_1(z)^{m_1} \cdots p_s(z)^{m_s}$, where $p_1(z), \ldots, p_s(z)$ are the eigenfactors of $\varphi$. From (b) we get $\mu_\psi(z) = a^{\deg(p_1(z))} p_1(\frac{1}{a}(z-b))^{m_1} \cdots a^{\deg(p_s(z))} p_s(\frac{1}{a}(z-b))^{m_s}$. Each polynomial $a^{\deg(p_i(z))} p_i(\frac{1}{a}(z-b))^{m_i}$ is monic. It is also irreducible, since a decomposition of it would determine a decomposition of $p_i(z)$ by substituting $z$ with $az + b$. By combining these observations, we obtain the claim.                  $\square$

The next result is a classical tool in Linear Algebra. To prove it, we use the notion of the adjugate of a square matrix which is defined as follows. Given a ring $R$ and a matrix $A \in \mathrm{Mat}_d(R)$, we let $A_{ij}$ be the matrix obtained from $A$ by deleting the $i$-th row and the $j$-th column. Furthermore we let $c_{ij} = \det(A_{ij})$ and call $\mathrm{adj}(A) = ((-1)^{i+j} c_{ij})^{\mathrm{tr}}$ the **adjugate matrix** (or the **classical adjoint matrix**) of $A$. Its main property is expressed by $\mathrm{adj}(A) \cdot A = A \cdot \mathrm{adj}(A) = \det(A) \cdot I_d$.

**Theorem 1.2.5 (Cayley-Hamilton)**
*Let $\varphi : V \longrightarrow V$ be a $K$-linear map. Then the minimal polynomial $\mu_\varphi(z)$ of $\varphi$ is a divisor of its characteristic polynomial $\chi_\varphi(z)$.*

*Proof* Let $A \in \mathrm{Mat}_d(K)$ be a matrix which represents $\varphi$, let $\chi_\varphi(z) = \sum_{i=0}^d c_i z^i$ with $c_i \in K$, and let $B = \mathrm{adj}(z I_d - A) \in \mathrm{Mat}_d(K[z])$. Notice that the entries of $B$ are polynomials in $K[z]$ of degree at most $d - 1$. By expanding $B$, we obtain a representation $B = \sum_{i=0}^{d-1} B_i z^i$ with $B_i \in \mathrm{Mat}_d(K)$. If in $\chi_\varphi(z) I_d = \det(z I_d - A) I_d = (z I_d - A) B$ we substitute $\chi_\varphi(z)$ with $\sum_{i=0}^d c_i z^i$ and $B$ with $\sum_{i=0}^{d-1} B_i z^i$, we get

$$\sum_{i=0}^d c_i I_d z^i = \sum_{i=0}^{d-1} B_i z^{i+1} - \sum_{i=0}^{d-1} A B_i z^i = -A B_0 + \sum_{i=1}^{d-1} (B_{i-1} - A B_i) z^i + B_{d-1} z^d .$$

By comparing coefficients, we have the equalities $c_0 I_d = -A B_0$ and $I_d = B_{d-1}$ as well as $c_i I_d = B_{i-1} - A B_i$ for $i = 1, \ldots, d - 1$. By multiplying the latter equality by $A^i$ on the left, we get $c_i A^i = A^i B_{i-1} - A^{i+1} B_i$. Summing up everything, we obtain

$$\sum_{i=0}^d c_i A^i = -A B_0 + \left( A B_0 - A^2 B_1 \right) + \cdots + \left( A^{d-1} B_{d-2} - A^d B_{d-1} \right) + A^d B_{d-1} = 0$$

as was to be shown. $\qquad\square$

This theorem implies that the eigenfactors of $\varphi$ are also irreducible factors of $\chi_\varphi(z)$. In fact, the converse is true, too. In order to prove this we need a further ingredient.

**Proposition 1.2.6** *Let $p(z) \in K[z] \setminus K$. The following conditions are equivalent.*

(a) *The endomorphism $p(\varphi)$ has a non-trivial kernel.*
(b) *We have $\gcd(p(z), \chi_\varphi(z)) \neq 1$.*
(c) *We have $\gcd(p(z), \mu_\varphi(z)) \neq 1$.*

*Proof* The proof of (c) $\Rightarrow$ (b) follows immediately from Theorem 1.2.5.
  To prove the implication (b) $\Rightarrow$ (a), we let $g(z) = \gcd(p(z), \chi_\varphi(z))$. Since $g(z)$ divides $p(z)$, it suffices to prove that $g(\varphi)$ has a non-trivial kernel. Hence it suffices to prove the claim in the case $p(z) = g(z)$, i.e., in the case that $p(z)$ is a non-trivial factor of $\chi_\varphi(z)$. Let $A \in \mathrm{Mat}_d(K)$ be a matrix which represents $\varphi$. Then $p(\varphi)$ has a non-trivial kernel if and only if $\det(p(A)) = 0$. For a field extension $L \supseteq K$, the extended map $\varphi_L : V \otimes_K L \longrightarrow V \otimes_K L$ satisfies $\mathrm{Ker}(p(\varphi_L)) \neq \langle 0 \rangle$ if and only if $\mathrm{Ker}(p(\varphi)) \neq \langle 0 \rangle$. Therefore we may assume that $p(z)$ has a zero $\lambda \in K$. Then $\lambda$ is a zero of the polynomial $\chi_\varphi(z) = \det(z I_d - A)$, and hence $\mathrm{Ker}(\lambda I_d - A) \neq \langle 0 \rangle$. Now we write $p(z) = (z - \lambda) q(z)$ with $q(z) \in K[z]$ and observe that the claim follows from the inclusions $\langle 0 \rangle \subset \mathrm{Ker}(\lambda \, \mathrm{id}_V - \varphi) \subseteq \mathrm{Ker}(p(\varphi))$.

To conclude the proof, we show that (a) implies (c). Let $h(z) = \gcd(p(z), \mu_\varphi(z))$. Then we have $h(z) \mid p(z)$, and hence $\mathrm{Ker}(p(\varphi)) \subseteq \mathrm{Ker}(h(\varphi))$. By assumption we have $\mathrm{Ker}(p(\varphi)) \neq \langle 0 \rangle$. Thus we get $\mathrm{Ker}(h(\varphi)) \neq \langle 0 \rangle$, and therefore $h(z) \neq 1$.    □

**Remark 1.2.7** In the proof of (a) $\Rightarrow$ (c) of this proposition we have seen that $\mathrm{Ker}(p(\varphi)) \subseteq \mathrm{Ker}(h(\varphi))$. On the other hand, there exist two polynomials $a(z), b(z)$ such that $h(z) = a(z)p(z) + b(z)\mu_\varphi(z)$. Consequently, we have $h(\varphi) = a(\varphi)p(\varphi)$, and hence $\mathrm{Ker}(h(\varphi)) \subseteq \mathrm{Ker}(p(\varphi))$. Therefore we actually have the equality of kernels $\mathrm{Ker}(p(\varphi)) = \mathrm{Ker}(h(\varphi))$.

Now we are ready to state and prove the desired theorem about the irreducible factors of the minimal and characteristic polynomials of an endomorphism.

**Theorem 1.2.8 (Prime Factors of Minimal and Characteristic Polynomials)**
*Let $\varphi : V \longrightarrow V$ be a $K$-linear map, and let $\mu_\varphi(z) = p_1(z)^{m_1} \cdots p_s(z)^{m_s}$, where $m_i > 0$ and the polynomials $p_i(z)$ are the eigenfactors of $\varphi$. For $i = 1, \ldots, s$, we let $W_i = \mathrm{Gen}(\varphi, p_i(z)) = \mathrm{Ker}(p_i(\varphi)^{m_i})$.*

(a) *For every $i \in \{1, \ldots, s\}$, there exist an integer $a_i \geq m_i$ such that the characteristic polynomial of the restriction of $\varphi$ to $W_i$ satisfies $\chi_{\varphi|_{W_i}}(z) = p_i(z)^{a_i}$.*
(b) *We have $\chi_\varphi(z) = \chi_{\varphi|_{W_1}}(z) \cdots \chi_{\varphi|_{W_s}}(z) = p_1(z)^{a_1} \cdots p_s(z)^{a_s}$.*
(c) *The polynomials $\mu_\varphi(z)$ and $\chi_\varphi(z)$ have the same irreducible factors, and hence the same squarefree part.*

*Proof* Let us prove claim (a). By Theorem 1.1.17.b, the polynomial $p_i(z)^{m_i}$ is the minimal polynomial of $\varphi|_{W_i}$. By the Cayley-Hamilton Theorem 1.2.5, it divides the characteristic polynomial of $\varphi|_{W_i}$ which is therefore of the form $p_i(z)^{a_i} q(z)$ with $a_i \geq m_i$ and a polynomial $q(z) \in K[z]$ such that $p_i(z)$ and $q(z)$ are coprime. We want to show that $q(z)$ is constant.

For a contradiction, assume that $\deg(q(z)) > 0$. There exist $f(z), g(z) \in K[z]$ such that $1 = f(z)p_i(z)^{m_i} + g(z)q(z)$. Since the polynomial $q(z)$ divides the characteristic polynomial of $\varphi|_{W_i}$, we can apply the preceding proposition and find a non-zero vector $v \in W_i$ such that we have $q(\varphi)(v) = 0$. Therefore we obtain $v = f(\varphi)p_i(\varphi)^{m_i}(v) + g(\varphi)q(\varphi)(v) = 0$, a contradiction.

Next we show (b). By Theorem 1.1.17.a, the vector subspaces $W_i$ are $\varphi$-invariant and $V$ is their direct sum. Therefore, if we choose bases for the vector subspaces $W_i$ and combine them to a basis of $V$, the matrix representing $\varphi$ with respect to that basis will be block diagonal. This yields the first equality in (b), and the second one follows from (a). Clearly, claim (c) follows from (b).    □

**Remark 1.2.9** By Remark 1.2.2 and the theorem, the indeterminate $z$ is a factor of $\mu_\varphi(z)$ if and only if $\varphi$ is not invertible.

The following continuation of Example 1.1.4 illustrates the theorem.

**Example 1.2.10** In the setting of Example 1.1.4, we have $\mu_\varphi(z) = z^6 - z^2 = (z+1)(z-1)(z^2+1)z^2$, while $\chi_\varphi(z) = (z+1)(z-1)(z^2+1)z^4$. To compute the direct sum decomposition of Theorem 1.1.17, we use the exponents of the factors of $\mu_\varphi(z)$ and get $\mathrm{Ker}(\varphi + \mathrm{id}_V) = (v_1)$, $\mathrm{Ker}(\varphi - \mathrm{id}_V) = (v_2)$, $\mathrm{Ker}(\varphi^2 + \mathrm{id}_V) = (v_3, v_4)$, and $\mathrm{Ker}(\varphi^2) = (v_5, v_6, v_7, v_8)$, where

$$
\begin{aligned}
v_1 &= (0,0,1,0,0,-1,1,-1) & v_2 &= (0,0,-1,0,0,-1,-1,-1) \\
v_3 &= (0,0,1,0,0,0,-1,0) & v_4 &= (0,0,0,0,0,1,0,-1) \\
v_5 &= (0,0,0,1,0,-1,0,0) & v_6 &= (0,0,0,0,1,0,0,-1) \\
v_7 &= (1,0,-1,0,0,0,0,0) & v_8 &= (0,1,0,0,0,0,-1,0)
\end{aligned}
$$

Then the tuple $(v_1, v_2, v_3, v_4, v_5, v_6, v_7, v_8)$ is a $K$-basis of $V$, and we have

$$V = \mathrm{Ker}(\varphi + \mathrm{id}_V) \oplus \mathrm{Ker}(\varphi - \mathrm{id}_V) \oplus \mathrm{Ker}(\varphi^2 + \mathrm{id}_V) \oplus \mathrm{Ker}(\varphi^2)$$

Notice that $\mathrm{Ker}(\varphi) = (v_5, v_6)$ is properly contained in $\mathrm{Ker}(\varphi^2)$.

## 1.3 Nilpotent Endomorphisms and Multiplicities

> *What if everything is an illusion and nothing exists?*
> *In that case, I definitely overpaid for my carpet.*
> (Woody Allen)

In this section we have a closer look at nilpotent endomorphisms, i.e., endomorphisms $\varphi$ such that $\varphi^i = 0$ for some $i > 0$. In this case the minimal polynomial of $\varphi$ is clearly of the form $\mu_\varphi(z) = z^m$ for some $m > 0$, and therefore $\varphi$ has only one eigenfactor $p_1(z) = z$ and only one generalized eigenspace $\mathrm{Gen}(\varphi, z) = V$. The nice results of the previous sections become trivial or empty, and it appears that we definitively overpaid for their proofs.

Should we join this nihilistic view? Of course not! Instead we feel instigated to have a closer look and examine the finer structure of $\varphi$. More precisely, besides the eigenspace $\mathrm{Eig}(\varphi, z) = \mathrm{Ker}(\varphi)$, we have the kernels of the powers of $\varphi$ available until $\varphi^m = 0$ kills the game. Hence this section starts with a careful study of the dimensions of the components of the standard filtration of $V$.

Later on, we exploit the observation that, given an endomorphism $\varphi$ of $V$ and an eigenfactor $p(z)$ of $\varphi$, the endomorphism $p(\varphi)$ is nilpotent on $\mathrm{Gen}(\varphi, p(z))$. We use it to prove a number of properties of the algebraic, the geometric, and the minimal multiplicity of $p(z)$. A further application is the description of the behaviour of eigenspaces and generalized eigenspaces under base field extensions in the last part of the section.

Summarizing, we should not underestimate the power of a nilpotent endomorphism, even if this power is an illusion when it gets large enough.

> nilpotentate: *someone who gets more and more powerful*
> *until he gets killed.*

Let $K$ be a field, let $V$ be a $d$-dimensional $K$-vector space, where $d > 0$.

**Definition 1.3.1**  Let $\varphi : V \longrightarrow V$ be a $K$-linear map.

(a) The endomorphism $\varphi$ is called **nilpotent** if there exists a number $i \geq 1$ such that $\varphi^i = 0$. In other words, the endomorphism $\varphi$ is called nilpotent if $\mu_\varphi(z)$ is a power of $z$.
(b) If $\varphi$ is a nilpotent endomorphism, the smallest number $m \geq 1$ such that $\varphi^m = 0$ is called the **index of nilpotency** of $\varphi$.
(c) Given a nilpotent endomorphism $\varphi : V \longrightarrow V$ with index of nilpotency $m \geq 1$, the chain

$$\langle 0 \rangle = \operatorname{Ker}(\varphi^0) \subseteq \operatorname{Ker}(\varphi^1) \subseteq \cdots \subseteq \operatorname{Ker}(\varphi^m) = V$$

is called the **standard filtration** of $V$ with respect to $\varphi$.
(d) Let $k \in \{1, \ldots, m\}$. A vector $v \in \operatorname{Ker}(\varphi^k) \setminus \operatorname{Ker}(\varphi^{k-1})$ is called a **principal vector** of $\varphi$ of order $k$.

It is clear that an endomorphism $\varphi$ of $V$ is nilpotent if and only if we have $\operatorname{BigKer}(\varphi) = V$, that its minimal polynomial is $\mu_\varphi(z) = z^m$, and that its characteristic polynomial is $\chi_\varphi(z) = z^d$. Consequently, its index of nilpotency $m$ satisfies $m \leq d$. The following proposition collects some basic properties of the standard filtration of $V$ with respect to a nilpotent endomorphism.

**Proposition 1.3.2**  *Let* $\varphi : V \longrightarrow V$ *be a nilpotent endomorphism, let* $m$ *be the nilpotency index of* $\varphi$, *and let* $\delta_i = \dim_K(\operatorname{Ker}(\varphi^i)) - \dim_K(\operatorname{Ker}(\varphi^{i-1}))$ *for* $i = 1, \ldots, m$.

(a) *The standard filtration of $V$ has the form*

$$\langle 0 \rangle = \operatorname{Ker}(\varphi^0) \subset \operatorname{Ker}(\varphi^1) \subset \cdots \subset \operatorname{Ker}(\varphi^m) = V$$

*i.e., we have $\delta_i > 0$ for $i = 1, \ldots, m$.*
(b) *We have $\delta_1 + \cdots + \delta_m = d$.*
(c) *For every $i \in \{1, \ldots, m\}$, the endomorphism $\varphi$ induces an injective $K$-linear map $\vartheta_\varphi : \operatorname{Ker}(\varphi^{i+2}) / \operatorname{Ker}(\varphi^{i+1}) \longrightarrow \operatorname{Ker}(\varphi^{i+1}) / \operatorname{Ker}(\varphi^i)$.*

   *In particular, a vector $v \in \operatorname{Ker}(\varphi^i)$ is a principal vector of $\varphi$ of order $i$ if and only if $\varphi(v)$ is a principal vector of $\varphi$ of order $i - 1$.*
(d) *We have $\delta_1 \geq \cdots \geq \delta_m$.*

*Proof*  To show (a), we first assume that $\operatorname{Ker}(\varphi^i) = \operatorname{Ker}(\varphi^{i+1})$ for some $i \geq 0$ and prove $\operatorname{Ker}(\varphi^{i+1}) = \operatorname{Ker}(\varphi^{i+2})$. For $v \in \operatorname{Ker}(\varphi^{i+2})$, we have $\varphi(v) \in \operatorname{Ker}(\varphi^{i+1})$ and the hypothesis implies $\varphi(v) \in \operatorname{Ker}(\varphi^i)$. From this we get $v \in \operatorname{Ker}(\varphi^{i+1})$, as we wanted to show. Inductively, we get $\operatorname{Ker}(\varphi^i) = \operatorname{Ker}(\varphi^j)$ for all $j > i$. Then the claim follows from $\operatorname{Ker}(\varphi^m) = V$ and the definition of the index of nilpotency.

Since claim (b) follows by simply adding up the dimensions in the standard filtration, we prove claim (c) next. For every $i \geq 0$ and every vector $v \in \operatorname{Ker}(\varphi^{i+1})$, we have $\varphi(v) \in \mathrm{Ker}(\varphi^i)$. Consequently, the map $\varphi$ induces a $K$-linear map $\vartheta_\varphi$. Now we check that $\vartheta_\varphi$ is injective. Given a vector $v \in \mathrm{Ker}(\varphi^{i+2})$, this is a consequence of the observation that $\varphi(v + \mathrm{Ker}(\varphi^{i+1})) \subseteq \mathrm{Ker}(\varphi^i)$ implies $\varphi(v) \in \mathrm{Ker}(\varphi^i)$, and hence $v \in \mathrm{Ker}(\varphi^{i+1})$. Finally, claim (d) follows immediately from (c).    $\square$

Given an eigenvalue $\lambda \in K$ of an endomorphism $\varphi$ of $V$, Theorem 1.1.17.b says that the map $\varphi - \lambda \,\mathrm{id}_V$ restricts to a nilpotent endomorphism of $\mathrm{Gen}(\varphi, z - \lambda)$. Thus part (c) of the proposition yields the following corollary (see [22], Sect. 5).

**Corollary 1.3.3** *Let $\lambda \in K$ be an eigenvalue of $\varphi$, let $v \in \mathrm{Gen}(\varphi, z - \lambda)$, and let $k \geq 2$. Then the following conditions are equivalent.*

(a) *The vector $v$ is a principal vector of $\varphi - \lambda \,\mathrm{id}_V$ of order $k$.*
(b) *The vector $\varphi(v) - \lambda v$ is a principal vector of $\varphi - \lambda \,\mathrm{id}_V$ of order $k - 1$.*

If we have $\delta_1 = 1$ in the standard filtration of $V$ with respect to a nilpotent endomorphism $\varphi$, we obtain the following strong structural restrictions.

**Proposition 1.3.4** *Let $\varphi$ be a nilpotent endomorphism of $V$ with nilpotency index $m \geq 1$, and let $\delta_i = \dim_K(\mathrm{Ker}(\varphi^i)) - \dim_K(\mathrm{Ker}(\varphi^{i-1}))$ for $i = 1, \ldots, m$.*

(a) *If $\delta_1 = 1$, i.e. if $\dim_K(\mathrm{Ker}(\varphi)) = 1$, then we have $\delta_1 = \delta_2 = \cdots = \delta_m = 1$.*
(b) *If $\delta_1 = 1$ and $v \in V \setminus \mathrm{Ker}(\varphi^{m-1})$ then the tuple $(v, \varphi(v), \ldots, \varphi^{m-1}(v))$ is a $K$-basis of $V$.*
(c) *The equality $\delta_1 = 1$ holds if and only if $V$ is a cyclic $K[z]$-module via $\varphi$.*

*Proof* Claim (a) follows immediately from parts (a) and (c) of Proposition 1.3.2.

To prove (b), we let $v \in V \setminus \mathrm{Ker}(\varphi^{m-1})$. For every $i \in \{0, \ldots, m - 1\}$ we then have $\varphi^i(v) \in \mathrm{Ker}(\varphi^{m-i})$. Together with $\delta_1 = \cdots = \delta_m = 1$, this implies that we have $\varphi^i(v) \in \mathrm{Ker}(\varphi^{m-i}) \setminus \mathrm{Ker}(\varphi^{m-i-1})$ for $i = 0, \ldots, m - 1$. Hence $\{v, \varphi(v), \ldots, \varphi^{m-1}(v)\}$ generates $V$. By (a) and Proposition 1.3.2.b, we have $d = m$. This yields the claim.

Finally, we prove (c). If $\delta_1 = 1$, we get from (b) that $V$ is a cyclic $K[z]$-module via $\varphi$. Conversely, assume that there exists a vector $v \in V$ which generates $V$ as a $K[z]$-module. Then it follows that $\{v, \varphi(v), \ldots, \varphi^{m-1}(v)\}$ is a $K$-basis of $V$, since these are the only non-zero elements of the form $\varphi^i(v)$ and they generate $V$. Now we observe that we have $\varphi^i(v) \in \mathrm{Ker}(\varphi^{m-i})$, and using Proposition 1.3.2.a we conclude that $\varphi^i(v) \in \mathrm{Ker}(\varphi^{m-i}) \setminus \mathrm{Ker}(\varphi^{m-i-1})$ for $i = 0, \ldots, m - 1$. Altogether, we obtain $\delta_1 = \cdots = \delta_m = 1$.    $\square$

Given an eigenfactor $p_i(z)$ of an endomorphism $\varphi$ of $V$, Theorem 1.1.17.b shows that $p_i(\varphi)$ is a nilpotent endomorphism. In this case we can strengthen part (a) of the preceding proposition as follows.

**Proposition 1.3.5** *Let $\varphi \in \mathrm{End}_K(V)$, let $\mu_\varphi(z) = p_1(z)^{m_1} \cdots p_s(z)^{m_s}$, where the polynomials $p_1(z), \ldots, p_s(z)$ are the eigenfactors of $\varphi$, and let $i \in \{1, \ldots, s\}$.*

*Then the restriction of $p_i(\varphi)$ is a nilpotent endomorphism of $\mathrm{Gen}(\varphi, p_i(z)) = \mathrm{Ker}(p_i(\varphi)^{m_i})$, and for the standard filtration of $\mathrm{Gen}(\varphi, p_i(z))$ with respect to $p_i(\varphi)$, the number*

$$\delta_j = \dim_K\left(\mathrm{Ker}\left(p_i(\varphi)^j\right)\right) - \dim_K\left(\mathrm{Ker}\left(p_i(\varphi)^{j-1}\right)\right)$$

*is a multiple of $\deg(p_i(z))$ for $j = 1, \ldots, m_i$.*

*Proof* Notice that for $j \neq i$ the restriction of $p_i(\varphi)$ to a generalized eigenspace $\mathrm{Gen}(\varphi, p_j(z))$ is an isomorphism by Theorem 1.1.17.c. Thus it suffices to show that the $K$-vector space $W_{ij} = \mathrm{Ker}(p_i(\varphi)^j)/\mathrm{Ker}(p_i(\varphi)^{j-1})$ is a vector space over the field $L_i = K[z]/\langle p_i(z)\rangle$ for all $i = 1, \ldots, s$ and $j = 1, \ldots, m_i$. First we note that $W_{ij}$ is a $K[z]$-module via $p_i(\varphi)$, because the kernel $\mathrm{Ker}(p_i(\varphi)^k)$ is a $K[z]$-module via $p_i(\varphi)$ for every $k \geq 0$.

Given a vector $v \in \mathrm{Ker}(p_i(\varphi)^j)$, we have $p_i(\varphi)^{j-1}(p_i(\varphi)(v)) = p_i(\varphi)^j(v) = 0$, and thus $p_i(\varphi)(v) \in \mathrm{Ker}(p_i(\varphi)^{j-1})$. Hence the polynomials in $\langle p_i(z)\rangle$ operate trivially on $W_{ij}$, and we conclude that $W_{ij}$ is an $L_i$-vector space, as we wanted to show.                                                                                □

In the next part of this section we apply the results about the dimensions of the components of the standard filtration to prove many properties of the following types of multiplicities of eigenfactors.

**Definition 1.3.6**  Let $\mu_\varphi(z) = p_1(z)^{m_1} \cdots p_s(z)^{m_s}$ and $\chi_\varphi(z) = p_1(z)^{a_1} \cdots p_s(z)^{a_s}$, where the polynomials $p_1(z), \ldots, p_s(z)$ are the eigenfactors of $\varphi$ and $m_i, a_i \in \mathbb{N}$.

(a) For $i = 1, \ldots, s$, the number $m_i$ is called the **minimal multiplicity** of $p_i(z)$ and is denoted by $\mathrm{mmult}(\varphi, p_i(z))$.

  If $p_i(z)$ is of the form $p_i(z) = z - \lambda$ with $\lambda \in K$, we also write $\mathrm{mmult}(\varphi, \lambda)$ instead of $\mathrm{mmult}(\varphi, z - \lambda)$.

(b) For $i = 1, \ldots, s$, the number $a_i$ is called the **algebraic multiplicity** of $p_i(z)$ and is denoted by $\mathrm{amult}(\varphi, p_i(z))$.

  If $p_i(z)$ is of the form $p_i(z) = z - \lambda$ with $\lambda \in K$, we also write $\mathrm{amult}(\varphi, \lambda)$ instead of $\mathrm{amult}(\varphi, z - \lambda)$.

(c) For $i = 1, \ldots, s$, the number $\dim_K(\mathrm{Eig}(\varphi, p_i(z)))$ is called the **geometric multiplicity** of $p_i(z)$ and is denoted by $\mathrm{gmult}(\varphi, p_i(z))$.

  If $p_i(z)$ is of the form $p_i(z) = z - \lambda$ with $\lambda \in K$, we also write $\mathrm{gmult}(\varphi, \lambda)$ instead of $\mathrm{gmult}(\varphi, z - \lambda)$.

Let us collect some useful properties of these multiplicities.

**Proposition 1.3.7 (Basic Properties of Multiplicities)**
*Let $p(z)$ be an eigenfactor of $\varphi$.*

(a) *We have $\mathrm{amult}(\varphi, p(z)) \geq \mathrm{mmult}(\varphi, p(z))$.*
(b) *We have $\mathrm{amult}(\varphi, p(z)) \cdot \deg(p(z)) = \dim_K(\mathrm{Gen}(\varphi, p(z)))$.*
(c) *We have $\mathrm{amult}(\varphi, p(z)) \cdot \deg(p(z)) \geq \mathrm{gmult}(\varphi, p(z))$.*

(d) *Let $\varphi'$ be the restriction of $\varphi$ to $G = \text{Gen}(\varphi, p(z))$. The length of the standard filtration of $G$ with respect to $p(\varphi')$ is $m = \text{mmult}(\varphi, p(z))$.*
(e) *We have $\text{mmult}(\varphi, p(z)) = 1$ if and only if $\text{Eig}(\varphi, p(z)) = \text{Gen}(\varphi, p(z))$.*
(f) *The number $\text{gmult}(\varphi, p(z))$ is a multiple of $\deg(p(z))$.*
(g) *The equality $\text{amult}(\varphi, p(z)) = \text{mmult}(\varphi, p(z))$ holds if and only if we have $\text{gmult}(\varphi, p(z)) = \deg(p(z))$.*
(h) *If $\deg(p(z)) = 1$, then we have $\text{amult}(\varphi, p(z)) = \text{gmult}(\varphi, p(z))$ if and only if $\text{Eig}(\varphi, p(z)) = \text{Gen}(\varphi, p(z))$.*

*Proof* By Theorem 1.2.8.a, we have the inequality in (a) as well as $\chi_{\varphi|_G}(z) = p(z)^a$ for $a = \text{amult}(\varphi, p(z))$ and $G = \text{Gen}(\varphi, p(z))$. Therefore we have the equality $\deg(\chi_{\varphi|_G}(z)) = \text{amult}(\varphi, p(z)) \cdot \deg(p(z))$. Consequently, claim (b) follows from $\dim_K(G) = \deg(\chi_{\varphi|_G}(z))$. Claim (c) follows from $\text{Eig}(\varphi, p(z)) \subseteq \text{Gen}(\varphi, p(z))$ and (b), claim (d) follows from Theorem 1.1.17.b, and claim (e) follows from (d).

To prove (f), we note that the map $p(\varphi)$ operates trivially on the $K$-vector space $\text{Eig}(\varphi, p(z))$. This vector space is in fact a vector space over the field $L = K[z]/\langle p(z)\rangle$, and the claim follows from $\dim_K(L) = \deg(p(z))$.

Next we show (g). Consider the standard filtration

$$\langle 0 \rangle \subset \text{Ker}\big(p(\varphi)\big) \subset \text{Ker}\big(p(\varphi)^2\big) \subset \cdots \subset \text{Ker}\big(p(\varphi)^m\big) = \text{Gen}\big(\varphi, p(z)\big)$$

At every step of the filtration, the $K$-vector space dimension increases by at least $\deg(p(z))$. Hence we have

$$\text{amult}\big(\varphi, p(z)\big) \cdot \deg\big(p(z)\big) = \dim_K \big(\text{Gen}\big(\varphi, p(z)\big)\big)$$

$$\geq m \cdot \deg\big(p(z)\big) = \text{mmult}\big(\varphi, p(z)\big) \cdot \deg\big(p(z)\big)$$

Therefore, if $\text{amult}(\varphi, p(z)) = \text{mmult}(\varphi, p(z))$, then all those increases are equal to $\deg(p(z))$, and in particular we obtain $\text{gmult}(\varphi, p(z)) = \dim_K(\text{Ker}(p(\varphi))) = \deg(p(z))$. Conversely, if the first dimension in this standard filtration is $\deg(p(z))$, then Proposition 1.3.5 shows that all dimension increases are equal to $\deg(p(z))$ and the final dimension is $\dim_K(\text{Gen}(\varphi, p(z))) = \text{mmult}(\varphi, p(z)) \cdot \deg(p(z))$. Now it suffices to apply (b) to finish the proof. Finally, claim (h) follows immediately from (b). $\qquad\square$

The fact that there is no general inequality between the algebraic and the geometric multiplicity is illustrated by the following example.

**Example 1.3.8** Let $K = \mathbb{Q}$, let $M = \left(\begin{smallmatrix} 0 & 1 \\ -1 & 0 \end{smallmatrix}\right) \in \text{Mat}_2(K)$, and consider the two block matrices $A, B \in \text{Mat}_6(K)$ defined by

$$A = \begin{pmatrix} M & 0 & I_2 \\ 0 & M & 0 \\ 0 & 0 & M \end{pmatrix} \quad \text{and} \quad B = \begin{pmatrix} M & I_2 & 0 \\ 0 & M & I_2 \\ 0 & 0 & M \end{pmatrix}$$

For the $K$-linear map $\varphi : K^6 \longrightarrow K^6$ given by $A$, we have $\mu_\varphi(z) = (z^2 + 1)^2$ and $\chi_\varphi(z) = (z^2 + 1)^3$. Hence the map $\varphi$ has only one eigenfactor, namely $p(z) = z^2 + 1$. We calculate its geometric multiplicity and find $\text{gmult}(\varphi, p(z)) = 4$. On the other hand, the algebraic multiplicity is $\mathrm{amult}(\varphi, p(z)) = 3$. Notice that the inequality $\mathrm{amult}(\varphi, p(z)) \cdot \deg(p(z)) = 6 \geq \mathrm{gmult}(\varphi, p(z))$ of part (c) of the proposition holds.

Now we consider the $K$-linear map $\psi : K^6 \longrightarrow K^6$ given by $B$. Here we have $\mu_\psi(z) = \chi_\psi(z) = (z^2 + 1)^3$. Thus there is again just one eigenfactor $p(z) = z^2 + 1$. But here the geometric multiplicity turns out to be $\mathrm{gmult}(\varphi, p(z)) = 2$ which is smaller than $\mathrm{amult}(\varphi, p(z)) = 3$.

In the last part of this section we examine the behaviour of eigenspaces under base field extensions. Clearly, an arbitrary endomorphism $\varphi$ of $V$ need not have any eigenvalues. However, if we pass to an extension field $L$ of $K$, it may happen that there exists a zero $\lambda$ of $\mu_\varphi(z)$ in $L$.

Thus we now let $L \supseteq K$ be a field extension, we let $V_L = V \otimes_K L$, and we let $\varphi_L = \varphi \otimes_K L : V_L \longrightarrow V_L$ be the extension of $\varphi$. In the following we describe the relation between the eigenspaces and generalized eigenspaces of $\varphi$ and those of its extension $\varphi_L$. As above, let $\mu_\varphi(z) = p_1(z)^{m_1} \cdots p_s(z)^{m_s}$, where the polynomials $p_1(z), \ldots, p_s(z)$ are the eigenfactors of $\varphi$. Recall that a polynomial in $K[z]$ is called **separable** if it has pairwise distinct roots in the algebraic closure $\overline{K}$.

**Remark 1.3.9** Suppose that $L \supseteq K$ is a field extension such that $\mu_\varphi(z)$ splits in $L[z]$ into linear factors. Then every zero $\lambda \in L$ of $\mu_\varphi(z)$ is a zero of precisely one eigenfactor $p_i(z)$, and $p_i(z)$ is the minimal polynomial of $\lambda$ over $K$. Moreover, if $\mu_\varphi(z)$ is separable, the zeros of each $p_i(z)$ in $L$ are pairwise distinct. Hence in this case all zeros of all eigenfactors of $\varphi$ in $L$ are pairwise distinct, and their number is exactly $\deg(\mu_\varphi(z))$.

Keeping this remark in mind, we now describe the changes of eigenspaces and generalized eigenspaces when we pass to a field extension such that one of the eigenfactors of $\varphi$ splits into linear factors.

**Proposition 1.3.10 (Extension of Eigenspaces)**
*Let $p(z)$ be an eigenfactor of $\varphi$, and let $L \supseteq K$ be a field extension such that $p(z)$ splits in $L[z]$ into linear factors. Thus we can write $p(z) = (z - \lambda_1)^{c_1} \cdots (z - \lambda_k)^{c_k}$ with pairwise distinct $\lambda_i \in L$ and with $c_i > 0$.*

(a) *We have $\mathrm{Gen}(\varphi, p(z)) \otimes_K L = \bigoplus_{i=1}^{k} \mathrm{Gen}(\varphi_L, \lambda_i)$.*
(b) *We have $\mathrm{Eig}(\varphi, p(z)) \otimes_K L \supseteq \bigoplus_{i=1}^{k} \mathrm{Eig}(\varphi_L, \lambda_i)$.*
(c) *If $p(z)$ is separable, we have $c_1 = \cdots = c_k = 1$ and $k = \deg(p(z))$, and the inclusion in (b) is an equality.*
(d) *If $p(z)$ is separable, we have the equalities $\mathrm{amult}(\varphi, p(z)) = \mathrm{amult}(\varphi_L, \lambda_i)$ and $\mathrm{mmult}(\varphi, p(z)) = \mathrm{mmult}(\varphi_L, \lambda_i)$ for $i = 1, \ldots, k$.*

*Proof* First we show (a). Let $a = \mathrm{amult}(\varphi, p(z))$ and consider the calculation

$$\mathrm{Gen}(\varphi, p(z)) \otimes_K L = \mathrm{BigKer}(p(\varphi)) \otimes_K L = \mathrm{Ker}(p(\varphi)^a) \otimes_K L$$
$$= \mathrm{Ker}(p(\varphi_L)^a) = \mathrm{Ker}((\varphi_L - \lambda_1 \, \mathrm{id}_{V_L})^{c_1 a} \cdots (\varphi_L - \lambda_k \, \mathrm{id}_{V_L})^{c_k a})$$

$$\supseteq \mathrm{Ker}\big((\varphi_L - \lambda_1\,\mathrm{id}_{V_L})^{c_1 a}\big) + \cdots + \mathrm{Ker}\big((\varphi_L - \lambda_k\,\mathrm{id}_{V_L})^{c_k a}\big)$$

$$= \mathrm{Gen}(\varphi_L, \lambda_1) + \cdots + \mathrm{Gen}(\varphi_L, \lambda_k)$$

By Theorem 1.1.17.a, the vector subspaces $\mathrm{Gen}(\varphi_L, \lambda_i)$ form a direct sum. Hence we obtain the inclusion "$\supseteq$". Now the equality follows by comparing dimensions. By Proposition 1.3.7, we have the equalities $\dim_K(\mathrm{Gen}(\varphi, p(z))) = a \deg(p(z))$ and $\dim_L(\mathrm{Gen}(\varphi_L, \lambda_i)) = c_i a$.

The proof of (b) uses an analogous calculation based on the identification $\mathrm{Eig}(\varphi, p(z)) = \mathrm{Ker}(p(\varphi))$. Next we show (c) by proving

$$\mathrm{Eig}\big(\varphi, p(z)\big) \otimes_K L = \mathrm{Ker}\Big( \prod_{i=1}^{k} (\varphi_L - \lambda_i\,\mathrm{id}_{V_L}) \Big) = \bigoplus_{i=1}^{k} \mathrm{Ker}(\varphi_L - \lambda_i\,\mathrm{id}_{V_L})$$

The fact that a separable polynomial $p(z)$ splits into distinct linear factors in $L[z]$ implies the first equality. To prove the second equality, we take a vector $v$ in the kernel on the left-hand side and decompose it into a sum $v = v_1 + \cdots + v_k$ according to (a). By Theorem 1.1.17.a, each generalized eigenspace $\mathrm{Gen}(\varphi_L, \lambda_i)$ is invariant under the map $\prod_{j=1}^{k}(\varphi_L - \lambda_j\,\mathrm{id}_{V_L})$. Hence each vector $v_i$ is in the kernel of this map. Now Theorem 1.1.17.c implies $v_i \in \mathrm{Ker}(\varphi_L - \lambda_i\,\mathrm{id}_{V_L})$ for $i = 1, \ldots, k$. Therefore $v$ is contained in the right-hand side.

Finally we prove (d). For this we use that fact that, by Proposition 1.2.3, neither $\mu_\varphi(z)$ nor $\chi_\varphi(z)$ change when we extend the base field. By (c), we obtain the equality $p(z) = (z - \lambda_1) \cdots (z - \lambda_k)$, and the claim follows. $\square$

The final example in this section shows that the containment in part (b) of this proposition can be strict.

**Example 1.3.11** Let $K = \mathbb{F}_2(x)$, let $V = K^2$, and let $\varphi \in \mathrm{End}_K(V)$ be the endomorphism given by the matrix $A = \big(\begin{smallmatrix} 0 & x \\ 1 & 0 \end{smallmatrix}\big)$. Then the characteristic polynomial of $\varphi$ is $\chi_\varphi(z) = z^2 + x$. Since it is irreducible, it agrees with the minimal polynomial $\mu_\varphi(z)$. The only eigenspace of $\varphi$ is $\mathrm{Eig}(\varphi, z^2 + x) = \mathrm{Ker}(\varphi^2 + x\,\mathrm{id}_V) = \mathrm{Ker}(0) = V$, and this is obviously also the generalized eigenspace.

Now let us pass to the extension field $L = K(y)$ where $y = \sqrt{x}$. Then we have the factorization $z^2 + x = (z + y)^2$ in $L[z]$. Therefore the map $\varphi_L$ has again just one eigenspace $\mathrm{Eig}(\varphi_L, y) = \mathrm{Ker}(\varphi_L - y\,\mathrm{id}_{V_L}) = \langle (y, 1)^{\mathrm{tr}} \rangle$ which is 1-dimensional and properly contained in $\mathrm{Eig}(\varphi, z^2 + x) \otimes_K L = V_L$. The corresponding generalized eigenspace is $V_L$.

Together with part (a) of the proposition, this example shows that generalized eigenspaces are better behaved than eigenspaces with respect to extensions of the base field.

*When you transport something by car,*
*it is called a shipment.*
*But when you transport something by ship,*
*it is called cargo.*

## 1.4  The Module Structure Given by an Endomorphism

> *For every problem, there is one solution*
> *which is simple, neat, and wrong.*
> (Henry L. Mencken)

Given an endomorphism $\varphi$ of a finite dimensional vector space $V$ over a field $K$, there is a natural $K[\varphi]$-module structure on $V$ given by $\varphi \cdot v = \varphi(v)$ for all $v \in V$. In the following we examine this $K[\varphi]$-module structure more closely. A simple and neat way of doing this would be to define a $K[z]$-module structure on $V$ via the algebra homomorphism $K[z] \longrightarrow K[\varphi]$ given by $z \mapsto \varphi$ and to apply the structure theorem for finitely generated modules over principal ideal domains (see [10], Chap. 3). While not wrong, this would lead us far away from the topics of this chapter and the next, where we will study the module structure given by a family of endomorphisms. Therefore we prefer a more direct approach.

The first question that comes to mind is how much a single vector $v \in V$ generates via this module structure. The first observation is that $\dim_K(\langle v \rangle)$ can be at most $\dim_K(K[\varphi])$, and that this is the case if and only if $\mathrm{Ann}_{K[z]}(v) = \langle \mu_\varphi(z) \rangle$. More importantly, the question of how to find such a vector is answered by computing a principal vector of highest order for each eigenfactor of $\varphi$ and taking their sum (see Algorithm 1.4.6).

The second question that we analyze is what happens when there exists a $\varphi$-invariant vector subspace $W$ of $V$ and we want to know how much larger $\dim_K(W + \langle v \rangle)$ is than $\dim_K(W)$. Clearly, this is controlled by the *conductor ideal* $\mathfrak{C}(v, W) = \{f(z) \in K[z] \mid f(\varphi)(v) \in W\}$ and its monic generator, the *conductor* $\mathfrak{c}(v, W)$. The biggest increase in dimension can be achieved if $v$ is *well-separated* from $W$, i.e., if the conductor has the maximal possible degree. By carefully choosing vectors $w_1, \dots, w_r \in V$, we can actually compute a well-separated system of generators of $V$, i.e., a system of $K[\varphi]$-module generators such that $w_i$ is well-separated from $\langle w_1, \dots, w_{i-1} \rangle$ for $i = 1, \dots, r$ (see Algorithm 1.4.13).

Thirdly, we use a well-separated system of generators of $V$ to construct a system of $K[\varphi]$-module generators $(v_1, \dots, v_r)$ such that $V$ is the direct sum of the cyclic submodules $\langle v_i \rangle$ and such that $\langle \mu_\varphi(z) \rangle = \mathrm{Ann}_{K[z]}(v_i) \subseteq \cdots \subseteq \mathrm{Ann}_{K[z]}(v_r)$. This result is called the Cyclic Module Decomposition Theorem (see Theorem 1.4.15). In spite of its simple and neat formulation, the proof is rather tricky and long. However, it allows us to construct the desired system of generators $(v_1, \dots, v_r)$ explicitly, as Algorithm 1.4.17 shows.

As you can see from this brief overview, the following section is neither simple nor neat. Since it is not central for the remainder of the book, you may also skip it on the first reading. If you take the time to work through it, you should not expect to get that time back, but you may hope for a deep and detailed insight into the module structure given by an endomorphism.

> *Borrow money from a pessimist – if you can –*
> *he won't expect to get it back.*
> (John F. Haskett)

Let $K$ be a field, let $V$ be a finite dimensional $K$-vector space, let $d = \dim_K(V)$, and let $\varphi \in \mathrm{End}_K(V)$ be an endomorphism of $V$. The vector space $V$ acquires a module structure over the ring $K[\varphi]$ via the rule $\psi \cdot v = \psi(v)$ for all $\psi \in K[\varphi]$ and all $v \in V$.

**Remark 1.4.1** Using the $K$-algebra epimorphism $K[z] \longrightarrow K[\varphi]$ given by $z \mapsto \varphi$, it follows that $V$ acquires also a $K[z]$-module structure. This module structure is given by $f(z) \cdot v = f(\varphi)(v)$ for all $f(z) \in K[z]$ and $v \in V$. In the following we call it the $K[z]$-**module structure of** $V$ **via** $\varphi$. The $K[\varphi]$-module structure of $V$ is obtained from its $K[z]$-module structure by observing that $\mu_\varphi(z)$ acts trivially on $V$ and that we have $K[\varphi] \cong K[z]/\langle \mu_\varphi(z) \rangle$.

As in the previous sections, we write $\mu_\varphi(z) = p_1(z)^{m_1} \cdots p_s(z)^{m_s}$, where the polynomials $p_1(z), \ldots, p_s(z)$ are the eigenfactors of $\varphi$. Recall that $V$ is said to be a cyclic $K[\varphi]$-module if it is generated by a single vector. More generally, we now examine how large the $K[\varphi]$-submodule of $V$ generated by a single vector can be. We start this investigation with the following key lemma.

**Lemma 1.4.2** *Assume that we have $\mu_\varphi(z) = f_1(z) f_2(z)$ with coprime polynomials $f_1(z), f_2(z) \in K[z]$.*

(a) *For $1 = 1, 2$, let $w_i \in \mathrm{Ker}(f_i(\varphi))$, and let $w = w_1 + w_2$. Then $w_1$ and $w_2$ are multiples of $w$, i.e. we have $w_1, w_2 \in \langle w \rangle_{K[z]}$.*
(b) *Suppose that $f_1(z) = p_1(z)^{m_1}$ is a power of an eigenfactor of $\varphi$. Then there exists a vector $w_1 \in \mathrm{Ker}(f_1(\varphi))$ such that we have $\mathrm{Ann}_{K[z]}(w_1) = \langle f_1(z) \rangle$. In particular, we have $\dim_K(\langle w_1 \rangle_{K[z]}) = \deg(f_1(z)) = m_1 \cdot \deg(p_1(z))$.*

*Proof* First we prove (a). Since the polynomials $f_1(z)$ and $f_2(z)$ are coprime, there exist polynomials $q_1(z)$ and $q_2(z)$ such that $q_1(z) f_1(z) + q_2(z) f_2(z) = 1$. From this we obtain

$$w_1 = \big(q_1(\varphi) f_1(\varphi)\big)(w_1) + \big(q_2(\varphi) f_2(\varphi)\big)(w_1)$$
$$= \big(q_2(\varphi) f_2(\varphi)\big)(w_1) = \big(q_2(\varphi) f_2(\varphi)\big)(w)$$

Using the same argument, we also get $w_2 \in \langle w \rangle_{K[z]}$.

Now we show claim (b). We let $U = \mathrm{Ker}(f_1(\varphi))$. Since $f_1(\varphi)$ commutes with $\varphi$, Proposition 1.1.6 shows that $U$ is a $\varphi$-invariant subspace of $V$. Furthermore, by Lemma 1.1.16.d, we have $\mu_{\varphi|_U}(z) = f_1(z) = p_1(z)^{m_1}$. Consequently, there is a vector $w_1 \in U \setminus \mathrm{Ker}(p_1(\varphi)^{m_1-1})$. Next we let $g(z)$ be a generator of the ideal $\mathrm{Ann}_{K[z]}(w_1)$. Since $w_1 \in U$ means that we have $f_1(z) \cdot w_1 = 0$, we know that the polynomial $g(z)$ divides $f_1(z) = p_1(z)^{m_1}$. Since $w_1 \notin \mathrm{Ker}(p_1(\varphi)^{m_1-1})$, we know that $g(z)$ does not divide $p_1(z)^{m_1-1}$. Altogether, we get $g(z) = f_1(z)$.

The additional claim follows from $\dim_K(\langle w_1 \rangle_{K[z]}) = \dim_K(K[z]/\mathrm{Ann}_{K[z]}(w_1)) = \dim_K(K[z]/\langle f_1(z) \rangle) = \deg(f_1(z)) = m_1 \cdot \deg(p_1(z))$. $\square$

In part (a) of this lemma, the assumption that $\mu_\varphi(z) = f_1(z) f_2(z)$ with coprime polynomials $f_1(z), f_2(z)$ is essential, as the following example shows.

**Example 1.4.3** Let $K = \mathbb{Q}$, let $V = \mathbb{Q}^3$, and let $\varphi \in \text{End}_K(V)$ be defined by the matrix

$$A = \begin{pmatrix} 0 & 0 & 1 \\ 0 & 0 & 0 \\ 0 & 0 & 0 \end{pmatrix}$$

with respect to the canonical basis. We have $\mu_\varphi(z) = z^2$ and $\chi_\varphi(z) = z^3$.

Now we consider the vectors $w_1 = (1, 0, 0)$ and $w_2 = (0, 1, 0)$, and we let $w = w_1 + w_2$. Then we see that $w_1, w_2 \in \text{Ker}(\varphi) = \text{Ker}(f_i(\varphi))$ for $f_1(z) = f_2(z) = z$. However, the vectors $w_i$ are not multiples of $w$ because $z \cdot w = \varphi(w) = 0$ implies the equality $K[z] \cdot w = K \cdot w$.

Based on the preceding lemma we prove a key result about the structure of the vector space $V$ as a $K[z]$-module. It shows that there is always a cyclic submodule of dimension $\deg(\mu_\varphi(z))$.

**Proposition 1.4.4 (Cyclic Submodules of Maximal Dimension)**
*There exists a vector $w \in V$ such that the following two equivalent conditions hold:*

$$\dim_K\left(\langle w \rangle_{K[z]}\right) = \deg\left(\mu_\varphi(z)\right) \quad and \quad \text{Ann}_{K[z]}(w) = \langle \mu_\varphi(z) \rangle$$

*Proof* The two conditions are equivalent since $\mu_\varphi(\varphi)(w) = 0$ implies that $\langle \mu_\varphi(z) \rangle$ is contained in $\text{Ann}_{K[z]}(w)$.

The proof proceeds by induction on the number $s$ of eigenfactors of $\varphi$.

For $s = 1$, we let $f_1(z) = p_1(z)^{m_1}$ and apply Lemma 1.4.2.b.

For $s > 1$, we let $f_1(z) = p_1(z)^{m_1}$ and $f_2(z) = p_2(z)^{m_2} \cdots p_s(z)^{m_s}$. Clearly, the polynomials $f_1(z)$ and $f_2(z)$ are coprime. Using the lemma and the inductive assumption we get two vectors $w_1, w_2 \in V$ such that $\text{Ann}_{K[z]}(w_1) = \langle f_1(z) \rangle$ and $\text{Ann}_{K[z]}(w_2) = \langle f_2(z) \rangle$. In particular, we have $\dim_K(\langle w_i \rangle_{K[z]}) = \deg(f_i(z))$ for $i = 1, 2$. Now we let $w = w_1 + w_2$ and use part (a) of the lemma. It yields $\langle w \rangle_{K[z]} \supseteq \langle w_1 \rangle_{K[z]} + \langle w_2 \rangle_{K[z]}$.

Every polynomial $g(z) \in K[z]$ satisfies $g(z) \cdot w_1 = g(\varphi)(w_1) \in \text{Ker}(f_1(\varphi))$ since $f_1(\varphi)g(\varphi)(w_1) = g(\varphi)f_1(\varphi)(w_1) = g(\varphi)(0) = 0$. Therefore we have the inclusion $\langle w_1 \rangle_{K[z]} \subseteq \text{Ker}(f_1(\varphi))$, and similarly we get $\langle w_2 \rangle_{K[z]} \subseteq \text{Ker}(f_2(\varphi))$. Consequently, we can apply Lemma 1.1.16.c and conclude that the sum $\langle w_1 \rangle_{K[z]} + \langle w_2 \rangle_{K[z]}$ is direct. Altogether, we obtain $\dim_K(\langle w \rangle_{K[z]}) \geq \deg(f_1(z)) + \deg(f_2(z)) = \deg(\mu_\varphi(z))$.

On the other hand, every generator of the ideal $\text{Ann}_{K[z]}(w)$ divides $\mu_\varphi(z)$. This yields the inequality $\dim_K(\langle w \rangle_{K[z]}) \leq \deg(\mu_\varphi(z))$, and the proof is complete. $\square$

In particular, the vector $w$ in this proposition generates a $K[z]$-submodule of $V$ which is isomorphic to $K[\varphi]$.

**Corollary 1.4.5** *Let $\mu_\varphi(z) = p_1(z)^{m_1} \cdots p_s(z)^{m_s}$, where $m_i > 0$ and the polynomials $p_1(z), \ldots, p_s(z)$ are the eigenfactors of $\varphi(z)$. For $i = 1, \ldots, s$, let*

$$g_i(z) = p_1(z)^{m_1} \cdots p_{i-1}(z)^{m_{i-1}} \cdot p_i(z)^{m_i - 1} \cdot p_{i+1}(z)^{m_{i+1}} \cdots p_s(z)^{m_s}$$

*let $\psi_i = g_i(\varphi)$, and let $w \in V$.*

(a) *We have $\dim_K(\langle w \rangle_{K[z]}) = \deg(\mu_\varphi(z))$ if and only if $w \notin \bigcup_{i=1}^s \mathrm{Ker}(\psi_i)$.*
(b) *The set $V \setminus \bigcup_{i=1}^s \mathrm{Ker}(\psi_i)$ is not empty.*
(c) *Let $w \in V \setminus \bigcup_{i=1}^s \mathrm{Ker}(\psi_i)$. According to Theorem 1.1.17, there is a decomposition $w = w_1 + w_2 + \cdots + w_s$ where $w_i \in \mathrm{Ker}(p_i(\varphi)^{m_i})$ for $i = 1, \ldots, s$. Then every vector $w_i$ is a principal vector of order $m_i$, i.e. we have $w_i \notin \mathrm{Ker}(p_i(\varphi))^{m_i-1}$ for $i = 1, \ldots, s$.*

*Proof* First we prove the implication $\Rightarrow$ of (a). If there exists an index $i \in \{1, \ldots, s\}$ such that $w \in \mathrm{Ker}(\psi_i)$, then $g_i(z)$ annihilates $w$. Consequently, we have the inequality $\dim_K(\langle w \rangle_{K[z]}) < \deg(\mu_\varphi(z))$. To prove the converse implication, we note that $\mu_\varphi(z)$ annihilates $w$ and every proper divisor of $\mu_\varphi(z)$ divides one of the polynomials $g_i(z)$. Claim (b) follows from (a) and the proposition. To prove (c) we observe that if there exists an index $i \in \{1, \ldots, s\}$ such that $w_i \in \mathrm{Ker}(p_i(\varphi)^{m_i-1})$, then the polynomial $g_i(z)$ annihilates $w$. Hence the annihilator of $w$ properly divides $\mu_\varphi(z)$ which contradicts (a). $\square$

It is straightforward to turn the proof of the proposition into an algorithm for computing an element $w \in V$ which generates a cyclic submodule of dimension $\deg(\mu_\varphi(z))$.

---

**Algorithm 1.4.6 (Computing a Cyclic Module of Maximal Dimension)**
*Given an endomorphism $\varphi \in \mathrm{End}_K(V)$, consider the following sequence of instructions.*

(1) *Using Algorithm 1.1.8, compute the minimal polynomial $\mu_\varphi(z)$.*
(2) *Factorize $\mu_\varphi(z)$ and get its decomposition $\mu_\varphi(z) = p_1(z)^{m_1} \cdots p_s(z)^{m_s}$ into monic irreducible factors.*
(3) *For every $i \in \{1, \ldots, s\}$, compute vector space bases of $\mathrm{Ker}(p_i(\varphi)^{m_i-1})$ and $\mathrm{Ker}(p_i(\varphi)^{m_i})$ and choose a principal vector $w_i$ of order $m_i$.*
(4) *Return $w = w_1 + \cdots + w_s$.*

*This is an algorithm which computes a vector $w \in V$ with the property that $\dim_K(\langle w \rangle_{K[z]}) = \deg(\mu_\varphi(z))$. Equivalently, the vector $w$ satisfies the equality $\mathrm{Ann}_{K[z]}(w) = \langle \mu_\varphi(z) \rangle$.*

---

*Proof* For $s = 1$, the proof is contained in the proof of Lemma 1.4.2.b. For $s > 1$, we let $f_1(z) = p_1(z)^{m_1}$ and $f_2(z) = p_2(z)^{m_2} \cdots p_s(z)^{m_s}$ as in the proof of Proposition 1.4.4. Inductively, we may assume that $w_2 + \cdots + w_s$ is annihilated by $f_2(z)$. Then the proof of Proposition 1.4.4 shows that $w = w_1 + w_2 + \cdots + w_s$ is annihilated by $\mu_\varphi(z)$. Hence we obtain $\dim_K(\langle w \rangle_{K[z]}) = \deg(\mu_\varphi(z))$. $\square$

The following example illustrates this algorithm and the preceding corollary.

**Example 1.4.7** Let $K = \mathbb{Q}$, let $V = \mathbb{Q}^6$, and let $\varphi \in \mathrm{End}_{\mathbb{Q}}(V)$ be defined by the following matrix with respect to the canonical basis.

$$A = \begin{pmatrix} 1 & 7 & 7 & 7 & 5 & 4 \\ 0 & -14 & -15 & -15 & -12 & -8 \\ -1 & -5 & -5 & -5 & -3 & -4 \\ 1 & 20 & 21 & 21 & 16 & 12 \\ 0 & -1 & -1 & -1 & -1 & 0 \\ \frac{1}{2} & -1 & -\frac{1}{2} & -1 & -\frac{3}{2} & 1 \end{pmatrix}$$

The minimal polynomial of $\varphi$ is $\mu_\varphi(z) = z^2(z-1)^2$ while the characteristic polynomial is $\chi_\varphi(z) = z^3(z-1)^3$. Our goal is to find a vector $w \in V$ such that $\dim_{\mathbb{Q}}\langle w \rangle_{\mathbb{Q}[z]} = \deg(\mu_\varphi(z)) = 4$, i.e. a vector $w \in V$ such that $\mathrm{Ann}_{K[z]}(w) = \langle \mu_\varphi(z) \rangle$.

According to the algorithm, we have to choose a vector $w_1$ in $\mathrm{Ker}(\varphi^2) \setminus \mathrm{Ker}(\varphi)$ and a vector $w_2$ in $\mathrm{Ker}((\varphi - \mathrm{id}_V)^2) \setminus \mathrm{Ker}(\varphi - \mathrm{id}_V)$. A basis of $\mathrm{Ker}(\varphi)$ is given by the tuple $((-2, 3, 1, -3, -1, 0), (4, -8, -2, 10, 0, -1))$. A basis of $\mathrm{Ker}(\varphi^2)$ is $((-3/7, 6/7, 2/7, -1, 0, 0), (-5/7, 3/7, 1/7, 0, -1, 0), (-2/7, 4/7, 6/7, 0, 0, -1))$. So, for instance, we can choose $w_1 = (3/7, -6/7, -2/7, 1, 0, 0)$ and check that $w_1 \notin \mathrm{Ker}(\varphi)$.

A basis of $\mathrm{Ker}(\varphi - \mathrm{id}_V)$ is $((1, 1, -1, 0, 0, 0), (0, 1, 0, -1, 0, 0))$, and a basis of $\mathrm{Ker}((\varphi - \mathrm{id}_V)^2)$ is $((1, 1, -1, 0, 0, 0), (0, 1, 0, -1, 0, 0), (0, 0, 0, 0, 0, -1))$. So, for instance, we can choose $w_2 = (0, 0, 0, 0, 0, 1)$. Finally, we add the two vectors and get $w = w_1 + w_2 = (3/7, -6/7, -2/7, 1, 0, 1)$. It is easy to check that, indeed, we have $\dim_{\mathbb{Q}}\langle w \rangle_{\mathbb{Q}[z]} = 4$.

Another way to get such a vector is to use Corollary 1.4.5.a. First we compute a vector space basis of $\mathrm{Ker}(\varphi(\varphi - \mathrm{id}_V)^2)$ and get $((-1, 0, 0, 0, 0, 0),$ $(0, 1, -1, 0, 0, 0),$ $(0, 1, 0, -1, 0, 0),$ $(0, 1, 0, 0, -1, 0),$ $(0, 0, 0, 0, 0, -1))$. Then we compute a basis of $\mathrm{Ker}(\varphi^2(\varphi - \mathrm{id}_V))$ and obtain $((1, -1, 0, 0, 0, 0),$ $(2, 0, -1, 0, 0, 0), (1, 0, 0, -1, 0, 0), (0, 0, 0, 0, -1, 0), (2, 0, 0, 0, 0, -1))$. Now we pick a randomly chosen vector $w$ and verify that it is not contained in the two kernels. This is possible since the kernels are 5-dimensional vector subspaces of the 6-dimensional space $V$ over the infinite field $\mathbb{Q}$.

The next important theorem requires several preparatory results. The presentation of the following facts uses many ideas of [9], Chap. 7.

**Definition 1.4.8** Let $W$ be a $\varphi$-invariant vector subspace of $V$. For $v \in V$, the principal ideal $\{f(z) \in K[z] \mid f(\varphi)(v) \in W\}$ is called the **conductor ideal** of $v$ into $W$ and is denoted by $\mathfrak{C}(v, W)$. Its monic generator is called the **conductor** of $v$ into $W$ and denoted by $c_{v,W}(z)$.

Notice that the assumption that $W$ is $\varphi$-invariant is needed to conclude that $\mathfrak{C}(v, W)$ is a polynomial ideal. The following remark contains further useful observations about conductors and conductor ideals.

**Remark 1.4.9** Let $W$ be a $\varphi$-invariant vector subspace of $V$.

(a) The map $\varphi$ induces an endomorphism $\overline{\varphi} : V/W \longrightarrow V/W$. For $\bar{v} \in V/W$, we have $\mathrm{Ann}_{K[z]}(\bar{v}) = \mathfrak{C}(v, W)$.

(b) For every $v \in V$, the conductor $\mathfrak{c}_{v,W}(z)$ divides the minimal polynomial $\mu_{\overline{\varphi}}(z)$ of $\overline{\varphi}$ which, in turn, divides $\mu_{\varphi}(z)$. It follows that we have the inequalities $\deg(\mathfrak{c}_{v,W}(z)) \leq \deg(\mu_{\overline{\varphi}}(z)) \leq \deg(\mu_{\varphi}(z))$.

(c) Combining the preceding two parts with Proposition 1.4.4, it follows that the maximal degree of a conductor polynomial $\mathfrak{c}_{v,W}(z)$ with $v \in V$ is $\deg(\mu_{\overline{\varphi}}(z))$.

(d) Given two vector subspaces $U, W$ of $V$ such that $U \subseteq W$ and a vector $v \in V$, the conductor $c_{v,W}(z)$ divides $c_{v,U}(z)$.

(e) We have $\mathfrak{c}_{v,W}(\varphi)(v) = 0$ if and only if $W \cap \langle v \rangle_{K[z]} = \langle 0 \rangle$, i.e. if and only if the sum $W + \langle v \rangle_{K[z]}$ is direct.

The third part of this remark leads us to consider the following notion.

**Definition 1.4.10** Let $W$ be a $\varphi$-invariant vector subspace of the $K$-vector space $V$ and $\overline{\varphi} : V/W \to V/W$ the induced endomorphism. Then every $v \in V$ with $\deg(\mathfrak{c}_{v,W}(z)) = \deg(\mu_{\overline{\varphi}}(z))$ is called **well-separated** from $W$.

The next lemma yields a well-separated system of generators of $V$, viewed as a $K[z]$-module.

**Lemma 1.4.11** *There exist an integer $r \geq 1$ and vectors $w_1, w_2, \ldots, w_r \in V$ having the following properties.*

(a) *We have $V = \langle w_1 \rangle_{K[z]} + \cdots + \langle w_r \rangle_{K[z]} = \langle w_1, \ldots, w_r \rangle_{K[z]}$.*

(b) *For $i = 1, \ldots, r$, the vector subspace $\langle w_1 \rangle_{K[z]} + \cdots + \langle w_i \rangle_{K[z]}$ of $V$ is $\varphi$-invariant.*

(c) *The vector $w_1$ is well-separated from $\langle 0 \rangle$ and, for $i = 2, \ldots, r$, the vector $w_i$ is well-separated from $\langle w_1 \rangle_{K[z]} + \cdots + \langle w_{i-1} \rangle_{K[z]}$.*

*Proof* The vectors $w_1, \ldots, w_r \in V$ will be constructed inductively. According to Proposition 1.4.4, we can chose a vector $w_1$ with $\dim_K(\langle w_1 \rangle_{K[z]}) = \deg(\mu_{\varphi}(z))$. This means that $\mathrm{Ann}_{K[z]}(w_1) = \mathfrak{c}(v_1, \langle 0 \rangle) = \langle \mu_{\varphi}(z) \rangle$. By Remark 1.4.9.b, the degree of $\mu_{\varphi}(z)$ is the maximum degree of the conductors of vectors in $V$ with respect to $\langle 0 \rangle$.

If we now have $\langle w_1 \rangle_{K[z]} = V$, we are done. Otherwise, suppose that we have already constructed the vectors $w_1, \ldots, w_{i-1}$. Let $W = \langle w_1, \ldots, w_{i-1} \rangle_{K[z]}$, and let the vector $w_i \in V \setminus W$ be chosen such that $\deg(\mathfrak{c}(w_i, W))$ is maximal, i.e. such that $\deg(\mathfrak{c}_{w_i,W}) = \deg(\mu_{\overline{\varphi}}(z))$ where $\overline{\varphi}$ is the induced automorphism of $V/W$. Then $W + \langle w_i \rangle_{K[z]}$ is a $\varphi$-invariant subspace of $V$ and $w_i$ is well-separated from $W$. Continuing in this way, we eventually get $W = V$, since $V$ is finite dimensional. $\square$

To ease the formulation of the following results, we introduce a name for tuples of vectors as in the lemma.

**Definition 1.4.12**   A tuple $(w_1, \ldots, w_r)$ of vectors of $V$ is called a **well-separated system of $K[z]$-generators** of $V$ if the conditions of the lemma are satisfied, i.e. if $\{w_1, \ldots, w_r\}$ is a system of generators of the $K[z]$-module $V$ and if $w_i$ is well-separated from $\langle w_1, \ldots, w_{i-1} \rangle_{K[z]}$ for $i = 1, \ldots, r$.

Now we show how to turn the proof of the lemma into an algorithm for computing a well-separated system of $K[z]$-generators of $V$. The key ingredient turns out to be Remark 1.4.9.c.

---

**Algorithm 1.4.13 (Computing a Well-Separated System of Generators)**
*Given an endomorphism $\varphi \in \mathrm{End}_K(V)$, consider the following sequence of instructions.*

(1) *Let $W_0 = \langle 0 \rangle$ and $i = 0$.*
(2) *Repeat the following steps until $W_i = V$. Then return the tuples $(w_1, \ldots, w_i)$ and $(\mu_1(z), \ldots, \mu_i(z))$ and stop.*
(3) *Increase $i$ by one, and compute the minimal polynomial $\mu_i(z)$ of the endomorphism $\overline{\varphi}_i$ of $V/W_{i-1}$ induced by $\varphi$.*
(4) *Using Algorithm 1.4.6, compute $w_i \in V$ with $\dim_K(\langle w_i \rangle_{K[z]}) = \deg(\mu_i(z))$.*
(5) *Let $W_i = W_{i-1} + \langle w_i \rangle_{K[z]}$ and continue with Step (2).*

*This is an algorithm which computes a tuple $(w_1, \ldots, w_i)$ which is a well-separated system of $K[z]$-generators of $V$ as well as the tuple of conductor polynomials $(\mathfrak{c}_{w_1, W_0}(z), \ldots, \mathfrak{c}_{w_i, W_{i-1}}(z)) = (\mu_1(z), \ldots, \mu_i(z))$.*

---

*Proof* Clearly, the correctness of this algorithm follows the proof of the lemma. The choice of a vector $w_i$ such that $\deg(\mathfrak{c}(w_i, W))$ is maximal is achieved by using Remark 1.4.9.c.                                                                               $\square$

The next lemma provides a particular divisibility property of the conductor for vectors in a well-separated system of generators.

**Lemma 1.4.14**   *Let $(w_1, \ldots w_r)$ be a well-separated system of $K[z]$-generators of $V$. For $i = 1, \ldots, r$, denote the vector space $\langle w_1 \rangle_{K[z]} + \cdots + \langle w_i \rangle_{K[z]}$ by $W_i$.*

(a) *For every vector $v \in V$, there exist polynomials $g_1(z), \ldots, g_i(z) \in K[z]$ such that $c_{v, W_i}(\varphi)(v) = \sum_{j=1}^{i} g_j(\varphi)(w_j)$.*
(b) *In the setting of (a) the polynomial $c_{v, W_i}(z)$ divides each of the polynomials $g_1(z), \ldots, g_i(z)$.*

*Proof* Claim (a) follows from the definition of $c_{v, W_i}$ and that fact that $w_1, \ldots, w_i$ generate the $K[z]$-module $W_i$.

Next we prove (b). By performing a division with remainder of each polynomial $g_j(z)$ by $c_{v, W_i}$, we get equalities $g_j(z) = c_{v, W_i}(z) \cdot h_j(z) + r_j(z)$ and our goal is to show that $r_j(z) = 0$ for $j = 1, \ldots, i$. For a contradiction, assume that $j \in \{1, \ldots, i\}$ is the largest number such that $r_j(z) \neq 0$. Since $r_j(z)$ is a division remainder, it follows that

$$\deg(r_j(z)) < \deg(c_{v, W_i}(z)) \tag{1.1}$$

Now we consider the vector $u = v - \sum_{k=1}^{i} h_k(\varphi)(w_k)$. As $u$ and $v$ differ only by an element of $W_i$, we have $c_{v, W_i}(z) = c_{u, W_i}(z)$, and therefore

$$c_{u, W_i}(\varphi)(u) = c_{v, W_i}(\varphi)(u) = c_{v, W_i}(\varphi)(v) - \sum_{k=1}^{i} c_{v, W_i}(\varphi) \cdot h_k(\varphi)(w_k)$$

$$= \sum_{k=1}^{i} \big(g_k(\varphi)(w_k) - c_{v, W_i}(\varphi) \cdot h_k(\varphi)(w_k)\big) = \sum_{k=1}^{j} r_j(\varphi)(w_j) \tag{1.2}$$

Here we used the fact that $r_k(z) = 0$ for $k > j$ by the definition of $j$. By Remark 1.4.9.d, we know that the conductor $c_{u, W_i}(z)$ divides all polynomials $c_{u, W_k}(z)$ with $k \leq i$. Hence there exists a polynomial $q(z) \in K[z]$ such that

$$c_{u, W_{j-1}}(z) = q(z) \cdot c_{u, W_i}(z) \tag{1.3}$$

Using equalities (1.3) and (1.2), we deduce

$$c_{u, W_{j-1}}(\varphi)(u) = q(\varphi) \cdot c_{u, W_i}(\varphi)(u) = q(\varphi) \cdot r_j(\varphi)(w_j) + \sum_{k=1}^{j-1} q(\varphi) \cdot r_k(\varphi)(w_k)$$

Since both $c_{u, W_{j-1}}(\varphi)(u)$ and $\sum_{k=1}^{j-1} q(\varphi) \cdot r_k(\varphi)(w_k)$ are contained in $W_{j-1}$, we get

$$q(\varphi) \cdot r_j(\varphi)(w_j) \in W_{j-1}$$

Together with the definition of the conductor, this shows that we have the inequality $\deg(q(z) \cdot r_j(z)) \geq \deg(c_{w_j, W_{j-1}}(z))$. Since $(w_1, \ldots w_r)$ is a well-separated system of generators, we have the inequality $\deg(c_{w_j, W_{j-1}}(z)) \geq \deg(c_{u, W_{j-1}}(z))$. Using (1.3), we have the equality $\deg(c_{u, W_{j-1}}(z)) = \deg(q(z) \cdot c_{u, W_i}(z))$. Altogether, we get

$$\deg(q(z) \cdot r_j(z)) \geq \deg(c_{w_j, W_{j-1}}(z)) \geq \deg(c_{u, W_{j-1}}(z)) = \deg(q(z) \cdot c_{u, W_i}(z))$$

Hence we have $\deg(r_j(z)) \geq \deg(c_{u, W_i}(z))$, and this contradicts (1.1). Thus the proof is complete.                                                                               □

Finally we state and prove the fundamental result of this section.

**Theorem 1.4.15 (Cyclic Module Decomposition)**
*Let $\varphi \in \mathrm{End}_K(V)$ be an endomorphism of $V$. There exists a tuple $(v_1, v_2, \ldots, v_r)$ of vectors in $V$ such that*

(a) $V = \bigoplus_{i=1}^{r} \langle v_i \rangle_{K[z]}$,
(b) $\langle \mu_\varphi(z) \rangle = \mathrm{Ann}_{K[z]}(v_1) \subseteq \mathrm{Ann}_{K[z]}(v_2) \subseteq \cdots \subseteq \mathrm{Ann}_{K[z]}(v_r)$.

*Proof* By Lemma 1.4.11, there exists a well-separated system of $K[z]$-module generators $(w_1, \ldots, w_r)$ of $V$. For $i = 1, \ldots, r$, we let $W_i = \langle w_1, \ldots, w_i \rangle_{K[z]}$. We let $v_1 = w_1$ and note that $\langle \mu_\varphi(z) \rangle = \langle c_{w_1, \langle 0 \rangle}(z) \rangle = \mathrm{Ann}_{K[z]}(v_1)$. For $i = 2, \ldots, r$, we apply parts (a) and (b) of Lemma 1.4.14 to the vector $w_i$ and find polynomials $f_{ij}(z) \in K[z]$ such that

$$c_{w_i, W_{i-1}}(\varphi)(w_i) = \sum_{j=1}^{i-1} \big( c_{w_i, W_{i-1}}(\varphi) \cdot f_{ij}(\varphi) \big)(w_j) \tag{1.4}$$

Now we let $v_i = w_i - \sum_{j=1}^{i-1} f_{ij}(\varphi)(w_j)$ for $i = 2, \ldots, r$. Since the vectors $w_i$ and $v_i$ differ by an element of $W_{i-1}$, we have $c_{v_i, W_{i-1}}(z) = c_{w_i, W_{i-1}}(z)$. It is also clear that we have $V = \langle v_1 \rangle_{K[z]} + \cdots + \langle v_r \rangle_{K[z]}$. From (1.4) we deduce that $c_{v_i, W_{i-1}}(\varphi)(v_i) = 0$ for $i = 2, \ldots, r$ and thus, by Remark 1.4.9.e, the sum is direct. Thus proves (a).

It remains to show (b). The fact that $c_{v_i, W_{i-1}}(\varphi)(v_i) = 0$ for $i = 1, \ldots, r$ implies

$$\mathrm{Ann}_{K[z]}(v_i) = \big\langle c_{v_i, W_{i-1}}(z) \big\rangle \tag{1.5}$$

Using the equalities $c_{v_i, W_{i-1}}(\varphi)(v_i) = 0$ and $\sum_{j=1}^{i-1} c_{v_j, W_{j-1}}(\varphi)(v_j) = 0$, we get the equality $c_{v_i, W_{i-1}}(\varphi)(v_i) = \sum_{j=1}^{i-1} c_{v_j, W_{j-1}}(\varphi)(v_j)$. Now we use Lemma 1.4.14 together with (1.5) to deduce claim (b). □

**Remark 1.4.16** The annihilators $\mathrm{Ann}_{K[z]}(v_i)$ are principal ideals in $K[z]$. Their monic generators are usually called the **invariant factors** of $\varphi$.

Next we show how to turn the theorem into an algorithm for computing a decomposition of $V$ as a direct sum of cyclic $K[z]$-modules via $\varphi$.

**Algorithm 1.4.17 (Computing a Cyclic Module Decomposition)**
*Let $\varphi \in \mathrm{End}_K(V)$ be an endomorphism of the vector space $V$. Consider the following sequence of instructions.*

(1) *Using Algorithm 1.4.13 compute a well-separated system of $K[z]$-module generators $(w_1, \ldots, w_r)$ of $V$ and the tuple of conductor polynomials $(c_{w_1, W_0}(z), \ldots, c_{w_i, W_{i-1}}(z)) = (\mu_1(z), \ldots, \mu_i(z))$, where $W_i$ denotes the vector subspace $\langle w_1, \ldots, w_{i-1} \rangle_{K[z]}$ of $V$.*

(2) *Let $v_1 = w_1$.*
(3) *For every $i = 2, \ldots, r$, execute the following Steps (4)–(6).*
(4) *Compute the vector $\mathfrak{c}_{w_i, W_{i-1}}(\varphi)(w_i)$ and represent it using the basis $(w_1, \ldots, w_{i-1})$ of $W_{i-1}$. The result is of the form*

$$\mathfrak{c}_{w_i, W_{i-1}}(\varphi)(w_i) = \textstyle\sum_{j=1}^{i-1} g_j(\varphi)(w_j)$$

   *with polynomials $g_j(z) \in K[z]$.*
(5) *For $j = 1, \ldots, i - 1$, divide $g_j(z)$ by $\mathfrak{c}_{w_i, W_{i-1}}(z)$ and obtain a polynomial $f_{ij}(z)$ such that $g_j(z) = \mathfrak{c}_{w_i, W_{i-1}}(z) \cdot f_{ij}(z)$.*
(6) *Calculate $v_i = w_i - \sum_{j=1}^{i-1} f_{ij}(\varphi)(w_j)$.*
(7) *Return the tuple $(v_1, \ldots, v_r)$.*

   *This is an algorithm which computes a tuple of vectors $(v_1, \ldots, v_r)$ which satisfies the conditions of the theorem. In particular, we have the direct sum decomposition $V = \bigoplus_{i=1}^{r} \langle v_i \rangle_{K[z]}$.*

*Proof* The fact that $g_j(z)$ is divisible by $\mathfrak{c}_{w_i, W_{i-1}}(z)$ follows from Lemma 1.4.14. Hence all instructions can be executed. The correctness of the algorithm follows from the proof of the theorem.  □

The following example continues Example 1.4.7 and shows how one can use the theorem and the algorithm to represent $V$ explicitly as a direct sum of cyclic $K[z]$-modules via $\varphi$.

**Example 1.4.18**  Let $K = \mathbb{Q}$, let $V = \mathbb{Q}^6$, let $\varphi \in \mathrm{End}_K(V)$ be given by the matrix

$$A = \begin{pmatrix} 1 & 7 & 7 & 7 & 5 & 4 \\ 0 & -14 & -15 & -15 & -12 & -8 \\ -1 & -5 & -5 & -5 & -3 & -4 \\ 1 & 20 & 21 & 21 & 16 & 12 \\ 0 & -1 & -1 & -1 & -1 & 0 \\ \frac{1}{2} & -1 & -\frac{1}{2} & -1 & -\frac{3}{2} & 1 \end{pmatrix}$$

and let $v_1 = (3/7, -6/7, -2/7, 1, 0, 1)$. In Example 1.4.7 we showed that $v_1$ satisfies $\mathrm{Ann}_{K[z]}(v_1) = \mu_\varphi(z) = z^2(z - 1)^2$. Thus the vector $v_1$ can be chosen as the first vector of the desired tuple.

The $K[z]$-submodule $W_1$ of $V$ generated by $v_1$ is 4-dimensional and has the basis $(v_1, \varphi(v_1), \varphi^2(v_1), \varphi^3(v_1))$. We check that the canonical basis vector $e_1$ is not contained in this module and its conductor is $\mathfrak{c}_{e_1, W_1}(z) = z(z - 1)$. From $\dim_K(V/W_1) = 2$ it follows that $e_1$ is well-separated from $W_1$. At this point we know that $(v_1, e_1)$ is a well-separated system of $K[z]$-generators of $V$.

Next we compute $\varphi(\varphi - \mathrm{id}_V)(e_1)$ and represent it using the basis of $W_1$. The result is $\varphi(\varphi - \mathrm{id}_V)(e_1) = (\frac{1}{2}\varphi)(\varphi(\varphi - \mathrm{id}_V)(v_1))$. This shows that $f_{11}(z) = \frac{1}{2}z$ and we get

$$v_2 = e_1 - \frac{1}{2}\varphi(v_1) = (-5/7, 47/14, 13/7, -36/7, -1/14, -17/28)$$

As expected, we find $\varphi(\varphi - \mathrm{id}_V)(v_2) = 0$. Altogether, we have computed the decomposition $V = \langle v_1 \rangle_{K[z]} \oplus \langle v_2 \rangle_{K[z]}$ according to the theorem.

If we represent $\varphi$ with respect to the basis $(v_1, \varphi(v_1), \varphi^2(v_1), \varphi^3(v_1), v_2, \varphi(v_2))$ of $V$, we get the block diagonal matrix

$$B = \begin{pmatrix} 0 & 0 & 0 & 0 & 0 & 0 \\ 1 & 0 & 0 & -1 & 0 & 0 \\ 0 & 1 & 0 & 2 & 0 & 0 \\ 0 & 0 & 1 & 0 & 0 & 0 \\ 0 & 0 & 0 & 0 & 0 & 0 \\ 0 & 0 & 0 & 0 & 1 & 1 \end{pmatrix}$$

## 1.5 Commendable Endomorphisms

*I can resist anything except temptation.*
(Oscar Wilde)

In mathematics, if a certain property or notion comes up in a variety of contexts, or has a number of different characterizations, we have an indication that this property has a deeper meaning and may turn out to be useful in many ways. Thus we are tempted to introduce a new name for this property. A prime example of a property like this is that all eigenspaces of an endomorphism $\varphi$ have their minimal possible dimension. As we show in this section, it is equivalent to require that we have $\mu_\varphi(z) = \chi_\varphi(z)$, or that the vector space $V$ is a cyclic $K[z]$-module via $\varphi$. Later in the book we prove several further characterizations of this property. Hence it is rather remarkable that it has gone largely unnoticed by algebraists. Of course, we could not resist the temptation to give it a name and to study it carefully.

By now the readers who are versed in Numerical Analysis will have noticed that, in the case that all eigenfactors of $\varphi$ are linear, this property has been used in that field and called *non-derogatory*. But as we explained in the main introduction, we do not agree to call such a good property non-bad, and we have given in to the temptation to introduce a new term: a *commendable* endomorphism $\varphi$ is characterized by $\dim_K(\mathrm{Eig}(\varphi, p(z))) = \deg(p(z))$ for every eigenfactor $p(z)$. Besides the characterizations mentioned above, we prove that this notion is invariant under extensions of the base field.

Do commendable endomorphisms exist? How common are they? In the last part of the section we show that, for every given monic polynomial, we can use a suitable companion matrix to construct a commendable endomorphism having it as its minimal polynomial. Yielding to temptation once again, we claim that to be commendable is the best one can say about an endomorphism.

> *Lead me not into temptation;*
> *I can find the way myself.*
> (Rita Mae Brown)

Let $K$ be a field, let $V$ be a finite dimensional $K$-vector space, then let $d = \dim_K(V)$, and let $\varphi \in \text{End}_K(V)$. We write $\mu_\varphi(z) = p_1(z)^{m_1} \cdots p_s(z)^{m_s}$, where $p_1(z), \ldots, p_s(z)$ are the eigenfactors of $\varphi$, and where $m_i > 0$ for $i = 1, \ldots, s$.

Recall from Proposition 1.3.7.f and its proof that $\text{Eig}(\varphi, p_i(z))$ is a vector space over the field $K[z]/\langle p_i(z) \rangle$ and that its $K$-vector space dimension is therefore a multiple of $\deg(p_i(z))$. The $K[z]/\langle p_i(z) \rangle$-vector space structure on $\text{Eig}(\varphi, p_i(z))$ may have some surprising consequences, as the following example shows.

**Example 1.5.1**  Let $\varphi$ be the $\mathbb{Q}$-endomorphism of $\mathbb{Q}^4$ defined by the matrix

$$\begin{pmatrix} 0 & 2 & 0 & 0 \\ 1 & 0 & 0 & 0 \\ 0 & 0 & 0 & 2 \\ 0 & 0 & 1 & 0 \end{pmatrix}$$

Then we have $\mu_\varphi(z) = z^2 - 2$. Hence there is only one eigenfactor, and its kernel is the entire vector space $\mathbb{Q}^4$. Thus $\mathbb{Q}^4$ acquires the structure of a vector space over the field $L = \mathbb{Q}[z]/\langle z^2 - 2 \rangle$. With respect to this structure we have the equalities

$$\bar{z} \cdot e_1 = \varphi(e_1) = e_2$$
$$\bar{z} \cdot e_2 = \varphi(e_2) = 2e_1$$
$$\bar{z} \cdot e_3 = \varphi(e_3) = e_4$$
$$\bar{z} \cdot e_4 = \varphi(e_4) = 2e_3$$

Therefore the tuple $(e_1, e_3)$ is an $L$-basis of $\mathbb{Q}^4$, and the identification $L \cong \mathbb{Q}(\sqrt{2})$ yields equalities such as $\sqrt{2}e_1 = e_2$ and $\sqrt{2}e_2 = 2e_1$ which look quite odd.

A very important situation occurs when all eigenspaces $\text{Eig}(\varphi, p_i(z))$ of $\varphi$ have their minimal possible dimension $\deg(p_i(z))$. The following notion will be of crucial importance further on.

**Definition 1.5.2**  The $K$-linear map $\varphi : V \longrightarrow V$ is called **commendable** if, for every $i \in \{1, \ldots, s\}$, the eigenfactor $p_i(z)$ of $\varphi$ satisfies

$$\text{gmult}(\varphi, p_i(z)) = \deg(p_i(z))$$

Equivalently, we require that $\dim_{K[z]/\langle p_i(z) \rangle}(\text{Eig}(\varphi, p_i(z))) = 1$ for $i = 1, \ldots, s$.

Note that in the literature commendable endomorphisms with linear eigenfactors are called **non-derogatory**. The following remark shows that we can check effectively whether an endomorphism $\varphi$ is commendable.

**Remark 1.5.3** Let $B$ be a basis of $V$. For every $i \in \{1, \ldots, s\}$, let $A_i$ be a matrix which represents $p_i(\varphi)$ with respect to $B$. Then the endomorphism $\varphi$ is commendable if and only if $\text{rank}(A_i) = d - \deg(p_i(z))$ for $i = 1, \ldots, s$.

The following examples provide some endomorphisms which are commendable and some which are derogatory.

**Example 1.5.4** A nilpotent endomorphism $\varphi : V \longrightarrow V$ is commendable if and only if $\dim_K(\text{Ker}(\varphi)) = 1$, since its only eigenfactor is $p(z) = z$.

**Example 1.5.5** Let $A \in \text{Mat}_d(K)$ be a diagonal matrix and $\varphi \in \text{End}_K(K^d)$ the endomorphism defined by $A$. Then $\varphi$ is commendable if and only if the entries on the diagonal of $A$ are pairwise distinct. In particular, if $d > 1$ then the identity map, although very nice, is not commendable.

**Example 1.5.6** Let $\varphi, \psi : \mathbb{Q}^6 \longrightarrow \mathbb{Q}^6$ be the $\mathbb{Q}$-linear maps defined in Example 1.3.8. They have only one eigenfactor, namely $p(z) = z^2 + 1$. It turns $\mathbb{Q}^6$ into a vector space over $L = \mathbb{Q}(i)$. Now the equalities $\text{gmult}(\psi, p(z)) = 2$ and $\text{gmult}(\varphi, p(z)) = 4$ show that $\psi$ is commendable and $\varphi$ is not commendable.

In a special case, commendable endomorphisms have been considered and characterized in different ways before.

**Remark 1.5.7** Suppose that the minimal polynomial of $\varphi$ splits in $K[z]$ into linear factors. Under this hypothesis the endomorphism $\varphi$ is commendable if and only if we have $\text{gmult}(\varphi, \lambda) = 1$ for every eigenvalue $\lambda \in K$ of $\varphi$. In other words, the endomorphism $\varphi$ is commendable if and only if all of its eigenspaces are 1-dimensional vector spaces.

This is the definition which is typically found in the literature where the term *non-derogatory* is used instead of *commendable*. One of the main results of this section is that the endomorphism $\varphi$ is commendable if and only if its minimal and characteristic polynomials agree (see Theorem 1.5.8). This is another commonly used equivalent definition.

A third one is the requirement that a matrix $A$ in $\text{Mat}_d(K)$ which represents $\varphi$ and is in Jordan normal form has only one Jordan block per eigenvalue. This requirement is indeed an equivalent condition, since the subspace corresponding to a single Jordan block contains a 1-dimensional eigenspace.

By Example 1.5.4 and Proposition 1.3.4.c, we know that a nilpotent endomorphism $\varphi : V \longrightarrow V$ is commendable if and only if $V$ is a cyclic $K[z]$-module via $\varphi$. This equivalence holds in general, as the following theorem shows.

**Theorem 1.5.8 (First Characterization of Commendable Endomorphisms)**
*Let $\varphi : V \longrightarrow V$ be a $K$-linear map. Then the following conditions are equivalent.*

(a) *The endomorphism $\varphi$ is commendable.*
(b) *We have $\mu_\varphi(z) = \chi_\varphi(z)$.*
(c) *The vector space $V$ is a cyclic $K[z]$-module via $\varphi$.*

*Proof* Let $\mu_\varphi(z) = p_1(z)^{m_1} \cdots p_s(z)^{m_s}$, where $p_1(z), \ldots, p_s(z)$ are the eigenfactors of $\varphi$, and let $G_i = \mathrm{Gen}(\varphi, p_i(z))$ for $i = 1, \ldots, s$.

First we prove the equivalence of conditions (a) and (b). By Theorem 1.1.17.b and Theorem 1.2.8.b, we have $\mu_\varphi(z) = \chi_\varphi(z)$ if and only if $\mu_{\varphi|G_i}(z) = \chi_{\varphi|G_i}(z)$ for $i = 1, \ldots, s$. In view of Definition 1.5.2 it is therefore sufficient to show the claim in the case $\mu_\varphi(z) = p(z)^m$ and $\chi_\varphi(z) = p(z)^a$ with an irreducible polynomial $p(z)$ and integers $a \geq m \geq 1$. In this case the claim was proved in Proposition 1.3.7.g.

Now we prove that (b) implies (c). By Proposition 1.4.4, there exists a vector $w \in V$ such that $\dim_K \langle w \rangle_{K[z]} = \deg(\mu_\varphi(z))$. The assumption yields $\deg(\mu_\varphi(z)) = \deg(\chi_\varphi(z))$. Hence we have $\dim_K(\langle w \rangle_{K[z]}) = \deg(\chi_\varphi(z)) = \dim_K(V)$. Consequently, the vector $w$ generates the $K[z]$-module $V$.

Finally we show that (c) implies (b). Let $w \in V$ be such that $V = \langle w \rangle_{K[z]}$. Thus every polynomial $f(z) \in K[z]$ such that $f(z) \cdot w = 0$ annihilates $V$ and hence is a multiple of $\mu_\varphi(z)$. Therefore, as a $K[z]$-module, the vector space $V$ is isomorphic to $K[z]/\langle \mu_\varphi(z) \rangle$. Consequently, we obtain $\deg(\mu_\varphi(z)) = \dim_K(V) = \deg(\chi_\varphi(z))$. But $\mu_\varphi(z)$ divides $\chi_\varphi(z)$ by Theorem 1.2.5, so that we conclude $\mu_\varphi(z) = \chi_\varphi(z)$. $\square$

This theorem implies that the property of an endomorphism to be commendable is invariant under base field extensions. We use the abbreviations $V_L = V \otimes_K L$ and $\varphi_L = \varphi \otimes_K L$ to denote the extensions of $V$ and $\varphi$, respectively.

**Corollary 1.5.9** *Let $L \supseteq K$ be a field extension. Then the following conditions are equivalent.*

(a) *The endomorphism $\varphi$ of $V$ is commendable.*
(b) *The endomorphism $\varphi_L$ of $V_L$ is commendable.*

*Proof* This follows from Proposition 1.2.3 and condition (b) of the theorem. $\square$

Another consequence of the theorem is that the property of being commendable does not change if we add a multiple of the identity to an endomorphism, as the next corollary shows.

**Corollary 1.5.10** *Let $\varphi : V \longrightarrow V$ be a $K$-linear map, and let $c \in K$. Then the following conditions are equivalent.*

(a) *The endomorphism $\varphi$ is commendable.*
(b) *The endomorphism $\varphi + c\,\mathrm{id}_V$ is commendable.*

*Proof* This follows from Corollary 1.2.4 and condition (b) of the theorem.    □

Theorem 1.5.8 implies that, given any endomorphism $\varphi$ of $V$, the endomorphism of $K[\varphi]$ given by multiplication by $\varphi$ is always commendable. This observation will come in handy in later chapters. Let us see an example.

**Example 1.5.11**  Let $K$ be a field, let $V = K^3$, and let $\varphi \in \mathrm{End}_K(V)$ be defined by the following matrix with respect to the canonical basis.

$$A = \begin{pmatrix} 0 & 0 & 1 \\ 0 & 0 & 0 \\ 0 & 0 & 0 \end{pmatrix}$$

We have $\mu_\varphi(z) = z^2$ and $\chi_\varphi(z) = z^3$. Hence $\varphi$ is a derogatory endomorphism of $V$.

Now we consider the $K$-algebra $K[\varphi] \cong K[z]/\langle \mu_\varphi(z) \rangle$ and view it as a $K$-vector space. The multiplication by $\varphi$ is an endomorphism of $K[\varphi]$ which we denote by $\vartheta_\varphi$. With respect to the basis $(1, \varphi)$ of $K[\varphi]$, the map $\vartheta_\varphi$ is represented by the matrix $\begin{pmatrix} 0 & 0 \\ 1 & 0 \end{pmatrix}$. Since $\mu_{\vartheta_\varphi}(z) = \chi_{\vartheta_\varphi}(z) = z^2$, the endomorphism $\vartheta_\varphi$ is commendable.

In the last part of this section we discuss a classical tool which, among other things, allows us to construct commendable endomorphisms with a given minimal polynomial.

**Definition 1.5.12**  Let $d \geq 1$, let $c_0, \ldots, c_d \in K$ and let $p(z)$ be the monic polynomial $p(z) = z^d + c_{d-1}z^{d-1} + \cdots + c_1 z + c_0 \in K[z]$. Then the matrix

$$\begin{pmatrix} 0 & 0 & \cdots & 0 & -c_0 \\ 1 & 0 & \cdots & 0 & -c_1 \\ 0 & 1 & \cdots & 0 & -c_2 \\ \vdots & \vdots & & \vdots & \vdots \\ 0 & \cdots & 0 & 1 & -c_{d-1} \end{pmatrix}$$

is called the **companion matrix** of $p(z)$ and is denoted by $C_{p(z)}$.

Using induction on $d$ and Laplace expansion, it is straightforward to show that $\det(z \cdot I_d - C_{p(z)}) = p(z)$. In other words, the characteristic polynomial of the endomorphism defined by $C_{p(z)}$ is the given polynomial $p(z)$. Thus the following corollary provides us with a way to construct commendable endomorphisms having a given characteristic and minimal polynomial.

**Corollary 1.5.13**  *Let $p(z) \in K[z]$ be a monic polynomial of degree $d \geq 1$, let $V$ be a $K$-vector space of dimension $d$, and let $\varphi$ be the endomorphism of $V$ defined by $C_{p(z)}$ with respect to a basis of $V$.*

(a) *The endomorphism $\varphi$ is commendable.*
(b) *We have $\mu_\varphi(z) = \chi_\varphi(z) = p(z)$.*
(c) *The vector space $V$ is a cyclic $K[z]$-module via $\varphi$.*

*Proof* The equivalence of the three conditions is an immediate consequence of the theorem. Therefore it suffices to show that (c) holds. For this purpose we let $B = (v_1, v_2, \ldots, v_d)$ be the basis of $V$ with respect to which $\varphi$ is defined by $C_{p(z)}$. Then we have

$$\varphi(v_1) = v_2, \varphi^2(v_1) = \varphi(v_2) = v_3, \ldots, \varphi^{d-1}(v_1) = \cdots = \varphi(v_{d-1}) = v_d$$

and thus the element $v_1$ generates the $K[z]$-module $V$. □

Based on this corollary, we get another characterization of commendable endomorphisms.

**Corollary 1.5.14 (Second Characterization of Commendable Endomorphisms)**
*Let $\varphi$ be an endomorphism of $V$. Then the following conditions are equivalent.*

(a) *The endomorphism $\varphi$ is commendable.*
(b) *There exists a $K$-basis of $V$ such that $\varphi$ is represented by the companion matrix $C_{\mu_\varphi(z)}$ with respect to this basis.*
(c) *Every matrix representing $\varphi$ with respect to a $K$-basis of $V$ is similar to the matrix $C_{\mu_\varphi(z)}$.*

*Proof* Since conditions (b) and (c) are clearly equivalent, and since (b) implies (a) by the above corollary, it remains to prove that (b) follows from (a). So, let $\varphi$ be commendable. By the theorem, the vector space $V$ is a cyclic $K[z]$-module via $\varphi$. Let $v \in V$ be a generator of this module and let $d = \dim_K(V)$. Consequently, the set $B = \{v, \varphi(v), \ldots, \varphi^{d-1}(v)\}$ is a system of generators and thus a $K$-basis of $V$. Writing $\varphi^d(v) = -c_0 v - c_1\varphi(v) - \cdots - c_{d-1}\varphi^{d-1}(v)$ with $c_i \in K$, we see that the matrix of $\varphi$ with respect to this basis is the companion matrix of the polynomial $p(z) = z^d + c_{d-1}z^{d-1} + \cdots + c_0$. Hence the characteristic polynomial of $\varphi$ is $p(z)$. By Theorem 1.5.8, the polynomial $p(z)$ is also the minimal polynomial of $\varphi$. □

Let us see an example which illustrates this corollary.

**Example 1.5.15** Let $K = \mathbb{Q}$, let $V = K^6$ and let $\varphi$ be the endomorphism of $V$ defined by the matrix

$$A = \begin{pmatrix} 0 & 1 & 1 & 0 & 0 & 0 \\ -1 & 0 & 0 & 1 & 0 & 0 \\ 0 & 0 & 0 & 1 & 1 & 0 \\ 0 & 0 & -1 & 0 & 0 & 1 \\ 0 & 0 & 0 & 0 & 0 & 1 \\ 0 & 0 & 0 & 0 & -1 & 0 \end{pmatrix}$$

with respect to the canonical basis. We have $\chi_\varphi(z) = \mu_\varphi(z) = z^6 + 3z^4 + 3z^2 + 1 = (z^2 + 1)^3$. By Corollary 1.4.5.a, to get a generator of the $K[\varphi]$-module $V$, we have to pick a vector $v$ in $V \setminus \text{Ker}((\varphi^2 + \text{id}_V)^2)$, for instance $v = e_5$. As in the proof of the preceding corollary, the tuple $B = (v, \varphi(v), \varphi^2(v), \varphi^3(v), \varphi^4(v), \varphi^5(v))$ is a $K$-basis of $V$ and the matrix

$$\begin{pmatrix} 0 & 0 & 1 & 0 & 0 & -1 \\ 1 & 0 & 0 & 0 & 0 & 0 \\ 0 & 1 & 0 & 0 & 0 & -3 \\ 0 & 0 & 1 & 0 & 0 & 0 \\ 0 & 0 & 0 & 1 & 0 & -3 \\ 0 & 0 & 0 & 0 & 1 & 0 \end{pmatrix}$$

representing $\varphi$ in this basis $B$ is the companion matrix of $\mu_\varphi(z)$.

To conclude this section, we show that every endomorphism of $V$ can be represented as a direct sum of commendable endomorphisms.

**Theorem 1.5.16 (Direct Sum of Commendable Endomorphisms)**
*Let $\varphi$ be an endomorphism of V and let $v_1, \ldots, v_r \in V$ be vectors such that we have $V = \bigoplus_{i=1}^r \langle v_i \rangle_{K[z]}$. For $i = 1, \ldots, r$, let $\varphi_i$ be the endomorphism of the $\varphi$-invariant vector subspace $U_i = \langle v_i \rangle_{K[z]}$ induced by $\varphi$.*

(a) *The endomorphism $\varphi$ is the direct sum of the endomorphisms $\varphi_1, \ldots, \varphi_r$.*
(b) *For $i = 1, \ldots, r$, the endomorphism $\varphi_i$ is commendable.*
(c) *For $i = 1, \ldots, r$, let $\mu_i(z)$ be the minimal polynomial of $v_i$, viewed as a vector of $U_i$, and let $d_i = \deg(\mu_i(z))$. Then the matrix representing $\varphi_i$ with respect to the basis $(v_i, \varphi(v_i), \ldots, \varphi^{d_i-1}(v_i))$ of $U_i$ is the companion matrix of $\mu_i(z)$.*

*Proof* Claim (a) is clearly true and claim (b) follows from Theorem 1.5.8.c. Finally, claim (c) is a consequence of Corollary 1.5.14.                                            □

Thus, if we pick the basis $(v_i, \varphi_i(v_i), \ldots, \varphi_i^{d_i-1}(v_i))$ for each $U_i$ and join these bases together, we get a basis of $V$ such that the matrix representing $\varphi$ with respect to this basis is a block diagonal matrix whose diagonal blocks are companion matrices. A matrix having a decomposition like this has been given at the end of Example 1.4.18.

Pinocchio: *"My nose is about to grow!"*

## 1.6  Other Special Endomorphisms

*You boys line up, alphabetically*
*according to your height.*
(Edgar Allen Diddle)

To get some order into the many different kinds of endomorphisms of a vector space, it is a classical and well-researched topic to line them up according to certain

normal forms of their matrices. The first and most well-known type are diagonalizable endomorphisms, i.e. endomorphisms for which there exists a basis of the vector
space consisting of eigenvectors. Next in line are triangularizable endomorphisms.
Their matrices can be made upper triangular by a suitable choice of basis. Thirdly,
we can decompose a trigularizable endomorphism uniquely as a sum of a diagonalizable and a nilpotent endomorphism. The hypothesis which makes all three of
these results possible is that all eigenfactors of the endomorphism have to be linear,
and this property is what makes them special.

One normal form is conspicuously absent from this section, namely the Jordan
normal form. Although it, too, requires the eigenfactors to be linear, do not hold
your breath waiting for it. It is not going to happen, because starting from the next
section we are mainly interested in families of commuting endomorphisms, and they
do not have joint Jordan normal forms, as it turns out.

> *Some people want it to happen,*
> *some wish it would happen,*
> *others make it happen.*
> (Michael Jordan)

Let $K$ be a field, let $V$ be a finite dimensional $K$-vector space, and let $\varphi$ be an
endomorphism of $V$. Recall that $\varphi$ is called **diagonalizable** if there exists a basis
of $V$ in which $\varphi$ is represented by a diagonal matrix. Equivalently, we are looking for
a basis of $V$ consisting of eigenvectors of $\varphi$. The following proposition characterizes
diagonalizable endomorphisms.

**Proposition 1.6.1 (Characterization of Diagonalizable Endomorphisms)**
*For a $K$-linear map $\varphi : V \longrightarrow V$, the following conditions are equivalent.*

(a) *The map $\varphi$ is diagonalizable.*
(b) *All eigenfactors of $\varphi$ are linear, and for every eigenfactor $z - \lambda$ we have*
$\mathrm{gmult}(\varphi, \lambda) = \mathrm{amult}(\varphi, \lambda)$.
(c) *All eigenfactors of $\varphi$ are linear, and for every eigenfactor $z - \lambda$ we have the*
*equality $\mathrm{Eig}(\varphi, \lambda) = \mathrm{Gen}(\varphi, \lambda)$.*
(d) *The minimal polynomial of $\varphi$ splits in $K[z]$ into distinct linear factors.*

*Proof* The equivalence of (b), (c), and (d) follows from Proposition 1.3.7. To show
(a) $\Rightarrow$ (d), we argue as follows. If $\varphi$ is diagonalizable, then its distinct eigenvalues
$\lambda_1, \ldots, \lambda_r \in K$ are the elements contained in the main diagonal of a diagonal matrix
representing $\varphi$. Consequently, we have $(\varphi - \lambda_1 \mathrm{id}_V) \cdots (\varphi - \lambda_r \mathrm{id}_V) = 0$, and hence
$\mu_\varphi(z)$ splits in $K[z]$ into distinct linear factors.

For the proof of (c) $\Rightarrow$ (a), we let $\lambda_1, \ldots, \lambda_r \in K$ be the eigenvalues of $\varphi$. Then
we have $V = \mathrm{Gen}(\varphi, \lambda_1) \oplus \cdots \oplus \mathrm{Gen}(\varphi, \lambda_r)$ by Theorem 1.1.17.a. Hence we get the
equality $V = \mathrm{Eig}(\varphi, \lambda_1) \oplus \cdots \oplus \mathrm{Eig}(\varphi, \lambda_r)$. Thus $V$ has a $K$-basis which consists
of $\lambda_i$-eigenvectors of $\varphi$. $\qquad\square$

Let us have a look at an example.

**Example 1.6.2**  Let $K = \mathbb{Q}$, let $V = K^6$, and let $\varphi \in \mathrm{End}_K(V)$ be defined by

$$
A = \begin{pmatrix}
-8 & -\frac{8}{3} & 8 & 20 & 0 & 12 \\
0 & 1 & 0 & 0 & 0 & 0 \\
6 & 2 & -5 & -15 & 0 & 12 \\
-6 & -\frac{8}{3} & 8 & 18 & 0 & 12 \\
0 & 0 & 0 & 0 & -2 & 0 \\
3 & \frac{2}{3} & -2 & -5 & 0 & 1
\end{pmatrix}
$$

with respect to the canonical basis. We have

$$\mu_\varphi(z) = z^4 - 6z^3 + 3z^2 + 26z - 24 \qquad\qquad = (z-3)(z-4)(z+2)(z-1)$$
$$\chi_\varphi(z) = z^6 - 5z^5 - 5z^4 + 41z^3 - 4z^2 - 76z + 48 = (z-3)(z-4)(z+2)^2(z-1)^2$$

A $K$-basis of $\mathrm{Ker}(\varphi - 3\,\mathrm{id}_V)$ is $(v_1)$, where $v_1 = (4, 0, -3, 4, 0, -1)$. A $K$-basis of $\mathrm{Ker}(\varphi - 4\,\mathrm{id}_V)$ is $(v_2)$, where $v_2 = (-1, 0, 1, -1, 0, 0)$. A $K$-basis of $\mathrm{Ker}(\varphi + 2\,\mathrm{id}_V)$ is $(v_3, v_4)$, where $v_3 = -e_5$ and $v_4 = (4, 0, -3, 3, 0, -1)$. A $K$-basis of $\mathrm{Ker}(\varphi - \mathrm{id}_V)$ is $(v_5, v_6)$, where $v_5 = (0, -3, -1, 0, 0, 0)$ and $v_6 = (4, 12, 0, 4, 0, -1)$. Joining these bases, we get the $K$-basis $B = (v_1, v_2, v_3, v_4, v_5, v_6)$ of $V$. With respect to the basis $B$, the endomorphism $\varphi$ is represented by the diagonal matrix

$$
D = \begin{pmatrix}
-2 & 0 & 0 & 0 & 0 & 0 \\
0 & -2 & 0 & 0 & 0 & 0 \\
0 & 0 & 1 & 0 & 0 & 0 \\
0 & 0 & 0 & 1 & 0 & 0 \\
0 & 0 & 0 & 0 & 3 & 0 \\
0 & 0 & 0 & 0 & 0 & 4
\end{pmatrix}
$$

If $\varphi$ is not diagonalizable, we can at least try to triangularize it. Recall that an endomorphism $\varphi : V \longrightarrow V$ is called **triangularizable** if there exists a basis of $V$ such that $\varphi$ is represented by an upper triangular matrix in this basis. In this case, the upper triangular matrix is also called a **Schur form** of $\varphi$. The following proposition characterizes triangularizable endomorphisms.

**Proposition 1.6.3 (Characterization of Triangularizable Endomorphisms)**
*For a $K$-linear map $\varphi : V \longrightarrow V$, the following conditions are equivalent.*

(a) *The map $\varphi$ is triangularizable.*
(b) *The eigenfactors of $\varphi$ are linear.*

*Proof*  To show that (a) implies (b), it suffices to represent $\varphi$ by an upper triangular matrix $A$ and to compute $\chi_\varphi(z) = \det(zI_d - A)$. Thus it remains to prove that (b) implies (a). By Theorem 1.1.17.a, we have $V = \mathrm{Gen}(\varphi, \lambda_1) \oplus \cdots \oplus \mathrm{Gen}(\varphi, \lambda_s)$ where $\lambda_1, \ldots, \lambda_s \in K$ are the distinct eigenvalues of $\varphi$. Since the direct summands are $\varphi$-invariant, it suffices to prove the claim in the case $V = \mathrm{Gen}(\varphi, \lambda)$ with an eigenvalue $\lambda \in K$ of $\varphi$. In this case the map $\varphi - \lambda \, \mathrm{id}_V$ is nilpotent and we have $\chi_\varphi(z) = (z - \lambda)^d$ by Theorem 1.2.8.a. We consider the chain of strict inclusions

$$\langle 0 \rangle \subset \mathrm{Ker}(\varphi - \lambda \, \mathrm{id}_V) \subset \cdots \subset \mathrm{Ker}(\varphi - \lambda \, \mathrm{id}_V)^m = V$$

By repeated extensions along this chain, we now construct a $K$-basis of $V$ with respect to which $\varphi$ is represented by an upper triangular matrix. For this purpose we start with a $K$-basis of $\mathrm{Eig}(\varphi, \lambda) = \mathrm{Ker}(\varphi - \lambda \, \mathrm{id}_V)$. It suffices to show that for $i \in \{2, \ldots, m\}$ and every vector $v \in \mathrm{Ker}(\varphi - \lambda \, \mathrm{id}_V)^i$ we have $\varphi(v) = \lambda v + w$ with a vector $w \in \mathrm{Ker}(\varphi - \lambda \, \mathrm{id}_V)^{i-1}$. This fact follows from Corollary 1.3.3, and the proof is complete.  $\square$

The following example shows an endomorphism which cannot be diagonalized, but is still triangularizable.

**Example 1.6.4**  Let $K = \mathbb{Q}$, let $V = K^4$, and let $\varphi$ be the endomorphism of $V$ defined by the matrix

$$A = \begin{pmatrix} -506 & -16 & -360 & -368 \\ 1185 & 11 & -24 & -2281 \\ 306 & 34 & 403 & 782 \\ 36 & 4 & 90 & -270 \end{pmatrix}$$

with respect to the canonical basis. We have

$$\mu_\varphi(z) = z^3 - 98283z + 11859482 \qquad\qquad = (z + 362)(z - 181)^2$$
$$\chi_\varphi(z) = z^4 + 362z^3 - 98283z^2 - 23718964z + 4293132484 = (z + 362)^2(z - 181)^2$$

A $K$-basis of $\mathrm{Ker}(\varphi + 362 \, \mathrm{id}_V)$ is $(v_1, v_2)$, where $v_1 = (31/8, -99/8, -1, 0)$ and $v_2 = (5, -22, 0, -1)$. A $K$-basis of $\mathrm{Ker}(\varphi - 181 \, \mathrm{id}_V)$ is $(v_3)$, where we have $v_3 = (4, 85/2, -17/2, -1)$. A $K$-basis of $\mathrm{Ker}(\varphi - 181 \, \mathrm{id}_V)^2$ is $(v_3, v_4)$, where $v_4 = e_2$. According to the proposition, with respect to the basis $B = (v_1, v_2, v_3, v_4)$ of $V$, the endomorphism $\varphi$ is represented by the upper-triangular matrix

$$\begin{pmatrix} -362 & 0 & 0 & 0 \\ 0 & -362 & 0 & 0 \\ 0 & 0 & 181 & -4 \\ 0 & 0 & 0 & 181 \end{pmatrix}$$

It is clear that a triangularizable endomorphism can be decomposed into a diagonalizable endomorphism $\delta$ and a nilpotent endomorphism $\eta$, since it suffices to define $\delta$ by the diagonal part of the upper triangular matrix and to let $\eta$ be represented by the matrix whose non-zero entries are given by the part of the upper triangular matrix above the main diagonal. However, in this case the maps $\delta$ and $\eta$ will, in general, not commute. The next proposition says that we can achieve a decomposition

$\varphi = \delta + \eta$ into a diagonal endomorphism $\delta$ and a nilpotent endomorphism $\eta$ which commute if the eigenfactors of $\varphi$ are linear. In fact, this decomposition is unique in the following sense.

**Proposition 1.6.5 (Diagonal-Nilpotent Decomposition)**
*Let $\varphi : V \longrightarrow V$ be a $K$-linear map such that the eigenfactors of $\varphi$ are linear.*

(a) *There exist a diagonalizable endomorphism $\delta$ and a nilpotent endomorphism $\eta$ of $V$ such that $\varphi = \delta + \eta$ and such that $\delta$ and $\eta$ commute.*
(b) *The endomorphisms $\delta$ and $\eta$ in (a) are uniquely determined by $\varphi$.*
(c) *The endomorphisms $\delta$ and $\eta$ are contained in $K[\varphi]$.*

*Proof* To prove (a), i.e. to find $\delta$ and $\eta$, we write $\mu_\varphi(z) = (z - \lambda_1)^{m_1} \cdots (z - \lambda_s)^{m_s}$ with $\lambda_i \in K$ and $m_i > 0$, and we let $G_i = \mathrm{Gen}(\varphi, \lambda_i)$. On $G_i$, we define $\delta_i = \lambda_i \, \mathrm{id}_V$ and $\eta_i = \varphi - \delta_i$. Then the map $\eta_i$ is nilpotent, since $(\varphi - \lambda_i \, \mathrm{id}_V)^{m_i} = 0$ by Theorem 1.1.17.d. The direct sums $\delta = \delta_1 \oplus \cdots \oplus \delta_s$ and $\eta = \eta_1 \oplus \cdots \oplus \eta_s$ provide a decomposition of $\varphi$ of the desired kind, since $\delta_i = \lambda_i \, \mathrm{id}_V$ and $\eta_i$ clearly commute.

Next we show (b), i.e. the uniqueness of this decomposition. Let $\varphi = \delta' + \eta'$ be another decomposition as in (a). Then $\delta' \eta' = \eta' \delta'$ shows $\delta' \varphi = \delta'(\delta' + \eta') = \varphi \delta'$ as well as $\eta' \varphi = \varphi \eta'$. By Proposition 1.1.6, it follows that $G_i$ is $\delta'$-invariant and $\eta'$-invariant, and therefore it suffices to prove the claimed uniqueness on $G_i$.

Now let us assume that there exists an element $\lambda \in K$ such that $V = \mathrm{Gen}(\varphi, \lambda)$. We choose a $K$-basis $B$ of $V$ in which $\delta'$ is represented by a diagonal matrix. In this basis also $\delta = \lambda \, \mathrm{id}_V$ is represented by a diagonal matrix. Hence the matrix of the map $\delta' - \delta = \eta - \eta'$ is both diagonal and nilpotent, and therefore zero.

To prove (c), we let $f_i(z) = \prod_{j \neq i}(z - \lambda_j)^{m_j}$ for $j = 1, \ldots, s$. Since no factor $(z - \lambda_k)$ divides all of them, they generate the unit ideal in $K[z]$, i.e. there are polynomials $g_i(z) \in K[z]$ such that $f_1(z)g_1(z) + \cdots + f_s(z)g_s(z) = 1$. For $i = 1, \ldots, s$, we now let $h_i(z) = f_i(z)g_i(z)$ so that $h_1(\varphi) + \cdots + h_s(\varphi) = \mathrm{id}_V$. For $v \in G_i$, we have $h_j(\varphi)(v) = g_j(\varphi)f_j(\varphi)(v) = g_j(\varphi)(0) = 0$ for $j \neq i$ hence $h_i(\varphi)(v) = \mathrm{id}_V(v) = v$. Hence the map $\lambda_1 h_1(\varphi) + \cdots + \lambda_k h_k(\varphi)$ agrees with $\delta$ on each $G_i$, and therefore on $V$ since we can write $V = \mathrm{Gen}(\varphi, \lambda_1) \oplus \cdots \oplus \mathrm{Gen}(\varphi, \lambda_s)$. Thus the map $\delta$ is contained in $K[\varphi]$, and therefore also $\eta = \varphi - \delta$.                    $\square$

If we drop the requirement that the maps $\delta$ and $\eta$ in this proposition commute, the decomposition $\varphi = \delta + \eta$ is not necessarily unique any more, as the following example shows.

**Example 1.6.6** Let $K = \mathbb{Q}$ and $V = K^2$. Consider the endomorphism $\varphi$ of $V$ which is given by the matrix $\left(\begin{smallmatrix} 1 & 0 \\ 0 & 2 \end{smallmatrix}\right)$ with respect to the canonical basis. Then $\varphi$ is clearly diagonalizable and has a decomposition $\varphi = \varphi + 0$ into the sum of a diagonalizable and a nilpotent map.

However, if we let $\delta \in \mathrm{End}_K(V)$ be given by the matrix $\left(\begin{smallmatrix} 1 & 1 \\ 0 & 2 \end{smallmatrix}\right)$ and $\eta \in \mathrm{End}_K(V)$ by the matrix $\left(\begin{smallmatrix} 0 & -1 \\ 0 & 0 \end{smallmatrix}\right)$, then we have $\varphi = \delta + \eta$, the map $\delta$ is diagonalizable, and $\eta$ is nilpotent. Notice that we have $\delta\eta \neq \eta\delta$ in this case.

If the minimal polynomial of $\varphi$ splits into linear factors, there is another normal form, namely the Jordan normal form of $\varphi$, about which we have the following to say.

**Remark 1.6.7** By carefully combining the preceding proposition, Proposition 1.3.2, and Corollary 1.3.3, one can deduce that, if $\mu_\varphi(z)$ splits in $K[z]$ into linear factors, then there exists a basis of $V$ with respect to which $\varphi$ is represented by a matrix having the usual Jordan normal form (see [9], Chap. 7). Since we are not going to use this here, we leave it to the interested reader to spell out the details. The main reason why we do not use it is that we do not have a useful generalization of the Jordan decomposition to families of commuting endomorphisms (see Example 2.6.4).

# Chapter 2
# Families of Commuting Endomorphisms

> Family: *a social unit where*
> *the father is concerned with parking space,*
> *the children with outer space,*
> *and the mother with closet space.*
> (Evan Esar)

Given a finite dimensional vector space $V$ over a field $K$, a commuting family is a mathematical unit where every member is an endomorphism of $V$, and its only concern is to commute with the other family members. More precisely, a commuting family $\mathscr{F}$ is a $K$-algebra consisting of pairwise commuting endomorphisms of $V$. Since $\mathscr{F}$ is a commutative ring, and since the vector space $V$ is a finitely generated $\mathscr{F}$-module via the operation $\varphi \cdot v = \varphi(v)$ for $\varphi \in \mathscr{F}$ and $v \in V$, the study of commuting families involves a considerable dose of commutative algebra.

After fixing a system of $K$-algebra generators $\Phi = (\varphi_1, \ldots, \varphi_n)$ of $\mathscr{F}$, we can present $\mathscr{F}$ in the form $\mathscr{F} = P/\operatorname{Rel}_P(\Phi)$, where $P = K[x_1, \ldots, x_n]$ and $\operatorname{Rel}_P(\Phi)$ is the ideal of algebraic relations of $\Phi$. Thus we can also consider $V$ as a $P$-module. Notice that this point of view depends on the choice of $\Phi$, and differs from the one in Chap. 1, where we started from a single endomorphism $\varphi$ rather than the commuting family $K[\varphi]$. Moreover, notice that the ideal $\operatorname{Rel}_P(\Phi)$ generalizes the principal ideal $\langle \mu_\varphi(z) \rangle$ generated by the minimal polynomial of a single endomorphism, while the characteristic polynomial of $\varphi$ does not generalize nicely to commuting families. Thus one of the key tools for computations involving commuting families is the Buchberger-Möller Algorithm for Matrices 2.1.4 which calculates $\operatorname{Rel}_P(\Phi)$ efficiently.

> *Ideals have a place in the kernel of the soul.*

**Electronic supplementary material** The online version of this chapter
(DOI:10.1007/978-3-319-43601-2_2) contains supplementary material, which is available to authorized users.

M. Kreuzer and L. Robbiano, *Computational Linear and Commutative Algebra*,
DOI 10.1007/978-3-319-43601-2_2

In the second section we continue our commutative algebra approach and examine ideals in a commuting family. Like fairy tales, they have a kernel of truth, and we are looking to find it. When we mentioned the definition of the kernel of an ideal to some colleagues, they would answer something like "If, in an exam, a student speaks about the kernel of an ideal, I would immediately fail him, since he is all mixed up about the basic definitions." Nevertheless, by letting the kernel of an ideal in $\mathscr{F}$ be the intersection of the kernels $\mathrm{Ker}(\varphi)$ with $\varphi \in \mathscr{F}$ and by defining its big kernel analogously as the intersection of the individual big kernels, we arrive at the soul of the chapter: the joint eigenspaces and generalized eigenspaces of $\mathscr{F}$ are, respectively, the kernels and big kernels of the maximal ideals of $\mathscr{F}$. Here a joint eigenspace of $\mathscr{F}$ is a vector subspace of $V$ which is the intersection of the eigenspaces $\mathrm{Eig}(\varphi, p_\varphi(z))$ for all $\varphi \in \mathscr{F}$. Rivalling the wonders and marvels of outer space, the vector space $V$ amazingly turns out to be the direct sum of the joint generalized eigenspaces $\mathrm{BigKer}(\mathfrak{m}_i)$ for $i = 1, \ldots, s$, where the ideals $\mathfrak{m}_1, \ldots, \mathfrak{m}_s$ are the maximal ideals of $\mathscr{F}$ (see Theorem 2.4.3).

The relationship between joint generalized eigenspaces and maximal ideals of $\mathscr{F}$ becomes even more intriguing when we consider the annihilator ideals $\mathfrak{q}_i = \mathrm{Ann}_\mathscr{F}(\mathrm{BigKer}(\mathfrak{m}_i))$, i.e., the ideals of all $\varphi \in \mathscr{F}$ such that $\varphi$ vanishes on $\mathrm{BigKer}(\mathfrak{m}_i)$. As we shall see in Sect. 2.4, the intersection of the ideals $\mathfrak{q}_i$ is the zero ideal. They are powers of the corresponding ideals $\mathfrak{m}_i$, and they yield a decomposition $\mathscr{F} \cong \prod_i \mathscr{F}/\mathfrak{q}_i$ into local rings. Thus the joint generalized eigenspaces of $\mathscr{F}$ are also related to the *primary components* $\mathfrak{q}_i$ of its zero ideal.

> *The purpose of computing is insight, not numbers.*
> (Richard Hamming)

In order to compute these ideals $\mathfrak{q}_i$, it would be nice if we could represent them as principal ideals and describe a generator explicitly. To do this, we need a further ingredient. Given a maximal ideal $\mathfrak{m}$ of $\mathscr{F}$, the monic generator $p_{\mathfrak{m},\varphi}(z)$ of the kernel of the $K$-algebra homomorphism $K[z] \longrightarrow \mathscr{F}/\mathfrak{m}$ given by $z \mapsto \varphi + \mathfrak{m}$ is one of the eigenfactors of $\varphi$. This assignment is surjective, but not in general injective. If an eigenfactor $p(z)$ of $\varphi$ arises from just one maximal ideal $\mathfrak{m}_i$ of $\mathscr{F}$, then the corresponding ideal $\mathfrak{q}_i$ is given by $\mathfrak{q}_i = \langle p(\varphi)^m \rangle$, where $m$ is the multiplicity of $p(z)$ in the minimal polynomial of $\varphi$. If the assignment $\mathfrak{m} \mapsto p_{\mathfrak{m},\varphi}(z)$ is bijective, we say that $\varphi$ is a splitting endomorphism. In this case we can compute all primary components $\mathfrak{q}_i$ as principal ideals.

Therefore we study the existence of splitting endomorphisms in Sect. 2.5. In particular, we show that the set of splitting endomorphisms in $\mathscr{F}$ contains a Zariski open subset if $K$ is infinite. This means that a generically chosen endomorphism $\varphi$ is splitting, and the calculation of the primary components $\mathfrak{q}_i$ is essentially reduced to the calculation and factorization of $\mu_\varphi(z)$.

Finally, to top off the chapter, we treat a couple of leftover topics. In the last section we characterize simultaneously diagonalizable and trigonalizable commuting families.

> *I read that you can make chocolate fondue from chocolate leftovers.*
> *I am confused. What are chocolate leftovers?*

## 2.1  Commuting Families

> *A commuting family is a space*
> *where part of the family waits*
> *until the rest of the family brings back the car.*

Morally speaking, matrices should not commute. Setting aside moral problems, commuting matrices do occur in many natural settings, as we shall for instance see in Chap. 4. So, let us not wait for them to show up, but let us get into the driver's seat and start studying them. Given $n$ pairwise commuting endomorphisms $\varphi_1, \ldots, \varphi_n$ of a finite dimensional vector space $V$ over a field $K$, we can form the *commuting family* $\mathcal{F} = K[\varphi_1, \ldots \varphi_n]$, where a constant $a \in K$ is interpreted as $a \cdot \mathrm{id}_V$. Since $\mathcal{F}$ is a zero-dimensional commutative ring, we can revisit and utilize many methods of Computational Commutative Algebra. In particular, the Buchberger-Möller algorithm for Matrices 2.1.4 allows us to compute a presentation $\mathcal{F} \cong K[x_1, \ldots, x_n]/\operatorname{Rel}_P(\Phi)$, where $\operatorname{Rel}_P(\Phi)$ is the ideal of algebraic relations of $\Phi = (\varphi_1, \ldots, \varphi_n)$. A further excursion leads to the insight that we can restrict a commuting family to an invariant subspace $U$ of $V$ and that we can find a presentation of the restricted family by calculating the annihilator of $U$ in $\mathcal{F}$. Finally, we have a short tour of the possible values of the dimension $\dim_K(\mathcal{F})$. Here we have to drive carefully, since we should not assume accidentally that this dimension equals $\dim_K(V)$, although it will do so in important special cases. In general, the dimension of $\mathcal{F}$ can be both larger and smaller than that of $V$. With the necessary patience and care, our commuting families will get going happily and safely.

> *Be careful! 80 % of people are caused by accidents.*

In the following we let $K$ be a field, let $V$ be a finite dimensional $K$-vector space, and let $d = \dim_K(V)$. Assume that we are given a set of commuting endomorphisms of $V$, in other words a set $S = \{\varphi_i\}_{i \in \Sigma}$ in $\operatorname{End}_K(V)$ indexed by a set $\Sigma$ and having the property that $\varphi_i \circ \varphi_j = \varphi_j \circ \varphi_i$ for all $i, j \in \Sigma$. Notice that also polynomial combinations of the elements of $S$ commute. Thus the $K$-subalgebra $\mathcal{F}$ of $\operatorname{End}_K(V)$ generated by $S$, i.e. the $K$-algebra $\mathcal{F} = K[S]$, is a commutative ring. To define the $K$-algebra structure of $\mathcal{F}$, we identify an element $c \in K$ with the endomorphism $c \operatorname{id}_V$ in $\operatorname{End}_K(V)$.

**Definition 2.1.1**   Given a set of commuting endomorphisms $S$ of the vector space $V$, we let $\mathcal{F} = K[S]$ be the commutative $K$-subalgebra of $\operatorname{End}_K(V)$ generated by $S$ and call it the **family of commuting endomorphisms**, or simply the **commuting family**, generated by $S$.

**Remark 2.1.2**   Notice that the vector space $V$ carries a natural structure of a module over the $K$-algebra $\mathcal{F}$ given by $\varphi \cdot v = \varphi(v)$ for all $\varphi \in \mathcal{F}$ and $v \in V$. We have $\operatorname{Ann}_{\mathcal{F}}(V) = \langle 0 \rangle$, i.e., the vector space $V$ is a faithful $\mathcal{F}$-module, since a map $\varphi \in \mathcal{F}$ with $\varphi(v) = 0$ for all $v \in V$ is necessarily the zero map.

In this section we study general properties of commuting families. Later on, special commuting families and their properties will be examined. Since the family $\mathscr{F}$ is a vector subspace of $\mathrm{End}_K(V)$, it is a finite dimensional $K$-vector space. It follows that $\mathscr{F}$ is a zero-dimensional commutative $K$-algebra (see for instance [15], Proposition 3.7.1).

Given a system of $K$-algebra generators $\Phi = (\varphi_1, \ldots, \varphi_n)$ of $\mathscr{F}$, we form the polynomial ring $P = K[x_1, \ldots, x_n]$ and introduce the following ideal.

**Definition 2.1.3** Let $\Phi = (\varphi_1, \ldots, \varphi_n)$ be a system of $K$-algebra generators of $\mathscr{F}$, and let the $K$-algebra epimorphism $\pi : P \longrightarrow \mathscr{F}$ be defined by $\pi(x_i) = \varphi_i$ for every $i \in \{1, \ldots, n\}$. Then the kernel of $\pi$ is called the **ideal of algebraic relations** of $\Phi$ and denoted by $\mathrm{Rel}_P(\Phi)$.

Clearly, if $\Phi$ consists of a single endomorphism $\varphi$, the ideal of algebraic relations of $\Phi$ is the principal ideal generated by the minimal polynomial of $\varphi$.

Given the tuple $\Phi$ of endomorphisms of $V$, the ideal $\mathrm{Rel}_P(\Phi)$ can be computed effectively, as the following algorithms shows. In particular, given a matrix $A = (a_{ij}) \in \mathrm{Mat}_d(K)$, we order the entries according to the lexicographic ordering of their indices $(i, j)$ and thereby transform (a.k.a. "flatten") the matrix to a vector in $K^{d^2}$.

**Algorithm 2.1.4 (The Buchberger-Möller Algorithm for Matrices)**
*Let $\Phi = (\varphi_1, \ldots, \varphi_n)$ be a system of $K$-algebra generators of $\mathscr{F}$. For every $i \in \{1, \ldots, n\}$, let $M_i \in \mathrm{Mat}_d(K)$ be a matrix representing $\varphi_i$ with respect to a fixed $K$-basis of $V$, and let $\sigma$ be a term ordering on $\mathbb{T}^n$. Consider the following sequence of instructions.*

(1) *Let $G = \emptyset$, $\mathscr{O} = \emptyset$, $S = \emptyset$, $\mathscr{N} = \emptyset$, and $L = \{1\}$.*

(2) *If $L = \emptyset$, return the pair $(G, \mathscr{O})$ and stop. Otherwise let $t = \min_\sigma(L)$ and delete it from $L$.*

(3) *Compute $t(M_1, \ldots, M_n)$ and reduce it against $\mathscr{N} = (N_1, \ldots, N_k)$ to obtain*

$$R = t(M_1, \ldots, M_n) - \textstyle\sum_{i=1}^k c_i N_i \quad \text{with } c_i \in K$$

(4) *If $R = 0$, append the polynomial $t - \sum_{i=1}^k c_i s_i$ to $G$, where $s_i$ denotes the $i$-th element of $S$. Remove from $L$ all multiples of $t$. Continue with Step (2).*

(5) *Otherwise, we have $R \neq 0$. Append $R$ to $\mathscr{N}$ and $t - \sum_{i=1}^k c_i s_i$ to $S$. Append the term $t$ to $\mathscr{O}$, and append to $L$ those elements of $\{x_1 t, \ldots, x_n t\}$ which are neither multiples of a term in $L$ nor in $\mathrm{LT}_\sigma(G)$. Continue with Step (2).*

> *This is an algorithm which computes the reduced $\sigma$-Gröbner basis of $\mathrm{Rel}_P(\Phi)$ and a list of terms $\mathcal{O}$ whose residue classes form a vector space basis of $P/\mathrm{Rel}_P(\Phi)$.*

The proof of this algorithm uses the theory of Gröbner bases. A sketch can be found in [16], Tutorial 91.j, and a complete proof is contained in [14], Theorem 4.1.7. Notice that the result of this algorithm does not depend on the choice of the basis of $V$. The following proposition shows the invariance of $\mathrm{Rel}_P(\Phi)$ with respect to field extensions.

**Proposition 2.1.5** *Let $\Phi = (\varphi_1, \ldots, \varphi_n)$ be a system of $K$-algebra generators of the family $\mathscr{F}$, let $L \supseteq K$ be a field extension, let $V_L = V \otimes_K L$. Then let $(\varphi_i)_L = \varphi_i \otimes_K L : V_L \longrightarrow V_L$ be the extension of $\varphi_i$ for $i = 1, \ldots, n$, and let $\Phi_L = ((\varphi_1)_L, \ldots, (\varphi_n)_L)$. Then the ideal $\mathrm{Rel}_{L[x_1,\ldots,x_n]}(\Phi_L)$ is the extension ideal of $\mathrm{Rel}_P(\Phi)$ to $L[x_1, \ldots, x_n]$.*

*Proof* A $K$-basis $B$ of $V$ is also an $L$-basis of $V_L$, and the matrices which represent the endomorphisms $\varphi_i$ with respect to $B$ also represent $(\varphi_i)_L$. Now the claim follows by applying the above algorithm for $\Phi_L$ and noticing that all of its operations are performed over $K$. Hence the resulting polynomials will have coefficients in $K$. $\square$

In analogy to Definition 1.1.5, we define invariant subspaces for a commuting family as follows.

**Definition 2.1.6** A $K$-vector subspace $U$ of $V$ is called an **invariant subspace** for the family $\mathscr{F}$ if we have $\varphi(U) \subseteq U$ for all $\varphi \in \mathscr{F}$.

Given an invariant subspace $U$ for the family $\mathscr{F}$, we can consider the **restricted family** $\mathscr{F}_U = \{\varphi|_U \mid \varphi \in \mathscr{F}\}$. Then the vector space $U$ carries the structure of an $\mathscr{F}_U$-module. As a consequence of the Buchberger-Möller Algorithm for Matrices we get the following algorithm for computing annihilators of invariant subspaces.

**Algorithm 2.1.7 (The Annihilator of an Invariant Subspace)**
*Let $\Phi = (\varphi_1, \ldots, \varphi_n)$ be a system of $K$-algebra generators of the family $\mathscr{F}$, let $U \subseteq V$ be an invariant subspace for $\mathscr{F}$, and let $\Phi_U = (\varphi_1|_U, \ldots, \varphi_n|_U)$ be the corresponding system of generators of the restricted family. The following instructions compute $\mathrm{Ann}_{\mathscr{F}}(U)$.*

(1) *Using the Buchberger-Möller Algorithm for Matrices, compute $\mathrm{Rel}_P(\mathscr{F}_U)$.*
(2) *Return $\pi(\mathrm{Rel}_P(\mathscr{F}_U))$, where $\pi : P \longrightarrow \mathscr{F}$ is given by $x_i \mapsto \varphi_i$ for every $i \in \{1, \ldots, n\}$.*

*Proof* Step (1) computes a set of generators of the ideal in $P$ whose members are the polynomials $f$ such that $f(\varphi_1|_U, \ldots, \varphi_n|_U) = 0$. Its image in $\mathscr{F}$, computed in Step (2), is clearly the annihilator of $U$.                                             $\square$

Let us apply this algorithm to a concrete example.

**Example 2.1.8** Let $K = \mathbb{Q}$, let $V = K^6$ and let $\varphi_1, \varphi_2 \in \operatorname{End}_K(V)$ be defined by the matrices

$$A_1 = \begin{pmatrix} 0 & 0 & 0 & 0 & 0 & 0 \\ 0 & 0 & 0 & 0 & 0 & 0 \\ 1 & 0 & 0 & 0 & 0 & 0 \\ 0 & 1 & 0 & 0 & 0 & 0 \\ 0 & 0 & 1 & 0 & 1 & 0 \\ 0 & 0 & 0 & 1 & 0 & 1 \end{pmatrix} \quad \text{and} \quad A_2 = \begin{pmatrix} 0 & 2 & 0 & 0 & 0 & 0 \\ 1 & 0 & 0 & 0 & 0 & 0 \\ 0 & 0 & 0 & 2 & 0 & 0 \\ 0 & 0 & 1 & 0 & 0 & 0 \\ 0 & 0 & 0 & 0 & 0 & 2 \\ 0 & 0 & 0 & 0 & 1 & 0 \end{pmatrix}$$

respectively. Since $A_1$ and $A_2$ commute, we have a commuting family $\mathscr{F} = K[\varphi_1, \varphi_2]$. It is easy to check that the vectors $u_1 = e_1 - e_3$, $u_2 = e_2 - e_4$, $u_3 = e_1 - e_5$, and $u_4 = e_2 - e_6$ are $K$-linearly independent. Let $U = \langle u_1, u_2, u_3, u_4 \rangle$. We calculate

$$\varphi(u_1) = u_3 - u_1, \quad \varphi(u_2) = u_4 - u_2, \quad \varphi(u_3) = u_3 - u_1, \quad \varphi(u_4) = u_4 - u_2$$
$$\varphi_2(u_1) = u_2, \qquad \varphi_2(u_2) = 2u_1, \qquad \varphi_2(u_3) = u_4, \qquad \varphi_2(u_4) = 2u_3$$

and conclude that $U$ is an invariant subspace for $\mathscr{F}$.

The matrices associated to $\varphi_1|_U$ and $\varphi_2|_U$ with respect to the basis $(u_1, u_2, u_3, u_4)$ are

$$M_1 = \begin{pmatrix} -1 & 0 & -1 & 0 \\ 0 & -1 & 0 & -1 \\ 1 & 0 & 1 & 0 \\ 0 & 1 & 0 & 1 \end{pmatrix} \quad \text{and} \quad M_2 = \begin{pmatrix} 0 & 2 & 0 & 0 \\ 1 & 0 & 0 & 0 \\ 0 & 0 & 0 & 2 \\ 0 & 0 & 1 & 0 \end{pmatrix}$$

Using the Buchberger-Möller Algorithm for Matrices 2.1.4 applied to the pair $(M_1, M_2)$, we get the ideal $\operatorname{Rel}_P(\mathscr{F}_U) = \langle x_1^2, x_2^2 - 2 \rangle$. Consequently, we obtain $\operatorname{Ann}_{\mathscr{F}}(U) = \langle \varphi_1^2 \rangle$, since $\varphi_2^2 - \operatorname{id}_V = 0$ and $\varphi_1^2 \neq 0$.

Clearly, the **annihilator** in $\mathscr{F}$ can be defined for any vector subspace of $V$, not just an invariant one. The following remark explains how to compute it.

**Remark 2.1.9** Let $(\psi_1, \ldots, \psi_\delta)$ be a $K$-basis of $\mathscr{F}$, where $\delta = \dim_K(\mathscr{F})$, let $U$ be a $K$-vector subspace of $V$, and let $(u_1, \ldots, u_e)$ be a $K$-basis of $V$. Then we can calculate the annihilator $\operatorname{Ann}_{\mathscr{F}}(U) = \{u \in U \mid \varphi(u) = 0 \text{ for all } \varphi \in \mathscr{F}\}$ as follows.

Introduce new indeterminates $y_1, \ldots, y_\delta$ and write down the linear system of equations

$$\psi_1(u_1)y_1 + \cdots + \psi_\delta(u_1)y_\delta = 0$$

$$\vdots$$

$$\psi_1(u_e)y_1 + \cdots + \psi_\delta(u_e)y_\delta = 0$$

Compute a $K$-basis $\{(c_{i1}, \ldots, c_{i\delta}) \mid i = 1, \ldots, \varepsilon\}$ of its solution space. Then the endomorphisms $c_{i1}\psi_1 + \cdots + c_{i\delta}\psi_\delta$ with $i \in \{1, \ldots, \varepsilon\}$ form a $K$-basis of $\mathrm{Ann}_{\mathscr{F}}(U)$.

As a consequence of Algorithm 2.1.7, we get the following description of the restriction of the family $\mathscr{F}$ to an invariant subspace.

**Corollary 2.1.10** *Let $U \subseteq V$ be an invariant subspace for $\mathscr{F}$. Then the restriction map $\varrho : \mathscr{F} \longrightarrow \mathscr{F}_U$ induces an isomorphism of $K$-algebras $\mathscr{F} / \mathrm{Ann}_{\mathscr{F}}(U) \cong \mathscr{F}_U$.*

*Proof* In Algorithm 2.1.7, we have the inclusion $\mathrm{Rel}(\mathscr{F}) \subseteq \mathrm{Rel}(\mathscr{F}_U)$, and therefore $\mathscr{F}_U \cong P / \mathrm{Rel}(\mathscr{F}_U) \cong (P / \mathrm{Rel}(\mathscr{F})) / (\mathrm{Rel}(\mathscr{F}_U) / \mathrm{Rel}(\mathscr{F})) \cong \mathscr{F} / \mathrm{Ann}_{\mathscr{F}}(U)$. $\qquad\square$

Based on the Buchberger-Möller Algorithm for Matrices, we have an alternative to Algorithm 1.1.8 for computing the minimal polynomial of an element of $\mathscr{F}$.

**Algorithm 2.1.11 (The Minimal Polynomial of a Family Member)**
*Let $\Phi = (\varphi_1, \ldots, \varphi_n)$ be a system of $K$-algebra generators of the family $\mathscr{F}$, let $f(x_1, \ldots, x_n) \in P$, and let $\psi = f(\varphi_1, \ldots, \varphi_n)$. The following instructions compute the minimal polynomial of $\psi$.*

(1) *In the ring $P[z]$ form the ideal $I_\psi = \langle z - f(x_1, \ldots, x_n) \rangle + \mathrm{Rel}_P(\Phi) \cdot P[z]$ and compute $J_\psi = I_\psi \cap K[z]$.*
(2) *Return the monic generator of $J_\psi$.*

*Proof* By [15], Proposition 3.6.2, these instructions compute the monic generator of the kernel of the $K$-algebra homomorphism $K[z] \longrightarrow \mathscr{F}$ defined by $z \mapsto \psi$. This polynomial is exactly the minimal polynomial of $\psi$. $\qquad\square$

Recall that a field $K$ is called **perfect** if either its characteristic is 0 or its characteristic is $p > 0$ and we have $K = K^p$ (see also [15], Definition 3.7.7). It is known that finite fields are perfect (see [10], Sect. 4.4). Moreover, the **nilradical** of $\mathscr{F}$ is $\mathrm{Rad}_{\mathscr{F}}(0) = \{\psi \in \mathscr{F} \mid \psi^i = 0 \text{ for some } i \geq 1\}$, i.e., the set of nilpotent endomorphisms in $\mathscr{F}$. We can compute the nilradical of $\mathscr{F}$ as follows.

**Remark 2.1.12** Let $\Phi = (\varphi_1, \ldots, \varphi_n)$ be a system of $K$-algebra generators of $\mathscr{F}$. Using the Buchberger-Möller Algorithm for Matrices 2.1.4, we compute the ideal

$\text{Rel}_P(\Phi)$. Then we use Algorithm 1.1.8 or Algorithm 2.1.11 to compute the minimal polynomial $\mu_{\varphi_i}(x_i)$ for every $i \in \{1, \ldots, n\}$.

If $K$ is a perfect field, we can use the algorithms in [15], 3.7.9 and 3.7.12 to compute the squarefree parts $f_i(x_i) = \text{sqfree}(\mu_{\varphi_i}(x_i))$ for $i = 1, \ldots, n$. Then the nilradical of $\mathscr{F}$ is given by $\text{Rad}_{\mathscr{F}}(0) = \langle f_1(\varphi_1, \ldots, \varphi_n), \ldots, f_t(\varphi_1, \ldots, \varphi_n) \rangle$. Later we will see an improvement of this method (see 5.4.2). If $K$ is not perfect, we can use [13], Alg. 7 to compute the radical of $\text{Rel}_P(\Phi)$ and get $\text{Rad}_{\mathscr{F}}(0)$ as above.

In the last part of this section we study the dimension of a commuting family.

**Definition 2.1.13** Let $\mathscr{F}$ be a family of commuting endomorphisms of $V$. The number $\dim_K(\mathscr{F})$ is called the **dimension** of the family $\mathscr{F}$.

The dimension of a commuting family can be larger than, equal to, or smaller than the dimension of $V$, as the following examples show.

**Example 2.1.14** If $\varphi \in \text{End}_K(V)$ and $\mathscr{F} = K[\varphi]$. Then the dimension of $\mathscr{F}$ is given by $\dim_K(\mathscr{F}) = \deg(\mu_\varphi(z))$. Thus it is smaller than or equal to $d$, with equality if and only if $\varphi$ is commendable (see Theorem 1.5.8).

The maximal dimension of a commuting family was determined by J. Schur (see [26]) and N. Jacobson (see [12]) a long time ago: it coincides with $\lfloor d^2/4 \rfloor + 1$ for $d = \dim_K(V)$. In the next example $\dim_K(\mathscr{F})$ surpasses $\dim_K(V)$ maximally.

**Example 2.1.15** Let $V = K^6$, and let $\mathscr{F}$ be the $K$-algebra generated by $\{\text{id}_V\}$ and the set of all endomorphisms of $V$ whose matrix with respect to the canonical basis of $V$ is of the form $\begin{pmatrix} 0 & A \\ 0 & 0 \end{pmatrix}$ with a matrix $A$ of size $3 \times 3$. The family $\mathscr{F}$ is commuting, since $\begin{pmatrix} 0 & A \\ 0 & 0 \end{pmatrix}\begin{pmatrix} 0 & B \\ 0 & 0 \end{pmatrix} = \begin{pmatrix} 0 & B \\ 0 & 0 \end{pmatrix}\begin{pmatrix} 0 & A \\ 0 & 0 \end{pmatrix} = \begin{pmatrix} 0 & 0 \\ 0 & 0 \end{pmatrix}$ for all matrices $A$, $B$ of size $3 \times 3$. Here we have $\dim_K(V) = 6$ and $\dim_K(\mathscr{F}) = 10 = 6^2/4 + 1$, the maximal possible dimension.

The special case of a family generated by two commuting matrices was considered by Gerstenhaber (see [8]) who proved that in this case the sharp upper bound for the dimension of $\mathscr{F}$ is $d$. A sharp upper bound for the dimension of a family generated by three commuting matrices is apparently not known.

## 2.2 Kernels and Big Kernels of Ideals

> *A serious and good philosophical work*
> *could be written consisting entirely of jokes.*
> (Ludwig Wittgenstein)

Dear Reader, although the notion of the kernel of an ideal may initially sound like a joke, let us assure you that it is a serious and philosophically sound idea. Let us meditate a bit.

The eigenspaces and generalized eigenspaces of an endomorphism $\varphi$ are defined as the kernels and big kernels, respectively, of maps of the form $p_i(\varphi)$, where $p_i(z)$ is an eigenfactor of $\varphi$. It is a standard fact in Linear Algebra that commuting endomorphisms have joint eigenspaces if their eigenfactors are linear. But what about the general case? How can we define a "joint" kernel or big kernel of many endomorphisms? Naturally, we can take the intersection of the individual kernels or big kernels. But will this intersection be non-zero in interesting cases?

The purpose of this section is to convince you that it can be done! We define the kernel and the big kernel of an ideal $I$ in a commuting family and show that it is non-zero whenever $I$ is a proper ideal. To prove that this definition is no joke, we have to subject you to a crash course on zero-dimensional rings in the first subsection. Then things get really serious in the second part of the section where we study properties of kernels and big kernels of ideals in a commuting family. Without further ado, but with the advice to take everything we write with a grain of salt, let us conclude this introduction with the following utterance of a famous philosopher.

*When you get used to sugar,*
*you do not accept salt anymore.*
(Giovanni Trapattoni)

## 2.2.A Properties of Zero-Dimensional Rings

*Never make fun of someone who speaks broken English.*
*It means they know another language.*
(H. Jackson Brown, Jr.)

Before we continue our study of commuting families, we need additional tools. So far we have only talked about Linear Algebra. Now we need to learn to speak the language of Commutative Algebra, and in particular the part of it which deals with zero-dimensional rings. We assume that the reader knows already the terms ideal, prime ideal, and maximal ideal, and their basic properties. Here we introduce a few further words such as comaximal ideals, zero radical, and prime decomposition. More precisely, we show that a zero-dimensional ring has only finitely many maximal ideals, that their intersection is the zero radical of the ring, and that the ring modulo the zero radical is isomorphic to the product of its residue class fields modulo the maximal ideals. Another important notion is the annihilator of an element. In fact, as we shall see, every maximal ideal of a zero-dimensional algebra over a field can be represented as the annihilator of an element. After reaching this basic level, we announce that the continuation of our language course for the Commutative Algebra of zero-dimensional rings is contained in Chap. 4.

*I speak two languages: Body and English.*
(Mae West)

Let $R$ be a commutative ring. Recall that an ideal $I$ in $R$ is called **proper** if $I \neq R$, and two ideals $I_1, I_2$ in $R$ are called **comaximal** if $I_1 + I_2 = R$. The following results are classical tools in Commutative Algebra. We include them here for the convenience of the reader.

**Theorem 2.2.1 (Chinese Remainder Theorem)**
*Let $R$ be a ring, let $I_1, \dots, I_t$ be comaximal ideals in $R$, and let $\pi$ denote the canonical $R$-linear map $\pi : R/(I_1 \cap \cdots \cap I_t) \longrightarrow \prod_{i=1}^{t} R/I_i$.*

(a) *For $i = 1, \dots, s$, the ideals $I_i$ and $J_i = \prod_{j \neq i} I_j$ are comaximal.*
(b) *We have $I_1 \cap \cdots \cap I_s = I_1 \dots I_s$.*
(c) *The map $\pi$ is an isomorphism of rings.*

*Proof* To show (a), we argue by contradiction. Suppose that there exists a maximal ideal $\mathfrak{m}$ of $R$ containing both $I_i$ and $J_i$. Since $\mathfrak{m}$ is prime, it then contains one of the factors $I_j$, and this contradicts the hypothesis $I_i + I_j = R$.

Claim (b) follows from (a) by induction on $s$. For $s = 2$, it suffices to note that

$$I_1 \cap I_2 = (I_1 \cap I_2)(I_1 + I_2) = (I_1 \cap I_2)I_1 + (I_1 \cap I_2)I_2 \subseteq I_1 I_2$$

and that the other inclusion is obviously true.

Now we prove claim (c). The map $\pi$ is clearly injective, so we need to prove its surjectivity. Claim (a) implies that for every $i \in \{1, \dots, t\}$ there are elements $a_i \in I_i$ and $b_i \in J_i$ such that $a_i + b_i = 1$. Now it is easy to check that an element $(c_1 + I_1, \dots, c_t + I_t)$ of $\prod_{i=1}^{t} R/I_i$ is the image of the residue class of the element $c_1 b_1 + \cdots + c_t b_t = c_1(1 - a_1) + \cdots + c_t(1 - a_t)$ under $\pi$.  □

A first application of this theorem is given by the following corollary. It can be interpreted as a special case of Lemma 1.1.11. Recall that the **minimal polynomial** $\mu_r(z)$ of an element $r$ of a zero-dimensional $K$-algebra $R$ is the monic polynomial in $K[z]$ of least degree such that $\mu_r(r) = 0$.

**Corollary 2.2.2** *Let $K$ be a field, let $P = K[x_1, \dots, x_n]$, let $I_1, \dots, I_s$ be pairwise comaximal ideals in $P$, let $J = I_1 \cap \cdots \cap I_s$, and assume that $P/J$ is a zero-dimensional $K$-algebra. Given $f \in P/J$ we denote the residue classes of $f$ in $P/I_i$ by $f_i$. Then we have $\mu_f(z) = \mathrm{lcm}(\mu_{f_1}(z), \dots, \mu_{f_s}(z))$.*

*Proof* The theorem yields an isomorphism $\pi : P/J \longrightarrow \prod_{i=1}^{s} P/I_i$. Then we let $\varepsilon : K[z] \longrightarrow P/J$ be the $K$-algebra homomorphism which maps $z$ to $f$, and, for $i = 1, \dots, s$, we let $\varepsilon_i : K[z] \longrightarrow P/I_i$ be the $K$-algebra homomorphism which maps $z$ to $f_i$. In view of the definition of $\pi$, we see that we have $\pi \circ \varepsilon = \prod_{i=1}^{s} \varepsilon_i$.

Hence we deduce

$$\langle \mu_f(z) \rangle = \mathrm{Ker}(\varepsilon) = \mathrm{Ker}(\pi \circ \varepsilon) = \mathrm{Ker}\left(\prod_{i=1}^{s} \varepsilon_i\right)$$

$$= \bigcap_{i=1}^{s} \mathrm{Ker}(\varepsilon_i) = \bigcap_{i=1}^{s} \langle \mu_{f_i}(z) \rangle = \left\langle \mathrm{lcm}\big(\mu_{f_1}(z), \dots, \mu_{f_s}(z)\big)\right\rangle$$

and the proof is complete.                                                              □

Let us illustrate this corollary with an example.

**Example 2.2.3**  In the ring $P = \mathbb{Q}[x_1, x_2]$ we consider the ideal

$$J = \langle x_1^2 - 3x_1 + 2, \ x_2^3 - x_1 x_2 - 2x_2^2 + 2x_1 \rangle$$

Furthermore, we define the ideals $I_1 = \langle x_1 - 1 \rangle + J = \langle x_1 - 1, x_2^3 - 2x_2^2 - y_2 + 2 \rangle$ and $I_2 = \langle x_1 - 2 \rangle + J = \langle x_1 - 2, x_2^3 - 2x_2^2 - 2x_2 + 4 \rangle$. They are comaximal ideals and satisfy $J = I_1 \cap I_2$.

For the residue class $f = x_2 + J$ in $P/J$ and for $i = 1, 2$, we let $f_i$ denote the residue class of $f$ in $P/I_i$. Then we have $\mu_{f_1}(z) = z^3 - 2z^2 - z + 2 = (z + 1)(z - 1)(z - 2)$ and $\mu_{f_2}(z) = z^3 - 2z^2 - 2z + 4 = (z - 2)(z^2 - 2)$. Now an application of the corollary yields $\mu_f(z) = \mathrm{lcm}(\mu_{f_1}(z), \mu_{f_2}(z)) = (z + 1)(z - 1) \times (z - 2)(z^2 - 2)$, in agreement with the result of the direct calculation of $\mu_f(z)$.

Using the Chinese Remainder Theorem, we can study the ideals in a zero-dimensional algebra by examining suitable direct products.

**Proposition 2.2.4**  *Let $R$ be a zero-dimensional $K$-algebra, and let $d = \dim_K(R)$. Then there are only finitely many maximal ideals in $R$, say $\mathfrak{m}_1, \dots, \mathfrak{m}_s$, and $s \leq d$.*

*Proof*  For a contradiction, suppose that there are more that $d$ maximal ideals in $R$, and let $\mathfrak{m}_1, \dots, \mathfrak{m}_{d+1}$ be $d + 1$ of them. Since they are pairwise comaximal, Theorem 2.2.1 implies that the map $\varphi : R/(\bigcap_{i=1}^{d+1} \mathfrak{m}_i) \longrightarrow \prod_{i=1}^{d+1} R/\mathfrak{m}_i$ is an isomorphism. We deduce that $d \geq \dim_K(R/(\bigcap_{i=1}^{d+1} \mathfrak{m}_i)) = \sum_{i=1}^{d+1} \dim_K(R/\mathfrak{m}_i) \geq d + 1$, a contradiction.                                                              □

Another application of the Chinese Remainder Theorem is the following proposition which describes the nilpotent elements of a zero-dimensional $K$-algebra.

**Proposition 2.2.5 (Prime Decomposition of the Radical)**
*Let $R$ be a zero-dimensional $K$-algebra, let $\mathfrak{m}_1, \dots, \mathfrak{m}_s$ be its maximal ideals, and let $d = \dim_K(R)$.*

(a) *For every ideal $J$ in $R$, there exists $t \geq 1$ such that $J^t = J^u$ for every $u \geq t$.*
(b) *If $J \subseteq \bigcap_{i=1}^{s} \mathfrak{m}_i$, there exists a number $t \geq 1$ such that $J^t = \langle 0 \rangle$.*
(c) *We have $\mathrm{Rad}(0) = \bigcap_{i=1}^{s} \mathfrak{m}_i = \prod_{i=1}^{s} \mathfrak{m}_i$. This description is also called the* **prime decomposition** *of the* **zero radical** $\mathrm{Rad}(0)$.

*Proof* To prove (a), we observe that the chain $R \supset J \supseteq J^2 \supseteq \cdots \supseteq J^3 \supseteq \cdots$ becomes stationary, since $R$ is a finite dimensional $K$-vector space. Let $t \geq 1$ be such that $J^t = J^{t+1}$. Then we have $J^{t+2} = J \cdot J^{t+1} = J \cdot J^t = J^{t+1} = J^t$. Continuing to argue in this way, we get $J^u = J^t$ for all $u \geq t$.

To prove (b), we use (a) and let $t \geq 1$ be such that $J^t = J^u$ for every $u \geq t$. Then we let $b_1, \ldots, b_r \in R$ such that $J^t = \langle b_1, \ldots, b_r \rangle$. From $J^t = J \cdot J^t$ we deduce that there exist elements $a_{ij} \in J$ such that $b_i = \sum_{j=1}^r a_{ij} b_j$ for $i = 1, \ldots, r$. Thus the matrix $A = (a_{ij}) \in \mathrm{Mat}_r(R)$ satisfies $(I_r - A) \cdot (b_1, \ldots, b_r)^{\mathrm{tr}} = 0$. By multiplying this equation on the left by the adjoint matrix of $I_r - A$, we get $\det(I_r - A) \cdot b_i = 0$ for $i = 1, \ldots, r$. The determinant has the form $1 - a$ with $a \in J$. Hence we have $(1 - a)b_i = 0$ for $i = 1, \ldots, r$. Next we observe that $1 - a$ is not contained in any maximal ideal $\mathfrak{m}_i$ of $R$, because otherwise we would have $1 = a + (1 - a) \in \mathfrak{m}_i$. Therefore the element $1 - a$ is a unit of $R$. It follows that $b_1 = \cdots = b_r = 0$, and claim (b) is proved.

Finally, we prove claim (c). The inclusion $\mathrm{Rad}(0) \subseteq \bigcap_{i=1}^s \mathfrak{m}_i$ is clearly true. Claim (b) implies that there exists a number $t \geq 1$ such that $(\bigcap_{i=1}^s \mathfrak{m}_i)^t = \langle 0 \rangle$. This implies the inclusion $\bigcap_{i=1}^s \mathfrak{m}_i \subseteq \mathrm{Rad}(0)$. Finally, the equality $\bigcap_{i=1}^s \mathfrak{m}_i = \prod_{i=1}^s \mathfrak{m}_i$ follows from Theorem 2.2.1.b.                                                                                      □

**Corollary 2.2.6** *Let $R$ be a zero-dimensional $K$-algebra, and assume that $R$ is an integral domain. Then $R$ is a field.*

*Proof* Let $\mathfrak{m}_1, \ldots, \mathfrak{m}_s$ be the maximal ideals of $R$. By assumption, the only nilpotent element in $R$ is 0. We deduce from part (c) of the proposition that $\langle 0 \rangle = \bigcap_{i=1}^s \mathfrak{m}_i$. Then the Chinese Remainder Theorem 2.2.1 implies that there is an isomorphism $R \cong \prod_{i=1}^s R/\mathfrak{m}_i$. The ring on the right-hand side is an integral domain if and only if $s = 1$. This concludes the proof.                                                                                      □

Recall that the **Krull dimension** of a commutative ring is the length $\ell$ of the longest proper chain of prime ideals $\mathfrak{p}_0 \subset \mathfrak{p}_1 \subset \cdots \subset \mathfrak{p}_\ell$. The following corollary says that the name "zero-dimensional algebra" given to a $K$-algebra $R$ with $\dim_K(R) < \infty$ in [15], Definition 3.7.2, is justified.

**Corollary 2.2.7** *Every prime ideal in a zero-dimensional $K$-algebra $R$ is maximal. In other words, the Krull dimension of $R$ is zero.*

*Proof* Let $\mathfrak{m}_1, \ldots, \mathfrak{m}_s$ be the maximal ideals of $R$, and let $\mathfrak{p}$ be a prime ideal of $R$. Since the intersection $\mathfrak{m}_1 \cap \cdots \cap \mathfrak{m}_s$ consists of nilpotent elements by the proposition, it is contained in $\mathfrak{p}$. Hence the residue class ideal of $\mathfrak{p}$ corresponds under the isomorphism $R/(\mathfrak{m}_1 \cap \cdots \cap \mathfrak{m}_s) \longrightarrow R/\mathfrak{m}_1 \times \cdots \times R/\mathfrak{m}_s$ to a prime ideal in a product of fields. Therefore this residue class ideal, and consequently the ideal $\mathfrak{p}$ itself, is a maximal ideal.                                                                                      □

Next we draw some conclusions about the shape of a maximal ideal in a zero-dimensional $K$-algebra.

**Proposition 2.2.8 (Maximal Ideals as Annihilators)**
*Let $R$ be a zero-dimensional $K$-algebra, and let $\mathfrak{m}$ be a maximal ideal in $R$. Then there exists an element $f \in R \setminus \{0\}$ such that $\mathfrak{m} = \operatorname{Ann}_R(f)$.*

*Proof* If $\mathfrak{m} = \langle 0 \rangle$, we can take $f = 1$. So, let $\langle 0 \rangle \subset \mathfrak{m}$. Proposition 2.2.5 implies that $\operatorname{Rad}(0) = \prod_{i=1}^{s} \mathfrak{m}_i$ and that there exists a number $t \geq 1$ such that $(\prod_{i=1}^{s} \mathfrak{m}_i)^t = \langle 0 \rangle$, where $\mathfrak{m}_1, \ldots, \mathfrak{m}_s$ are the maximal ideals of $R$. We may assume that $\mathfrak{m} = \mathfrak{m}_1$. Next we let $u \geq 0$ be the maximal exponent such that we have $J = \mathfrak{m}^u \cdot (\prod_{i=2}^{s} \mathfrak{m}_i)^t \neq \langle 0 \rangle$. For every $f \in J \setminus \{0\}$, we get $\mathfrak{m} \cdot f = \langle 0 \rangle$, and therefore $\mathfrak{m} \subseteq \operatorname{Ann}_R(f)$. The fact that the ideal $\operatorname{Ann}_R(f)$ is a proper ideal of $R$ implies the equality $\mathfrak{m} = \operatorname{Ann}_R(f)$, as was to be shown. $\qquad\square$

## 2.2.B Properties of Kernels and Big Kernels of Ideals

> *I can't understand why people*
> *are frightened of new ideas.*
> *I'm frightened of the old ones.*
> (John Cage)

After this crash course in Commutative Algebra, we introduce a key new idea for combining Linear Algebra and Commutative Algebra: the kernel of an ideal $I$ in a commuting family $\mathscr{F}$. It is defined as the intersection of the kernels of the endomorphisms in $I$. Similarly, the big kernel of $I$ is the intersection of the big kernels of the endomorphisms in $I$. One of the basic properties of this kernel is that it can be computed as the intersection of the kernels of a system of generators of $I$. This leads to a straightforward algorithm later in the subsection.

However, what is not so clear is whether this intersection of kernels is non-zero. At this point our grass-roots work in the preceding subsection comes to the rescue: from the fact that a maximal ideal $\mathfrak{m}$ in $\mathscr{F}$ is the annihilator of an endomorphism $\varphi$ and from a non-zero vector in the image of $\varphi$ we get a non-zero element in the kernel of $\mathfrak{m}$. Since every ideal is contained in a maximal ideal, we see that every proper ideal of $\mathscr{F}$ has a non-zero kernel, and a more refined analysis based on the prime decomposition of the zero radical of $\mathscr{F}/I$ allows us to estimate its size.

So, absurd as it may seem at the first glance, kernels and big kernels of ideals of $\mathscr{F}$ are very much for real. In fact they will become the central objects when we extend the theory of eigenspaces and generalized eigenspaces to commuting families a little further down the road in this chapter.

> *If at first the idea is not absurd,*
> *then there is no hope for it.*
> (Albert Einstein)

In the following we let $K$ be a field, let $V$ be a finite dimensional $K$-vector space, and let $\mathscr{F}$ be a family of commuting endomorphisms of $V$. We want to examine ideals in the $K$-algebra $\mathcal{F}$ and how their properties are related to properties of the endomorphisms they contain.

**Definition 2.2.9** Let $I$ be an ideal in the commuting family $\mathcal{F}$.

(a) The $K$-vector subspace $\mathrm{Ker}(I) = \bigcap_{\varphi \in I} \mathrm{Ker}(\varphi)$ of $V$ is called the **kernel** of $I$.
(b) The $K$-vector subspace $\mathrm{BigKer}(I) = \bigcap_{\varphi \in I} \mathrm{BigKer}(\varphi)$ of $V$ is called the **big kernel** of $I$.

The following proposition collects some properties of kernels and big kernels of ideals in $\mathcal{F}$.

**Proposition 2.2.10 (Basic Properties of Kernels and Big Kernels of Ideals)**
*Let $I$ be an ideal in the commuting family $\mathcal{F}$.*

(a) *The vector space $\mathrm{Ker}(I)$ has a natural structure of an $\mathcal{F}/I$-module.*
(b) *For $\varphi \in \mathcal{F}$ and the principal ideal $I = \langle \varphi \rangle$, we have $\mathrm{Ker}(I) = \mathrm{Ker}(\varphi)$ and $\mathrm{BigKer}(I) = \mathrm{BigKer}(\varphi)$.*
(c) *For $\varphi_1, \ldots, \varphi_n \in \mathcal{F}$ and the ideal $I = \langle \varphi_1, \ldots, \varphi_n \rangle$, let $a_1, \ldots, a_n \in \mathbb{N}$ be such that $\mathrm{BigKer}(\varphi_i) = \mathrm{Ker}(\varphi_i^{a_i})$, and let $J = \langle \varphi_1^{a_1}, \ldots, \varphi_n^{a_n} \rangle$. Then we have*

$$\mathrm{Ker}(I) = \bigcap_{j=1}^{n} \mathrm{Ker}(\varphi_j) \quad and \quad \mathrm{BigKer}(I) = \bigcap_{j=1}^{n} \mathrm{BigKer}(\varphi_j) = \mathrm{Ker}(J)$$

(d) *There is a number $m \geq 1$ such that for all $t \geq m$ we have $\mathrm{BigKer}(I) = \mathrm{Ker}(I^t)$.*
(e) *The vector spaces $\mathrm{Ker}(I)$ and $\mathrm{BigKer}(I)$ are invariant for $\mathcal{F}$.*

*Proof* Claim (a) follows from the definition of $\mathrm{Ker}(I)$. Claim (b) is a consequence of the observations that $\mathrm{Ker}(\psi\varphi) \supseteq \mathrm{Ker}(\varphi)$ and $\mathrm{BigKer}(\psi\varphi) \supseteq \mathrm{BigKer}(\varphi)$. To show the first formula in (c), it suffices to check the inclusion "$\supseteq$". This follows by writing an arbitrary element $\psi \in I$ in the form $\psi = \eta_1\varphi_1 + \cdots + \eta_n\varphi_n$ with $\eta_i \in \mathcal{F}$ and noting that $\varphi_1(v) = \cdots = \varphi_n(v) = 0$ implies $\psi(v) = 0$ for every $v \in V$. Similarly, to prove the inclusion "$\supseteq$" in the first part of the second formula, we write $\psi \in I$ in the form $\psi = \eta_1\varphi_1 + \cdots + \eta_n\varphi_n$, and for every vector $v$ contained in the right-hand side, we choose $a_i \geq 1$ such that $\varphi_i^{a_i}(v) = 0$. Then we have $\psi^b(v) = 0$ for $b \geq a_1 + \cdots + a_n - n + 1$, and thus $v \in \mathrm{BigKer}(\psi)$. The second equality follows from Proposition 1.1.2.b.

Next, we prove (d). Let $J = \langle \varphi_1^{a_1}, \ldots, \varphi_n^{a_n} \rangle$ be the ideal which satisfies the second formula in (c). If we let $m = a_1 + \cdots + a_n - n + 1$, it follows that every genera-tor $\varphi_1^{c_1} \cdots \varphi_n^{c_n}$ of $I^m$ satisfies $c_i \geq a_i$ for at least one index $i$. Using (c), we see that $\mathrm{BigKer}(I) = \mathrm{Ker}(J) \subseteq \mathrm{Ker}(I^m) \subseteq \mathrm{BigKer}(I^m) = \mathrm{BigKer}(I)$. This shows the equality $\mathrm{BigKer}(I) = \mathrm{Ker}(I^m)$. For $t \geq m$, the equality $\mathrm{BigKer}(I) = \mathrm{Ker}(I^t)$ is a consequence of $\mathrm{Ker}(\varphi_j^{b_j}) = \mathrm{Ker}(\varphi_j^{a_j})$ for $b_j \geq a_j$.

Finally, the proof of (e) follows from (c) and Proposition 1.1.6.                                        $\square$

Part (c) of this proposition can be easily generalized to sums of ideals.

**Corollary 2.2.11**  *Let $I_1, \ldots, I_s$ be ideals in $\mathscr{F}$, and let $I = I_1 + \cdots + I_s$. Then we have*

$$\mathrm{Ker}(I) = \bigcap_{j=1}^{s} \mathrm{Ker}(I_j) \quad and \quad \mathrm{BigKer}(I) = \bigcap_{j=1}^{s} \mathrm{BigKer}(I_j)$$

*Proof* This follows from part (c) of the proposition by choosing systems of generators $\{\varphi_{j1}, \ldots, \varphi_{jk_j}\}$ for the ideals $I_j$ and combining them to form a system of generators of $I$. $\qquad\square$

Now we provide an important building block for the theory of joint eigenspaces developed later.

**Proposition 2.2.12 (Kernels and Big Kernels of Comaximal Ideals)**
*Let $I_1, \ldots, I_s$ be pairwise comaximal ideals in the family $\mathscr{F}$, and let $I = I_1 \cap \cdots \cap I_s$.*

(a)  *We have $\mathrm{Ker}(I) = \mathrm{Ker}(I_1) \oplus \cdots \oplus \mathrm{Ker}(I_s)$.*
(b)  *We have $\mathrm{BigKer}(I) = \mathrm{BigKer}(I_1) \oplus \cdots \oplus \mathrm{BigKer}(I_s)$.*

*Proof* Using Theorem 2.2.1.a and induction on $s$, it suffices to treat the case $s = 2$. To prove that the sum in (a) is direct we observe that Corollary 2.2.11 yields the equalities $\mathrm{Ker}(I_1) \cap \mathrm{Ker}(I_2) = \mathrm{Ker}(I_1 + I_2) = \langle 0 \rangle$. Since $I \subseteq I_1$ and $I \subseteq I_2$ imply $\mathrm{Ker}(I) \supseteq \mathrm{Ker}(I_1) + \mathrm{Ker}(I_2)$, it remains to show the reverse inclusion.

Let $v \in \mathrm{Ker}(I) = \mathrm{Ker}(I_1 \cdot I_2)$. Then we have $\varphi_1(v) \in \mathrm{Ker}(I_2)$ and $\varphi_2(v) \in \mathrm{Ker}(I_1)$ for all $\varphi_1 \in I_1$ and all $\varphi_2 \in I_2$. From $I_1 + I_2 = \mathscr{F}$ we get $\varphi_1 \in I_1$ and $\varphi_2 \in I_2$ such that $\varphi_1 + \varphi_2 = \mathrm{id}_V$. Thus we obtain $v = \varphi_1(v) + \varphi_2(v) \in \mathrm{Ker}(I_2) + \mathrm{Ker}(I_1)$, as had to be shown.

To prove (b) we use Proposition 2.2.10.d in order to get $\mathrm{BigKer}(I) = \mathrm{Ker}(I^t)$ and $\mathrm{BigKer}(I_j) = \mathrm{Ker}(I_j^t)$ for $j = 1, 2$ and some $t > 0$. Notice that also $I_1^t$ and $I_2^t$ are comaximal, as follows by taking a sufficiently high power of $\mathrm{id}_V = \varphi_1 + \varphi_2$. Moreover, since $I = I_1 \cdot I_2$ implies $I^t = I_1^t \cdot I_2^t = I_1^t \cap I_2^t$, the conclusion follows from (a). $\qquad\square$

Next we derive some information about the size of the kernel of a maximal ideal in a commuting family $\mathscr{F}$.

**Proposition 2.2.13 (Kernels of Maximal Ideals)**
*Let $\mathfrak{m}$ be a maximal ideal in a commuting family $\mathscr{F}$.*

(a)  *There exists a non-zero map $\varphi \in \mathscr{F}$ such that $\mathfrak{m} = \mathrm{Ann}_{\mathscr{F}}(\varphi)$.*
(b)  *For a map $\varphi$ as in (a) and for every $v \in \mathrm{Im}(\varphi) \setminus \{0\}$, we have $\mathrm{Ann}_{\mathscr{F}}(v) = \mathfrak{m}$.*
      *In particular, we have $v \in \mathrm{Ker}(\mathfrak{m}) \neq \langle 0 \rangle$.*
(c)  *We have $\dim_K(\mathrm{Ker}(\mathfrak{m})) = \dim_{\mathscr{F}/\mathfrak{m}}(\mathrm{Ker}(\mathfrak{m})) \cdot \dim_K(\mathscr{F}/\mathfrak{m})$. In particular, if we have $\dim_K(\mathrm{Ker}(\mathfrak{m})) = 1$ then $\mathscr{F}/\mathfrak{m} \cong K$.*

*Proof* The first claim follows immediately from Proposition 2.2.8. To show (b) we note that the annihilator of $K \cdot v$ is a proper ideal of $\mathscr{F}$ which contains $\mathrm{Ann}_{\mathscr{F}}(\varphi)$. Therefore it coincides with the maximal ideal $\mathfrak{m}$. To prove claim (c) it suffices to use Proposition 2.2.10.a.                                                                    □

Notice that the annihilator $\mathrm{Ann}_{\mathscr{F}}(\varphi)$ in (a) is the annihilator of an element of the ring $\mathscr{F}$, whereas the annihilator $\mathrm{Ann}_{\mathscr{F}}(v)$ in (b) is the annihilator of an element of an $\mathscr{F}$-module. The following example shows that the reverse implication in the special case of part (c) does not hold.

**Example 2.2.14**   Let $K = \mathbb{Q}$, let $V = \mathbb{Q}^3$, and let $\varphi$ be the endomorphism of $V$ defined by the matrix

$$\begin{pmatrix} 0 & 0 & 1 \\ 0 & 0 & 0 \\ 0 & 0 & 0 \end{pmatrix}$$

with respect to the canonical basis. Since $\mu_{\varphi}(z) = z^2$ and $\chi_{\varphi}(z) = z^3$, the endomorphism $\varphi$ is not commendable by Theorem 1.5.8. Now we consider the commuting family $\mathscr{F} = K[\varphi] \cong K[z]/\langle z^2 \rangle$. It has only one maximal ideal $\mathfrak{m} = \langle \varphi \rangle$ and satisfies $\dim_K(\mathscr{F}/\mathfrak{m}) = 1$, but $\dim_K(\mathrm{Ker}(\mathfrak{m})) = \dim_K(\mathrm{Ker}(\varphi)) = 2$.

At this point we are ready to estimate the size of the kernel of an arbitrary ideal in $\mathscr{F}$.

**Proposition 2.2.15 (Kernels of Proper Ideals)**
*Let $I = \langle \psi_1, \ldots, \psi_s \rangle$ be an ideal of $\mathscr{F}$, and let $\mathfrak{m}_1, \ldots, \mathfrak{m}_\ell$ be the maximal ideals of $\mathscr{F}$ containing $I$.*

(a) *The ideal $I$ is proper if and only if we have $\mathrm{Ker}(I) \neq \langle 0 \rangle$.*
(b) *We have $\dim_K(\mathrm{Ker}(I)) \geq \sum_{i=1}^{\ell} \dim_K(\mathscr{F}/\mathfrak{m}_i)$.*

*Proof* First we show claim (a). If $I$ is a proper ideal of $\mathscr{F}$, it is contained in a maximal ideal $\mathfrak{m}$. Hence we have $\mathrm{Ker}(I) \supseteq \mathrm{Ker}(\mathfrak{m}) \neq \langle 0 \rangle$ by Proposition 2.2.13.b. Conversely, if $I = \mathscr{F}$ then the ideal $I$ contains the map $\mathrm{id}_V$ whose kernel is zero.

To prove claim (b) we observe that $\mathrm{Rad}(I) = \bigcap_{i=1}^{\ell} \mathfrak{m}_i$, and since we have the inequality $\dim_K(\mathrm{Ker}(I)) \geq \dim_K(\mathrm{Ker}(\mathrm{Rad}(I)))$, the conclusion follows by applying Propositions 2.2.12.a and 2.2.13.c to $\mathrm{Rad}(I)$.                                                           □

In the final part of this section we illustrate a technique for computing kernels of ideals in $\mathscr{F}$.

**Definition 2.2.16** Let $A_1, \ldots, A_s \in \mathrm{Mat}_d(K)$, and let $\mathrm{Col}(A_1, \ldots, A_s) \in \mathrm{Mat}_{sd,d}(K)$ be the block matrix

$$\mathrm{Col}(A_1, \ldots, A_s) = \begin{pmatrix} A_1 \\ A_2 \\ \vdots \\ A_s \end{pmatrix}$$

Then we say that $\mathrm{Col}(A_1, \ldots, A_s)$ is the **block column matrix** of $(A_1, \ldots, A_s)$.

Using a suitable block column matrix, we can compute the kernel of an ideal as follows.

---

**Algorithm 2.2.17 (The Kernel of an Ideal in a Commuting Family)**
Let $I = \langle \psi_1, \ldots, \psi_s \rangle$ be an ideal of $\mathscr{F}$, and for every $i \in \{1, \ldots, s\}$ let $A_i \in \mathrm{Mat}_d(V)$ be the matrix representing $\psi_i$ with respect to a fixed $K$-basis of $V$. Then the following algorithm computes the kernel of $I$.

(1) Form the block column matrix $C = \mathrm{Col}(A_1, \ldots, A_s)$.
(2) Compute a system of generators of $\mathrm{Ker}(C)$ and return the corresponding vectors of $V$.

---

*Proof* Clearly, the vectors in the intersection $\bigcap_{i=1}^s \mathrm{Ker}(\psi_i)$ correspond to the tuples in $\mathrm{Ker}(\mathrm{Col}(A_1, \ldots, A_s))$. Hence the correctness of this algorithm follows from the fact that $\mathrm{Ker}(I) = \bigcap_{i=1}^s \mathrm{Ker}(\psi_i)$ by Proposition 2.2.10. $\qquad \square$

Finally we apply the above proposition to obtain an effective criterion for checking whether an ideal of $\mathscr{F}$ is proper.

**Corollary 2.2.18** *Let $I = \langle \psi_1, \ldots, \psi_s \rangle$ be an ideal of $\mathscr{F}$, and for $i \in \{1, \ldots, s\}$ let $M_i \in \mathrm{Mat}_d(V)$ be the matrix representing $\psi_i$ with respect to a fixed $K$-basis of $V$. Then the ideal $I$ is proper in $\mathscr{F}$ if and only if $\mathrm{rank}(\mathrm{Col}(A_1, \ldots, A_s)) < d$.*

*Proof* By Proposition 2.2.15.a, the ideal $I$ is a proper ideal in the ring $\mathscr{F}$ if and only if $\mathrm{Ker}(I) \neq \langle 0 \rangle$. Then Algorithm 2.2.17 implies that $\mathrm{Ker}(I) \neq \langle 0 \rangle$ if and only if the matrix $\mathrm{Col}(A_1, \ldots, A_s) \in \mathrm{Mat}_{sd,d}(K)$ has a non-trivial kernel. This happens if and only if its rank is smaller than the number of its columns. $\qquad \square$

Let us see an example for the application of the preceding algorithm and the corollary.

**Example 2.2.19** Let $K = \mathbb{Q}$, let $V = K^4$ and let $\varphi_1, \varphi_2$ be the endomorphisms of $V$ represented by the matrices

$$A_1 = \begin{pmatrix} 0 & \frac{5}{2} & 0 & 0 \\ 1 & 0 & 1 & 1 \\ 0 & 0 & 0 & 0 \\ 0 & \frac{5}{2} & 0 & 0 \end{pmatrix} \quad \text{and} \quad A_2 = \begin{pmatrix} 0 & 2 & 0 & 1 \\ 0 & 1 & 0 & 0 \\ 1 & 0 & 0 & -1 \\ 0 & 0 & 1 & 1 \end{pmatrix}$$

respectively. From $A_1 A_2 = A_2 A_1$ it follows that $\mathscr{F} = K[\varphi_1, \varphi_2]$ is a commuting family. Let $I$ be the ideal of $\mathscr{F}$ generated by $\{\varphi_1^2 - 5\,\mathrm{id}_V, \varphi_2 - \mathrm{id}_V\}$. The matrices which represent the two generators of $I$ are

$$M_1 = \begin{pmatrix} -\frac{5}{2} & 0 & \frac{5}{2} & \frac{5}{2} \\ 0 & 0 & 0 & 0 \\ 0 & 0 & -5 & 0 \\ \frac{5}{2} & 0 & \frac{5}{2} & -\frac{5}{2} \end{pmatrix} \quad \text{and} \quad M_2 = \begin{pmatrix} -1 & 0 & 0 & 1 \\ 0 & 0 & 0 & 0 \\ 1 & 0 & -1 & -1 \\ 0 & 0 & 1 & 0 \end{pmatrix}$$

We form the block column matrix $C = \mathrm{Col}(M_1, M_2)$. This is an matrix of size $8 \times 4$ whose rank is 2. Hence $I$ is a proper ideal in $\mathscr{F}$ and we have $\mathrm{Ker}(I) = \mathrm{Ker}(C) = \langle e_2, e_1 + e_4 \rangle$.

## 2.3 Eigenfactors

> *"What do you get when you combine a joke with a rhetorical question?"*
> *"What if there were no hypothetical questions?"*

The preceding section contains the first hints that the maximal ideals of a commuting family $\mathscr{F}$, and in particular their kernels and big kernels, are fundamental for generalizing the theory of eigenvalues and eigenvectors to arbitrary endomorphisms, i.e., to endomorphisms whose eigenfactors are not assumed to be linear. This invites several rhetorical questions.

*Are the maximal ideals of $\mathscr{F}$ somehow related to the eigenfactors of an endomorphism $\varphi$ in $\mathscr{F}$?* Surprise, surprise! They are indeed related: associated to every maximal ideal $\mathfrak{m}$ of $\mathscr{F}$ there is an eigenfactor of $\varphi$, namely the monic generator $p_{\mathfrak{m},\varphi}(z)$ of the kernel of the $K$-algebra homomorphism $K[z] \longrightarrow \mathscr{F}/\mathfrak{m}$ given by $z \mapsto \varphi + \mathfrak{m}$. It is called the $\mathfrak{m}$-eigenfactor of $\varphi$.

*Is every eigenfactor of $\varphi$ of the form $p_{\mathfrak{m},\varphi}(z)$ for some maximal ideal $\mathfrak{m}$?* Yes, it is, but the correspondence is in general not 1–1, since two maximal ideals may correspond to the same eigenfactor.

*Is the kernel of $\mathfrak{m}$ related to the eigenfactor $p_{\mathfrak{m},\varphi}(z)$?* Of course! It is contained in the corresponding eigenspace $\mathrm{Eig}(\varphi, p_{\mathfrak{m},\varphi}(z))$ for every endomorphism $\varphi$ in $\mathscr{F}$.

If you feel nostalgic and long back to the setting in Chap. 1, where we treated a single endomorphism instead of a commuting family, you can specialize everything to the case of the family $\mathscr{F} = K[\varphi]$, as we do in Proposition 2.3.4. Among others, you will discover that the maximal ideals of $\mathscr{F}$ are precisely the principal ideals $\langle p(\varphi) \rangle$, where $p(z)$ is an eigenfactor of $\varphi$. And if you feel even more nostalgic and wish yourself back to the good old days when all eigenfactors were assumed to be linear, you can look at the last part of the section where we characterize linear maximal ideals. But should you really forego the pleasure of studying eigenfactors of arbitrary endomorphisms? Is this a joke or a rhetorical question?

*Nostalgia isn't what it used to be.*

Let $K$ be a field, let $V$ be a finite dimensional $K$-vector space, and let $\mathscr{F}$ be a commuting family of endomorphisms of $V$. In the following we want to relate the minimal polynomials of elements of $\mathscr{F}$ and their eigenfactors to properties of the family. The following notion provides the first link.

**Definition 2.3.1**   For an endomorphism $\varphi$ of the family $\mathscr{F}$ and a maximal ideal $\mathfrak{m}$ of $\mathscr{F}$, the monic generator of the kernel of the $K$-algebra homomorphism from $K[z]$ to $\mathscr{F}/\mathfrak{m}$ defined by $z \mapsto \varphi + \mathfrak{m}$ is called the $\mathfrak{m}$-**eigenfactor** of $\varphi$ and is denoted by $p_{\mathfrak{m},\varphi}(z)$.

Observe that, by definition, we have $p_{\mathfrak{m},\varphi}(\varphi) \in \mathfrak{m}$. Further properties of the $\mathfrak{m}$-eigenfactors of $\varphi$ are collected in the next proposition.

**Proposition 2.3.2 (Basic Properties of $\mathfrak{m}$-Eigenfactors)**
*Let $\varphi \in \mathscr{F}$, and let $\mathfrak{m}$ be a maximal ideal of $\mathscr{F}$. Then the following facts hold true.*

(a) *If $\varphi \in \mathfrak{m}$ then we have $p_{\mathfrak{m},\varphi}(z) = z$.*
(b) *For $\varphi \in \mathscr{F}$ and $\psi \in \mathfrak{m}$, we have $p_{\mathfrak{m},\varphi}(z) = p_{\mathfrak{m},\varphi+\psi}(z)$.*
(c) *The polynomial $p_{\mathfrak{m},\varphi}(z)$ is an eigenfactor of $\varphi$.*
(d) *We have $\mathrm{Ker}(\mathfrak{m}) \subseteq \mathrm{Ker}(p_{\mathfrak{m},\varphi}(\varphi)) = \mathrm{Eig}(\varphi, p_{\mathfrak{m},\varphi}(z))$.*
(e) *We have $\mathrm{BigKer}(\mathfrak{m}) \subseteq \mathrm{BigKer}(p_{\mathfrak{m},\varphi}(\varphi)) = \mathrm{Gen}(\psi, p_{\mathfrak{m},\varphi}(z))$.*
(f) *Let $p(z)$ be an eigenfactor of $\varphi$. Then there exists a maximal ideal $\mathfrak{n}$ of $\mathscr{F}$ containing $p(\varphi)$ and such that $p(z) = p_{\mathfrak{n},\varphi}(z)$.*

*Proof* Claims (a) and (b) follow immediately from the definition. To prove (c) we first observe that $\mu_\varphi(z)$ is in $\mathrm{Ker}(\varepsilon)$. Since $\mathscr{F}/\mathfrak{m}$ is a field, the kernel of $\varepsilon$ is a prime ideal. Therefore the polynomial $p_{\mathfrak{m},\varphi}(z)$ is an irreducible factor of $\mu_\varphi(z)$. The proof of the inclusion in (d) follows from the fact that $p_{\mathfrak{m},\varphi}(\varphi)$ is in $\mathfrak{m}$. Claim (e) follows similarly, so let us prove (f).

Since the polynomial $p(z)$ divides the minimal polynomial of $\varphi$, its image $p(\varphi)$ in $\mathscr{F}$ is zero or a zero divisor of $\mathscr{F}$. Hence there exists a maximal ideal $\mathfrak{n}$ of $\mathscr{F}$ containing $p(\varphi)$. Then $p(z)$ is an element of the kernel of the $K$-algebra homomorphism $\varepsilon : K[z] \longrightarrow \mathscr{F}/\mathfrak{n}$ defined by $z \mapsto \varphi + \mathfrak{n}$. Since this kernel is a prime ideal and since $p(z)$ is a monic irreducible polynomial, it follows that we have $p(z) = p_{\mathfrak{n},\varphi}(z)$.                                                            $\square$

Given explicit generators of $\mathfrak{m}$, we can compute the $\mathfrak{m}$-eigenfactor of an endomorphism as follows.

**Algorithm 2.3.3 ($\mathfrak{m}$-Eigenfactors of Endomorphisms)**
*Let $\Phi = (\varphi_1, \ldots, \varphi_n)$ be a system of $K$-algebra generators of the family $\mathcal{F}$, let $P = K[x_1, \ldots, x_n]$, let $f(x_1, \ldots, x_n) \in P$, let $\psi = f(\varphi_1, \ldots, \varphi_n) \in \mathcal{F}$, and let $\mathfrak{m}$ be a maximal ideal of $\mathcal{F}$ given by an explicit set of generators of the preimage $\mathfrak{M}$ of $\mathfrak{m}$ in the ring $P$. The following algorithm computes the $\mathfrak{m}$-eigenfactor of $\psi$.*

(1) *In ring $P[z]$ form the ideal $I_{\mathfrak{m}, \psi} = \langle z - f(x_1, \ldots, x_n) \rangle + \mathfrak{M} \cdot P[z]$ and compute the elimination ideal $J_{\mathfrak{m}, \psi} = I_{\mathfrak{m}, \psi} \cap K[z]$.*
(2) *Return the monic generator of $J_{\mathfrak{m}, \psi}$.*

*Proof* The correctness of this algorithm follows from [15], Proposition 3.6.2, since $\mu_\psi(z)$ is the monic generator of the kernel of the $K$-algebra homomorphism $K[z] \longrightarrow \mathcal{F}$ given by $z \mapsto \psi$, and the $\mathfrak{m}$-eigenfactor of $\psi$ is the kernel of the composition of $\alpha$ and the canonical epimorphism $\mathcal{F} \longrightarrow \mathcal{F}/\mathfrak{m}$. $\qquad\square$

Our next proposition collects some properties of $\mathfrak{m}$-eigenfactors of endomorphisms in the family generated by a single endomorphism.

**Proposition 2.3.4 (Further Properties of $\mathfrak{m}$-Eigenfactors)**
*Let $\varphi$ be an endomorphism of $V$, let $\mathcal{F} = K[\varphi]$, let $f(z) \in K[z]$, and let $\psi = f(\varphi)$.*

(a) *The eigenfactors $p(z)$ of $\varphi$ are in 1–1 correspondence with the maximal ideals $\mathfrak{m} = \langle p(\varphi) \rangle$ of $\mathcal{F}$. In particular, to every eigenfactor of $\varphi$ we have an associated eigenfactor $p_{\mathfrak{m}, \psi}(z)$ of $\psi$.*
(b) *For every maximal ideal $\mathfrak{m}$ of $\mathcal{F}$, the $\mathfrak{m}$-eigenfactor of $\varphi$ divides $p_{\mathfrak{m}, \psi}(f(z))$.*
(c) *For every maximal ideal $\mathfrak{m}$ of $\mathcal{F}$, we have $\mathrm{Eig}(\varphi, p_{\mathfrak{m}, \varphi}(z)) \subseteq \mathrm{Eig}(\psi, p_{\mathfrak{m}, \psi}(z))$ and $\mathrm{Gen}(\varphi, p_{\mathfrak{m}, \varphi}(z)) \subseteq \mathrm{Gen}(\psi, p_{\mathfrak{m}, \psi}(z))$.*
(d) *Let $\mathfrak{m}$ be a maximal ideal of $\mathcal{F}$ and assume that also $\psi$ is a generator of the $K$-algebra $\mathcal{F}$. Then we have $\mathrm{Eig}(\varphi, p_{\mathfrak{m}, \varphi}(z)) = \mathrm{Eig}(\psi, p_{\mathfrak{m}, \psi}(z))$ as well as $\mathrm{Gen}(\varphi, p_{\mathfrak{m}, \varphi}(z)) = \mathrm{Gen}(\psi, p_{\mathfrak{m}, \psi}(z))$.*

*Proof* Claim (a) follows from the definition. To prove (b), we consider the composition of the multiplication homomorphism $m_f : K[z] \longrightarrow K[z]$ given by $z \mapsto f(z)$ and the $K$-algebra homomorphism $\varepsilon : K[z] \longrightarrow \mathcal{F}/\mathfrak{m}$ which is defined as above by $z \mapsto \varphi + \mathfrak{m}$. This composition agrees with the $K$-algebra homomorphism $\delta : K[z] \longrightarrow \mathcal{F}/\mathfrak{m}$ which satisfies $\delta(z) = \psi + \mathfrak{m}$. Then the polynomial $p_{\mathfrak{m}, \varphi}(z)$ is the monic generator of $\mathrm{Ker}(\varepsilon)$ and $p_{\mathfrak{m}, \psi}(z)$ is the monic generator of $\mathrm{Ker}(\delta)$. Thus the claim follows from $\mathrm{Ker}(\delta) = m_f^{-1}(\mathrm{Ker}(\varepsilon))$.

Claim (c) follows from the fact that (b) implies the inclusion $\mathrm{Ker}(p_{\mathrm{m},\varphi}(\varphi)) \subseteq$ $\mathrm{Ker}(p_{\mathrm{m},\psi}(f(\varphi)))$ and from the observation that

$$\mathrm{Eig}\big(\varphi, p_{\mathrm{m},\varphi}(z)\big) = \mathrm{Ker}\big(p_{\mathrm{m},\varphi}(\varphi)\big)$$
$$\subseteq \mathrm{Ker}\big(p_{\mathrm{m},\psi}\big(f(\varphi)\big)\big) = \mathrm{Ker}\big(p_{\mathrm{m},\psi}(\psi)\big) = \mathrm{Eig}\big(\psi, p_{\mathrm{m},\psi}(z)\big)$$

as well as the analogous observation for $\mathrm{Gen}(\varphi, p_{\mathrm{m},\varphi}(z))$. Claim (d) is an immediate consequence of (c). $\qquad\qquad\qquad\qquad\qquad\qquad\qquad\qquad\qquad\qquad\qquad\qquad\qquad\square$

The following example illustrates this proposition.

**Example 2.3.5** As in Example 1.1.10, let $K = \mathbb{Q}$, let $V = K^4$, and let $\varphi$ be the $K$-endomorphism of $V$ defined by the matrix

$$A = \begin{pmatrix} 0 & 0 & 0 & -1 \\ 1 & 0 & 0 & 0 \\ 0 & 1 & 0 & 1 \\ 0 & 0 & 1 & -1 \end{pmatrix}$$

with respect to the canonical basis. Notice that $A$ is the companion matrix of the polynomial $z^4 + z^3 - z^2 + 1 = (z+1)(z^3 - z + 1)$ (see Definition 1.5.12). By Corollary 1.5.13, we have $\mu_\varphi(z) = \chi_\varphi(z) = z^4 + z^3 - z^2 + 1$. We observe that this is the product of two irreducible polynomials, i.e., that there are two maximal ideals $\mathfrak{m}_1 = \langle \varphi + 1 \rangle$ and $\mathfrak{m}_2 = \langle \varphi^3 - \varphi + 1 \rangle$ in the family $\mathscr{F} = K[\varphi]$.

Now let us consider the endomorphism $\psi = \varphi^2 - \varphi$ of $K^4$. In Example 1.1.10 we computed its minimal polynomial and obtained

$$\mu_\psi(z) = z^4 - 4z^3 + z^2 + 5z + 2 = (z-2)(z^3 - 2z^2 - 3z - 1)$$

Thus we have $\dim_K(K[\psi]) = \dim_K(K[\varphi]) = 4$, which implies that the inclusion $K[\psi] \subseteq K[\varphi]$ is an equality and hence that $\psi$ is a $K$-algebra generator of $\mathscr{F}$. To express $\varphi$ explicitly as an element of $K[\psi]$, we use the Subalgebra Membership Test (see [15], Corollary 3.6.7). The result is $\varphi = \frac{3}{7}\psi^3 - \frac{13}{7}\psi^2 + \frac{5}{7}\psi + \frac{11}{7}\mathrm{id}_V$.

Starting with $\mathfrak{m}_1 = \langle \varphi + 1 \rangle$ and the $\mathfrak{m}_1$-eigenfactor $p_{\mathfrak{m}_1,\varphi}(z) = z + 1$ of $\varphi$, we can use Algorithm 2.3.3 to compute $p_{\mathfrak{m}_1,\psi}(z)$. The result is $p_{\mathfrak{m}_1,\psi}(z) = z - 2$, and indeed, as predicted by part (b) of the proposition, we find that $p_{\mathfrak{m}_1,\varphi}(z) = z + 1$ divides $p_{\mathfrak{m}_1,\psi}(z^2 - z) = z^2 - z - 2 = (z+1)(z-2)$.

Next we compute the polynomial $p_{\mathfrak{m}_2,\psi}(z)$ using Algorithm 2.3.3. For this we form the polynomial ring $K[x, z]$ and the ideal $I_{\mathfrak{m},\psi} = \langle x^3 - x + 1, z - (x^2 - x) \rangle$ in $K[x, z]$. Then we compute the monic generator of $I_{\mathfrak{m}_2,\psi} \cap K[z]$ and get the eigenfactor $p_{\mathfrak{m}_2,\psi}(z) = z^3 - 2z^2 - 3z - 1$ of $\psi$. Again, let us verify that the polynomial $p_{\mathfrak{m}_2,\varphi}(z) = z^3 - z + 1$ divides $p_{\mathfrak{m}_2,\psi}(z^2 - z)$ here. This follows from $p_{\mathfrak{m}_2,\psi}(z^2 - z) = (z^2 - z)^3 - 2(z^2 - z)^2 - 3(z^2 - z) - 1 = (z^3 - z + 1)(z^3 - 3z^2 + 2z - 1)$.

If we compute the two eigenspaces $\mathrm{Eig}(\varphi, p_{\mathfrak{m}_2,\varphi}(z)) = \mathrm{Ker}(\varphi^3 - \varphi + \mathrm{id}_V)$ and $\mathrm{Eig}(\psi, p_{\mathfrak{m}_2,\psi}(z)) = \mathrm{Ker}(\psi^3 - 2\psi^2 - 3\psi - \mathrm{id}_V)$, we get the same subspace of $V$, namely the subspace $\langle e_1 + e_2, e_1 - e_3, e_1 + e_4 \rangle$, in agreement with part (d) of the proposition.

Notice that we did not claim in Proposition 2.3.4.c that $\mathrm{Gen}(\varphi, p_{\mathfrak{m},\varphi}(z))$ equals the generalized eigenspace of $\psi = f(\varphi)$ for the eigenfactor $p_{\mathfrak{m},\psi}(z)$. In fact, this is in general not true, as the following example shows.

**Example 2.3.6** Let $K = \mathbb{Q}$ and $V = K^2$. The map $\varphi \in \mathrm{End}_K(V)$ given by the matrix $\begin{pmatrix} 1 & 0 \\ 0 & -1 \end{pmatrix}$ satisfies $\mu_\varphi(z) = \chi_\varphi(z) = z^2 - 1$. It has two 1-dimensional generalized eigenspaces corresponding to the eigenvalues $\lambda_1 = 1$ and $\lambda_2 = -1$. For the map $\psi = \varphi^2 - \mathrm{id}_V = 0$, we have one eigenvalue $\lambda' = 0$ and the corresponding generalized eigenspace is $V$.

In this setting the family $\mathscr{F} = K[\varphi]$ has two maximal ideals $\mathfrak{m}_1 = \langle \varphi - 1 \rangle$ and $\mathfrak{m}_2 = \langle \varphi + 1 \rangle$. Clearly, the corresponding eigenfactors are $p_{\mathfrak{m}_1,\varphi}(z) = z - 1$, $p_{\mathfrak{m}_2,\varphi}(z) = z + 1$, and $p_{\mathfrak{m}_1,\psi}(z) = p_{\mathfrak{m}_2,\psi}(z) = z$. As predicted by part (b) of the proposition, both $p_{\mathfrak{m}_1,\varphi}(z)$ and $p_{\mathfrak{m}_2,\varphi}(z)$ divide $p_{\mathfrak{m}_i,\psi}(z^2 - 1) = z^2 - 1$. Moreover, since we have $\mathrm{Eig}(\varphi, p_{\mathfrak{m}_1,\varphi}(z)) = \mathrm{Gen}(\varphi, p_{\mathfrak{m}_1,\varphi}(z)) = \langle e_1 \rangle$ and $\mathrm{Eig}(\psi, p_{\mathfrak{m}_1,\psi}(z)) = \mathrm{Gen}(\psi, p_{\mathfrak{m}_1,\psi}(z)) = V$, the inclusions in part (c) of the proposition are proper inclusions. The same holds for the maximal ideal $\mathfrak{m}_2$.

A particularly important case occurs when all $\mathfrak{m}$-eigenfactors are linear.

**Definition 2.3.7** A maximal ideal $\mathfrak{m}$ of a $K$-algebra $R$ is called **linear** if $R/\mathfrak{m} \cong K$.

By Proposition 2.2.13.c, if we have $\dim_K(\mathrm{Ker}(\mathfrak{m})) = 1$ then $\mathfrak{m}$ is linear. The following proposition characterizes linear maximal ideals of $\mathscr{F}$.

**Proposition 2.3.8 (Characterization of Linear Maximal Ideals)**
*Let $\Phi = (\varphi_1, \ldots, \varphi_n)$ be a system of $K$-algebra generators of $\mathscr{F}$, and let $\mathfrak{m}$ be a maximal ideal of $\mathscr{F}$. Then the following conditions are equivalent.*

(a) *The maximal ideal $\mathfrak{m}$ of $\mathscr{F}$ is linear.*
(b) *We have $\mathfrak{m} = \langle \varphi_1 - \lambda_1 \mathrm{id}_V, \ldots, \varphi_n - \lambda_n \mathrm{id}_V \rangle$ for some $\lambda_1, \ldots, \lambda_n \in K$.*
(c) *For $i = 1, \ldots, n$, the $\mathfrak{m}$-eigenfactor of $\varphi_i$ is of the form $z - \lambda_i$ with $\lambda_i \in K$.*

*Proof* The equivalence of (a) and (b) is clear. The implication (b) $\Rightarrow$ (c) follows from the definition of an $\mathfrak{m}$-eigenfactor. To show that (c) implies (b), it suffices to observe that $\mathfrak{m}$ contains the right-hand side by the definition of the $\mathfrak{m}$-eigenfactor $p_{\mathfrak{m},\varphi_i}(z) = z - \lambda_i$, and that the right-hand side is a maximal ideal of $\mathscr{F}$.  $\square$

The equivalence of conditions (b) and (c) in this proposition will be interpreted in terms of the solutions of polynomial systems in later chapters.

**Corollary 2.3.9** *Let $\mathfrak{m} = \langle \varphi_1 - \lambda_1 \mathrm{id}_V, \ldots, \varphi_n - \lambda_n \mathrm{id}_V \rangle$ be a linear maximal ideal of $\mathscr{F}$, and let $\psi \in \mathscr{F}$ be of the form $\psi = f(\varphi_1, \ldots, \varphi_n)$ with $f \in K[x_1, \ldots, x_n]$.*

(a) *We have $p_{\mathfrak{m},\psi}(z) = z - f(\lambda_1, \ldots, \lambda_n)$.*
(b) *We have $\mathrm{Eig}(\psi, f(\lambda_1, \ldots, \lambda_n)) \supseteq \mathrm{Ker}(\mathfrak{m}) = \bigcap_{i=1}^{n} \mathrm{Eig}(\varphi_i, \lambda_i)$.*

*Proof* From $f(x_1, \ldots, x_n) - f(\lambda_1, \ldots, \lambda_n) \in \langle x_1 - \lambda_1, \ldots, x_n - \lambda_n \rangle$ we deduce the relation $\psi - f(\lambda_1, \ldots, \lambda_n) \, \mathrm{id}_V \in \langle \varphi_1 - \lambda_1 \, \mathrm{id}_V, \ldots, \varphi_n - \lambda_n \, \mathrm{id}_V \rangle = \mathfrak{m}$. Thus the $\mathfrak{m}$-eigenfactor of $\psi$ divides the polynomial $z - f(\lambda_1, \ldots, \lambda_n)$. Since this polynomial is irreducible, the equality follows. This shows (a). Part (b) follows from Proposition 2.3.2.d and Proposition 2.2.10.c.                                     $\square$

In general, we have the inclusion $\langle p_{\mathfrak{m}, \varphi_1}(\varphi_1), \ldots, p_{\mathfrak{m}, \varphi_n}(\varphi_n) \rangle \subseteq \mathfrak{m}$. This is not always an equality, as we see in the next example.

**Example 2.3.10** Let $K = \mathbb{Q}$, let $V = K^4$, and let $\varphi_1$ and $\varphi_2$ be the endomorphisms of $V$ defined by the matrices

$$A_1 = \begin{pmatrix} 0 & 2 & 0 & 0 \\ 1 & 0 & 0 & 0 \\ 0 & 0 & 0 & 2 \\ 0 & 0 & 1 & 0 \end{pmatrix} \quad \text{and} \quad A_2 = \begin{pmatrix} 0 & 0 & 8 & 0 \\ 0 & 0 & 0 & 8 \\ 1 & 0 & 0 & 0 \\ 0 & 1 & 0 & 0 \end{pmatrix}$$

respectively. Later we will interpret $A_1$ and $A_2$ as the multiplication matrices of the $K$-algebra $R = K[x_1, x_2]/\langle x_1^2 - 2, x_2^2 - 8 \rangle$ with respect to the $K$-basis given by the residue classes of $(1, x_1, x_2, x_1 x_2)$. In particular, the endomorphisms $\varphi_1$ and $\varphi_2$ commute and generate a commuting family $\mathscr{F} = K[\varphi_1, \varphi_2]$.

The minimal polynomials $\mu_{\varphi_1}(z) = z^2 - 2$ and $\mu_{\varphi_2}(z) = z^2 - 8$ are irreducible. Hence they are the only eigenfactors of $\varphi_1$ and $\varphi_2$, respectively. However, the zero ideal $\langle 0 \rangle = \langle \mu_{\varphi_1}(\varphi_1), \mu_{\varphi_2}(\varphi_2) \rangle$ is not a maximal ideal of $\mathscr{F}$, since it is the intersection of the two maximal ideals $\mathfrak{m}_1 = \langle \varphi_2 - 2\varphi_1 \rangle$ and $\mathfrak{m}_2 = \langle \varphi_2 + 2\varphi_1 \rangle$.

## 2.4 Joint Eigenspaces

> *The trouble with eating Italian food is*
> *that five or six days later, you're hungry again.*
> (George Miller)

It has been cooking for quite some time. Now, in this section, we take the lid off and pull the rabbit out of the pot: the kernels of the maximal ideals of a commuting family $\mathscr{F}$ are nothing but the joint eigenspaces of the endomorphisms in $\mathscr{F}$! By a joint eigenspace of $\mathscr{F}$ we mean a nonzero intersection $\bigcap_{\varphi \in \mathscr{F}} \mathrm{Eig}(\varphi, p_\varphi(z))$, where $p_\varphi(z)$ is an eigenfactor of $\varphi$ for every $\varphi \in \mathscr{F}$. Similarly, a joint generalized eigenspace is a nonzero intersection $\bigcap_{\varphi \in \mathscr{F}} \mathrm{Gen}(\varphi, p_\varphi(z))$. The first and very satisfying theorem of this section says that, given the maximal ideals $\mathfrak{m}_1, \ldots, \mathfrak{m}_s$ of $\mathscr{F}$, the kernels $\mathrm{Ker}(\mathfrak{m}_i)$ are the joint eigenspaces of $\mathscr{F}$, the big kernels $\mathrm{BigKer}(\mathfrak{m}_i)$ are the joint generalized eigenspaces of $\mathscr{F}$, and $V$ is always the direct sum of the joint generalized eigenspaces of $\mathscr{F}$.

However, this is only the equivalent of the *antipasto* of the Italian meal provided here. The *primo* is pasta *al dente* based on the annihilators $\mathfrak{q}_i = \mathrm{Ann}_{\mathscr{F}}(\mathrm{BigKer}(\mathfrak{m}_i))$, where $i = 1, \ldots, s$. They are the *primary components* of the zero ideal of $\mathscr{F}$ and satisfy $\mathfrak{q}_1 \cap \cdots \cap \mathfrak{q}_s = \langle 0 \rangle$. They yield an isomorphism $\mathscr{F} \cong \mathscr{F}/\mathfrak{q}_1 \times \cdots \times \mathscr{F}/\mathfrak{q}_s$, and they are given as powers $\mathfrak{q}_i = \mathfrak{m}_i^{a_i}$ of the maximal ideals of $\mathscr{F}$.

Next comes the *secondo*. It details the relation between the joint eigenspaces of $\mathscr{F}$ and the eigenspaces of the individual family members. More precisely, if $\mathfrak{m}_1, \ldots, \mathfrak{m}_\ell$ are the maximal ideals of $\mathscr{F}$ with $p_{\mathfrak{m}_i, \varphi}(z) = p(z)$ for a given eigenfactor $p(z)$ of an endomorphism $\varphi$, then $\mathrm{Gen}(\varphi, p(z))$ equals the direct sum of joint generalized eigenspaces $\bigoplus_{i=1}^{\ell} \mathrm{BigKer}(\mathfrak{m}_i)$.

Finally, we offer a very sweet *dolce:* if we have $p_{\mathfrak{m}_j, \varphi}(z) \neq p_{\mathfrak{m}_i, \varphi}(z)$ for $j \neq i$, then the primary component $\mathfrak{q}_i$ is in fact the principal ideal generated by $p_{\mathfrak{m}_i, \varphi}(\varphi)^{m_i}$, where $m_i$ is the minimal multiplicity of the eigenfactor $p_{\mathfrak{m}_i, \varphi}(z)$. No doubt, reading this section will provide you with a feeling of satiety. After consuming all this pasta and antipasto, you should have enough pure energy to be fully charged for five or six days.

> *If you ate pasta and antipasto,*
> *would you still be hungry?*

Let $K$ be a field, and let $V$ be a finite dimensional $K$-vector space. Given a family $\mathscr{F}$ of commuting endomorphisms of $V$, we want to introduce joint eigenspaces and joint generalized eigenspaces of $\mathscr{F}$. Loosely speaking, we are looking for a non-zero vector subspace which is contained in an eigenspace of $\varphi$ for every $\varphi \in F$. More precisely, we introduce the following terminology.

**Definition 2.4.1**  For every endomorphism $\varphi \in \mathscr{F}$, let $p_\varphi(z)$ be an eigenfactor of $\varphi$, and let $\mathscr{C} = \{ p_\varphi(z) \mid \varphi \in \mathscr{F} \}$.

(a) The set $\mathscr{C}$ is called a **coherent family of eigenfactors** of $\mathscr{F}$ if we have a non-trivial intersection $\bigcap_{\varphi \in \mathscr{F}} \mathrm{Eig}(\varphi, p_\varphi(z)) \neq \langle 0 \rangle$.

(b) If for every $\varphi \in \mathscr{F}$ there exists an element $\lambda_\varphi \in K$ such that $p_\varphi(z) = z - \lambda_\varphi$ and if $\mathscr{C}$ is a coherent family of eigenfactors, we say that $\{ \lambda_\varphi \mid \varphi \in \mathscr{F} \}$ is a **coherent family of eigenvalues** of $\mathscr{F}$.

(c) Given a coherent family of eigenfactors $\mathscr{C}$ of $\mathscr{F}$, the corresponding intersection $\bigcap_{\varphi \in \mathscr{F}} \mathrm{Eig}(\varphi, p_\varphi(z)) \neq \langle 0 \rangle$ is called a **joint eigenspace** of $\mathscr{F}$. Its non-zero elements are called **joint eigenvectors** of $\mathscr{F}$.

(d) If the intersection $\bigcap_{\varphi \in \mathscr{F}} \mathrm{Gen}(\varphi, p_\varphi(z))$ is non-zero, it is called a **joint generalized eigenspace** of $\mathscr{F}$.

The next theorem is the main result about joint eigenspaces and generalized eigenspaces. It says that the vector space $V$ is the direct sum of the joint generalized eigenspaces of $\mathscr{F}$, and that every joint generalized eigenspace of $\mathscr{F}$ contains a joint eigenspace. The following lemma will simplify the proof.

**Lemma 2.4.2** *For every $\varphi \in \mathscr{F}$, let $p_\varphi(z)$ be an eigenfactor of $\varphi$.*

(a) *If $\mathscr{C} = \{p_\varphi(z) \mid \varphi \in \mathscr{F}\}$ is a coherent family of eigenfactors of $\mathscr{F}$, i.e. if the intersection $\bigcap_{\varphi \in \mathscr{F}} \mathrm{Eig}(\varphi, p_\varphi(z))$ is non-zero, then there exists a maximal ideal $\mathfrak{m}$ of $\mathscr{F}$ such that $p_\varphi(z) = p_{\mathfrak{m},\varphi}(z)$ for all $\varphi \in \mathscr{F}$.*
(b) *If the intersection $\bigcap_{\varphi \in \mathscr{F}} \mathrm{Gen}(\varphi, p_\varphi(z))$ is non-zero, then there exists a maximal ideal $\mathfrak{m}$ of $\mathscr{F}$ such that $p_\varphi(z) = p_{\mathfrak{m},\varphi}(z)$ for all $\varphi \in \mathscr{F}$.*

*Proof* Let $\mathfrak{m}_1, \dots, \mathfrak{m}_s$ be the maximal ideals of $\mathscr{F}$. To prove (a), we argue by contradiction. Assume that there is no index $i$ such that $p_\varphi(z) = p_{\mathfrak{m}_i,\varphi}(z)$ for all $\varphi \in \mathscr{F}$. This implies that, for every $i \in \{1, \dots, s\}$, there exists an endomorphism $\psi_i \in \mathscr{F}$ for which $p_{\psi_i}(z)$ is not equal to $p_{\mathfrak{m}_i,\psi_i}(z)$. Then it follows that we have $p_{\psi_i}(\psi_i) \notin \mathfrak{m}_i$, because otherwise $p_{\psi_i}(z)$ would be a monic irreducible polynomial in the kernel of the map $K[z] \longrightarrow \mathscr{F}/\mathfrak{m}_i$ and hence generate that kernel.

Therefore the ideal $\langle p_{\psi_1}(\psi_1), \dots, p_{\psi_s}(\psi_s) \rangle$ is not contained in any maximal ideal of $\mathscr{F}$. Hence it contains $\mathrm{id}_V$, i.e. we can find elements $\eta_1, \dots, \eta_s \in \mathscr{F}$ such that we have $\mathrm{id}_V = \sum_{i=1}^{s} \eta_i \cdot p_{\psi_i}(\psi_i)$. For every vector $v \in \bigcap_{\varphi \in \mathscr{F}} \mathrm{Eig}(\varphi, p_\varphi(z))$, this shows $v = (\sum_{i=1}^{s} \eta_i \cdot p_{\psi_i}(\psi_i))(v) = 0$, in contradiction to the hypothesis.

Next we prove (b). Assume that no index $i$ exists such that $p_\varphi(z) = p_{\mathfrak{m}_i,\varphi}(z)$ for all $\varphi \in \mathscr{F}$. We choose endomorphisms $\psi_1, \dots, \psi_s \in \mathscr{F}$ as above so that we can construct an equality $\mathrm{id}_V = \sum_{i=1}^{s} \eta_i \cdot p_{\psi_i}(\psi_i)$ with $\eta_1, \dots, \eta_s \in \mathscr{F}$. For every vector $v \in \bigcap_{\varphi \in F} \mathrm{Gen}(\varphi, p_\varphi(z))$ and every $i \in \{1, \dots, s\}$, we find a number $a_i > 0$ such that $(p_{\psi_i}(\psi_i))^{a_i}(v) = 0$. Letting $m = a_1 + \cdots + a_s - s + 1$, it follows that we have $v = (\mathrm{id}_V)^m(v) = (\sum_{i=1}^{s} \eta_i \cdot p_{\psi_i}(\psi_i))^m(v) = 0$, and this contradicts the hypothesis. $\square$

Now we are ready to formulate and prove the aforementioned theorem.

**Theorem 2.4.3 (Joint Generalized Eigenspace Decomposition)**
*Let $\mathscr{F}$ be a family of commuting endomorphisms of $V$, and let $\mathfrak{m}_1, \dots, \mathfrak{m}_s$ be the maximal ideals of $\mathscr{F}$.*

(a) *We have $V = \bigoplus_{i=1}^{s} \mathrm{BigKer}(\mathfrak{m}_i)$.*
(b) *The joint eigenspaces of $\mathscr{F}$ are $\mathrm{Ker}(\mathfrak{m}_1), \dots, \mathrm{Ker}(\mathfrak{m}_s)$.*
(c) *The joint generalized eigenspaces of $\mathscr{F}$ are $\mathrm{BigKer}(\mathfrak{m}_1), \dots, \mathrm{BigKer}(\mathfrak{m}_s)$.*

*Proof* Recall that the $K$-algebra $\mathscr{F}$ is zero-dimensional and that its nilradical is $\mathrm{Rad}(0) = \bigcap_{i=1}^{s} \mathfrak{m}_i$ (see Proposition 2.2.5.c). Moreover, since the ideals $\mathfrak{m}_1, \dots, \mathfrak{m}_s$ are pairwise comaximal, we have $\bigcap_{i=1}^{s} \mathfrak{m}_i = \prod_{i=1}^{s} \mathfrak{m}_i$ by Theorem 2.2.1.b. Therefore we obtain $\langle 0 \rangle = \prod_{i=1}^{s} \mathfrak{m}_i^\alpha$ for every sufficiently large $\alpha \geq 1$. Using Proposition 2.2.12, we get $V = \mathrm{Ker}(0) = \bigoplus_{i=1}^{s} \mathrm{Ker}(\mathfrak{m}_i^\alpha) = \bigoplus_{i=1}^{s} \mathrm{BigKer}(\mathfrak{m}_i)$, and (a) is proved.

Now we show (b). Let $W = \bigcap_{\varphi \in \mathscr{F}} \mathrm{Eig}(\varphi, p_\varphi(z)) = \bigcap_{\varphi \in \mathscr{F}} \mathrm{Ker}(p_\varphi(\varphi))$ be a joint eigenspace of $\mathscr{F}$. Using the lemma, we find an index $i \in \{1, \dots, s\}$ such that $p_\varphi(z) = p_{\mathfrak{m}_i,\varphi}(z)$ for all $\varphi \in \mathscr{F}$. By definition, we know that for an $\mathfrak{m}_i$-eigenfactor $p_{\mathfrak{m}_i,\varphi}(z)$ of an endomorphism $\varphi \in \mathscr{F}$, we have $p_{\mathfrak{m}_i,\varphi}(\varphi) \in \mathfrak{m}_i$. Hence the ideal of $\mathcal{F}$ generated by the endomorphisms $p_{\mathfrak{m}_i,\varphi}(\varphi)$ is contained in $\mathfrak{m}_i$, and thus $\mathrm{Ker}(\mathfrak{m}_i) \subseteq W$. Given a set of generators $\{\psi_1, \ldots, \psi_t\}$ of $\mathfrak{m}_i$, we have $\mathrm{Ker}(\mathfrak{m}_i) = \bigcap_{j=1}^{t} \mathrm{Ker}(\psi_j)$ by Corollary 2.2.11. Using Proposition 2.3.2.a, we see that the $\mathfrak{m}_i$-eigenfactor of $\psi_j$ is $z$, and hence $p_{\mathfrak{m}_i,\psi_j}(\psi_j) = \psi_j$ yields $\mathrm{Ker}(\mathfrak{m}_i) = \bigcap_{j=1}^{t} \mathrm{Ker}(\psi_j) \supseteq W$. Altogether, it follows that $W = \mathrm{Ker}(\mathfrak{m}_i)$.

Finally we prove (c). Let $W = \bigcap_{\varphi \in \mathcal{F}} \mathrm{Gen}(\varphi, p_\varphi(z))$ be a joint generalized eigenspace of $\mathcal{F}$. By the lemma, there exists an index $i \in \{1, \ldots, s\}$ such that $p_\varphi(z) = p_{\mathfrak{m}_i,\varphi}(z)$ for all $\varphi \in \mathcal{F}$. Consequently, we get $\mathrm{BigKer}(\mathfrak{m}_i) \subseteq W$. The reverse inclusion follows as in the proof of (b).                                    $\square$

The following corollary is an immediate consequence of the theorem. In Chap. 5 it will be used to improve some computations.

**Corollary 2.4.4** *Let* $\mathfrak{m}_1, \ldots, \mathfrak{m}_\ell$ *be the maximal ideals of* $\mathcal{F}$ *such that we have the equality* $\sum_{i=1}^{\ell} \dim_K(\mathrm{BigKer}(\mathfrak{m}_i)) = \dim_K(V)$. *Then* $\mathfrak{m}_1, \ldots, \mathfrak{m}_\ell$ *are all maximal ideals of* $\mathcal{F}$. *In particular, the vector spaces* $\mathrm{BigKer}(\mathfrak{m}_1), \ldots, \mathrm{BigKer}(\mathfrak{m}_\ell)$ *are all joint generalized eigenspaces of* $\mathcal{F}$.

The theorem allows us to choose a basis of $V$ such that all endomorphisms in $\mathcal{F}$ are represented by block matrices as follows.

**Corollary 2.4.5** *Let* $\mathcal{F}$ *be a family of commuting endomorphisms of* $V$, *and let* $\mathfrak{m}_1, \ldots, \mathfrak{m}_s$ *be the maximal ideals of* $\mathcal{F}$. *For* $i = 1, \ldots, s$, *let* $B_i$ *be a* $K$-*basis of* $\mathrm{BigKer}(\mathfrak{m}_i)$, *and let* $B$ *be the concatenation of* $B_1, \ldots, B_s$. *With respect to* $B$, *every endomorphism in* $\mathcal{F}$ *is represented by a block diagonal matrix having* $s$ *blocks.*

*Proof* This follows from part (a) of the theorem, because the vector subspaces $\mathrm{BigKer}(\mathfrak{m}_i)$ are invariant for $\mathcal{F}$ by Proposition 2.2.10.e.                                    $\square$

The following proposition describes the annihilators of the joint eigenspaces and of the joint generalized eigenspaces of a commuting family.

**Proposition 2.4.6 (Annihilators of Joint Eigenspaces)**
*Let* $\mathcal{F}$ *be a commuting family, let* $\mathfrak{m}_1, \ldots, \mathfrak{m}_s$ *be the maximal ideals of* $\mathcal{F}$, *and for* $i = 1, \ldots, s$ *let* $\mathfrak{q}_i = \mathrm{Ann}_{\mathcal{F}}(\mathrm{BigKer}(\mathfrak{m}_i))$.

(a) *We have* $\mathrm{Ann}_{\mathcal{F}}(\mathrm{Ker}(\mathfrak{m}_i)) = \mathfrak{m}_i$ *for* $i = 1, \ldots, s$.

(b) *The ideals* $\mathfrak{q}_1, \ldots, \mathfrak{q}_s$ *are pairwise comaximal and satisfy* $\mathfrak{q}_1 \cap \cdots \cap \mathfrak{q}_s = \langle 0 \rangle$.

(c) *We have* $\mathcal{F} \cong \prod_{i=1}^{s} \mathcal{F}/\mathfrak{q}_i$ *and* $\mathcal{F}/\mathfrak{q}_i \cong \mathcal{F}_{\mathrm{BigKer}(\mathfrak{m}_i)}$ *for* $i = 1, \ldots, s$.

(d) *Let* $\psi_{i,1}, \ldots, \psi_{i,n_i}$ *be generators of the ideal* $\mathfrak{m}_i$ *and let* $a_{i,1}, \ldots, a_{i,n_i}$ *be such that* $\mathrm{BigKer}(\psi_{i,j}) = \mathrm{Ker}(\psi_{i,j}^{a_{i,j}})$ *for* $i = 1, \ldots, s$ *and* $j = 1, \ldots, n_i$. *Then we have the equality* $\mathfrak{q}_i = \langle \psi_{i,1}^{a_{i,1}}, \ldots, \psi_{i,n_i}^{a_{i,n_i}} \rangle$. *In particular, we have* $\mathrm{Rad}(\mathfrak{q}_i) = \mathfrak{m}_i$ *and* $\mathrm{Ker}(\mathfrak{q}_i) = \mathrm{BigKer}(\mathfrak{m}_i)$.

(e) *For every* $i \in \{1, \ldots, s\}$, *there exists a number* $a_i \geq 1$ *such that* $\mathfrak{q}_i = \mathfrak{m}_i^{b_i}$ *for every* $b_i \geq a_i$.

*Proof* Clearly, we have $\mathfrak{m}_i \subseteq \text{Ann}_{\mathscr{F}}(\text{Ker}(\mathfrak{m}_i))$. The latter is a proper ideal by Proposition 2.2.13.b, so claim (a) is proved.

To prove (b), we note that, for every $i \in \{1, \ldots, s\}$, there exists a number $r_i \geq 1$ such that $\text{BigKer}(\mathfrak{m}_i) = \text{Ker}(\mathfrak{m}_i^{a_i})$. Therefore we have $\mathfrak{m}_i^{r_i} \subseteq \mathfrak{q}_i$, and thus $\mathfrak{m}_i$ is the only maximal ideal of $F$ containing $\mathfrak{q}_i$. This implies the first claim.

Next we show the second claim. Every map $\varphi \in \mathfrak{q}_1 \cap \cdots \cap \mathfrak{q}_s$ satisfies the relation $\varphi \in \text{Ann}_{\mathscr{F}}(\bigoplus_{i=1}^s \text{BigKer}(\mathfrak{m}_i))$. Since Theorem 2.4.3.a yields the equality $\bigoplus_{i=1}^s \text{BigKer}(\mathfrak{m}_i) = V$, we get $\varphi = 0$.

The first part of (c) follows from (b) and the Chinese Remainder Theorem 2.2.1. The second part is a special case of Corollary 2.1.10.

Next we prove (d). We let $J_i = \langle \psi_{i,1}^{a_{i,1}}, \ldots, \psi_{i,n_i}^{a_{i,n_i}} \rangle$ and use Proposition 2.2.10.c. We get the equality $\text{Ker}(J_i) = \text{BigKer}(\mathfrak{m}_i)$. Hence we obtain $\mathfrak{q}_i \supseteq J_i$. Now we apply the isomorphism of (c) and observe that both ideals correspond to the unit ideal in $\mathscr{F}/\mathfrak{q}_j$ for $j \neq i$ and to the zero ideal in $\mathscr{F}/\mathfrak{q}_i$. Therefore we get $\mathfrak{q}_i = J_i$, and it follows that $\text{Rad}(\mathfrak{q}_i) = \mathfrak{m}_i$ as well as $\text{Ker}(\mathfrak{q}_i) = \text{Ker}(J_i) = \text{BigKer}(\mathfrak{m}_i)$.

Finally, using Proposition 2.2.10.d, we get the proof of (e) in an analogous way to the proof of (d). $\square$

Given a system of $K$-algebra generators $\Phi = (\varphi_1, \ldots, \varphi_n)$ of $\mathscr{F}$ and a $K$-vector space basis of $\text{Ker}(\mathfrak{m}_i)$, we can compute matrices representing the restrictions of the generators $\varphi_j$ to $\text{Ker}(\mathfrak{m}_i)$. Then Algorithm 2.1.7 allows us to determine $\text{Ann}_{\mathscr{F}}(\text{Ker}(\mathfrak{m}_i))$, and hence $\mathfrak{m}_i$ by part (a) of the proposition. Similarly, given a $K$-vector space basis of $\text{BigKer}(\mathfrak{m}_i)$, we can compute the ideals $\mathfrak{q}_i$. Note that in the setting of Chap. 4, the ideals $\mathfrak{q}_i$ will turn out to be the **primary components** of $\langle 0 \rangle$.

A particularly simple situation occurs for a reduced family. Recall that a ring is said to be **reduced** if $\text{Rad}(0) = \langle 0 \rangle$, i.e. if the only nilpotent element is zero.

**Corollary 2.4.7** *Let $\mathscr{F}$ be a reduced commuting family, let $\mathfrak{m}_1, \ldots, \mathfrak{m}_s$ be the maximal ideals of $\mathscr{F}$, and let $\mathfrak{q}_i = \text{Ann}_{\mathscr{F}}(\text{BigKer}(\mathfrak{m}_i))$ for $i = 1, \ldots, s$.*

(a) *We have $\mathfrak{m}_i = \mathfrak{q}_i = \mathfrak{m}_i^a$ for $i = 1, \ldots, s$ and every $a \geq 1$.*
(b) *We have $\mathscr{F} \cong \prod_{i=1}^s \mathscr{F}/\mathfrak{m}_i$.*
(c) *For $i = 1, \ldots, s$, we have $\text{BigKer}(\mathfrak{m}_i) = \text{Ker}(\mathfrak{m}_i)$.*

*Proof* First we prove (a). By part (d) of the proposition, we have $\text{Rad}(\mathfrak{q}_i) = \mathfrak{m}_i$ for $i = 1, \ldots, s$. This implies that $\mathfrak{q}_i \supseteq \mathfrak{m}_i^a$ for $a \gg 0$. Hence it suffices to prove that $\mathfrak{m}_i = \mathfrak{m}_i^a$ for $a \geq 1$. To do this, we let $\mathfrak{n}_i = \bigcap_{j \neq i} \mathfrak{m}_j$ and observe that $\mathfrak{m}_i^a$ and $\mathfrak{n}_i$ are comaximal. So, there exist $\varphi \in \mathfrak{m}_i^a$ and $\psi_j \in \mathfrak{n}_i$ such that $\text{id}_V = \varphi + \psi$. Therefore, for every $\varrho \in \mathfrak{m}_i$, we have $\varrho = \varrho\varphi + \varrho\psi$, and hence $\varrho - \varrho\varphi \in \mathfrak{m}_i \cap \mathfrak{n}_i = \langle 0 \rangle$. This implies $\varrho = \varrho\varphi \in \mathfrak{m}_i^a$. The other inclusion is clear.

The proof of (b) follows from (a) and the isomorphism $\mathscr{F} \cong \prod_{i=1}^s \mathscr{F}/\mathfrak{q}_i$ of part (c) of the proposition. In order to show (c), we note that Proposition 2.2.10.d yields $\text{BigKer}(\mathfrak{m}_i) = \text{Ker}(\mathfrak{m}_i^a)$ for some $a \geq 1$. Together with (a), this proves the claim. $\square$

Let us apply Theorem 2.4.3 is a concrete example and compute the joint generalized eigenspace decomposition.

**Example 2.4.8**  Let us consider the two matrices

$$
A_1 = \begin{pmatrix} 0 & 0 & 0 & 0 & 0 & 0 \\ 1 & 0 & 0 & 0 & 0 & 0 \\ 0 & 0 & 0 & 0 & 0 & 0 \\ 0 & 1 & 0 & 1 & 0 & 0 \\ 0 & 0 & 1 & 0 & 0 & 0 \\ 0 & 0 & 0 & 0 & 1 & 1 \end{pmatrix} \quad \text{and} \quad A_2 = \begin{pmatrix} 0 & 0 & 2 & 0 & 0 & 0 \\ 0 & 0 & 0 & 0 & 2 & 0 \\ 1 & 0 & 0 & 0 & 0 & 0 \\ 0 & 0 & 0 & 0 & 0 & 2 \\ 0 & 1 & 0 & 0 & 0 & 0 \\ 0 & 0 & 0 & 1 & 0 & 0 \end{pmatrix}
$$

For $i = 1, 2$, let $\varphi_i$ be the endomorphism of $V = \mathbb{Q}^6$ represented by $A_i$ with respect to the canonical basis, and let $\Phi = (\varphi_1, \varphi_2)$. The two matrices commute and the ideal of algebraic relations of $\Phi$ is the ideal $\mathrm{Rel}_P(\Phi) = \langle x_1^3 - x_1^2, x_2^2 - 2\rangle$ in the polynomial ring $P = \mathbb{Q}[x_1, x_2]$. Later we will interpret $A_1$ and $A_2$ as multiplication matrices of $\mathscr{F}$ with respect to the basis $(\mathrm{id}_V, \varphi_1, \varphi_2, \varphi_1^2, \varphi_1\varphi_2, \varphi_1^2\varphi_2)$.

When we compute the radical of the ideal of algebraic relations of $\Phi$ using [15], Corollary 3.7.16, we get $\mathrm{Rad}(\mathrm{Rel}_P(\Phi)) = \langle x_1^2 - x_1, x_2^2 - 2\rangle$. Clearly, this ideal is of the form $\mathrm{Rad}(\mathrm{Rel}_P(\Phi)) = \mathfrak{M}_1 \cap \mathfrak{M}_2$ where $\mathfrak{M}_1 = \langle x_1, x_2^2 - 2\rangle$ and $\mathfrak{M}_2 = \langle x_1 - 1, x_2^2 - 2\rangle$ are maximal ideals of $P$. Hence the family $\mathscr{F}$ has two maximal ideals, namely $\mathfrak{m}_1 = \langle \varphi_1\rangle$ and $\mathfrak{m}_2 = \langle \varphi_1 - \mathrm{id}_V\rangle$. By the theorem, it follows that $V$ is the direct sum of the two joint generalized eigenspaces $\mathrm{BigKer}(\mathfrak{m}_1)$ and $\mathrm{BigKer}(\mathfrak{m}_2)$. Let us have a closer look at this decomposition.

In view of Proposition 1.1.2, we calculate $\mathrm{BigKer}(\mathfrak{m}_1) = \mathrm{Ker}(\varphi_1^2)$. This is the 4-dimensional vector subspace of $V$ given by

$$
\mathrm{BigKer}(\mathfrak{m}_1) = \big\langle (1, 0, 0, -1, 0, 0), (0, 1, 0, -1, 0, 0),
$$
$$
(0, 0, 1, 0, 0, -1), (0, 0, 0, 0, 1, -1)\big\rangle
$$

The corresponding joint eigenspace is $\mathrm{Ker}(\mathfrak{m}_1) = \mathrm{Ker}(\varphi_1)$ which is the 2-dimensional vector subspace of $V$ given by

$$
\mathrm{Ker}(\mathfrak{m}_1) = \big\langle (0, 1, 0, -1, 0, 0), (0, 0, 0, 0, 1, -1)\big\rangle
$$

To compute the second joint generalized eigenspace $\mathrm{BigKer}(\mathfrak{m}_2)$, we proceed similarly and find that it agrees with the joint eigenspace $\mathrm{Ker}(\mathfrak{m}_2) = \mathrm{Ker}(\varphi_1 - \mathrm{id}_V)$. Thus the result is the 2-dimensional vector subspace

$$
\mathrm{BigKer}(\mathfrak{m}_2) = \mathrm{Ker}(\mathfrak{m}_2) = \big\langle (0, 0, 0, 1, 0, 0), (0, 0, 0, 0, 0, 1)\big\rangle
$$

At this point it is easy to see that, indeed, the vector space $V$ has a direct sum decomposition $V = \mathrm{BigKer}(\mathfrak{m}_1) \oplus \mathrm{BigKer}(\mathfrak{m}_2)$.

Let us have a look at the situation when we choose a couple of specific endomorphisms in the family of Example 2.4.8.

**Example 2.4.9**   Let $\mathscr{F} = \mathbb{Q}[\varphi_1, \varphi_2]$ be the family of commuting endomorphisms of $V = \mathbb{Q}^6$ introduced in Example 2.4.8.

For the endomorphism $\psi = \varphi_1^3 - 3\varphi_2 + 5\,\mathrm{id}_V$, we compute its minimal polynomial using Algorithm 1.1.8 and get $\mu_\psi(z) = z^4 - 22z^3 + 145z^2 - 264z + 126$. A computation of the eigenfactors of $\psi$ using Algorithm 2.3.3 yields

$$p_{\mathfrak{m}_1, \psi}(z) = z^2 - 10z + 7 \quad\text{and}\quad p_{\mathfrak{m}_2, \psi}(z) = z^2 - 12z + 18$$

and then we have $\mu_\psi(z) = p_{\mathfrak{m}_1, \psi}(z) \cdot p_{\mathfrak{m}_2, \psi}(z)$.

Starting with the eigenfactor $p_{\mathfrak{m}_1, \psi}(z)$, we compute the kernel of $\psi^2 - 10\psi + 7$ and find $\mathrm{Eig}(\psi, p_{\mathfrak{m}_1, \psi}(z)) = \mathrm{BigKer}(\mathfrak{m}_1)$. It follows that also $\mathrm{Gen}(\psi, p_{\mathfrak{m}_1, \psi}(z))$ agrees with $\mathrm{BigKer}(\mathfrak{m}_1)$. Now let us consider the eigenfactor $z^2 - 12z + 18$ of $\psi$. If we compute the big kernel of $\psi^2 - 12\psi + 18\,\mathrm{id}_V$, we see that the generalized eigenspace $\mathrm{Gen}(\psi, p_{\mathfrak{m}_2, \psi}(z))$ is equal to $\mathrm{BigKer}(\mathfrak{m}_2)$. Thus the decomposition of $V$ into generalized eigenspaces of $\psi$ agrees with the decomposition of $V$ into joint generalized eigenspaces in this case.

On the other hand, for the endomorphism $\eta = \varphi_1^2 - \varphi_1 \in \mathscr{F}$, the minimal polynomial is $\mu_\eta(z) = z^2$. Thus there is only one eigenfactor $p_{\mathfrak{m}_1, \eta}(z) = p_{\mathfrak{m}_2, \eta}(z) = z$, and the associated generalized eigenspace is $\mathrm{Gen}(\eta, p_{\mathfrak{m}_i, \eta}(z)) = V$ for $i = 1, 2$. In other words, for this member of the family $\mathscr{F}$, there is one generalized eigenspace and it is the direct sum of the two joint generalized eigenspaces of $\mathscr{F}$.

The next proposition provides a connection between the joint generalized eigenspaces of a commuting family $\mathscr{F}$ and the generalized eigenspaces of the individual endomorphisms in $\mathscr{F}$. In particular, it is shown that every generalized eigenspace is the direct sum of joint generalized eigenspaces.

**Proposition 2.4.10 (Eigenspaces of an Endomorphism in a Family)**
*Let $\mathscr{F}$ be a family of commuting endomorphisms of $V$, let $\mathfrak{m}_1, \ldots, \mathfrak{m}_s$ be the maximal ideals of $\mathscr{F}$, let $\varphi$ be an element of $\mathscr{F}$, and let $p(z)$ be an eigenfactor of $\varphi$. Without loss of generality, let $\mathfrak{m}_1, \ldots, \mathfrak{m}_\ell$ be those maximal ideals of $\mathscr{F}$ for which we have $p_{\mathfrak{m}_i, \varphi}(z) = p(z)$.*

(a) *We have $\mathrm{Gen}(\varphi, p(z)) = \bigoplus_{i=1}^\ell \mathrm{BigKer}(\mathfrak{m}_i)$.*
(b) *We have $\mathrm{Eig}(\varphi, p(z)) \supseteq \bigoplus_{i=1}^\ell \mathrm{Ker}(\mathfrak{m}_i)$.*

*Proof* Let $p_1(z), \ldots, p_s(z)$ be the eigenfactors of $\varphi$, and for every $i \in \{1, \ldots, s\}$, let $\mathfrak{m}_{i1}, \ldots, \mathfrak{m}_{i\ell_i}$ be the maximal ideals of $\mathscr{F}$ such that $p_i(z) = p_{\mathfrak{m}_{ij}, \varphi}(z)$ for every $j = 1, \ldots, \ell_i$.

To prove (a), we let $W_i = \bigoplus_{j=1}^{\ell_i} \mathrm{BigKer}(\mathfrak{m}_{ij})$. By Theorem 1.1.17.a, we know that $V = \bigoplus_{i=1}^s \mathrm{Gen}(\varphi, p_i(z))$, and Theorem 2.4.3 implies $V = \bigoplus_{i=1}^s W_i$. Furthermore, Proposition 2.3.2.e shows that we have $W_i \subseteq \mathrm{Gen}(\varphi, p_i(z))$ for $i = 1, \ldots, s$. Putting these facts together, we get $W_i = \mathrm{Gen}(\varphi, p_i(z))$ for $i = 1, \ldots, s$, and the proof of (a) is complete. Claim (b) follows immediately from Proposition 2.3.2.d. $\square$

The next example shows that the containment in part (b) of this proposition can be strict.

**Example 2.4.11** Let $K = \mathbb{Q}$, let $V = K^3$, and let $\varphi_1$ and $\varphi_2$ be the endomorphisms of $V$ represented with respect to the canonical basis by the matrices

$$A_1 = \begin{pmatrix} 0 & 1 & 0 \\ 0 & 0 & 0 \\ 0 & 0 & 0 \end{pmatrix} \quad \text{and} \quad A_2 = \begin{pmatrix} 0 & 0 & 1 \\ 0 & 0 & 0 \\ 0 & 0 & 0 \end{pmatrix}$$

respectively. Then $\mathcal{F} = K[\varphi_1, \varphi_2]$ is a commuting family. It is isomorphic to $K[x, y]/\langle x^2, xy, y^2\rangle$. There is only one maximal ideal $\mathfrak{m} = \langle \varphi_1, \varphi_2\rangle$, and we have $\mathrm{Ker}(\mathfrak{m}) = \langle e_1\rangle$. An element $\psi \in \mathcal{F}$ can be represented as $\psi = c_0\,\mathrm{id}_V + c_1\varphi_1 + c_2\varphi_2$ with $c_0, c_1, c_2 \in K$. It has only one eigenfactor, namely $p(z) = z - c_0$. Since the associated matrix has at least two zero rows, we see that $\dim_K \mathrm{Ker}(\psi - c_0\,\mathrm{id}_V) \geq 2$. Therefore we have $\mathrm{Eig}(\psi, z - c_0) \supset \mathrm{Ker}(\mathfrak{m})$.

The following theorem will become very useful later on. Given an endomorphism $\varphi \in \mathcal{F}$, we recall that for every eigenfactor $p(z)$ of $\varphi$ there exists at least one maximal ideal $\mathfrak{m}$ in $\mathcal{F}$ such that $p(z) = p_{\mathfrak{m},\varphi}(z)$ (see Proposition 2.3.2.f). If there are several maximal ideals $\mathfrak{m}_i$ like that, the intersection of the corresponding ideals $\mathfrak{q}_i = \mathrm{Ann}_{\mathcal{F}}(\mathrm{BigKer}(\mathfrak{m}_i))$ (as defined in Proposition 2.4.6) can be described as follows.

### Theorem 2.4.12 (Primary Components as Principal Ideals)

*Let $\mathcal{F}$ be a family of commuting endomorphisms of $V$, let $\mathfrak{m}_1, \ldots, \mathfrak{m}_s$ be the maximal ideals of $\mathcal{F}$, and denote $\mathrm{Ann}_{\mathcal{F}}(\mathrm{BigKer}(\mathfrak{m}_i))$ by $\mathfrak{q}_i$ for $i = 1, \ldots, s$. Given an endomorphism $\varphi \in \mathcal{F}$, let $p(z)$ be an eigenfactor of $\varphi$, let $m = \mathrm{mmult}(p(z))$, and w.l.o.g. let $\mathfrak{m}_1, \ldots, \mathfrak{m}_\ell$ be those maximal ideals of $\mathcal{F}$ for which $p_{\mathfrak{m}_i,\varphi}(z) = p(z)$.*

(a) *We have $\mathfrak{q}_1 \cap \cdots \cap \mathfrak{q}_\ell = \langle p(\varphi)^m\rangle$.*
(b) *Let $i \in \{1, \ldots, s\}$ be such that $p_{\mathfrak{m}_j,\varphi}(z) \neq p_{\mathfrak{m}_i,\varphi}(z)$ for every $j \in \{1, \ldots, s\}\setminus\{i\}$, and let $m_i = \mathrm{mmult}(p_{\mathfrak{m}_i,\varphi}(z))$. Then we have $\mathfrak{q}_i = \langle p_{\mathfrak{m}_i,\varphi}(\varphi)^{m_i}\rangle$.*

*Proof* It is clear that (b) follows immediately from (a), so let us prove (a). For simplicity we denote $\bigcap_{i=1}^{\ell} \mathfrak{q}_i$ by $\mathfrak{a}$. The preceding Proposition 2.4.10 implies that $\mathrm{Gen}(\varphi, p(z)) = \bigoplus_{i=1}^{\ell} \mathrm{BigKer}(\mathfrak{m}_j)$. From this we get

$$\mathfrak{a} = \bigcap_{i=1}^{\ell} \mathrm{Ann}_{\mathcal{F}}\big(\mathrm{BigKer}(\mathfrak{m}_i)\big) = \mathrm{Ann}_{\mathcal{F}}\big(\bigoplus_{i=1}^{\ell} \mathrm{BigKer}(\mathfrak{m}_i)\big)$$

$$= \mathrm{Ann}_{\mathcal{F}}\big(\mathrm{Gen}(\varphi, p(z))\big)$$

By Theorem 1.1.17.d, we have $\mathrm{Gen}(\varphi, p(z)) = \mathrm{Ker}(p(\varphi)^m)$, and therefore we obtain $\mathfrak{a} = \mathrm{Ann}_{\mathcal{F}}(\mathrm{Ker}(p(\varphi)^m))$. In particular, we have $p(\varphi)^m \in \mathfrak{a}$, and hence the inclusion $\langle p(\varphi)^m\rangle \subseteq \mathfrak{a}$ holds.

Consequently, using the isomorphism $\mathscr{F} \cong \prod_{i=1}^{s} \mathscr{F}/\mathfrak{q}_i$ of Proposition 2.4.6, it suffices to show that the two ideals $\mathfrak{a}$ and $\langle p(\varphi)^m \rangle$ have the same image in each component $\mathscr{F}/\mathfrak{q}_i$ in order to get the claimed equality. In view of the above inclusion, we see that both ideals reduce to the zero ideal modulo $\mathfrak{q}_j$ for every $j = 1, \ldots, \ell$. Now let $\ell < i \le s$. By Proposition 2.4.6.b, the ideals $\mathfrak{q}_1, \ldots, \mathfrak{q}_s$ are pairwise comaximal. Hence also the ideals $\mathfrak{a}$ and $\mathfrak{q}_i$ are pairwise comaximal. Therefore the image of $\mathfrak{a}$ in $\mathscr{F}/\mathfrak{q}_i$ is the unit ideal. By assumption, we have $p(\varphi) \notin \mathfrak{m}_i$. Thus also the ideals $\langle p(\varphi)^m \rangle$ and $\mathfrak{q}_i$ are comaximal. Consequently, the image of $\langle p(\varphi)^m \rangle$ in $\mathscr{F}/\mathfrak{q}_i$ is the unit ideal. By combining these observations we get the claimed equality.            $\square$

A map $\varphi$ for which we have $\ell < s$ in this theorem corresponds to a partially splitting endomorphism in the next section. Such maps play an important role in Chap. 5.

## 2.5  Splitting Endomorphisms

> Brain, n. *An apparatus with which we think that we think.*
>                                                         (Ambrose Bierce)

At the end of the preceding section we saw that the primary components of the zero ideal in a commuting family are principal ideals and can be computed easily, if there exists an endomorphism in the family whose eigenfactors corresponding to different maximal ideals are distinct. Pondering about this result, we realize that the latter property is a useful one in several other respects. For instance, it means that the generalized eigenspaces of the endomorphism agree with the joint generalized eigenspaces of the family. Thus we say that an element $\varphi$ of a commuting family $\mathscr{F}$ is a *splitting* endomorphism if $p_{\mathfrak{m}_i,\varphi}(z) \ne p_{\mathfrak{m}_j,\varphi}(z)$ for maximal ideals $\mathfrak{m}_i \ne \mathfrak{m}_j$. Continuing the thinking processes, we examine the question if and when such an endomorphism exists. If it doesn't, we can still think about a *partially splitting* endomorphism, i.e., an endomorphism for which there exist maximal ideals $\mathfrak{m}_i \ne \mathfrak{m}_j$ such that $p_{\mathfrak{m}_i,\varphi}(z) \ne p_{\mathfrak{m}_j,\varphi}(z)$. The result of these deliberations is that a partially splitting endomorphism exists in every commuting family, while for the existence of a splitting endomorphism we need to assume that the base field has at least as many elements as there are maximal ideals in $\mathscr{F}$. These insights lead us to the question of how we can actually find a splitting endomorphism, if it exists. The traditional answer "Take a random element of $\mathscr{F}$" makes only sense if the base field is infinite and the *splitting locus*, i.e., the set of splitting endomorphisms in $\mathscr{F}$, contains a non-empty Zariski open subset of $\mathscr{F}$.

The goal of the second part of the section is to prove rigorously that this is the case. A valuable idea, versions of which will reappear in later chapters, is the construction of a *generically extended* endomorphism. This is a linear combination of a $K$-basis of $\mathscr{F}$ with indeterminate coefficients. Since it is defined over a field of the form $K(y_1, \ldots, y_\delta)$, we have enough leeway to show that it is a splitting endomorphism for the extended family, and we can cogitate about writing down polynomials in the indeterminates $y_1, \ldots, y_\delta$ such that the complement of their zero set is a nonempty Zariski open subset of the splitting locus. This section is garnished with a particularly rich set of examples which should keep you thinking for quite some time.

*Your brain works faster than you think.*

As before, we let $K$ be a field, $V$ a finite dimensional $K$-vector space, and $\mathscr{F}$ a family of commuting endomorphisms of $V$. We start with a definition.

**Definition 2.5.1**  Let $\mathfrak{m}_1, \ldots, \mathfrak{m}_s$ be the maximal ideals of $\mathscr{F}$, and let $\varphi \in \mathscr{F}$.

(a) The map $\varphi$ is called a **partially splitting endomorphism** for $\mathscr{F}$ if there exist two distinct maximal ideals $\mathfrak{m}_i$ and $\mathfrak{m}_j$ such that $p_{\mathfrak{m}_i,\varphi}(z) \neq p_{\mathfrak{m}_j,\varphi}(z)$. Equivalently, the map $\varphi$ is a partially splitting endomorphism for $\mathscr{F}$ if and only if it has more than one eigenfactor.

(b) The map $\varphi$ is called a **splitting endomorphism** for $\mathscr{F}$ if $p_{\mathfrak{m}_i,\varphi}(z) \neq p_{\mathfrak{m}_j,\varphi}(z)$ for $i \neq j$. Equivalently, the map $\varphi$ is a splitting endomorphism for $\mathscr{F}$ if and only if it has $s$ eigenfactors.

(c) If we have the equalities $\mathrm{Eig}(\varphi, p_{\mathfrak{m}_i,\varphi}(z)) = \mathrm{Ker}(\mathfrak{m}_i)$ for $i = 1, \ldots, s$ then $\varphi$ is called a **strongly splitting endomorphism** for $\mathscr{F}$.

Partially splitting endomorphism will be studied and used in Chap. 5. If $s = 1$, i.e. if the family $\mathscr{F}$ has only one maximal ideal $\mathfrak{m}$, then every element $\varphi \in \mathscr{F}$ is a splitting endomorphism, because we have $V = \mathrm{Gen}(\varphi, p_{\mathfrak{m},\varphi}(z))$. The next proposition clarifies some aspects of the above definition.

**Proposition 2.5.2 (Basic Properties of Splitting Endomorphisms)**
*Let $\mathscr{F}$ be a commuting family, let $\mathfrak{m}_1, \ldots, \mathfrak{m}_s$ be the maximal ideals of $\mathscr{F}$, and let $\varphi \in \mathscr{F}$.*

(a) *The map $\varphi$ is a splitting endomorphism for $\mathscr{F}$ if and only if we have the equalities $\mathrm{Gen}(\varphi, p_{\mathfrak{m}_i,\varphi}(z)) = \mathrm{BigKer}(\mathfrak{m}_i)$ for $i = 1, \ldots, s$.*

(b) *If $\varphi$ is a strongly splitting endomorphism for $\mathscr{F}$ then it is a splitting endomorphism for $\mathscr{F}$.*

(c) *If $\mathscr{F} = K[\varphi]$ then $\varphi$ is a splitting endomorphism for $\mathscr{F}$.*

(d) *Given $c \in K$, the map $\varphi$ is a splitting endomorphism if and only if $\varphi + c\,\mathrm{id}_V$ is a splitting endomorphism.*

*Proof*  Claim (a) follows immediately from Proposition 2.4.10.a. To prove (b), we use Proposition 2.4.10.b to deduce that $\mathfrak{m}_i$ is the only maximal ideal for which $p_{\mathfrak{m}_i,\varphi}(z)$ is the corresponding eigenfactor of $\varphi$. Now part (a) of the same proposition implies the claim. Claim (c) is a consequence of claim (a) and the observation that the isomorphism $\mathscr{F} \cong K[z]/\langle \mu_\varphi(z) \rangle$ yields a 1–1 correspondence between the maximal ideals of $\mathscr{F}$ and the eigenfactors of $\varphi$. Finally, claim (d) follows from Corollary 1.2.4 and the definition, since the number of eigenfactors does not change under the substitution $z \mapsto z - c$. $\qquad\square$

Our next corollary is a useful tool for computing primary decompositions in the last chapter.

**Corollary 2.5.3 (Primary Components of a Splitting Endomorphism)**
*Let $\varphi \in \mathcal{F}$ be a splitting endomorphism for the family $\mathcal{F}$. For every $i = 1, \ldots, s$, let $q_i = \mathrm{Ann}_{\mathcal{F}}(\mathrm{BigKer}(\mathfrak{m}_i))$, and let $\mu_\varphi(z) = \prod_{i=1}^s p_{\mathfrak{m}_i, \varphi}(z)^{m_i}$. Then we have the equality $q_i = \langle p_{\mathfrak{m}_i, \varphi}(\varphi)^{m_i} \rangle$ for $i = 1, \ldots, s$.*

*Proof* This is a straightforward consequence of Theorem 2.4.12. $\qquad\square$

If the base field has sufficiently many elements, splitting endomorphisms always exist, as our next proposition and its corollary show. For a finite field $K$, we let card$(K)$ be its **cardinality**, and if $K$ is infinite, we let card$(K) = \infty$.

**Proposition 2.5.4 (Existence of Splitting Endomorphisms)**
*Let $\mathfrak{m}_1, \ldots, \mathfrak{m}_s$ be the maximal ideals of the commuting family $\mathcal{F}$, let $i \in \{1, \ldots, s\}$, and let $q_j = \mathrm{Ann}_{\mathcal{F}}(\mathrm{BigKer}(\mathfrak{m}_j))$ for $j = 1, \ldots, s$.*

(a) *If $s > 1$, there exists $\varphi \in \mathcal{F}$ such that $\varphi \in \mathfrak{m}_i$ and $\varphi - \mathrm{id}_V \in \mathfrak{m}_j$ for $j \neq i$. In this case we have $\mu_\varphi(z) = z^a \cdot (z-1)^b$ with $a, b \in \mathbb{N}_+$, the equalities $\mathrm{Gen}(\varphi, z) = \mathrm{BigKer}(\mathfrak{m}_i)$, and $q_i = \langle \varphi^a \rangle$. In particular, the map $\varphi$ is a partially splitting endomorphism for $\mathcal{F}$.*

(b) *Suppose that $s \leq \mathrm{card}(K)$, and let $(c_1, \ldots, c_s)$ be a tuple of distinct elements of $K$. Then there exists an endomorphism $\varphi \in \mathcal{F}$ such that $\mu_\varphi(z) = \prod_{i=1}^s (z - c_i)^{a_i}$ for some $a_i > 0$, and we have the equalities $\mathrm{Gen}(\varphi, z - c_i) = \mathrm{BigKer}(\mathfrak{m}_i)$ for $i = 1, \ldots, s$. In particular, the map $\varphi$ is a splitting endomorphism for $\mathcal{F}$ and $q_i = \langle (\varphi - c_i \, \mathrm{id}_V)^{a_i} \rangle$ for $i = 1, \ldots, s$.*

*Proof* First we prove (a). Let $\pi : \mathcal{F}/(\mathfrak{m}_1 \cap \cdots \cap \mathfrak{m}_s) \longrightarrow \bigoplus_{i=1}^s \mathcal{F}/\mathfrak{m}_i$ be the isomorphism which follows from the Chinese Remainder Theorem 2.2.1, and let $\varphi \in \mathcal{F}$ be such that $\pi(\bar\varphi) = (\overline{\mathrm{id}}_V, \ldots, \overline{\mathrm{id}}_V, 0, \overline{\mathrm{id}}_V, \ldots, \overline{\mathrm{id}}_V)$. Then we have $\varphi \in \mathfrak{m}_i$, and thus the $\mathfrak{m}_i$-eigenfactor of $\varphi$ is $z$. Likewise, we see that the $\mathfrak{m}_j$-eigenfactor of $\varphi$ is $z - 1$ for $j \neq i$. Consequently, Lemma 1.1.11 implies that there exist integers $a, b > 0$ such that $\mu_\varphi(z) = z^a \cdot (z-1)^b$. Moreover, as we have $\varphi \in \mathfrak{m}_i$ and $\varphi \notin \mathfrak{m}_j$ for $j \neq i$, the equality $\mathrm{Gen}(\varphi, z) = \mathrm{BigKer}(\mathfrak{m}_i)$ follows from Proposition 2.4.10. The equality $q_i = \langle \varphi^a \rangle$ follows from Theorem 2.4.12.

The proof of (b) proceeds similarly. If $s = 1$, the claim is obviously true. If $s > 1$, we construct an endomorphism $\varphi \in \mathcal{F}$ such that $\pi(\varphi) = (c_1 \overline{\mathrm{id}}_V, \ldots, c_s \overline{\mathrm{id}}_V)$ and note that $\varphi - c_i \, \mathrm{id}_V \in q_i \subseteq \mathfrak{m}_i$ implies that the $\mathfrak{m}_i$-eigenfactor of $\varphi$ is $z - c_i$. Hence there are $s$ distinct eigenfactors, the minimal polynomial of $\varphi$ has the desired form, the generalized eigenspaces of $\varphi$ are precisely the joint generalized eigenspaces of $\mathcal{F}$, and $\varphi$ is a splitting endomorphism for $\mathcal{F}$. For every $i \in \{1, \ldots, s\}$, the equality $q_i = \langle (\varphi - c_i \, \mathrm{id}_V)^{a_i} \rangle$ follows from Corollary 2.5.3. $\qquad\square$

The following corollary provides another condition on the cardinality of $K$ which implies the existence of a splitting endomorphism.

**Corollary 2.5.5** *If we have* $\mathrm{card}(K) \geq \dim_K(\mathcal{F})$ *then there exists a splitting endomorphism for* $\mathcal{F}$.

*Proof* By Proposition 2.2.4, the family $\mathcal{F}$ has at most $\dim_K(\mathcal{F})$ maximal ideals. Hence the claim follows from part (b) of the proposition.                    □

The following example shows that a splitting endomorphism may exist even if $\mathrm{card}(K)$ is smaller than the number of maximal ideals in $\mathcal{F}$, and hence smaller than $\dim_K(\mathcal{F})$.

**Example 2.5.6** Let $K = \mathbb{F}_2$, let $V = K^4$, and let $\mathcal{F} = K[\varphi_1, \varphi_2]$ be the commuting family generated by the two endomorphisms represented by

$$
A_1 = \begin{pmatrix} 0 & 0 & 0 & 0 \\ 1 & 1 & 0 & 0 \\ 0 & 1 & 0 & 1 \\ 0 & 0 & 1 & 1 \end{pmatrix} \quad \text{and} \quad A_2 = \begin{pmatrix} 0 & 0 & 0 & 0 \\ 0 & 0 & 0 & 0 \\ 1 & 0 & 1 & 0 \\ 0 & 1 & 0 & 1 \end{pmatrix}
$$

Using the Buchberger Möller Algorithm for Matrices 2.1.4, we get $\mathcal{F} \cong K[x, y]/I$, where $I = \langle x^4 + x, y + x^2 + x \rangle = \langle x, y \rangle \cap \langle x + 1, y \rangle \cap \langle x^2 + x + 1, y + 1 \rangle$. The family $\mathcal{F}$ has three maximal ideals and $\mu_{\varphi_1}(z) = z^4 + z = z(z + 1)(z^2 + z + 1)$ has three eigenfactors. Therefore $\varphi_1$ is a splitting endomorphism, although there are more maximal ideals in $\mathcal{F}$ than elements in the field $K$.

In the case of a reduced commuting family, we even get strongly splitting endomorphisms, as our next corollary shows.

**Corollary 2.5.7** *Let $\mathcal{F}$ be a reduced commuting family, and let $\mathfrak{m}_1, \ldots, \mathfrak{m}_s$ be its maximal ideals.*

(a) *For every $i \in \{1, \ldots, s\}$, there exists an endomorphism $\varphi \in \mathcal{F}$ such that we have $\varphi \in \mathfrak{m}_i$ and $\varphi - \mathrm{id}_V \in \mathfrak{m}_j$ for all $j \neq i$. It satisfies $\mu_\varphi(z) = z \cdot (z - 1)$ and $\mathrm{Eig}(\varphi, z) = \mathrm{Ker}(\mathfrak{m}_i)$.*

(b) *If $s \leq \mathrm{card}(K)$ and $(c_1, \ldots, c_s)$ is a tuple of distinct elements of $K$ then there exists a map $\varphi \in \mathcal{F}$ such that $\mu_\varphi(z) = \prod_{i=1}^{s}(z - c_i)$. For $i = 1, \ldots, s$, it satisfies the equalities $\mathrm{Eig}(\varphi, z - c_i) = \mathrm{Ker}(\mathfrak{m}_i)$ for $i = 1, \ldots, s$. In particular, the map $\varphi$ is a strongly splitting endomorphism.*

*Proof* To prove (a) we note that the proposition implies $\mathrm{Gen}(\varphi, z) = \mathrm{BigKer}(\mathfrak{m}_i)$. From Corollary 2.4.7.b we deduce that $\varphi(\varphi - \mathrm{id}_V)$ is the zero map. Hence we have $\mu_\varphi(z) = z \cdot (z - 1)$, and therefore Theorem 1.1.17.d yields $\mathrm{Gen}(\varphi, z) = \mathrm{Eig}(\varphi, z)$. On the other hand, Corollary 2.4.7.c shows $\mathrm{BigKer}(\mathfrak{m}_i) = \mathrm{Ker}(\mathfrak{m}_i)$ for $i = 1, \ldots, s$. Thus we get (a), and (b) follows analogously.                    □

Example 2.4.11 shows that a strongly splitting endomorphism may not exist (see also Example 6.1.8). On the other hand, the preceding proposition and its corollaries say that splitting endomorphisms exist if the base field has enough elements. In view of the next example, this extra assumption is essential.

**Example 2.5.8**   Let $K = \mathbb{F}_2$, let $V = K^3$, and let $\mathscr{F} = K[\varphi_1, \varphi_2]$ be the commuting family generated by the two endomorphisms given by

$$A_1 = \begin{pmatrix} 0 & 0 & 0 \\ 1 & 1 & 0 \\ 0 & 0 & 0 \end{pmatrix} \quad \text{and} \quad A_2 = \begin{pmatrix} 0 & 0 & 0 \\ 0 & 0 & 0 \\ 1 & 0 & 1 \end{pmatrix}$$

respectively. Using the Buchberger Möller Algorithm for Matrices 2.1.4 we find that $\mathscr{F} \cong K[x, y]/I$ where $I = \langle x^2 + x, xy, y^2 + y \rangle$, and that $\mathscr{O} = (1, x, y)$ represents a $K$-basis of $\mathscr{F}$. It is easy to check that $I = \langle x, y \rangle \cap \langle x + 1, y \rangle \cap \langle x, y + 1 \rangle$. Thus the family $\mathscr{F}$ is reduced.

Since we have $\dim_K(\mathscr{F}) = 3$, the family $\mathscr{F}$ has eight elements. Besides the zero map and the identity, the other endomorphisms in $\mathscr{F}$ are $\varphi_1, \varphi_2, \varphi_1 + \mathrm{id}_V, \varphi_2 + \mathrm{id}_V$, $\varphi_1 + \varphi_2$, and $\varphi_1 + \varphi_2 + \mathrm{id}_V$. They all have minimal polynomial $z(z + 1)$. Hence no element in $\mathscr{F}$ is a splitting endomorphism.

Even if the base field is large enough, we do not yet know how to find a splitting endomorphism. The following example provides a case where the two generators of the family are not splitting endomorphisms, but their sum is.

**Example 2.5.9**   Let us go back to Example 2.3.10, i.e., to $K = \mathbb{Q}$, the vector space $V = K^4$, and the commuting family $\mathscr{F} = K[\varphi_1, \varphi_2]$ generated by the endomorphisms represented by the matrices

$$A_1 = \begin{pmatrix} 0 & 2 & 0 & 0 \\ 1 & 0 & 0 & 0 \\ 0 & 0 & 0 & 2 \\ 0 & 0 & 1 & 0 \end{pmatrix} \quad \text{and} \quad A_2 = \begin{pmatrix} 0 & 0 & 8 & 0 \\ 0 & 0 & 0 & 8 \\ 1 & 0 & 0 & 0 \\ 0 & 1 & 0 & 0 \end{pmatrix}$$

respectively. We have already seen that $\mu_{\varphi_1}(z) = z^2 - 2$ is irreducible. Similarly, we get that $\mu_{\varphi_2}(z) = z^2 - 8$ is irreducible. However, these two endomorphisms are not splitting, as we show in the following.

Let us consider the endomorphism $\psi = \varphi_1 + \varphi_2 \in \mathscr{F}$. It is represented by the matrix

$$B = \begin{pmatrix} 0 & 2 & 8 & 0 \\ 1 & 0 & 0 & 8 \\ 1 & 0 & 0 & 2 \\ 0 & 1 & 1 & 0 \end{pmatrix}$$

and satisfies $\mu_\psi(z) = z^4 - 20z^2 + 36 = (z^2 - 2)(z^2 - 18)$. This minimal polynomial splits into quadratic eigenfactors $p_1(z) = z^2 - 2$ and $p_2(z) = z^2 - 18$. The corresponding eigenspaces are $\mathrm{Eig}(\psi, p_1(z)) = \mathrm{Ker}(\psi^2 - 2\,\mathrm{id}_V) = \langle v_1, v_2 \rangle$, where $v_1 = (0, 2, -1, 0)$ and $v_2 = (4, 0, 0, -1)$, and $\mathrm{Eig}(\psi, p_2(z)) = \mathrm{Ker}(\psi^2 - 18\,\mathrm{id}_V) =$

$\langle w_1, w_2 \rangle$, where $w_1 = (0, 2, 1, 0)$ and $w_2 = (4, 0, 0, 1)$. The dimensions of the two eigenspaces of $\psi$ add up to four which is the dimension of $V$. Therefore these eigenspaces are also the generalized eigenspaces of $\psi$. Next we check that there are no 1-dimensional joint eigenspaces of $\mathcal{F}$ (for instance, using Algorithm 6.1.1). Hence the map $\psi$ is a splitting endomorphism for the family $\mathcal{F}$, and because of $\mathrm{Eig}(\psi, p_i(z)) = \mathrm{Gen}(\psi, p_i(z))$ for $i = 1, 2$ it is even a strongly splitting endomorphism for $\mathcal{F}$. Notice that neither $\varphi_1$ nor $\varphi_2$ is a splitting endomorphism.

The next example shows that, in general, it is not enough to use only linear combinations of the generators of the family in order to find a splitting endomorphism.

**Example 2.5.10**  Let $K = \mathbb{F}_2$, let $V = K^8$, and let $\mathcal{F} = K[\varphi_1, \varphi_2]$ be the commuting family generated by the two endomorphisms $\varphi_1, \varphi_2$ given by

$$A_1 = \begin{pmatrix} 0 & 1 & 0 & 0 & 0 & 0 & 0 & 0 \\ 1 & 1 & 0 & 0 & 0 & 0 & 0 & 0 \\ 0 & 0 & 0 & 1 & 0 & 0 & 0 & 0 \\ 0 & 0 & 1 & 1 & 0 & 0 & 0 & 0 \\ 0 & 0 & 0 & 0 & 0 & 1 & 0 & 0 \\ 0 & 0 & 0 & 0 & 1 & 1 & 0 & 0 \\ 0 & 0 & 0 & 0 & 0 & 0 & 0 & 1 \\ 0 & 0 & 0 & 0 & 0 & 0 & 1 & 1 \end{pmatrix} \text{ and } A_2 = \begin{pmatrix} 0 & 0 & 0 & 0 & 0 & 0 & 1 & 0 \\ 0 & 0 & 0 & 0 & 0 & 0 & 0 & 1 \\ 1 & 0 & 0 & 0 & 0 & 0 & 1 & 0 \\ 0 & 1 & 0 & 0 & 0 & 0 & 0 & 1 \\ 0 & 0 & 1 & 0 & 0 & 0 & 0 & 0 \\ 0 & 0 & 0 & 1 & 0 & 0 & 0 & 0 \\ 0 & 0 & 0 & 0 & 1 & 0 & 0 & 0 \\ 0 & 0 & 0 & 0 & 0 & 1 & 0 & 0 \end{pmatrix}$$

respectively. The minimal polynomials of $\varphi_1$ and $\varphi_2$ are $\mu_{\varphi_1}(z) = z^2 + z + 1$ and $\mu_{\varphi_2}(z) = z^4 + z + 1$. Both polynomials are irreducible. The non-trivial linear combinations of $\mathrm{id}_V$, $\varphi_1$ and $\varphi_2$ are

$$\mathrm{id}_V, \ \varphi_1, \ \varphi_2, \ \mathrm{id}_V + \varphi_1, \ \mathrm{id}_V + \varphi_2, \ \varphi_1 + \varphi_2, \ \mathrm{id}_V + \varphi_1 + \varphi_2$$

Their minimal polynomials are

$$z + 1, \ z^2 + z + 1, \ z^4 + z + 1, \ z^2 + z + 1, \ z^4 + z + 1, \ z^4 + z + 1, \ z^4 + z + 1$$

respectively. All of them are irreducible. This would indicate that $\mathcal{F}$ has only one maximal ideal.

However, the minimal polynomial of $\psi = \varphi_1 \varphi_2$ is $\mu_\psi(z) = z^8 + z^4 + z^2 + 1$ which is the product of $p_1(z) = z^4 + z^3 + z^2 + z + 1$ and $p_2(z) = z^4 + z^3 + 1$. The corresponding eigenspaces are $\mathrm{Eig}(\psi, p_1(z)) = \mathrm{Ker}(\psi_1^4 + \psi^3 + \psi^2 + \psi + \mathrm{id}_V) = \langle v_1, v_2, v_3, v_4 \rangle$, where $v_1 = (1, 1, 1, 0, 1, 0, 0, 0)$, $v_2 = (1, 0, 0, 1, 0, 1, 0, 0)$, $v_3 = (1, 1, 0, 1, 0, 0, 1, 0)$, and $v_4 = (1, 0, 1, 1, 0, 0, 0, 1)$, and, furthermore, $\mathrm{Eig}(\psi, p_2(z)) = \mathrm{Ker}(\psi^4 + \psi^3 + \mathrm{id}_V) = \langle w_1, w_2, w_3, w_4 \rangle$, where we have $w_1 = (0, 1, 1, 0, 1, 0, 0, 0)$, $w_2 = (1, 1, 0, 1, 0, 1, 0, 0)$, $w_3 = (0, 1, 1, 1, 0, 0, 1, 0)$, and $w_4 = (1, 1, 1, 0, 0, 0, 0, 1)$. Since the dimensions of these two eigenspaces add up to $\dim_K(V) = 8$, they are the generalized eigenspaces of $\psi$.

So, the map $\psi$ seems to indicate that $\mathcal{F}$ has two maximal ideals. Let us verify that this is indeed the case. Using the Buchberger Möller Algorithm for Matrices 2.1.4, we get the ideal $\mathrm{Rel}_P(\varphi_1, \varphi_2) = \langle x^2 + x + 1, y^4 + y + 1 \rangle$ in $P = K[x, y]$. Then we use any of the methods in Chap. 5 to show that $\mathrm{Rel}_P(\varphi_1, \varphi_2) = \mathfrak{M}_1 \cap \mathfrak{M}_2$, where $\mathfrak{M}_1 = \langle x^2 + x + 1, y^2 + x + y \rangle$ and $\mathfrak{M}_2 = \langle x^2 + x + 1, y^2 + x + y + 1 \rangle$ are maximal ideals in $P$. Hence the eigenspaces of $\psi$ are the joint eigenspaces of $\mathscr{F}$ as well as the joint generalized eigenspaces of $\mathscr{F}$, and $\psi$ is a strictly splitting endomorphism for $\mathscr{F}$.

Proposition 2.5.4.b and Corollary 2.5.5 suggest that "generic" elements of $\mathscr{F}$ are splitting endomorphisms if the base field is large enough. In the following we want to clarify what the term "generic" means here.

**Definition 2.5.11**   Let $\mathfrak{m}_1, \ldots, \mathfrak{m}_s$ be the maximal ideals of the commuting family $\mathscr{F}$. Then the set

$$\mathrm{Spl}(\mathscr{F}) = \left\{ \varphi \in \mathscr{F} \mid p_{\mathfrak{m}_i, \varphi}(z) \neq p_{\mathfrak{m}_j, \varphi}(z) \text{ for } i \neq j \right\}$$

is called the **splitting locus** of $\mathscr{F}$. Moreover, the set $\mathrm{NSpl}(\mathscr{F}) = \mathscr{F} \setminus \mathrm{Spl}(\mathscr{F})$ is called the **non-splitting locus** of $\mathscr{F}$.

By definition, the elements of $\mathrm{Spl}(\mathscr{F})$ are precisely the splitting endomorphisms for $\mathscr{F}$. In order to describe the splitting locus of $\mathscr{F}$, we use an extension of the family $\mathscr{F}$ which is defined as follows.

**Definition 2.5.12**   Let $\delta = \dim_K(\mathscr{F})$, let $y_1, \ldots, y_\delta$ be new indeterminates, and let $K(y_1, \ldots, y_\delta)$ be the quotient field of $K[y_1, \ldots, y_\delta]$.

(a) By extending the base field, we form the vector space $V \otimes_K K(y_1, \ldots, y_\delta)$. The family $\mathscr{F} \otimes_K K(y_1, \ldots, y_\delta)$ of commuting endomorphisms of this vector space is called the **generically extended family** of $\mathscr{F}$.

(b) For a $K$-basis $\Psi = (\psi_1, \ldots, \psi_\delta)$ of $\mathscr{F}$, the element $\Theta_{y,\psi} = y_1 \psi_1 + \cdots + y_\delta \psi_\delta$ in $\mathscr{F} \otimes_K K(y_1, \ldots, y_\delta)$ is called the **generically extended endomorphism** associated to $\Psi$.

Notice that the canonical map $\iota : \mathscr{F} \longrightarrow \mathscr{F} \otimes_K K(y_1, \ldots, y_\delta)$ is injective. Extended families are examined in a more general setting in Sect. 3.6. The generically extended endomorphism is a splitting endomorphism and can be used to determine the number of maximal ideals and joint generalized eigenspaces of $\mathscr{F}$, as our next proposition shows.

**Proposition 2.5.13 (Properties of Generically Extended Endomorphisms)**
*Let $\mathscr{F}$ be a commuting family, let $\delta = \dim_K(\mathscr{F})$, let $y_1, \ldots, y_\delta$ be new indeterminates, let $\Psi = (\psi_1, \ldots, \psi_\delta)$ be a $K$-basis of $\mathscr{F}$, let $\Theta_{y,\psi}$ be the generically extended endomorphism, and let $\iota : \mathscr{F} \longrightarrow \mathscr{F} \otimes_K K(y_1, \ldots, y_\delta)$ be the canonical map.*

(a) *The families $\mathscr{F}$ and $\mathscr{F} \otimes_K K(y_1, \ldots, y_\delta)$ have the same number of maximal ideals. More precisely, maps given by $\mathfrak{m} \mapsto \iota(\mathfrak{m}) = \mathfrak{m} \otimes_K K(y_1, \ldots, y_\delta)$*

*and* $\mathfrak{M} \mapsto \iota^{-1}(\mathfrak{M})$ *define a 1–1 correspondence between the maximal ideals of* $\mathscr{F}$ *and those of* $\mathscr{F} \otimes_K K(y_1, \ldots, y_\delta)$.

(b) *The map* $\Theta_{y,\psi}$ *is a splitting endomorphism for* $\mathscr{F} \otimes_K K(y_1, \ldots, y_\delta)$.

(c) *The number of eigenfactors of* $\Theta_{y,\psi}$ *is precisely the number of maximal ideals of* $\mathscr{F}$.

*Proof* First we prove (a). A basis of the $K$-vector space $\mathscr{F}$ is also a basis of the $K(y_1, \ldots, y_\delta)$-vector space $\mathscr{F} \otimes_K K(y_1, \ldots, y_\delta)$. Therefore the generically extended family $\mathscr{F} \otimes_K K(y_1, \ldots, y_\delta)$ is a zero-dimensional $K(y_1, \ldots, y_\delta)$-algebra (see [15], Proposition 3.7.1). Let $\mathfrak{m}_1, \ldots, \mathfrak{m}_s$ be the maximal ideals of $\mathscr{F}$ and for $i = 1, \ldots, s$ let $\mathfrak{q}_i = \mathrm{Ann}_{\mathscr{F}}(\mathrm{BigKer}(\mathfrak{m}_i))$. From Proposition 2.4.6 we obtain an isomorphism of $K$-algebras $\pi : \mathscr{F} \cong \mathscr{F}/\mathfrak{q}_1 \times \cdots \times \mathscr{F}/\mathfrak{q}_s$. Consequently, we have an isomorphism $\mathscr{F} \otimes_K K(y_1, \ldots, y_\delta) \cong \prod_{i=1}^{s}(\mathscr{F}/\mathfrak{q}_i) \otimes_K K(y_1, \ldots, y_\delta)$. Since the ring $\mathscr{F}/\mathfrak{q}_i$ has only one maximal ideal, namely $\mathfrak{m}_i/\mathfrak{q}_i$, it suffices to prove that $(\mathscr{F}/\mathfrak{q}_i) \otimes_K K(y_1, \ldots, y_\delta)$ has only one maximal ideal, namely the residue class ideal of $\mathfrak{m}_i \otimes_K K(y_1, \ldots, y_\delta)$.

Consequently, we may assume that $\mathscr{F}$ has only one maximal ideal $\mathfrak{m}$. Since $\mathfrak{m}$ is a nilpotent ideal by Proposition 2.2.5.b, the inverse image under $\iota$ of every maximal ideal of $\mathscr{F} \otimes_K K(y_1, \ldots, y_\delta)$ contains $\mathfrak{m}$. Hence, to show the claim, we may pass to the induced map $\bar{\iota} : \mathscr{F}/\mathfrak{m} \longrightarrow (\mathscr{F} \otimes_K K(y_1, \ldots, y_\delta))/(\mathfrak{m} \otimes_K K(y_1, \ldots, y_\delta))$ and show that the ring on the right-hand side is a field. In other words, we may assume that $\mathscr{F}$ is a field. In this case it follows from [11], Theorem 8.47, that $\mathscr{F} \otimes_K K(y_1, \ldots, y_\delta)$ is an integral domain. Now Corollary 2.2.6 implies that $\mathscr{F} \otimes_K K(y_1, \ldots, y_\delta)$ is a field, and the proof of claim (a) is complete.

From (a) and Definition 2.5.1 we deduce that conditions (b) and (c) are equivalent. Therefore it suffices prove (c). The vector subspaces $G_i = \mathrm{BigKer}(\mathfrak{m}_i)$ with $i \in \{1, \ldots, s\}$ are the joint generalized eigenspaces of $\mathscr{F}$. By concatenating $K$-bases of the spaces $G_i$, we obtain a $K$-basis $C$ of $V$ such that the matrices of all elements of $\mathscr{F}$ are block diagonal with respect to this basis.

For $i = 1, \ldots, s$, we now choose a $K$-basis $B_i$ of the preimage of $\mathscr{F}/\mathfrak{q}_i$ under the isomorphism $\pi$ with the property that its first element is mapped by $\pi$ to the tuple $(0, \ldots, 0, \bar{1}, 0, \ldots, 0)$ with $\bar{1} \in \mathscr{F}/\mathfrak{q}_i$. Notice that the matrix of an element of $B_i$ in the basis $C$ is block diagonal and only the $i$-th block is non-zero. Next we concatenate the bases $B_i$ to a $K$-basis $B = (\beta_1, \ldots, \beta_\delta)$ of $\mathscr{F}$. Then there exists an invertible matrix $A \in \mathrm{Mat}_\delta(K)$ such that $A \cdot (\beta_1, \ldots, \beta_\delta)^{\mathrm{tr}} = (\psi_1, \ldots, \psi_\delta)^{\mathrm{tr}}$, and we have

$$\Theta_{y,\psi} = y_1\psi_1 + \cdots + y_\delta\psi_\delta = (y_1, \ldots, y_\delta) \cdot A \cdot (\beta_1, \ldots, \beta_\delta)^{\mathrm{tr}}$$

$$= (y_1', \ldots, y_\delta') \cdot (\beta_1, \ldots, \beta_\delta)^{\mathrm{tr}} = y_1'\beta_1 + \cdots + y_\delta'\beta_\delta = \Theta_{y',B}$$

with $y' = (y_1', \ldots, y_\delta') = (y_1, \ldots, y_\delta) \cdot A$. We note that the tuple $(y_1', \ldots, y_\delta')$ contains homogeneous linear polynomials which are algebraically independent and generate $K[y_1, \ldots, y_\delta]$.

Since the generically extended endomorphism $\Theta_{y',B} = y_1'\beta_1 + \cdots + y_\delta'\beta_\delta$ of the $K(y_1', \ldots, y_\delta')$-vector space $V' = V \otimes_K K(y_1', \ldots, y_\delta')$ agrees with $\Theta_{y,\psi}$, it suffices to prove that the number of eigenfactors $t$ of $\Theta_{y',B}$ is exactly $s$. By (a) and Proposition 2.3.2.f, we have $t \leq s$. With respect to the indeterminate $z$, the polynomial $\chi_{\Theta_{y',B}}(z) \in K[y'_1, \ldots, y'_\delta][z]$ is monic and has degree $d = \dim_K(V)$. Since the ring $K[y'_1, \ldots, y'_\delta][z]$ is a unique factorization domain (cf. [15], Proposition 1.2.12), we can factorize $\chi_{\Theta_{y',B}}(z)$ and get $t$ factors which are monic in $z$ and whose coefficients are polynomials in $K[y'_1, \ldots, y'_\delta]$.

Clearly, the tuple $C$ is also a $K(y'_1, \ldots, y'_\delta)$-basis of $V'$. From the fact that the matrix of an element of $B_i$ in the basis $C$ has non-zero entries only in the $i$-th diagonal block it follows that the matrix of $\Theta_{y',B}$ in the basis $C$ is block diagonal and that the entries of the $i$-th diagonal block only involve the indeterminates from the $i$-th subtuple of $(y'_1, \ldots, y'_\delta)$, i.e., the subtuple corresponding to the elements of $B_i$. Let $j$ be the first index corresponding to an element of $B_i$. By construction, the element $\beta_j$ corresponds to the identity on $G_i$. Hence the indeterminate $y'_j$ appears in the matrix of $\Theta_{y',B}$ with coefficient 1 everywhere on the main diagonal of the $i$-th diagonal block and nowhere else. Therefore, letting $\delta_i = \dim_K(G_i)$, we see that the term $(y'_j)^{\delta_i}$ appears in the support of the $i$-th factor of the characteristic polynomial of $\Theta_{y',B}$. Consequently, this characteristic polynomial has at least $s$ distinct factors, i.e., we have $t \geq s$, and the proof is complete.                               $\square$

In the proof of the following theorem we describe a non-empty Zariski open subset of the splitting locus of a commuting family. Let us recall what this means. Using zero-sets as the closed sets, we define the **Zariski topology** on $K^\delta$ for $\delta \geq 1$ (see [15], Tutorial 27 and [16], Sect. 5.5.C). Given a finite dimensional $K$-vector space $W$ and an isomorphism $\iota : W \longrightarrow K^\delta$, we can then define **Zariski open** subsets of $W$ as those subsets whose images under $\iota$ are Zariski open in $K^\delta$. Notice that this definition does not depend on the choice of the isomorphism $\iota$ (see [16], Remark 5.5.22). The following definition slightly extends Definition 5.5.19.c in [16] and will be used to formulate the next theorem.

**Definition 2.5.14** Let $W$ be finite dimensional $K$-vector space, and let $\mathscr{P}$ be a property of elements of $W$. If there exists a non-empty Zariski open subset $U$ of $W$ such that every endomorphism in $U$ has property $\mathscr{P}$, we say that a **generic element** of $W$ has property $\mathscr{P}$ or that $\mathscr{P}$ holds **for a generic element** of $W$.

Now we are ready to state the theorem.

**Theorem 2.5.15 (Genericity of the Splitting Locus)**
*Let $K$ be an infinite field, and let $\mathscr{F}$ be a family of commuting endomorphisms of a finitely generated $K$-vector space $V$. Then the splitting locus $\mathrm{Spl}(\mathscr{F})$ contains a non-empty Zariski open subset of $\mathscr{F}$. In other words, a generic element of $\mathscr{F}$ is a splitting endomorphism for $\mathscr{F}$.*

*Proof* Let $\delta = \dim_K(\mathscr{F})$, let $y_1, \ldots, y_\delta$ be new indeterminates, let $\Psi = (\psi_1, \ldots, \psi_\delta)$ be a $K$-basis of $\mathscr{F}$, and let $\Theta_\Psi$ be the generically extended endomorphism. Let $p_1(y_1, \ldots, y_\delta, z), \ldots, p_s(y_1, \ldots, y_\delta, z)$ denote the eigenfactors of $\Theta_\psi$. As we saw in the proof of the preceding proposition, they are polynomials in $K[y_1, \ldots, y_\delta, z]$. By part (b) of that proposition, they correspond 1–1 to the maximal ideals $\mathfrak{M}_1, \ldots, \mathfrak{M}_s$ of $\mathscr{F} \otimes_K K(y_1, \ldots, y_\delta)$, and by part (a), the maximal ideals of $\mathscr{F} \otimes_K K(y_1, \ldots, y_\delta)$ are of the form $\mathfrak{M}_i = \mathfrak{m}_i \otimes_K K(y_1, \ldots, y_\delta)$ where $\mathfrak{m}_1, \ldots, \mathfrak{m}_s$ are the maximal ideals of $\mathscr{F}$.

For every tuple $a = (a_1, \ldots, a_\delta) \in K^\delta$, we denote the endomorphism obtained by substituting $y_i \mapsto a_i$ in $\Theta_\psi$ by $\vartheta_a$. In other words, the endomorphism $\vartheta_a$ is given by $\vartheta_a = a_1 \psi_1 + \cdots + a_\delta \psi_\delta$. Notice that every element of $\mathscr{F}$ is of the form $\vartheta_a$ for a unique tuple $a \in K^\delta$.

Given indices $i, j \in \{1, \ldots, s\}$ with $i \neq j$, the polynomials $p_i(y_1, \ldots, y_\delta, z)$ and $p_j(y_1, \ldots, y_\delta, z)$ are coprime elements of $K(y_1, \ldots, y_\delta)[z]$. Hence there exist elements $f_i(y_1, \ldots, y_\delta, z), f_j(y_1, \ldots, y_\delta, z) \in K(y_1, \ldots, y_\delta)[z]$ such that

$$f_i(y_1, \ldots, y_\delta, z) p_i(y_1, \ldots, y_\delta, z) + f_j(y_1, \ldots, y_\delta, z) p_j(y_1, \ldots, y_\delta, z) = 1$$

Let $g_i, g_j \in K[y_1, \ldots, y_\delta]$ be the common denominators of $f_i, f_j$, respectively, and let $U_{ij} \subseteq K^\delta$ be the set of all tuples $a = (a_1, \ldots, a_\delta)$ such that we have $g_i(a) \neq 0$ and $g_j(a) \neq 0$. Since $K$ is infinite, the set $U_{ij}$ is a non-empty Zariski open subset of $K^\delta$ (see for instance [16], Proposition 5.5.21.b). For all $a \in U_{ij}$, we have $f_i(a, z) p_i(a, z) + f_j(a, z) p_j(a, z) = 1$. Therefore the polynomials $p_i(a, z)$ and $p_j(a, z)$ are coprime for these tuples $a$. Altogether, the set $U = \bigcap_{i \neq j} U_{ij}$ is a non-empty Zariski open subset of $K^\delta$, and the polynomials $p_1(a, z), \ldots, p_s(a, z)$ are pairwise coprime for $a \in U$.

The substitution $y_i \mapsto a_i$ sends an element of $\mathfrak{M}_j = \mathfrak{m}_j \otimes_K K(y_1, \ldots, y_\delta)$ into $\mathfrak{m}_j$. Therefore we have $p_j(a, \vartheta_a) \in \mathfrak{m}_j$, i.e. the polynomial $p_j(a, z)$ is a multiple of the eigenfactor $p_{\mathfrak{m}_j, \vartheta_a}(z)$. For $a \in U$, the facts that there are $s$ pairwise coprime factors $p_j(a, z)$ and $s$ maximal ideals $\mathfrak{m}_j$ imply that $p_j(a, z)$ is a power of $p_{\mathfrak{m}_j, \vartheta_a}(z)$. Using the observation that the substitution $y_i \mapsto a_i$ maps $\chi_{\Theta_\psi}(z)$ to $\chi_{\vartheta_a}(z)$, it follows that the element $\vartheta_a$ is contained in $\mathrm{Spl}(\mathscr{F})$ for $a \in U$.                    $\square$

In the case of an infinite base field $K$, this theorem suggests that a *random* choice of the coefficients $a_i$ produces a splitting endomorphism $\psi$ *with probability one*, since the chance of hitting the Zariski closed, proper subset $\mathrm{NSpl}(\mathscr{F})$ is zero. If $K$ is finite, there may be no splitting endomorphism, as we saw in Example 2.5.8, but if $K$ has sufficiently many elements, the set $\mathrm{Spl}(\mathscr{F})$ is non-empty (see Proposition 2.5.4 and its corollaries), although it is not always easy to find one of its elements (see Example 2.5.10).

> *We're not splitting atoms here;*
> *we're trying to entertain people.*
> (Boomer Esiason)

The following example illustrates the preceding theorem. It shows that $\mathrm{Spl}(\mathscr{F})$ contains a Zariski open subset of $\mathscr{F}$, but it is not necessarily Zariski open itself. In the language of algebraic geometry, it is what is called a **constructible set**.

**Example 2.5.16** Let $K = \mathbb{Q}$, let $V = K^4$, let the endomorphisms $\varphi_1$ and $\varphi_2$ of $V$ be given by the matrices

$$A_1 = \begin{pmatrix} 0 & 0 & 0 & 4 \\ 1 & 0 & 0 & 2 \\ 0 & 0 & 0 & -4 \\ 0 & 1 & 0 & -2 \end{pmatrix} \quad \text{and} \quad A_2 = \begin{pmatrix} 0 & 0 & 0 & 0 \\ 0 & 0 & 0 & 0 \\ 1 & 0 & 1 & 0 \\ 0 & 0 & 0 & 0 \end{pmatrix}$$

and let $\mathscr{F} = K[\varphi_1, \varphi_2]$. It is easy to check that $\mathscr{F}$ is a commuting family. If we apply the Buchberger-Möller Algorithm for Matrices 2.1.4 to $(A_1, A_2)$, we get $\mathrm{Rel}_P(\varphi_1, \varphi_2) = \langle x^3 + 2x^2 - 2x + 4y - 4, xy, y^2 - y \rangle$. Moreover, we have $\mu_{\varphi_1}(z) = z^4 + 2z^3 - 2z^2 - 4z = z(z + 2)(z^2 - 2)$ and $\mu_{\varphi_2}(z) = z^2 - z = z(z - 1)$.

Using the algorithms in Chap. 5, we find that $\mathscr{F}$ is reduced, has one non-linear maximal ideal $\mathfrak{m}_1 = \langle \varphi_1^2 - 2\,\mathrm{id}_V, \varphi_2 \rangle$ and two linear maximal ideals $\mathfrak{m}_2 = \langle \varphi_1 + 2\,\mathrm{id}_V, \varphi_2 \rangle$ and $\mathfrak{m}_3 = \langle \varphi_1, \varphi_2 - \mathrm{id}_V \rangle$.

Next we consider the isomorphism $\pi : \mathscr{F} \cong \bigoplus_{i=1}^3 \mathscr{F}/\mathfrak{m}_i$ and compute a basis $B = (B_1, B_2, B_3)$ such that $\pi(B_i)$ is a $K$-basis of $\mathscr{F}/\mathfrak{m}_i$ for $i = 1, 2, 3$. Clearly, a basis of $\mathscr{F}/\mathfrak{m}_1$ is given by $\{\mathrm{id}_V + \mathfrak{m}_1, \varphi_1 + \mathfrak{m}_1\}$, a basis of $\mathscr{F}/\mathfrak{m}_2$ is $\{\mathrm{id}_V + \mathfrak{m}_2\}$, and a basis of $\mathscr{F}/\mathfrak{m}_3$ is given by $\{\mathrm{id}_V + \mathfrak{m}_3\}$. By lifting these bases to $\mathscr{F}$ via the Chinese Remainder Theorem 2.2.1, we find

$$B_1 = \left( -\frac{1}{2}\varphi_1^2 - 2\varphi_2 + 2\,\mathrm{id}_V, \; \varphi_1^2 + \varphi_1 + 2\varphi_2 - 2\,\mathrm{id}_V \right)$$

$$B_2 = \left( \frac{1}{2}\varphi_1^2 + \varphi_2 - \mathrm{id}_V \right)$$

$$B_3 = (\varphi_2)$$

Therefore we can write a given endomorphism $\vartheta_a$ in $\mathscr{F}$ as

$$\vartheta_a = a_1\left( -\frac{1}{2}\varphi_1^2 - 2\varphi_2 + 2\,\mathrm{id}_V \right) + a_2\left(\varphi_1^2 + \varphi_1 + 2\varphi_2 - 2\,\mathrm{id}_V\right)$$

$$+ a_3\left( \frac{1}{2}\varphi_1^2 + \varphi_2 - \mathrm{id}_V \right) + a_4\varphi_2$$

with $a = (a_1, a_2, a_3, a_4) \in K^4$. If we replace the numbers $a_i$ by indeterminates $y_i$, the resulting generically extended endomorphism $\Theta_B$ satisfies

$$\chi_{\Theta_B}(z) = z^4 + (-2y_1 - y_3 - y_4)z^3 + \left(y_1^2 - 2y_2^2 + 2y_1 y_3 + 2y_1 y_4\right)z^2$$

$$+ \left(-y_1^2 y_3 + 2y_2^2 y_3 - y_1^2 y_4 + 2y_2^2 y_4 - 2y_1 y_3 y_4\right)z + \left(y_1^2 y_3 y_4 - 2y_2^2 y_3 y_4\right)$$

which factorizes as $\chi_{\Theta_B}(z) = (z^2 - 2y_1 z + y_1^2 - 2y_2^2)(z - y_3)(z - y_4)$. The three factors correspond 1–1 to the maximal ideals $\mathfrak{m}_1$, $\mathfrak{m}_2$, and $\mathfrak{m}_3$ of $\mathscr{F}$ (see Proposition 2.5.13.c). Notice that each factor uses only the indeterminates related to the corresponding maximal ideal.

Now let us look at the splitting locus in this case. To compute $\mathrm{Spl}(\mathscr{F})$, we need to find the values of the coefficients $a_i$ such that two irreducible factors of $\chi_{\vartheta_a}(z)$ coincide. First of all, to check when the quadratic factor splits into a product of linear factors, we consider its discriminant and get $4y_1^2 - 4(y_1^2 - 2y_2^2) = 8y_2^2$. Since 8 is not a square in $\mathbb{Q}$, this factor splits if and only if $a_2 = 0$, and in this case it splits as $(z - a_1)^2$. Next we check when two factors become equal and we get three cases:

(1) $a_2 = 0$ and $a_1 = a_3$,
(2) $a_2 = 0$ and $a_1 = a_4$,
(3) $a_3 = a_4$.

Therefore, if we have $a_2 \neq 0$ and $a_3 \neq a_4$, the map $\psi$ is a splitting endomorphism, and the Zariski open set $\{\psi \mid a_2 \neq 0, a_3 - a_4 \neq 0\}$ is contained in $\mathrm{Spl}(\mathscr{F})$.

Finally, let us mention what happens it we consider this example over the base field $\mathbb{Q}(\sqrt{2})$. In this case, the family $\mathscr{F}$ has four maximal ideals, the polynomial $\chi_{\Theta_B}(z)$ splits as $\chi_{\vartheta_B}(z) = (z - y_1 - \sqrt{2}y_2)(z - y_1 + \sqrt{2}y_2)(z - y_3)(z - y_4)$ into linear factors, and the splitting locus is the intersection of six non-empty Zariski open sets.

Notice that in Example 2.5.8 no element of $\mathscr{F}$ is a splitting endomorphism. However, a generically extended endomorphism is a splitting endomorphism by Proposition 2.5.13.b. Let us check this in the mentioned example.

**Example 2.5.17** In the following we continue the discussion of Example 2.5.8. Recall that the maps $\varphi_1, \varphi_2$ generate a 3-dimensional commuting family $\mathscr{F}$ over $\mathbb{F}_2$ and satisfy $\mu_{\varphi_i}(z) = z^2 + z$. The family $\mathscr{F}$ has three maximal ideals $\mathfrak{m}_1 = \langle \varphi_1, \varphi_2 \rangle$, $\mathfrak{m}_2 = \langle \varphi_1 + \mathrm{id}_V, \varphi_2 \rangle$, and $\mathfrak{m}_3 = \langle \varphi_1, \varphi_2 + \mathrm{id}_V \rangle$.

Every element $\vartheta_a$ in the family $\mathscr{F}$ is of the form $\vartheta_a = a_0 \mathrm{id}_V + a_1 \varphi_1 + a_2 \varphi_2$ with a tuple $a = (a_0, a_1, a_2) \in \mathbb{F}_2^3$. After replacing the numbers $a_i$ by indeterminates $y_i$ and factoring the corresponding characteristic polynomial, we get

$$\chi_{\Theta_B}(z) = z^3 + (y_0 + y_1 + y_2)z^2 + \left(y_0^2 + y_1 y_2\right)z + y_0^3 + y_0^2 y_1 + y_0^2 y_2 + y_0 y_1 y_2$$

$$= (z + y_0)(z + y_0 + y_1)(z + y_0 + y_2)$$

In agreement with part (c) of Proposition 2.5.13, there are three irreducible factors. The conditions for two factors to become equal after the substitution $y_i \mapsto a_i$ are $a_1 = 0$ or $a_2 = 0$ or $a_1 = a_2$. Consequently, the splitting locus $\mathrm{Spl}(\mathscr{F})$ is given by $\mathrm{Spl}(\mathscr{F}) = \{\vartheta_a \in \mathscr{F} \mid a_1 \neq 0, a_2 \neq 0, a_1 \neq a_2\}$. Since the base field has only two elements, this set is empty. In agreement with Example 2.5.8, this means that there is no splitting endomorphism in $\mathscr{F}$.

In the next example we continue the discussion of Example 2.5.10.

**Example 2.5.18** Let $K = \mathbb{F}_2$, let $V = \mathbb{F}_2^8$ and let $\mathscr{F} = K[\varphi_1, \varphi_2]$ be the commuting family introduced in Example 2.5.10. We extend the base field and use $L = \mathbb{F}_2(a, b)$

where $a, b$ are indeterminates. Then we use the same matrices $A_1$, $A_2$ to define endomorphisms $\psi_1$, $\psi_2$ of the $L$-vector space $V_L = L^8$.

In Example 2.5.10 we saw that no linear combination of $\varphi_1$ and $\varphi_2$ is a splitting endomorphism for $\mathscr{F}$. Now, if we let $\vartheta = a\psi_1 + b\psi_2$, we get

$$\mu_\vartheta(z) = z^8 + \left(a^4 + ab^3\right)z^4 + b^6z^2 + \left(a^4b^3 + ab^6\right)z + a^8 + a^4b^4 + a^2b^6 + ab^7 + b^8$$

which factorizes as the product of the two irreducible factors

$$z^4 + \left(a^2 + ab\right)z^2 + \left(a^2b + ab^2 + b^3\right)z + a^4 + a^3b + b^4 \quad \text{and}$$

$$z^4 + \left(a^2 + ab\right)z^2 + \left(a^2b + ab^2 + b^3\right)z + a^4 + a^3b + a^2b^2 + ab^3 + b^4$$

Using similar calculations as in Exercise 2.5.10, one can show that the family $\mathscr{F}_L = L[\psi_1, \psi_2]$ has two maximal ideals. Therefore the map $\vartheta$ is a splitting endomorphism for $\mathscr{F}_L$. In other words, over the extension field $L = K(a, b)$, a linear combination of the extensions of $\varphi_1, \varphi_2$ is a splitting endomorphism.

## 2.6 Simultaneous Diagonalization and Triangularization

> *A lot of people are afraid of heights.*
> *Not me, I'm afraid of widths.*
> (Steven Wright)

Given a family of commuting endomorphisms $\mathscr{F}$, it is a classical question to decide when they are simultaneously diagonalizable or triangularizable. Let us try to keep the width of the set of non-zero entries of a family of commuting matrices as small as possible! Clearly, one necessary condition is that every generator of $\mathscr{F}$ is diagonalizable or triangularizable, respectively. By Propositions 1.6.1 and 1.6.3, this requires at least that all of their eigenfactors are linear. Using the techniques developed in this chapter, we can extend these propositions to characterizations of simultaneous diagonalizablity and triangularizability.

Under the hypothesis that all eigenfactors are linear, one can also transform a single matrix to Jordan normal form. Also for this shape the set of non-zero elements is rather slim. However, as we are going to show at the end of this section, not all endomorphisms of a commuting family can be forced to diet this much simultaneously.

> *Middle age is when your age*
> *starts to show around the middle.*
> (Bob Hope)

Let $K$ be a field, let $V$ be a finite dimensional $K$-vector space, and let $\mathscr{F}$ be a family of commuting endomorphisms of $V$. Recall that $\mathscr{F}$ is called **simultaneously diagonalizable** if there exists a $K$-basis of $V$ such that every endomorphism of $\mathscr{F}$ is represented by a diagonal matrix with respect to this basis. Simultaneously diagonalizable commuting families can be characterized as follows.

**Proposition 2.6.1 (Simultaneously Diagonalizable Families)**
*Let $\mathscr{F}$ be a family of commuting endomorphisms of $V$, let $\Phi = (\varphi_1, \ldots, \varphi_n)$ be a system of generators of $\mathscr{F}$, and let $\mathfrak{m}_1, \ldots, \mathfrak{m}_s$ be the maximal ideals of $\mathscr{F}$. Then the following conditions are equivalent.*

(a) *The family $\mathscr{F}$ is simultaneously diagonalizable.*
(b) *The endomorphisms $\varphi_1, \ldots, \varphi_n$ are simultaneously diagonalizable.*
(c) *Every endomorphism in $\mathscr{F}$ is diagonalizable.*
(d) *For $i = 1, \ldots, n$, the endomorphism $\varphi_i$ is diagonalizable.*
(e) *For $i = 1, \ldots, s$, the maximal ideal $\mathfrak{m}_i$ is linear and $\mathrm{Ker}(\mathfrak{m}_i) = \mathrm{BigKer}(\mathfrak{m}_i)$.*
(f) *For $i = 1, \ldots, s$, the maximal ideal $\mathfrak{m}_i$ is linear and $V = \bigoplus_{i=1}^{s} \mathrm{Ker}(\mathfrak{m}_i)$.*
(g) *For $i = 1, \ldots, s$, the maximal ideal $\mathfrak{m}_i$ is linear and $\mathscr{F}$ is reduced.*
(h) *For $i = 1, \ldots, n$, all eigenfactors of $\varphi_i$ are linear and $\mathscr{F}$ is reduced.*

*Proof* Since it is obvious that (a) implies (b), let us prove first that (b) implies (a). Let $\psi \in \mathscr{F}$ be of the form $\psi = f(\varphi_1, \ldots, \varphi_n)$ with $f \in K[x_1, \ldots, x_n]$. Let $B$ be the basis of $V$ such that $\varphi_i$ is represented by a diagonal matrix $D_i$ for $i = 1, \ldots, n$. Then $\psi$ is represented in the basis $B$ by the diagonal matrix $f(D_1, \ldots, D_n)$.

As it is clear that (a) implies (c) and (c) implies (d), we next prove that (d) implies (e). By Proposition 1.6.1, the minimal polynomials of $\varphi_1, \ldots, \varphi_n$ split into distinct linear factors. This implies that, for every maximal ideal $\mathfrak{m}_i$, the corresponding eigenfactor of $\varphi_j$ is of the form $z - \lambda_{ij}$ with $\lambda_{ij} \in K$. Consequently, the ideal $\langle \varphi_1 - \lambda_{i1} \mathrm{id}_V, \ldots, \varphi_n - \lambda_{in} \mathrm{id}_V \rangle$ is contained in $\mathfrak{m}_i$, and hence coincides with $\mathfrak{m}_i$. Therefore we get $\mathrm{BigKer}(\mathfrak{m}_i) = \bigcap_{j=1}^{n} \mathrm{BigKer}(\varphi_j - \lambda_{ij} \mathrm{id}_V)$. Now $\mathrm{mmult}(\varphi_j, \lambda_{ij}) = 1$ and Theorem 1.1.17.b imply that we have $\mathrm{BigKer}(\varphi_j - \lambda_{ij} \mathrm{id}_V) = \mathrm{Ker}(\varphi_j - \lambda_{ij} \mathrm{id}_V)$, and thus also $\mathrm{BigKer}(\mathfrak{m}_i) = \mathrm{Ker}(\mathfrak{m}_i)$.

Condition (f) follows from (e) by Theorem 2.4.3 and (a) is a consequence of (f) and Corollary 2.3.9. It remains to prove that condition (g) and (h) are equivalent to the other conditions. Since it follows from Corollary 2.4.7 that (g) implies (e), we now prove that (e) implies (g). For $i = 1, \ldots, s$, the hypothesis implies $\mathrm{Ann}_{\mathscr{F}}(\mathrm{Ker}(\mathfrak{m}_i)) = \mathrm{Ann}_{\mathscr{F}}(\mathrm{BigKer}(\mathfrak{m}_i))$. By parts (a) and (b) of Proposition 2.4.6, we have $\mathfrak{m}_1 \cap \cdots \cap \mathfrak{m}_s = \langle 0 \rangle$, and then an application of Proposition 2.2.5.c finishes the proof. Finally, the equivalence of (g) and (h) follows from Proposition 2.3.8. $\square$

Notice that the implication (c) $\Rightarrow$ (a) in this proposition is non-trivial and depends strongly on the fact that the endomorphisms in $\mathscr{F}$ commute. Let us look at an example which shows how one can use this proposition to simultaneously diagonalize a commuting family.

**Example 2.6.2** Let $K = \mathbb{Q}$ and $V = \{(a_1, a_2, a_3, a_4) \in K^4 \mid a_1 + a_2 + a_3 + a_4 = 0\}$. Then $V$ is a $K$-vector subspace of $K^4$ with basis $B = (v_1, v_2, v_3)$, where we have $v_1 = (-1, 1, 0, 0)$, $v_2 = (-1, 0, 1, 0)$, and $v_3 = (-1, 0, 0, 1)$. Let $\varphi_1$ and $\varphi_2$ be the endomorphisms of $V$ defined by the matrices

$$A_1 = \begin{pmatrix} 1 & 0 & 0 \\ 0 & 1 & 0 \\ -1 & -1 & -1 \end{pmatrix} \quad \text{and} \quad A_2 = \begin{pmatrix} 0 & 1 & 0 \\ 1 & 0 & 0 \\ 0 & 0 & 1 \end{pmatrix}$$

with respect to $B$, respectively. Since $\varphi_1$ and $\varphi_2$ commute, we have a commuting family $\mathscr{F} = K[\varphi_1, \varphi_2]$. Since we have $\mu_{\varphi_1}(z) = \mu_{\varphi_2}(z) = (z+1)(z-1)$, Proposition 1.6.1 shows that condition (d) of the above characterization is satisfied. Hence the family $\mathscr{F}$ is simultaneously diagonalizable. It is straightforward to check that $\langle v_1 + v_2 - v_3 \rangle$, $\langle v_1 - v_2 \rangle$, and $\langle v_3 \rangle$ are three 1-dimensional joint eigenspaces of $\mathscr{F}$. Hence every endomorphism in $\mathscr{F}$ is represented by a diagonal matrix with respect to the basis $(v_1 + v_2 - v_3, v_1 - v_2, v_3)$ of $V$.

Of course, not every family of commuting endomorphisms is simultaneously diagonalizable. However, an important property of commuting families is that they admit simultaneous triangularizations if each individual endomorphism is triangularizable. Here a family $\mathscr{F}$ of endomorphisms of $V$ is called **simultaneously triangularizable** if there exists a $K$-basis of $V$ such that every endomorphism in $\mathscr{F}$ is represented by an upper triangular matrix with respect to this basis. Let us spell out the characterization of simultaneously triangularizable families in detail.

**Proposition 2.6.3 (Simultaneous Triangularization)**
*Let $\mathscr{F}$ be a family of commuting endomorphisms of $V$, let $\Phi = (\varphi_1, \dots, \varphi_n)$ be a system of generators of the $K$-algebra $\mathscr{F}$, and let $\mathfrak{m}_1, \dots, \mathfrak{m}_s$ be the maximal ideals of $\mathscr{F}$. Then the following conditions are equivalent.*

(a) *The family $\mathscr{F}$ is simultaneously triangularizable.*
(b) *The endomorphisms $\varphi_1, \dots, \varphi_n$ are simultaneously triangularizable.*
(c) *Every endomorphism in $\mathscr{F}$ is triangularizable.*
(d) *For $i = 1, \dots, n$, the endomorphism $\varphi_i$ is triangularizable.*
(e) *For $i = 1, \dots, n$, the eigenfactors of $\varphi_i$ are linear.*
(f) *For $i = 1, \dots, s$, the maximal ideal $\mathfrak{m}_i$ of $\mathscr{F}$ is linear.*

*Proof* It is clear that (a) implies (b). To prove the converse, let $B$ be a $K$-basis of $V$ such that $\varphi_i$ is represented by an upper triangular matrix $U_i$ in this basis. Let $\psi \in \mathscr{F}$ be of the form $\psi = f(\varphi_1, \dots, \varphi_n)$ with $f \in K[x_1, \dots, x_n]$. Then $\varphi$ is represented by the upper triangular matrix $f(U_1, \dots, U_n)$ in the basis $B$.

Furthermore, it is clear that (a) implies (c), that (c) implies (d), and that conditions (d) and (e) are equivalent by Proposition 1.6.3. Since conditions (e) and (f) are equivalent by Proposition 2.3.8, it remains to prove that (e) and (f) imply (a). By Theorem 2.4.3, it is sufficient to prove the claim in the case where $V$ is a joint generalized eigenspace of the family $\mathscr{F}$. We proceed by induction on $d = \dim_K(V)$. If $d = 1$, the vector space $V$ is a joint eigenspace for $\mathscr{F}$ and the claim holds for any choice of a non-zero basis vector in $V$.

Now let $d > 1$. For $i \in \{1, \ldots, s\}$ and a non-zero joint eigenvector $v \in \mathrm{Ker}(\mathfrak{m}_i)$, the vector subspace $\langle v \rangle$ is $\varphi_i$-invariant for $i = 1, \ldots, n$. Let $W$ be the residue class vector space $W = V/\langle v \rangle$. Then $W$ satisfies $\dim_K(W) = d - 1$ and the fact that $\langle v \rangle$ is $\varphi_i$-invariant for every $i \in \{1, \ldots, n\}$ implies that we get an induced family of commuting endomorphisms $\overline{\mathscr{F}}$ consisting of induced endomorphisms $\overline{\psi} : W \longrightarrow W$ for each $\psi \in \mathscr{F}$.

Now the fact that the eigenfactors of $\varphi$ are linear implies the also the eigenfactors of $\overline{\varphi}_i$ are linear, since $\mu_{\overline{\varphi}_i}(z)$ divides $\mu_{\varphi_i}(z)$. Consequently, also the maps $\overline{\varphi}_i$ are triangularizable. By the induction hypothesis, there is a basis $\overline{B}$ of $W$ such that all maps $\overline{\psi}$ are represented by upper triangular matrices with respect to this basis. Now we represent the vectors in $\overline{B}$ by vectors in $V$ and append them to $v$ to get a basis of $V$. Clearly, all endomorphisms in $\mathscr{F}$ are represented by upper triangular matrices with respect to this basis.                                                                                                $\square$

Another way to obtain a particularly simple matrix representing a given endomorphism, if its minimal polynomial splits into linear factors, is the Jordan normal form. Hence it is a natural question whether there is a kind of "simultaneous Jordan form" for endomorphisms in a commuting family. Dashing any hopes that this may be possible, the next example indicates that, in general, there is no obvious way to define a simultaneous Jordan decomposition for a commuting family of matrices.

**Example 2.6.4**  Let $K = \mathbb{Q}$, let $V = K^3$, and let the endomorphisms $\varphi_1$ and $\varphi_2$ of $V$ be given by the matrices

$$A_1 = \begin{pmatrix} 0 & 0 & 0 \\ 1 & 0 & 0 \\ 0 & 0 & 0 \end{pmatrix} \quad \text{and} \quad A_2 = \begin{pmatrix} 0 & 0 & 0 \\ 0 & 0 & 0 \\ 1 & 0 & 0 \end{pmatrix}$$

respectively. It is easy to check that $\mathscr{F} = K[\varphi_1, \varphi_2]$ is a commuting family. If we apply the Buchberger-Möller Algorithm for Matrices 2.1.4 to $\Phi = (\varphi_1, \varphi_2)$, we get $\mathrm{Rel}_P(\Phi) = \mathfrak{m}^2$, where $\mathfrak{m}$ is the maximal ideal $\mathfrak{m} = \langle x, y \rangle$ in $P = K[x, y]$. Using Algorithm 2.2.17, we compute the joint eigenspace $\mathrm{Ker}(\mathfrak{m})$ of $\mathscr{F}$ and get $\mathrm{Ker}(\mathfrak{m}) = \langle e_2, e_3 \rangle$.

Next we construct a $K$-basis of $\mathrm{Ker}(\mathfrak{m})$ and extend it to a basis of $V$. Let $a_1, a_2, a_3, a_4 \in K$, let $v_1 = a_1 e_2 + a_2 e_3$, and let $v_2 = a_3 e_2 + a_4 e_3$. Then the pair $(v_1, v_2)$ is a $K$-basis of $\mathrm{Ker}(\mathfrak{m})$ if and only if $a_2 a_3 - a_1 a_4 \neq 0$. To extend this basis to a basis of $V$, we let $b_1, b_2, b_3 \in K$ and $v_3 = (b_1, b_2, b_3)$. Then the triple $B = (v_1, v_2, v_3)$ is a $K$-basis of $V$ if and only if $b_1(a_2 a_3 - a_1 a_4) \neq 0$.

Now let us examine whether it is possible to choose the basis $B$ such that the matrices of both endomorphisms $\varphi_1 + \varphi_2$ and $\varphi_1 - \varphi_2$ are in Jordan normal form. These matrices are given by

$$\begin{pmatrix} 0 & 0 & \frac{(a_3 - a_4)b_1}{a_2 a_3 - a_1 a_4} \\ 0 & 0 & \frac{(a_2 - a_1)b_1}{a_2 a_3 - a_1 a_4} \\ 0 & 0 & 0 \end{pmatrix} \quad \text{and} \quad \begin{pmatrix} 0 & 0 & \frac{(a_3 + a_4)b_1}{a_2 a_3 - a_1 a_4} \\ 0 & 0 & \frac{-(a_1 + a_2)b_1}{a_2 a_3 - a_1 a_4} \\ 0 & 0 & 0 \end{pmatrix}$$

respectively. In agreement with the preceding proposition, they are upper triangular. However, there is no way to choose the coefficients $(a_1, a_2, a_3, a_4, b_1, b_2, b_3)$ such that both matrices have a zero entry in position $(1, 3)$. Consequently, there is no way to construct a basis $B$ of $V$ in which all endomorphisms of the family are in Jordan normal form.

After being forced to abandon any hope of getting all matrices of a commuting family simultaneously into a very sparse and slim form, how can we continue? Should we start dieting?

*I'm on a seafood diet.*
*I see food, I eat it.*
(Dolly Parton)

# Chapter 3
# Special Families of Endomorphisms

*WiFi went down for five minutes,*
*so I had to talk to my family.*
*They seem like nice people.*

In this chapter we continue our study of commuting families, identifying several
different types: unigenerated families, commendable families, local families, dual
families, and extended families, to name just a few. All of them seem to be quite
nice, while commendable families deserve the highest praise. What are these special
families good for? Since each of them has its own merits, let us discuss them one by
one.

The first type we consider comprises commuting families $\mathscr{F}$ over which the vector space $V$ is a cyclic module. This condition is, for instance, satisfied for the
families of multiplication endomorphisms considered later in Chap. 4. One of its
main features is that it can be checked algorithmically via the Cyclicity Test 3.1.4.
Another type comprises unigenerated families, i.e., commuting families which are
generated as a $K$-algebra by a single endomorphism. This is clearly a case where
most of the results of the first chapter apply almost directly. By combining the first
two properties, we get a third type, namely the commuting families containing a
commendable endomorphism (see Theorem 3.2.4).

This last property is a very powerful one. A commendable endomorphism is the
leader and provider of the family. In real life, the provider is the person who carries
pictures where money used to be. In a commuting family, a commendable endomorphism generates it, makes the vector space $\mathscr{F}$-cyclic, is a splitting endomorphism,
and provides a host of further useful properties.

Can any family make do without a special member like this? Apparently yes.
As we show in Sect. 3.3, commendable families are a special type with many good

**Electronic supplementary material** The online version of this chapter
(DOI:10.1007/978-3-319-43601-2_3) contains supplementary material, which is available to
authorized users.

traits. They are defined by the condition that the joint eigenspaces have the minimal possible dimensions $\dim_K(\mathscr{F}/\mathfrak{m})$, where $\mathfrak{m}$ runs through the maximal ideals of $\mathscr{F}$. Many good things can be said about them, not least that they are the stars in the Duality Theorem discussed below. They get by just fine, even if they contain no commendable endomorphism.

Home is where you can say anything, because nobody listens to you anyway. Is this also true for local families? This type is characterized by the fact that they have only one maximal ideal. So, there is only one joint eigenspace, namely $\mathrm{Ker}(\mathfrak{m})$, and the only joint generalized eigenspace is $V$. Can we say more? Indeed we can! For instance, in the special case when $\mathscr{F}$ is a field, the existence of a commendable endomorphism in the family is equivalent to the Primitive Element Theorem.

Batman: *"That's one trouble with dual identities, Robin. Dual responsibilities."*

Are you still paying attention? We hope so, because the next type of commuting family is a rather important one: the dual family $\mathscr{F}^{\vee}$ is the commuting family on the dual vector space $V^*$ consisting of the duals of the endomorphisms in $\mathscr{F}$. The family and its dual family are intrinsically related in a deep and striking way. The Duality Theorem 3.5.12 states that a family $\mathscr{F}$ is commendable if and only if $V^*$ is a cyclic $\mathscr{F}^{\vee}$-module. Analogously, the family $\mathscr{F}^{\vee}$ is commendable if and only if $V$ is a cyclic $\mathscr{F}$-module. This seemingly abstract property has profound ramifications in later chapters. A first application, given at the end of Sect. 3.5, is that we can test effectively whether $\mathscr{F}$ is a commendable family by applying the Cyclicity Test 3.1.4 to $\mathscr{F}^{\vee}$.

Finally, the last type of commuting families we examine here are the extended families. According to Wikipedia, an extended family is a family that extends beyond the nuclear family, consisting of parents, aunts, uncles, and cousins, all living nearby or in the same household. In our setting, we can extend a commuting family by enlarging the base field: if $K \subseteq L$ is a field extension, then $\mathscr{F}_L = \mathscr{F} \otimes_K L$ is a commuting family of endomorphisms of $V_L = V \otimes_K L$. Miraculously, several good properties are mirrored in both a family and its extension. For instance, the family $\mathscr{F}$ is commendable if and only if $\mathscr{F}_L$ is, and the vector space $V$ is $\mathscr{F}$-cyclic if and only if $V_L$ is $\mathscr{F}_L$-cyclic.

Summing up, it is fair to claim that families are like fudge—mostly sweet, but with a few nuts. Please turn off your WiFi and get to know them!

## 3.1   $\mathscr{F}$-Cyclic Vector Spaces

> We only believe what we see.
> That's why we believe everything
> since TV was invented.
> (Dieter Hildebrand)

To commence our study of special types of commuting families, we look at the case when the vector space $V$ is a cyclic module over the commuting family $\mathscr{F}$ via the natural operation $\varphi \cdot v = \varphi(v)$ for $\varphi \in \mathscr{F}$ and $v \in V$. Since we do not believe that every commuting family has this property, we need a way to see whether it really holds. Luckily, we can recycle the result of Tutorial 91 of the second and last volume [16] of this trilogy and get a nice Cyclicity Test 3.1.4. If the vector space $V$ is the direct sum of invariant subspaces $U_i$, the task can be further simplified by restricting $\mathscr{F}$ to the invariant subspaces and checking whether $U_i$ is a cyclic $\mathscr{F}|_{U_i}$-module for all $i$. Lastly, we note that $\mathscr{F}$ is always a cyclic $\mathscr{F}$-module via the multiplication $\varphi \cdot \psi = \varphi \circ \psi$, and that it is isomorphic to $V$ in the cyclic case. This observation will be recycled later to reduce some algorithms to the cyclic case.

*This message was sent using 100 % recycled spam mails.*

Let $K$ be a field, let $V$ be a finite dimensional $K$-vector space, and let $\mathscr{F}$ be a commuting family of endomorphisms of $V$. Recall that $V$ has a natural $\mathscr{F}$-module structure given by $\varphi \cdot v = \varphi(v)$ for all $\varphi \in \mathscr{F}$ and $v \in V$. In the following we examine the case when $V$ is a cyclic $\mathscr{F}$-module, i.e., when $V$ is generated as an $\mathscr{F}$-module by a single vector. In this case we also say that $V$ is $\mathscr{F}$-**cyclic**. For a family $\mathscr{F}$ generated by one endomorphism, cyclic $\mathscr{F}$-modules were characterized in Theorem 1.5.8.

The following proposition collects some important properties of an $\mathscr{F}$-cyclic vector space.

**Proposition 3.1.1 (Basic Properties of Cyclic $\mathscr{F}$-Modules)**
*Let $\mathscr{F}$ be a commuting family of endomorphisms of $V$.*

(a) *If $V$ is $\mathscr{F}$-cyclic and $v \in V$ is an $\mathscr{F}$-module generator of $V$ then $\mathrm{Ann}_{\mathscr{F}}(v) = 0$.*
(b) *For a vector $v \in V$, the following conditions are equivalent.*

    (1) *The vector space $V$ is a cyclic $\mathscr{F}$-module with generator $v$.*
    (2) *The $K$-linear map $\eta : \mathscr{F} \longrightarrow V$ which sends $\varphi$ to $\varphi(v)$ for every $\varphi \in \mathscr{F}$ is an isomorphism.*

(c) *If $V$ is $\mathscr{F}$-cyclic then we have $\dim_K(\mathscr{F}) = \dim_K(V)$.*
(d) *If $\mathscr{F}$ is a field then $V$ is $\mathscr{F}$-cyclic if and only if $\dim_{\mathscr{F}}(V) = 1$.*

*Proof* To show (a), we suppose that $\varphi \in \mathscr{F}$ satisfies $\varphi(v) = 0$. Let $w \in V$. Since $V$ is $\mathscr{F}$-cyclic, there exists an endomorphism $\psi \in \mathscr{F}$ such that $w = \psi(v)$. Hence we have $\varphi(w) = \varphi(\psi(v)) = \psi(\varphi(v)) = 0$. Thus we have $\varphi = 0$.

Next we show (b). To prove that (1) implies (2), we observe that the $K$-linear map $\eta$ is injective by (a). To prove the surjectivity, we let $w \in V$ and observe that there exists a map $\psi \in \mathscr{F}$ with $w = \psi(v) = \eta(\psi)$. The other implication follows from the definitions. Finally, claim (c) is a consequence of (b), and (d) follows from the definitions.                                                                                                       $\square$

Part (b) of this proposition has the following important interpretation.

**Remark 3.1.2** Let $V$ be $\mathscr{F}$-cyclic with generator $v \in V$, and let the $K$-linear map $\eta : \mathscr{F} \longrightarrow V$ be defined by $\varphi \mapsto \varphi(v)$. Then $V$ acquires the structure of a $K$-algebra via $\eta$. The multiplication on $V$ can be described as follows. For $v_1, v_2 \in V$, there exist unique maps $\varphi_1, \varphi_2 \in \mathscr{F}$ such that $v_1 = \varphi_1(v)$ and $v_2 = \varphi_2(v)$. Then the multiplication on $V$ is given by $v_1 v_2 = (\varphi_1 \circ \varphi_2)(v)$.

Let us consider the setting of Example 2.4.11 in the present context.

**Example 3.1.3**  Let $K = \mathbb{Q}$, and let $\varphi_1, \varphi_2$ be the endomorphisms of $V = \mathbb{Q}^3$ given by

$$A_1 = \begin{pmatrix} 0 & 1 & 0 \\ 0 & 0 & 0 \\ 0 & 0 & 0 \end{pmatrix} \quad \text{and} \quad A_2 = \begin{pmatrix} 0 & 0 & 1 \\ 0 & 0 & 0 \\ 0 & 0 & 0 \end{pmatrix}$$

respectively with respect to the canonical basis. In Example 2.4.11 we showed that $A_1^2 = A_2^2 = A_1 \cdot A_2 = 0$. Consequently, there exists an isomorphism $\mathscr{F} \cong K[x, y]/\langle x^2, xy, y^2 \rangle$. Moreover, we have $\dim_K(\mathscr{F}) = \dim_K(V) = 3$.

Now let us verify that $V$ is not a cyclic $\mathscr{F}$-module. Let $v = (a_1, a_2, a_3) \in V$, where $a_1, a_2, a_3 \in \mathbb{Q}$. We have $\mathrm{id}_V(v) = v$, $\varphi_1(v) = (a_2, 0, 0)$, and $\varphi_2(v) = (a_3, 0, 0)$. Therefore we see that $\mathscr{F} \cdot v \subseteq \langle v, e_1 \rangle$. Hence the submodule generated by $v$ has at most vector space dimension 2 and is properly contained in $V$.

This example shows that the equality $\dim_K(\mathscr{F}) = \dim_K(V)$ is a necessary but not a sufficient condition for $V$ to be $\mathscr{F}$-cyclic. In fact, we have the following cyclicity test (see also [16], Tutorial 91).

---

**Algorithm 3.1.4 (Cyclicity Test)**
*Let $S = \{\varphi_1, \ldots, \varphi_r\} \subseteq \mathrm{End}_K(V)$ be a set of commuting endomorphisms, let $\mathscr{F}$ be the family generated by $S$, let $\Phi = (\varphi_1, \ldots, \varphi_r)$, let $B = (v_1, \ldots, v_d)$ be a basis of $V$, and let $A_i \in \mathrm{Mat}_d(K)$ be the matrix representing $\varphi_i$ in the basis $B$ for $i = 1, \ldots, r$. Consider the following instructions.*

(1) *Using the Buchberger-Möller Algorithm for Matrices (see 2.1.4) compute a tuple of terms $\mathscr{O} = (t_1, \ldots, t_s)$ whose residue classes form a $K$-basis of $P/\mathrm{Rel}_P(\Phi)$ where $P = K[x_1, \ldots, x_r]$.*
(2) *If $s \neq d$ then return "$\texttt{Not cyclic}$" and stop.*
(3) *Let $z_1, \ldots, z_d$ be indeterminates. Let $C \in \mathrm{Mat}_d(K[z_1, \ldots, z_d])$ be the matrix whose columns are $t_i(A_1, \ldots, A_d) \cdot (z_1, \ldots, z_d)^{\mathrm{tr}}$ for $i = 1, \ldots, d$.*
(4) *Compute the determinant $\Delta = \det(C)$. If $\Delta \neq 0$, find a tuple $(c_1, \ldots, c_d)$ in $K^d$ such that $\Delta(c_1, \ldots, c_d) \neq 0$. Then return "$\texttt{Cyclic}$" and the vector $v = c_1 v_1 + \cdots + c_d v_d$ and stop.*
(5) *Return "$\texttt{Not Cyclic}$" and stop.*

*This is an algorithm which checks whether $V$ is a cyclic $\mathscr{F}$-module and, in the affirmative case, computes a generator.*

*Proof* The Buchberger-Möller Algorithm for Matrices computes a list of terms $\mathscr{O}$ whose residue classes form a $K$-basis of $P/\mathrm{Rel}_P(\Phi) \cong \mathscr{F}$. If we get $s \neq d$ in Step (2), the vector space $V$ cannot be $\mathscr{F}$-cyclic by Proposition 3.1.1.c.

Now let us assume that $s = d$. The $\mathscr{F}$-module $V$ is cyclic if and only if we can find a vector $v = c_1 v_1 + \cdots + c_d v_d$ in $V$, where $c_i \in K$, such that the vectors in $\{t_i(\Phi)(v) \mid i = 1, \ldots, d\}$ are $K$-linearly independent, and therefore a $K$-basis of $V$. This is equivalent to finding substitutions $z_i \mapsto c_i$ such that the column vectors $\sum_{i=1}^{d} t_i(A_1, \ldots, A_d)(c_1, \ldots, c_d)^{\mathrm{tr}}$ are $K$-linearly independent. Equivalently, we have to find substitutions $z_i \mapsto c_i$ such that the determinant $\Delta$ is mapped to a non-zero element of $K$. This is the condition which is verified in Step (4).

To prove that the algorithm can always be executed, we still have to show that $\Delta \neq 0$ guaranties the existence of a tuple $(c_1, \ldots, c_d) \in K^d$ with $\Delta(c_1, \ldots, c_d) \neq 0$. If the field $K$ is infinite, this is clearly true. If $K$ is finite, we use the facts that such a tuple exists if and only if $V$ is $\mathscr{F}$-cyclic and that the latter property holds if and only if it holds after a base field extension (see Theorem 3.6.4 later in this chapter).   $\square$

Notice that the result of this algorithm does not depend on the choice of the basis $B$. The following remark suggests a way to implement Step (4).

**Remark 3.1.5**  Notice that Step (4) can be performed effectively. If $\Delta \neq 0$ and $K$ is infinite, then clearly there is a tuple $(c_1, \ldots, c_d)$ such that $\Delta(c_1, \ldots, c_d) \neq 0$. If $\Delta \neq 0$ and $K$ is finite with more than $d$ elements, than we can chose an indeterminate which shows up in the support of $\Delta$, substitute $c_i = 1$ for the remaining indeterminates, and then observe that the univariate polynomial that we get has degree at most $d$. Therefore at least one element of $K$ is not one of its roots.

If $\Delta \neq 0$ and $K$ is finite with less than $d + 1$ elements, one can, in principle, check all tuples in $K^d$. As shown above, this method succeeds even if $K$ is small. This means, for instance, that for the field $K = \mathbb{F}_2$, a polynomial such as $z_1^2 z_2 + z_2^2 z_1$ which vanishes for all pairs $(c_1, c_2) \in \mathbb{F}_2^2$ cannot occur as $\Delta$.

In the following example we apply the cyclicity test to an explicit case.

**Example 3.1.6**  Let us return to the setting of Example 2.6.2, i.e., let $K = \mathbb{Q}$, consider the $K$-vector space $V = \{(a_1, a_2, a_3, a_4) \in K^4 \mid a_1 + a_2 + a_3 + a_4 = 0\} \subset K^4$, and let $\mathscr{F} = K[\varphi_1, \varphi_2]$ be the commuting family generated by $\varphi_1$ and $\varphi_2$, where $\varphi_1(a_1, a_2, a_3, a_4) = (a_4, a_2, a_3, a_1)$ and $\varphi_2(a_1, a_2, a_3, a_4) = (a_1, a_3, a_2, a_4)$.

The matrices representing $\varphi_1$ and $\varphi_2$ with respect to the basis $B = (v_1, v_2, v_3)$ given by $v_1 = (-1, 1, 0, 0)$, $v_2 = (-1, 0, 1, 0)$, and $v_3 = (-1, 0, 0, 1)$ are

$$A_1 = \begin{pmatrix} 1 & 0 & 0 \\ 0 & 1 & 0 \\ -1 & -1 & -1 \end{pmatrix} \quad \text{and} \quad A_2 = \begin{pmatrix} 0 & 1 & 0 \\ 1 & 0 & 0 \\ 0 & 0 & 1 \end{pmatrix}$$

Let us apply the Cyclicity Test in this setting. First we use Buchberger's Algorithm for Matrices 2.1.4 to get the isomorphism $\mathscr{F} \cong K[x, y]/\mathrm{Rel}_P(\varphi_1, \varphi_2)$, where

$\mathrm{Rel}_P(\varphi_1, \varphi_2) = \langle x^2 - 1,\ y^2 - 1,\ xy - x - y + 1 \rangle$, and the triple $\mathcal{O} = (1, x, y)$ such that the residue classes of its elements form a $K$-basis of $\mathcal{F}$.

Next we compute the matrix $C$ and its determinant. We get

$$C = \begin{pmatrix} z_1 & z_1 & z_2 \\ z_2 & z_2 & z_1 \\ z_3 & -z_1 - z_2 - z_3 & z_3 \end{pmatrix}$$

and $\det(C) = z_1^3 + z_1^2 z_2 - z_1 z_2^2 - z_2^3 + 2 z_1^2 z_3 - 2 z_2^2 z_3$. Hence $V$ is a cyclic $\mathcal{F}$-module. As a generator, we can for instance use $v_1$ or $v_2$, but not $v_3$, since $(0, 0, 1)$ is a zero of $\det(C)$. To check directly that $v_1$ is a generator of $V$, we observe that $\mathrm{id}_V(v_1) = v_1$, $\varphi_1(v_1) = v_1 - v_3$, and $\varphi_2(v_1) = v_2$. Clearly, the tuple $(v_1, v_1 - v_3, v_2)$ is a $K$-basis of $V$.

If $V$ is a direct sum of invariant subspaces for $\mathcal{F}$, the task of checking whether $V$ is cyclic can be simplified.

**Proposition 3.1.7** *Let $U_1, \dots, U_s$ be $K$-vector subspaces of $V$ which are invariant for the family $\mathcal{F}$, and suppose that $V = U_1 \oplus \cdots \oplus U_s$. For every $i \in \{1, \dots, s\}$, let $I_i = \mathrm{Ann}_{\mathcal{F}}(U_i)$.*

(a) *If $V$ is $\mathcal{F}$-cyclic, then $U_i$ is $\mathcal{F}_{U_i}$-cyclic for every $i \in \{1, \dots, s\}$.*
(b) *If the ideals $I_1, \dots, I_s$ are pairwise comaximal, and if $U_i$ is $\mathcal{F}_{U_i}$-cyclic for every $i \in \{1, \dots, s\}$, then $V$ is $\mathcal{F}$-cyclic.*

*Proof* To prove (a) we assume that $V$ is $\mathcal{F}$-cyclic, and we let $v \in V$ be a generator. We write $v = u_1 + \cdots + u_s$ with $u_i \in U_i$. Then $\mathcal{F} \cdot u_i \subseteq U_i$ implies that $u_i$ is a generator of the $\mathcal{F}_{U_i}$-module $U_i$ for every $i \in \{1, \dots, s\}$.

Next we show (b). For every $i \in \{1, \dots, s\}$, let $u_i \in U_i$ be a generator of the cyclic $\mathcal{F}_{U_i}$-module $U_i$, and let $v = u_1 + \cdots + u_s$. We are going to prove that $v$ is a generator of the $\mathcal{F}$-module $V$.

First we show that $u_i$ is contained in $\mathcal{F} \cdot v$ for every $i \in \{1, \dots, s\}$. By assumption, the ideals $I_1, \dots, I_s$ are pairwise comaximal. If we let $J_i = \bigcap_{j \neq i} I_j$, it follows that $I_i$ and $J_i$ are comaximal for every $i \in \{1, \dots, s\}$. Hence there exist elements $\varphi_i \in I_i$ and $\psi_i \in J_i$ such that $\varphi_i + \psi_i = \mathrm{id}_V$. Then $u_i = \varphi_i(u_i) + \psi_i(u_i) = \psi_i(u_i) = \psi_i(v) \in \mathcal{F} \cdot v$, as we wanted to show.

Next we let $w \in V$ and write $w = w_1 + \cdots + w_s$ with $w_i \in U_i$. Since $U_i$ is a cyclic $\mathcal{F}_{U_i}$-module, there exists an endomorphism $\eta_i \in \mathcal{F}$ such that $w_i = \eta_i(u_i)$. By the above claim, we have $u_i = \psi_i(v)$ for some $\psi_i \in \mathcal{F}$. Altogether, we obtain

$$w = w_1 + \cdots + w_s = \eta_1(u_1) + \cdots + \eta_s(u_s) = (\eta_1 \psi_1)(v) + \cdots + (\eta_s \psi_s)(v) \in \mathcal{F} \cdot v$$

This finishes the proof of (b).                                                    $\square$

The next example shows that the extra assumption in part (b) of this proposition is necessary.

**Example 3.1.8** Let us consider the setting used in Example 1.4.3. Furthermore, let $\mathcal{F} = K[\varphi]$, let $U_1 = \langle e_1, e_3 \rangle = \langle e_3 \rangle_{K[z]}$, and let $U_2 = \langle e_2 \rangle$. Then $V$ is not $\mathcal{F}$-cyclic, since $\varphi$ is not commendable (cf. Theorem 1.5.8). However, both $U_1$ and $U_2$ are cyclic modules over the restricted families.

If the vector subspaces $U_i$ are the joint generalized eigenspaces of $\mathcal{F}$, then the ideals $I_1, \ldots, I_s$ in the preceding proposition are pairwise comaximal by Proposition 2.4.6.b, and the proposition simplifies as follows.

### Corollary 3.1.9 (Local Nature of $\mathcal{F}$-Cyclicity)
*Let $\mathfrak{m}_1, \ldots, \mathfrak{m}_s$ be the maximal ideals of a commuting family $\mathcal{F}$. Then $V$ is $\mathcal{F}$-cyclic if and only if $\mathrm{BigKer}(\mathfrak{m}_i)$ is $\mathcal{F}_{\mathrm{BigKer}(\mathfrak{m}_i)}$-cyclic for $i = 1, \ldots, s$.*

Notice that the restricted family $\mathcal{F}_{\mathrm{BigKer}(\mathfrak{m}_i)}$ has only one maximal ideal, namely the restriction of $\mathfrak{m}_i$. A family with this property is called a **local ring** in the language of Commutative Algebra. This explains the title of this corollary.

To end this section we include an observation that will be recalled in Chap. 4. Suppose that $V$ is $\mathcal{F}$-cyclic with generator $v \in V$. The $K$-linear map $\eta : \mathcal{F} \longrightarrow V$ defined by $\eta(\varphi) = \varphi(v)$ is an isomorphism by Proposition 3.1.1. This allows us to identify $\mathrm{End}_K(V)$ and $\mathrm{End}_K(\mathcal{F})$ in a natural way. The next proposition describes the image of an endomorphism $\varphi$ under this identification.

### Proposition 3.1.10 ($\mathcal{F}$-Cyclicity and Multiplication Endomorphisms)
*Suppose that $V$ is a cyclic $\mathcal{F}$-module with generator $v \in V$, let $\eta : \mathcal{F} \to V$ be the isomorphism defined by $\eta(\varphi) = \varphi(v)$, and let $\vartheta : \mathrm{End}_K(V) \longrightarrow \mathrm{End}_K(\mathcal{F})$ be defined by $\vartheta(\varphi) = \eta^{-1}\varphi\eta$.*

(a) *The map $\vartheta$ is an isomorphism of $K$-vector spaces.*
(b) *For every $\varphi \in \mathcal{F}$, the endomorphism $\vartheta(\varphi)$ is the multiplication by $\varphi$ on $\mathcal{F}$.*

*Proof* First we show (a). Since the map $\eta$ is an isomorphism, it follows that the map $\vartheta$ is injective. By Proposition 3.1.1.c, the vector spaces $\mathcal{F}$ and $V$ have the same dimension. Therefore also the vector spaces $\mathrm{End}_K(\mathcal{F})$ and $\mathrm{End}_K(V)$ have the same dimension. Consequently, the map $\vartheta$ is also surjective.

To prove (b), we let $\varphi, \psi \in \mathcal{F}$ and calculate $(\vartheta(\varphi))(\psi) = (\eta^{-1}\varphi\eta)(\psi) = (\eta^{-1}\varphi)(\psi(v)) = \eta^{-1}(\varphi\psi)(v) = \eta^{-1}\eta(\varphi\psi) = \varphi\psi$. This proves the claim. $\square$

## 3.2 Unigenerated Families

> *A celebrity is a person*
> *who works hard all his life to become well known,*
> *then wears dark glasses to avoid being recognized.*
> (Fred Allen)

In this section we study commuting families $\mathcal{F}$ which contain a special family member. As we know from the first chapter, the most celebrated property an endomorphism $\varphi$ in $\mathscr{F}$ can have is to be commendable. It turns out that the existence of a commendable family member $\varphi$ is a very powerful condition: the family is unigenerated by the commendable endomorphism $\varphi$, i.e., we have $\mathscr{F} = K[\varphi]$, the vector space $V$ is a cyclic $\mathscr{F}$-module, and $\varphi$ is a splitting endomorphism for $\mathscr{F}$. Conversely, if the family $\mathscr{F}$ is unigenerated by $\varphi$ and $V$ is a cyclic $\mathscr{F}$-module, then $\varphi$ is commendable.

In order to recognize a commendable endomorphism, we go on to prove several further characterizations. An important observation is that the property of being commendable is not local in the sense that it implies that the restrictions $\varphi|_{\mathrm{BigKer}(m_i)}$ to the generalized eigenspaces are commendable, but the converse is true only if $\varphi$ is a splitting endomorphism for $\mathscr{F}$. Finally, we examine the question when a commendable endomorphism $\varphi$ exists in $\mathscr{F}$. Besides the condition that $V$ has to be a cyclic $\mathscr{F}$-module, we need that the residue class rings $\mathscr{F}/q_i$, where $q_i = \mathrm{Ann}_{\mathscr{F}}(\mathrm{BigKer}(m_i))$, are of the form $K[z]/\langle p_i(z)^{m_i}\rangle$ with pairwise distinct, monic, irreducible polynomials $p_i(z)$. The latter condition is a rather subtle property, as we show using some examples. Its geometric interpretation will be explained in Sect. 4.3. For now we merely note that families containing a commendable endomorphism are anything but normal.

> *Now remember, as far as anyone knows,*
> *we're a nice, normal family.*
> (The Simpsons)

Let $K$ be a field, and let $V$ be a $d$-dimensional $K$-vector space.

**Definition 3.2.1**  Let $\mathscr{F}$ be a commuting family of endomorphisms of $V$.

(a) Let $\varphi, \psi \in \mathrm{End}_K(V)$. If there exists a polynomial $f(z) \in K[z]$ such that we have $\psi = f(\varphi)$, we say that $\psi$ is **polynomial in** $\varphi$.
(b) If the family $\mathscr{F}$ contains an endomorphism $\varphi$ such that all endomorphisms in $\mathscr{F}$ are polynomial in $\varphi$, it is called **unigenerated by** $\varphi$, and $\varphi$ is called a **generator** of $\mathscr{F}$.

In part (a) of this definition, we can bound the degree necessary to represent an endomorphism as a polynomial in another as follows.

**Remark 3.2.2**  Let $\varphi, \psi \in \mathrm{End}_K(V)$ and assume that $\psi$ is polynomial in $\varphi$. Then there exists a polynomial $f(z) \in K[z]$ of degree $\deg(f(z)) < \deg(\mu_\varphi(z))$ such that $\psi = f(\varphi)$. To see why this holds, it suffices to write $\psi = g(\varphi)$ with a polynomial $g(z) \in K[z]$ and to divide $g(z)$ with remainder by $\mu_\varphi(z)$.

Our next remark collects some basic properties of unigenerated families.

**Remark 3.2.3** Let $\mathscr{F}$ be a unigenerated family of commuting endomorphisms, and let $\varphi$ be a generator of $\mathscr{F}$.

(a) We have $\mathscr{F} = K[\varphi] \cong K[z]/\langle \mu_\varphi(z) \rangle$.
(b) The map $\varphi$ is a splitting endomorphism for $\mathscr{F}$ (see Proposition 2.5.2.c).
(c) The eigenspaces of $\varphi$ are precisely the joint eigenspaces of $\mathscr{F}$, because the joint eigenspaces of $\mathscr{F}$ are precisely the spaces $\mathrm{Ker}(\mathfrak{m})$, where $\mathfrak{m}$ is a maximal ideal of $\mathscr{F}$ (see Theorem 2.4.3) and the maximal ideals of $\mathscr{F}$ are the images of the ideals generated by the eigenfactors of $\varphi$ by (a).
(d) The generalized eigenspaces of $\varphi$ are the joint generalized eigenspaces of $\mathscr{F}$ for the same reasons.
(e) The endomorphism $\varphi$ is commendable if and only if we have $\dim_K(\mathrm{Ker}(\mathfrak{m})) = \dim_K(\mathscr{F}/\mathfrak{m})$ (equivalently $\dim_{\mathscr{F}/\mathfrak{m}}(\mathrm{Ker}(\mathfrak{m})) = 1$) for all maximal ideals $\mathfrak{m}$ of $\mathscr{F}$, as follows from (c) and the definition of a commendable endomorphism.

The following theorem provides a characterization of $\mathscr{F}$-cyclic modules over unigenerated families by the existence of a commendable endomorphism in the family. It shows that the existence of a commendable endomorphism has strong implications for the structure of the family $\mathscr{F}$ and of the $\mathscr{F}$-module $V$.

**Theorem 3.2.4 (Families Containing a Commendable Endomorphism)**
*Let $\mathscr{F}$ be a family of commuting endomorphisms of $V$, and let $\varphi \in \mathscr{F}$. Then the following conditions are equivalent.*

(a) *The endomorphism $\varphi$ is commendable.*
(b) *The $K$-algebra homomorphisms $K[z]/\langle \mu_\varphi(z) \rangle \longrightarrow \mathscr{F}$ induced by $z \mapsto \varphi$ and $K[z]/\langle \mu_\varphi(z) \rangle \longrightarrow K[z]/\langle \chi_\varphi(z) \rangle$ induced by $z \mapsto z$ are isomorphisms.*
(c) *The family $\mathscr{F}$ is unigenerated by $\varphi$ and the vector space $V$ is $\mathscr{F}$-cyclic.*

*Proof* First we prove (a) $\Rightarrow$ (c). By Theorem 1.5.8, we know that $V$ is a cyclic $K[z]$-module with respect to the module structure given by $z \cdot v = \varphi(v)$. Let $v \in V$ be a generator, let $\psi \in \mathscr{F}$, and let $g(z) \in K[z]$ be such that $\psi(v) = g(\varphi)(v)$. Now let $w \in V$, and let $f(z) \in K[z]$ be such that $w = f(\varphi)(v)$. Thus we get the equalities $\psi(w) = \psi(f(\varphi)(v)) = f(\varphi)(\psi(v)) = f(\varphi)(g(\varphi)(v)) = (g(\varphi)f(\varphi))(v) = g(\varphi)(w)$. Since $w$ was chosen arbitrarily, we conclude that $\psi = g(\varphi)$. Therefore we have $\mathscr{F} = K[\varphi]$, and the proof is complete.

To prove (c) $\Rightarrow$ (b), we note that part (a) of the preceding remark implies that there is an isomorphism $K[z]/\langle \mu_\varphi(z) \rangle \cong \mathscr{F}$ induced by $z \mapsto \varphi$. The further assumption that $V$ is $\mathscr{F}$-cyclic yields $\deg(\chi_\varphi(z)) = \dim_K(V) = \dim_K(\mathscr{F}) = \deg(\mu_\varphi(z))$. By the Cayley-Hamilton Theorem 1.2.5, it follows that $\mu_\varphi(z) = \chi_\varphi(z)$.

Finally, the implication (b) $\Rightarrow$ (a) follows from Theorem 1.5.8. $\qquad\square$

This theorem shows that the ideal of algebraic relations of a system of generators of $\mathscr{F}$ has a special shape. We note that this is precisely the shape of a Gröbner basis implied by the Shape Lemma (see [15], Theorem 3.7.25).

**Corollary 3.2.5** *Suppose that $\mathcal{F}$ is generated by endomorphisms $\varphi_1, \ldots, \varphi_n$ and that $\varphi_n$ is commendable. Then the ideal of algebraic relations of $\Phi = (\varphi_1, \ldots, \varphi_n)$ in $P = K[x_1, \ldots, x_n]$ is of the form*

$$\mathrm{Rel}_P(\Phi) = \langle x_1 - g_1(x_n), \ldots, x_{n-1} - g_{n-1}(x_n), \mu_{\varphi_n}(x_n) \rangle$$

*with $g_1(x_n), \ldots, g_{n-1}(x_n) \in K[x_n]$ such that $\deg(g_i(x_n)) < \deg(\mu_{\varphi_n}(x_n))$ for $i = 1, \ldots, n - 1$.*

**Corollary 3.2.6** *A commendable endomorphism $\varphi \in \mathcal{F}$ is a splitting endomorphism for $\mathcal{F}$.*

*Proof* This follows from the equivalence of (a) and (c) in the theorem and Proposition 2.5.2.c. □

Using the theorem and the Cyclic Module Decomposition 1.4.15, we can characterize a commendable endomorphism by the set of endomorphisms that commute with it in the following sense.

**Proposition 3.2.7 (Third Characterization of Commendable Endomorphisms)** *For $\varphi \in \mathrm{End}_K(V)$, the following conditions are equivalent.*

(a) *The endomorphism $\varphi$ is commendable.*
(b) *Every endomorphism $\psi \in \mathrm{End}_K(V)$ with $\psi \circ \varphi = \varphi \circ \psi$ is polynomial in $\varphi$.*

*Proof* To show that (a) implies (b), it suffices to consider the family $\mathcal{F} = K[\varphi, \psi]$ and to apply the theorem.

It remains to prove that (b) implies (a). Using Theorem 1.4.15, we can decompose $V$ into a direct sum of $\varphi$-invariant subspaces. One of these is of the form $W = \langle v_1 \rangle_{K[z]}$ with a vector $v_1 \in V$, where the annihilator of $W$ is $\langle \mu_\varphi(z) \rangle$. Now, if $W = V$, then $V$ is a cyclic $K[\varphi]$-module, and therefore $\varphi$ is commendable. Otherwise, using the above direct sum decomposition, there exists a non-zero $\varphi$-invariant subspace $U$ of $V$ such that $V = W \oplus U$. Let $\psi \in \mathrm{End}_K(V)$ be defined as the direct sum of the zero map on $W$ and the identity on $U$. Since it clearly commutes with $\varphi$, the hypothesis yields a polynomial $f(z) \in K[z]$ such that $\psi = f(\varphi)$. Then the equality $\psi(v_1) = 0$ implies that $f(z)$ is a multiple of $\mu_\varphi(z)$, and therefore we have $\psi = f(\varphi) = 0$. Since this contradicts $\psi|_U = \mathrm{id}_U$, the case $W \subset V$ cannot occur and the proof is complete. □

The following example (based on an example in [28], Sect. 2.2.2) illustrates the previous results.

**Example 3.2.8** Let $K = \mathbb{Q}$, let $V = K^4$, and let $\varphi_1$ and $\varphi_2$ be the $K$-endomorphisms of $V$ which are given by the matrices

$$
A_1 = \begin{pmatrix} 0 & 1 & 0 & 0 \\ 0 & 0 & 0 & 1 \\ -1 & 1 & 1 & 0 \\ 9 & -12 & -1 & 6 \end{pmatrix} \quad \text{and} \quad A_2 = \begin{pmatrix} 0 & 0 & 1 & 0 \\ -1 & 1 & 1 & 0 \\ -2 & 0 & 3 & 0 \\ -1 & 0 & 1 & 1 \end{pmatrix}
$$

respectively. Since we may check that $\varphi_1$ and $\varphi_2$ commute, we have a commuting family $\mathscr{F} = K[\varphi_1, \varphi_2]$.

To study $\varphi_1$, we compute $\chi_{\varphi_1}(z) = \mu_{\varphi_1}(z) = (z-2)^3(z-1)$. This implies that $\varphi_1$ is a commendable endomorphism. Hence Theorem 3.2.4 shows that the family $\mathscr{F}$ is unigenerated by $\varphi_1$ and that $\varphi_2$ is polynomial in $\varphi_1$. Using the Buchberger-Möller Algorithm for Matrices 2.1.4, we find the relation $\varphi_2 = -\varphi_1^3 + 6\varphi_1^2 + 12\varphi_1 - 9\,\mathrm{id}_V$.

According to Remark 3.2.3.c, the joint eigenspaces of $\mathscr{F}$ are the eigenspaces $\langle(1, 2, 1, 4)\rangle$ and $\langle(1, 1, 2, 1)\rangle$ of $\varphi_1$. Moreover, the joint generalized eigenspaces of $\mathscr{F}$ are $\langle(1, 2, 1, 4), (0, 1, 0, 4), (0, 0, 0, 1)\rangle$ and $\langle(1, 1, 2, 1)\rangle$.

Notice that, for the endomorphism $\varphi_2$, we have $\chi_{\varphi_2}(z) = (z-1)^3(z-2)$ and $\mu_{\varphi_2}(z) = (z-1)(z-2)$. Thus we see that $\varphi_2$ is not a commendable endomorphism.

In the following we show that commendability of an endomorphism implies commendability of its restrictions to the joint generalized eigenspaces of $\mathscr{F}$, but not vice versa. Therefore we introduce the following notion.

**Definition 3.2.9** Let $\mathscr{F}$ be a commuting family, let $\mathfrak{m}_1, \ldots, \mathfrak{m}_s$ be its maximal ideals, and let $\varphi \in \mathscr{F}$. The map $\varphi$ is called **locally commendable** if its restrictions $\varphi|_{\mathrm{BigKer}(\mathfrak{m}_i)}$ are commendable for $i = 1, \ldots, s$.

The next proposition provides another characterization of commendable endomorphisms.

**Proposition 3.2.10 (Fourth Characterization of Commendable Endomorphism)**
Let $\mathscr{F}$ be a commuting family, and let $\varphi \in \mathscr{F}$. Then the following conditions are equivalent.

(a) *The endomorphism $\varphi$ is commendable.*
(b) *The map $\varphi$ is a splitting endomorphism for $\mathscr{F}$ and $\varphi$ is locally commendable.*

*Proof* To prove the implication (a) $\Rightarrow$ (b), we first use Corollary 3.2.6 to see that $\varphi$ is a splitting endomorphism. By Theorem 3.2.4, we have $\mathscr{F} = K[\varphi] \cong K[z]/\langle\mu_\varphi(z)\rangle$. Hence the maximal ideals $\mathfrak{m}_1, \ldots, \mathfrak{m}_s$ of $\mathscr{F}$ are of the form $\mathfrak{m}_i = \langle p_i(\varphi)\rangle$ for every $i = 1, \ldots, s$, where $p_1(z), \ldots, p_s(z)$ are the eigenfactors of $\varphi$. Since $V$ is $\mathscr{F}$-cyclic by Theorem 3.2.4, we know from Corollary 3.1.9 that the generalized eigenspace $\mathrm{BigKer}(\mathfrak{m}_i)$ is $\mathscr{F}_{\mathrm{BigKer}(\mathfrak{m}_i)}$-cyclic for $i = 1, \ldots, s$. Using the fact that $\mathrm{BigKer}(\mathfrak{m}_i)$ is unigenerated by the restriction of $\varphi$, Theorem 3.2.4 implies that $\varphi|_{\mathrm{BigKer}(\mathfrak{m}_i)}$ is commendable.

Conversely, suppose that the conditions in (b) are satisfied. Since $\varphi$ is a splitting endomorphism, we have $\mathrm{BigKer}(\mathfrak{m}_i) = \mathrm{Gen}(\varphi, p_{\mathfrak{m}_i,\varphi}(z))$ for $i = 1, \ldots, s$ by Proposition 2.4.10 and the polynomials $p_{\mathfrak{m}_i,\varphi}(z)$ are pairwise distinct. The fact that $\varphi|_{\mathrm{BigKer}(\mathfrak{m}_i)}$ is commendable shows that the minimal polynomial of this map is of the form $p_{\mathfrak{m}_i,\varphi}(z)^{m_i}$ for some $m_i > 0$, and that we have the equality $\dim_K(\mathrm{BigKer}(\mathfrak{m}_i)) = m_i \cdot \deg(p_{\mathfrak{m}_i,\varphi}(z))$. By Lemma 1.1.11, it follows that $\mu_\varphi(z) = \prod_{i=1}^{s} p_{\mathfrak{m}_i,\varphi}(z)^{m_i}$, and Theorem 2.4.3 yields $\deg(\mu_\varphi(z)) = \dim_K(V) = \deg(\chi_\varphi(z))$. Hence Theorem 1.5.8 implies the claim. $\square$

The following example shows that commendability of an endomorphism is not a local property, i.e., that the implication (b) $\Rightarrow$ (a) in the preceding proposition does not hold if $\varphi$ is not a splitting endomorphism for $\mathscr{F}$.

**Example 3.2.11**  Let $K = \mathbb{Q}$, let $V = K^4$, and let $\varphi_1, \varphi_2 \in \mathrm{End}_K(V)$ be given by the matrices

$$A_1 = \begin{pmatrix} 0 & 0 & 0 & 0 \\ 0 & -\frac{1}{2} & -\frac{1}{2} & 0 \\ 0 & \frac{1}{2} & \frac{1}{2} & 0 \\ 1 & \frac{1}{2} & \frac{1}{2} & 0 \end{pmatrix} \quad \text{and} \quad A_2 = \begin{pmatrix} 0 & 0 & 0 & 0 \\ 1 & 0 & -\frac{1}{2} & -\frac{1}{2} \\ 0 & 1 & \frac{3}{2} & \frac{1}{2} \\ 0 & 0 & \frac{1}{2} & \frac{1}{2} \end{pmatrix}$$

respectively. We may verify that $\mathscr{F} = K[\varphi_1, \varphi_2]$ is a commuting family. Using the Buchberger Möller Algorithm for Matrices 2.1.4 with $\sigma = \mathrm{Lex}$, we calculate

$$\mathrm{Rel}_P(\varphi_1, \varphi_2) = \langle x - 2y^3 + 3y^2 - y,\ y^4 - 2y^3 + y^2 \rangle$$

From this result we deduce the isomorphism $\mathscr{F} \cong K[y]/\langle y^4 - 2y^3 + y^2 \rangle$ and the equalities $\mu_{\varphi_2}(z) = \chi_{\varphi_2}(z) = z^4 - 2z^3 + z^2 = z^2(z-1)^2$. Consequently, the endomorphism $\varphi_2$ is commendable, and there are two maximal ideals in the family $\mathscr{F}$, namely $\mathfrak{m}_1 = \langle \varphi_1, \varphi_2 \rangle$ and $\mathfrak{m}_2 = \langle \varphi_1, \varphi_2 - \mathrm{id}_V \rangle$.

What about the endomorphism $\varphi_1$? Since $\mu_{\varphi_1}(z) = z^2$, the map $\varphi_1$ is not a splitting endomorphism. Moreover, we have $\chi_{\varphi_1}(z) = z^4$, and this shows that $\varphi_1$ is not commendable, in agreement with Proposition 3.2.10.

Now let us check that $\varphi_1$ is locally commendable. First we compute the big kernels of the maximal ideals. For $\mathfrak{m}_1$, we get $\mathrm{BigKer}(\mathfrak{m}_1) = \mathrm{Ker}(\varphi_2^2) = \langle w_1, w_2 \rangle$, where $w_1 = (1, -2, 1, 0)$ and $w_2 = (1, -1, 0, 1)$. Analogously, for the ideal $\mathfrak{m}_2$, we obtain $\mathrm{BigKer}(\mathfrak{m}_2) = \mathrm{Ker}((\varphi_2 - \mathrm{id})^2) = \langle w_3, w_4 \rangle$, where $w_3 = (0, 0, 1, 0)$ and $w_4 = (0, 1, 0, -1)$. Notice that we have $V = \mathrm{BigKer}(\mathfrak{m}_1) \oplus \mathrm{BigKer}(\mathfrak{m}_2)$, as predicted by Theorem 2.4.3.

Next we let $U_1 = \mathrm{BigKer}(\mathfrak{m}_1)$ and $U_2 = \mathrm{BigKer}(\mathfrak{m}_2)$. Then $(w_1, w_2)$ is a $K$-basis of $U_1$ and $(w_3, w_4)$ is a $K$-basis of $U_2$. It is straightforward to check that $\varphi_1|_{U_1}$ and $\varphi_1|_{U_2}$ are represented with respect to these bases by

$$\frac{1}{2}\begin{pmatrix} -1 & -1 \\ 1 & 1 \end{pmatrix} \quad \text{and} \quad \frac{1}{2}\begin{pmatrix} 1 & 1 \\ -1 & -1 \end{pmatrix}$$

respectively. We have $\mu_{\varphi_1|_{U_1}} = z^2$ and $\mu_{\varphi_1|_{U_2}} = z^2$, and hence the maps $\varphi_1|_{U_1}$ and $\varphi_1|_{U_2}$ are commendable. Thus the map $\varphi_1$ is locally commendable.

Next, we consider a local version of the property of being unigenerated.

**Definition 3.2.12** Let $\mathscr{F}$ be a commuting family, and let $\mathfrak{m}_1, \ldots, \mathfrak{m}_s$ be the maximal ideals of $\mathscr{F}$. The family $\mathscr{F}$ is called **locally unigenerated** (or **weakly curvilinear**) if the restriction $\mathscr{F}_{\mathrm{BigKer}(\mathfrak{m}_i)}$ is unigenerated for every $i \in \{1, \ldots, s\}$.

The terminology *weakly curvilinear* will be explained in Sect. 4.3. Locally unigenerated families can be characterized as follows.

**Proposition 3.2.13 (Characterization of Locally Unigenerated Families)**
*Let $\mathscr{F}$ be a commuting family, let $\mathfrak{m}_1, \ldots, \mathfrak{m}_s$ be the maximal ideals of $\mathscr{F}$, and let $\mathfrak{q}_i = \mathrm{Ann}_{\mathscr{F}}(\mathrm{BigKer}(\mathfrak{m}_i))$ for $i = 1, \ldots, s$. Then the following conditions are equivalent.*

(a) *The family $\mathscr{F}$ is locally unigenerated.*
(b) *There exist monic irreducible polynomials $p_1(z), \ldots, p_s(z)$ in $K[z]$ and numbers $m_1, \ldots, m_s$ such that $\mathscr{F}/\mathfrak{q}_i \cong K[z]/\langle p_i(z)^{m_i}\rangle$ for $i = 1, \ldots, s$.*

*Proof* To prove (a) $\Rightarrow$ (b), we use Proposition 2.4.6.c to get $\mathscr{F}_{\mathrm{BigKer}(\mathfrak{m}_i)} \cong \mathscr{F}/\mathfrak{q}_i$ and Remark 3.2.3.a to get $\mathscr{F}_{\mathrm{BigKer}(\mathfrak{m}_i)} \cong K[z]/\langle p_i(z)^{m_i}\rangle$ for some monic irreducible polynomial $p_i(z) \in K[z]$ and an integer $m_i \geq 1$.

Conversely, the isomorphism in (b) implies that $\mathscr{F}/\mathfrak{q}_i$ is generated by the image of the residue class of $z$. An application of the isomorphism $\mathscr{F}_{\mathrm{BigKer}(\mathfrak{m}_i)} \cong \mathscr{F}/\mathfrak{q}_i$ finishes the proof. $\square$

The existence of a commendable endomorphism in a commuting family can be characterized as follows.

**Proposition 3.2.14 (Existence of Commendable Endomorphisms)**
*Let $\mathscr{F}$ be a commuting family with maximal ideals $\mathfrak{m}_1, \ldots, \mathfrak{m}_s$. Then the following conditions are equivalent.*

(a) *There exists a commendable endomorphism $\varphi$ in $\mathscr{F}$.*
(b) *The vector space $V$ is $\mathscr{F}$-cyclic, the family $\mathscr{F}$ is locally unigenerated, and the polynomials $p_1(z), \ldots, p_s(z)$ in condition 3.2.13.b can be chosen pairwise distinct.*

*Proof* The proof of (a) $\Rightarrow$ (b) follows immediately from the equivalence (a) $\Leftrightarrow$ (c) in Theorem 3.2.4. It remains to prove (b) $\Rightarrow$ (a). In the setting of Proposition 3.2.13.b, let $\mathscr{F}/\mathfrak{q}_i \cong K[z]/\langle p_i(z)^{m_i}\rangle$ for $i = 1, \ldots, s$. By Proposition 2.4.6.c and the Chinese Remainder Theorem 2.2.1, we have

$$\mathscr{F} \cong \prod_{i=1}^{s} \mathscr{F}/\mathfrak{q}_i \cong \prod_{i=1}^{s} K[z]/\langle p_i(z)^{m_i}\rangle \cong K[z]/\langle \textstyle\prod_{i=1}^{s} p_i(z)^{m_i}\rangle$$

Let $\varphi \in \mathscr{F}$ be a preimage of the residue class of $z$ under these isomorphisms. Since Proposition 2.4.6.c yields $\mathscr{F}/\mathfrak{q}_i \cong \mathscr{F}_{\mathrm{BigKer}(\mathfrak{m}_i)}$, the minimal polynomial of the restriction of $\varphi$ to $\mathrm{BigKer}(\mathfrak{m}_i)$ is $p_i(z)^{m_i}$ for $i = 1, \ldots, s$. Then Lemma 1.1.11 shows that the minimal polynomial of $\varphi$ is $\mu_\varphi(z) = \prod_{i=1}^s p_i(z)^{m_i}$. Using the hypothesis that the vector space $V$ is $\mathscr{F}$-cyclic, we get $\dim_K(V) = \dim_K(\mathscr{F}) = \deg(\mu_\varphi(z))$. Hence the endomorphism $\varphi$ is commendable.                                             $\square$

In Example 2.5.8 we saw a locally unigenerated family which does not contain a splitting endomorphism, and thus, by Proposition 3.2.10, no commendable endomorphism. If the field $K$ has enough elements, namely if $\mathrm{card}(K) \geq s$, then Proposition 2.5.4.b implies the existence of a splitting endomorphism in $\mathscr{F}$. This is not sufficient to guarantee the existence of a commendable endomorphism in $\mathscr{F}$, as the following example shows.

**Example 3.2.15** Let $K = \mathbb{F}_2$, let $V = K^4$, and let $\mathscr{F} = K[\varphi_1, \varphi_2]$ be the commuting family generated by the two endomorphisms given by

$$A_1 = \begin{pmatrix} 0 & 1 & 0 & 0 \\ 1 & 1 & 0 & 0 \\ 0 & 0 & 0 & 1 \\ 0 & 0 & 1 & 1 \end{pmatrix} \quad \text{and} \quad A_2 = \begin{pmatrix} 0 & 0 & 1 & 0 \\ 0 & 0 & 0 & 1 \\ 1 & 0 & 1 & 0 \\ 0 & 1 & 0 & 1 \end{pmatrix}$$

respectively. Using the Buchberger Möller Algorithm for Matrices 2.1.4, we find that $\mathscr{F} \cong K[x, y]/I$ where $I = \langle x^2 + x + 1, y^2 + y + 1 \rangle$. We have the decomposition $I = \mathfrak{M}_1 \cap \mathfrak{M}_1$ with $\mathfrak{M}_1 = \langle x^2 + x + 1, x + y \rangle$ and $\mathfrak{M}_2 = \langle x^2 + x + 1, x + y + 1 \rangle$. If we let $p(z) = z^2 + z + 1$, we get $\mathscr{F} \cong K[z]/\langle p(z) \rangle \oplus K[z]/\langle p(z) \rangle$. This shows that $\mathscr{F}$ is a reduced, locally unigenerated family.

Here we have $\mathrm{card}(K) = s$ and, in agreement with Corollary 2.5.7, the endomorphism $\varphi_1 + \varphi_2$, which is the preimage of $(0, 1)$ under the above isomorphism, is a strongly splitting endomorphism. It is not locally commendable, since the local minimal polynomials are $z$ and $z - 1$, while the restrictions to $\mathrm{BigKer}(\mathfrak{m}_i)$ are $K$-vector spaces of dimension 2 for $i = 1, 2$. This implies that $\varphi_1 + \varphi_2$ is not commendable, in agreement to Proposition 3.2.10.

Moreover, by computing the minimal polynomials of all elements in $\mathscr{F}$, it can be verified that there is no commendable endomorphism in $\mathscr{F}$. For instance, the endomorphism $\varphi_1$ is locally commendable, but it is not a splitting endomorphism, since the minimal polynomials of its restrictions to $\mathrm{BigKer}(\mathfrak{m}_i)$ coincide. This is in agreement with Proposition 3.2.10. Using the Cyclicity Test 3.1.4, we can check that $V$ is a cyclic $\mathscr{F}$-vector space.

The existence of pairwise distinct polynomials $p_i(z)$ in Proposition 3.2.14.b is a rather subtle property. In the preceding example, no linear change of coordinates is able to transform $p(z)$ into another polynomial. The next example shows that sometimes non-linear transformations can be used successfully, even if all linear transformations fail.

**Example 3.2.16** Let $K = \mathbb{F}_3$, let $V = K^9$, and let $\mathscr{F} = K[\varphi_1, \varphi_2]$ be the commuting family generated by the two endomorphisms given by

$$
A_1 = \begin{pmatrix}
0 & 0 & 0 & -1 & 0 & 0 & 0 & 0 & 0 \\
1 & 0 & 0 & 1 & 0 & 0 & 0 & 0 & 0 \\
0 & 0 & 0 & 0 & 0 & 0 & -1 & 0 & 0 \\
0 & 1 & 0 & 0 & 0 & 0 & 0 & 0 & 0 \\
0 & 0 & 1 & 0 & 0 & 0 & 1 & 0 & 0 \\
0 & 0 & 0 & 0 & 0 & 0 & 0 & 0 & -1 \\
0 & 0 & 0 & 0 & 1 & 0 & 0 & 0 & 0 \\
0 & 0 & 0 & 0 & 0 & 1 & 0 & 0 & 1 \\
0 & 0 & 0 & 0 & 0 & 0 & 0 & 1 & 0
\end{pmatrix}
\quad \text{and} \quad
A_2 = \begin{pmatrix}
0 & 0 & 0 & 0 & 0 & -1 & 0 & 0 & 0 \\
0 & 0 & 0 & 0 & 0 & 0 & 0 & -1 & 0 \\
1 & 0 & 0 & 0 & 0 & 1 & 0 & 0 & 0 \\
0 & 0 & 0 & 0 & 0 & 0 & 0 & 0 & -1 \\
0 & 1 & 0 & 0 & 0 & 0 & 0 & 1 & 0 \\
0 & 0 & 1 & 0 & 0 & 0 & 0 & 0 & 0 \\
0 & 0 & 0 & 1 & 0 & 0 & 0 & 0 & 1 \\
0 & 0 & 0 & 0 & 1 & 0 & 0 & 0 & 0 \\
0 & 0 & 0 & 0 & 0 & 0 & 1 & 0 & 0
\end{pmatrix}
$$

respectively. Using the Buchberger Möller Algorithm for Matrices 2.1.4, we find that $\mathscr{F} \cong K[x, y]/I$, where $I = \langle x^3 - x + 1, y^3 - y + 1 \rangle$. For instance, by applying the methods described in Chap. 5, we can calculate the decomposition

$$I = \left\langle x^3 - x + 1, x - y \right\rangle \cap \left\langle x^3 - x + 1, x - y + 1 \right\rangle \cap \left\langle x^3 - x + 1, x - y - 1 \right\rangle$$

and we get $\mathscr{F} \cong K[z]/\langle p(z) \rangle \oplus K[z]/\langle p(z) \rangle \oplus K[z]/\langle p(z) \rangle$ where $p(z) = z^3 - z + 1$. This shows that $\mathscr{F}$ is a reduced, locally unigenerated family. We have $s = \mathrm{card}(K)$ and, in agreement with Corollary 2.5.7, the preimage $\psi$ of $(0, 1, -1)$ under the above isomorphism is a strongly splitting endomorphism. We have $\psi = \varphi_1 - \varphi_2$ and the minimal polynomial of $\psi$ is $\mu_\psi(z) = (z + 1)z(z - 1)$. Since we have the inequality $\deg(\mu_\psi(z)) < \deg(\chi_\psi(z)) = 9$ we see that $\psi$ is not a commendable endomorphism of $V$.

Can we find a commendable endomorphism in the family $\mathscr{F}$? In order to make the three copies of $p(z)$ distinct, we cannot use a linear change of coordinates of the form $z \mapsto az + b$ with $a, b \in \mathbb{F}_3$, because under these changes $z^3 - z + 1$ is transformed either into itself or into $z^3 - z - 1$. However, using Algorithm 1.1.9, we check that the preimages of the elements $(\bar{z}^2 - \bar{z}, 0, 0)$, $(0, \bar{z}^2 - \bar{z} + 1, 0)$, and $(0, 0, \bar{z}^2 - \bar{z} - 1)$ in $\mathscr{F}$ have the minimal polynomials $p_1(z) = z^3 + z^2 - 1$, $p_2(z) = z^3 + z^2 + z - 1$, and $p_3(z) = z^3 + z^2 - z + 1$, respectively. These polynomials are monic, irreducible and distinct. Hence the preimage in $\mathscr{F}$ of the element $(\bar{z}^2 - \bar{z}, \bar{z}^2 - \bar{z} + 1, \bar{z}^2 - \bar{z} - 1)$ is an endomorphism $\varrho$ whose minimal polynomial is $\mu_\varrho(z) = z^9 + z^3 + z + 1 = p_1(z)p_2(z)p_3(z)$. It turns out that $\varrho = \varphi_1^2 - \varphi_2$. Then, we have the isomorphism $\mathscr{F} \cong K[z]/\langle p_1(z) \rangle \times K[z]/\langle p_2(z) \rangle \times K[z]/\langle p_3(z) \rangle$, and the endomorphism $\varrho$ is commendable. Moreover, we have

$$\mathscr{F} = K[\varphi_1, \varphi_2] \cong K[x, y]/\mathrm{Rel}(\varphi_1, \varphi_2)$$

$$\cong K[x, y]/\mathrm{Rel}\left(\varphi_1, \varphi_1^2 - \varphi_2\right) \cong K[x, y]/\left\langle x - \left(y^3 - y + 1\right), y^9 + y^3 + y + 1 \right\rangle$$

It is interesting to observe that we have a representation of the family of the form given in the Shape Lemma (see [15], Theorem 3.7.25), as predicted by Corollary 3.2.5.

In order to apply the Shape Lemma, we need that the ideal $I$ is in normal $x_n$-position. This property can be achieved by a linear change of coordinates if $\mathrm{card}(K) > \binom{\dim_K(\mathscr{F})}{2}$ (see [15], Proposition 3.7.22). In the current example we have card$(K) = 3$ and $\binom{\dim_K(\mathscr{F})}{2} = 36$, and we can check that no linear change of co-ordinates does the trick. However, as we saw, a non-linear change of coordinates transforms the ideal to the form given by the Shape Lemma.

## 3.3  Commendable Families

> *Why is lemon juice made with artificial flavor,*
> *and dishwashing liquid made with real lemons?*

Avoiding artificial definitions, we now go for the real thing, namely commendable families of commuting endomorphisms. A single endomorphism is called commendable if its eigenspaces have the minimal possible dimensions. Analogously, a commuting family is called commendable if its joint eigenspaces have the minimal possible dimensions. What is the minimum? Since a joint eigenspace Ker$(\mathfrak{m})$ of $\mathscr{F}$, where $\mathfrak{m}$ is a maximal ideal of $\mathscr{F}$, is an $\mathscr{F}/\mathfrak{m}$-vector space, its minimal $K$-dimension is $\dim_K(\mathscr{F}/\mathfrak{m})$. This yields the real definition and a small surprise: a commendable family need not contain a commendable endomorphism. In fact, as we shall see, an element $\varphi$ of $\mathscr{F}$ is commendable if and only if the family is commendable and unigenerated by $\varphi$.

In contrast to the property of a single endomorphism to be commendable, the notion of a commendable family is a local property. This means that the family $\mathscr{F}$ is commendable if and only if all restrictions $\mathscr{F}|_{\mathrm{BigKer}(\mathfrak{m}_i)}$ to its joint generalized eigenspaces BigKer$(\mathfrak{m}_i)$ are commendable. Thus the local nature of commendability corresponds very well to the local nature of the Miss Universe pageant.

> *The Miss Universe pageant is fixed.*
> *All the winners are from Earth.*

Let $K$ be a field, let $V$ be a finite dimensional $K$-vector space, let $\mathscr{F}$ be a family of commuting endomorphisms of $V$, and let $\mathfrak{m}_1, \ldots, \mathfrak{m}_s$ be the maximal ideals of $\mathscr{F}$.

By Proposition 2.2.13.c, the dimension of a joint eigenspace Ker$(\mathfrak{m}_i)$ is a multiple of the dimension of the corresponding residue class field $\mathscr{F}/\mathfrak{m}_i$. In Example 3.2.15 we met a commuting family $\mathscr{F}$ which contains no commendable endomorphism, but it is reduced and every joint eigenspace Ker$(\mathfrak{m}_i)$ has the minimal possible dimension $\dim_K(\mathscr{F}/\mathfrak{m}_i)$. This suggests the following definition.

**Definition 3.3.1**  In the above setting, the family $\mathscr{F}$ is called **commendable** if we have $\dim_K(\mathrm{Ker}(\mathfrak{m})) = \dim_K(\mathscr{F}/\mathfrak{m})$, or, equivalently, if $\dim_{\mathscr{F}/\mathfrak{m}}(\mathrm{Ker}(\mathfrak{m})) = 1$ for every maximal ideal $\mathfrak{m}$ of $\mathscr{F}$. If the family $\mathscr{F}$ is not commendable, we also say that it is **derogatory**.

The following example presents a non-reduced commendable family $\mathscr{F}$ whose joint eigenspaces are 1-dimensional and which does not contain a commendable endomorphism.

**Example 3.3.2** Let $K = \mathbb{Q}$, let $V = K^4$, and let $\mathcal{F} = K[\varphi_1, \varphi_2]$ with endomorphisms $\varphi_1, \varphi_2 \in \mathrm{End}_K(V)$ given by the matrices

$$A_1 = \begin{pmatrix} 0 & 0 & 0 & 0 \\ 1 & 0 & 0 & 0 \\ 0 & 0 & 0 & 0 \\ 0 & 0 & 1 & 0 \end{pmatrix} \quad \text{and} \quad A_2 = \begin{pmatrix} 0 & 0 & 0 & 0 \\ 0 & 0 & 0 & 0 \\ 1 & 0 & 0 & 0 \\ 0 & 1 & 0 & 0 \end{pmatrix}$$

respectively. Since we have $A_1 A_2 = A_2 A_1$, the $K$-algebra $\mathcal{F}$ is a commuting family.

If we let $P = K[x, y]$ and $\Phi = (\varphi_1, \varphi_2)$, the ideal of algebraic relations of $\Phi$ is given by $\mathrm{Rel}_P(\Phi) = \langle x^2, y^2 \rangle$. Consequently, we have $\mathcal{F} \cong P/\langle x^2, y^2 \rangle$ and the tuple $(\mathrm{id}_V, \varphi_1, \varphi_2, \varphi_1 \varphi_2)$ is a $K$-basis of $\mathcal{F}$. Moreover, the family $\mathcal{F}$ has only one maximal ideal, namely $\mathfrak{m} = \langle \varphi_1, \varphi_2 \rangle$. So, there is only one joint generalized eigenspace of $\mathcal{F}$, namely $\mathrm{BigKer}(\mathfrak{m}) = V$, and one joint eigenspace, namely the 1-dimensional vector space $\mathrm{Ker}(\mathfrak{m}) = \langle e_4 \rangle$. Hence $\mathcal{F}$ is a commendable family.

On the other hand, every endomorphism $\psi$ in $\mathcal{F}$ can be represented in the form $\psi = a_0 \, \mathrm{id}_V + a_1 \varphi_1 + a_2 \varphi_2 + a_3 \varphi_1 \varphi_2$ with $a_0, a_1, a_2, a_3 \in K$. The $\mathfrak{m}$-eigenfactor of $\psi$ is $z - a_0$. Hence the eigenspace of $\psi$ is $\mathrm{Ker}(\psi - a_0 \, \mathrm{id}_V)$. Moreover, the endomorphism $\psi - a_0 \, \mathrm{id}_V$ is represented by the matrix

$$B = \begin{pmatrix} 0 & 0 & 0 & 0 \\ a_1 & 0 & 0 & 0 \\ a_2 & 0 & 0 & 0 \\ a_3 & a_2 & a_1 & 0 \end{pmatrix}$$

The rank of $B$ is at most 2. Therefore we have $\dim_K(\mathrm{Eig}(\psi, z - a_0)) > 1$ which shows that no endomorphism in $\mathcal{F}$ is commendable.

In Example 3.2.15 we saw a case of a commendable family where the residue class fields $\mathcal{F}/\mathfrak{m}_i$ were proper extensions of $K$. Another example of this type is given by the following case.

**Example 3.3.3** Let $K = \mathbb{Q}$, let $V = K^4$, and let $\mathcal{F} = K[\varphi_1, \varphi_2]$ with endomorphisms $\varphi_1, \varphi_2 \in \mathrm{End}_K(V)$ given by the matrices

$$A_1 = \begin{pmatrix} 0 & 2 & 0 & 0 \\ 1 & 0 & 0 & 0 \\ 0 & 0 & 0 & 2 \\ 0 & 0 & 1 & 0 \end{pmatrix} \quad \text{and} \quad A_2 = \begin{pmatrix} 0 & 0 & 0 & 0 \\ 0 & 0 & 0 & 0 \\ 1 & 0 & 0 & 0 \\ 0 & 1 & 0 & 0 \end{pmatrix}$$

respectively. Since we have $A_1 A_2 = A_2 A_1$, the $K$-algebra $\mathcal{F}$ is a commuting family.

Letting $P = K[x, y]$ and $\Phi = (\varphi_1, \varphi_2)$, we calculate $\mathrm{Rel}_P(\Phi) = \langle x^2 - 2, y^2 \rangle$. Therefore the tuple $(\mathrm{id}_V, \varphi_1, \varphi_2, \varphi_1 \varphi_2)$ is a $K$-basis of $\mathcal{F}$. The family $\mathcal{F}$ has only one maximal ideal, namely $\mathfrak{m} = \langle \varphi_1^2 - 2 \, \mathrm{id}_V, \varphi_2 \rangle$. Thus there is only one joint generalized eigenspace, namely $\mathrm{BigKer}(\mathfrak{m}) = V$, and one joint eigenspace, namely the vector space $\mathrm{Ker}(\mathfrak{m}) = \langle e_3, e_4 \rangle$. Since $\dim_K(\mathscr{F}/\mathfrak{m}) = 2 = \dim_K(\mathrm{Ker}(\mathfrak{m}))$, we conclude that $\mathscr{F}$ is a commendable family.

An easy example of a derogatory family is obtained by reconsidering the setting of Example 2.6.4.

**Example 3.3.4** Let $K = \mathbb{Q}$, let $V = K^3$, and let $\varphi_1, \varphi_2$ be the endomorphisms of $V$ defined by the matrices

$$A_1 = \begin{pmatrix} 0 & 0 & 0 \\ 1 & 0 & 0 \\ 0 & 0 & 0 \end{pmatrix} \quad \text{and} \quad A_2 = \begin{pmatrix} 0 & 0 & 0 \\ 0 & 0 & 0 \\ 1 & 0 & 0 \end{pmatrix}$$

respectively. The family $\mathscr{F} = K[\varphi_1, \varphi_2]$ is a commuting family, and it is easy to check that $(\mathrm{id}_V, \varphi_1, \varphi_2)$ is a $K$-basis of $\mathscr{F}$. Moreover, the vector space $V$ is a cyclic $\mathscr{F}$-module generated by $e_1$, because we have $\varphi_1(e_1) = e_2$ and $\varphi_2(e_1) = e_3$. The family $\mathscr{F}$ has only one maximal ideal $\mathfrak{m} = \langle \varphi_1, \varphi_2 \rangle$, the 2-dimensional joint eigenspace $\langle e_2, e_3 \rangle$, and the residue class field $\mathscr{F}/\mathfrak{m} \cong K$. Therefore it is derogatory.

If we are given systems of generators of the maximal ideals of $\mathscr{F}$, we can use Algorithms 2.1.4, 2.2.17 and the Macaulay Basis Theorem (see [15], 1.5.7) to check whether a family is commendable via Definition 3.3.1. Later in this chapter we see a different method which does not require the knowledge of the maximal ideals of $\mathscr{F}$.

If a commuting family contains a commendable endomorphism, it is a commendable family, but with special properties. Remark 3.2.3 and Theorem 3.2.4 imply the following characterization.

**Proposition 3.3.5 (Fifth Characterization of Commendable Endomorphisms)**
*For a family $\mathscr{F}$ of commuting endomorphisms of $V$ and $\varphi \in \mathscr{F}$, the following conditions are equivalent.*

(a) *The endomorphism $\varphi$ is commendable.*
(b) *The family $\mathscr{F}$ is commendable and unigenerated by $\varphi$.*

The final proposition of this section says that commendability of a family is a local property in the following sense.

**Definition 3.3.6** Let $\mathscr{F}$ be a commuting family, and let $\mathfrak{m}_1, \ldots, \mathfrak{m}_s$ be the maximal ideals of $\mathscr{F}$. The family $\mathscr{F}$ is called **locally commendable** if $\mathscr{F}|_{\mathrm{BigKer}(\mathfrak{m}_i)}$ is a commendable family for every $i \in \{1, \ldots, s\}$.

Notice that this definition generalizes the corresponding definition for a single endomorphism (see Definition 3.2.9).

**Proposition 3.3.7 (Local Nature of Commendability)**
*Let $\mathscr{F}$ be a family of commuting endomorphisms of $V$. Then $\mathscr{F}$ is commendable if and only if it is locally commendable.*

*Proof* This follows from of Theorem 2.4.3 and Definition 3.3.1. □

Note that an analogous result does not hold for a single endomorphism, as Proposition 3.2.10 shows.

## 3.4 Local Families

> *"Have you lived in this village all your life?"*
> *"No, not yet."*

In this section we investigate commuting families which have a unique maximal ideal. In the language of ring theory, such families are local rings. This terminology is motivated by Algebraic Geometry, where the elements of a local ring are interpreted as germs of functions and the maximal ideal corresponds to the set of germs of functions whose value is zero at the given point. In our setting, a local family $\mathscr{F}$ has only one joint eigenspace, the only joint generalized eigenspace is the vector space itself, and the maximal ideal is nilpotent. Thus local families are rather special and we should not expect to have this property all the time.

However, one common type of local rings is given by fields. If $L$ is an extension field of $K$ and $\vartheta_a : L \longrightarrow L$ is the multiplication map given by $a \in L$, then the family $\mathscr{F} = K[\vartheta_a \mid a \in L]$ is a commuting family of endomorphisms of the $K$-vector space $L$ which is a field. When is one of these endomorphisms $\vartheta_\varphi$ commendable? As we shall see, this is the case if and only if $a$ is a primitive element for the extension $K \subseteq L$, i.e., if and only if $L = K(a)$. Thus we obtain the following version of the Primitive Element Theorem: if $K \subseteq L$ is a separable field extension, then the local family $\mathscr{F} = K[\vartheta_a \mid a \in L]$ contains a commendable endomorphism.

Another way in which we may exploit the fact that a local family $\mathscr{F}$ is very special is by constructing a particularly nice $K$-basis for it, namely a basis of the form $B = \{\varphi_i \tau_j \mid i = 1, \ldots, k \text{ and } j = 1, \ldots, m\}$ where the set of residue classes $\{\bar{\varphi}_1, \ldots, \bar{\varphi}_k\}$ is a $K$-basis of $\mathscr{F}/\mathfrak{m}$ and the set of residue classes $\{\bar{\tau}_1, \ldots, \bar{\tau}_k\}$ is a $K$-basis of the *graded ring* $\mathrm{gr}_\mathfrak{m}(\mathscr{F}) = \mathscr{F}/\mathfrak{m} \oplus \mathfrak{m}/\mathfrak{m}^2 \oplus \mathfrak{m}^2/\mathfrak{m}^3 \oplus \cdots$. Clearly, the graded ring is a new tool for examining a local commuting family. But is it really necessary to know about it? Should you care for that nice basis $B$? The answer to these questions is up to you.

> *"What's the difference between ignorance and apathy?"*
> *"I don't know and I don't care."*

Let $V$ be a $d$-dimensional vector space over a field $K$.

**Definition 3.4.1** A commuting family $\mathscr{F}$ of endomorphisms of $V$ is said to be **local** or a **local family** if it has a unique maximal ideal or, equivalently, if the vector space $V$ contains only one joint eigenspace of $\mathscr{F}$.

The following remark collects some easy properties of local families.

**Remark 3.4.2** Let $\mathscr{F}$ be a local family with maximal ideal $\mathfrak{m}$, let $\varphi \in \mathscr{F} \setminus \{0\}$, and let $\mathfrak{q} = \mathrm{Ann}_{\mathscr{F}}(\mathrm{BigKer}(\mathfrak{m}))$.

(a) By Proposition 2.3.2.g, the minimal polynomial $\mu_{\varphi}(z)$ is a power of an irreducible polynomial.
(b) By Proposition 2.4.6.b, we have $\mathfrak{q} = \langle 0 \rangle$.
(c) By Proposition 2.4.6.d, there exists a power of $\mathfrak{m}$ which is the zero ideal.

The first case of a local family is the case where $\mathscr{F}$ is a field. To examine it, we recall the following standard definitions from field theory.

**Definition 3.4.3** Let $K$ be a field, and let $K \subseteq L$ be a field extension.

(a) The extension $K \subseteq L$ is called a **finite field extension** if $\dim_K(L) < \infty$.
(b) An element $f \in L$ is called a **primitive element** for $K \subseteq L$ if we have the equality $L = K(f)$.

The following proposition characterizes the existence of a commendable multiplication endomorphism in a family which is a field.

**Proposition 3.4.4 (Commendable Multiplication Endomorphisms in a Field)**
*Let $K \subseteq L$ be a finite field extension. Given an element $a \in L$, let $\vartheta_a \in \mathrm{End}_K(L)$ be the multiplication by $a$, and let $\mathscr{F}$ be the family $\mathscr{F} = K[\vartheta_a \mid a \in L]$. For every $a \in L$, the following conditions are equivalent.*

(a) *The endomorphism $\vartheta_a \in \mathscr{F}$ is commendable.*
(b) *The element $a$ is a primitive element for $K \subseteq L$.*

*Proof* First we show that (a) implies (b). By Theorem 3.2.4, we have $\mathscr{F} = K[\vartheta_a]$. Hence, for element $b \in L$, there exists a polynomial $f \in K[x]$ such that $\vartheta_b = f(\vartheta_a)$, and therefore $b = f(a)$. This shows $L = K[a]$ and implies $L = K(a)$, i.e., that $a$ is a primitive element for $K \subseteq L$.

Conversely, assume that $a \in L$ satisfies $L = K(a)$. The ring $K[a] \subseteq L$ is a zero-dimensional $K$-algebra and an integral domain. Hence Corollary 2.2.6 shows that $K[a]$ is a field, and thus $L = K[a]$. Now it is straightforward to check the family $\mathscr{F}$ is unigenerated by $\vartheta_a$ and that $L$ is a cyclic $\mathscr{F}$-module. Consequently, an application of Theorem 3.2.4 finishes the proof.                                    $\square$

The family $\mathscr{F} = K[\vartheta_a \mid a \in L]$ is an example of a **multiplication family**. This notion will be studied in more detail in Chap. 4. Next we recall the following definition from field theory.

**Definition 3.4.5** Let $K$ be a field. A finite field extension $K \subseteq L$ is called **separable** if the minimal polynomial over $K$ of every element of $L$ has pairwise distinct roots in the algebraic closure.

An important class of separable field extensions if given as follows.

**Remark 3.4.6** In field theory it is shown that every finite field extension of a perfect field is separable (see for instance [30], Chap. II, Sect. 5), and has a primitive element (see for instance [10], Sect. 4.14).

From the above proposition we deduce that the existence of a commendable en-endomorphism is related to the existence of a primitive element of a finite field extension. The following classical example makes use of a non-separable field extension to provide a case where no primitive elements exist.

**Example 3.4.7** Let $p$ be a prime number, let $a, b, x, y$ be indeterminates, let $K$ be the field $K = \mathbb{F}_p(a, b)$, and let $L = K[x, y]/\mathfrak{M}$, where $\mathfrak{M} = \langle x^p - a, y^p - b \rangle$. Since the polynomial $x^p - a$ is irreducible over $K$, the ring $L' = K[x]/\langle x^p - a \rangle$ is a field. Since the polynomial $y^p - b$ is irreducible over $L'$, also $L = L'[y]/\langle y^p - b \rangle$ is a field. In particular, the ideal $\mathfrak{M}$ is a maximal ideal of $K[x, y]$.

On one hand, it is easy to check that $\dim_K(L) = p^2$. On the other hand, for every polynomial $f \in K[x, y]$, we have $f^p \in K[x^p, y^p]$. Therefore the residue class of $f^p$ modulo the ideal $\mathfrak{M} = \langle x^p - a, y^p - b \rangle$ is represented by an element of $K$. We conclude that every element $f \in L$ satisfies $f^p \in K$, and hence $\dim_K(K(f)) = \dim_K(K[f]) \leq p$. Altogether we see that there exists no primitive element for $K \subseteq L$. By the proposition, it follows that the multiplication family $\mathcal{F} = \{\vartheta_f \mid f \in L\}$ does not contain any commendable endomorphism.

By combining Propositions 3.1.1 and 3.5.9, one can see that the family $\mathcal{F}$ is commendable. In particular, like Example 3.3.2, it is another example of a commendable family which does not contain a commendable endomorphism.

Next we introduce a standard tool in commutative algebra, the associated graded ring which, in particular, can be used to construct special bases of $\mathcal{F}$. The following algebra associated to a local family is a special case of the general concept of the associated graded ring of a filtration (see [16], Definition 6.5.7).

**Lemma 3.4.8** *Let $(\mathcal{F}, \mathfrak{m})$ be a local family of commuting endomorphisms of $V$, let $r$ be the minimum integer such that $\mathfrak{m}^r = \langle 0 \rangle$, and let $L = \mathcal{F}/\mathfrak{m}$.*

(a) *There is a strictly increasing chain of $K$-vector subspaces*

$$\langle 0 \rangle = \mathfrak{m}^r \subset \mathfrak{m}^{r-1} \subset \cdots \subset \mathfrak{m} \subset \mathcal{F}$$

(b) *The set $\mathrm{gr}_\mathfrak{m}(\mathcal{F}) = \bigoplus_{i=0}^{r-1} \mathfrak{m}^i/\mathfrak{m}^{i+1}$ is a finitely generated algebra over the field $L = \mathfrak{m}^0/\mathfrak{m}^1 = \mathcal{F}/\mathfrak{m}$. It is called the **associated graded ring** of $(\mathcal{F}, \mathfrak{m})$.*
(c) *We have $\dim_K(\mathcal{F}) = \dim_K(\mathrm{gr}_\mathfrak{m}(\mathcal{F})) = \dim_L(\mathrm{gr}_\mathfrak{m}(\mathcal{F})) \cdot \dim_K(L)$.*

*Proof* First we prove (a). The strict inclusions in the claimed chain follow from the observation that $\mathfrak{m}^i = \mathfrak{m}^{i+1}$ implies $\mathfrak{m}^{i+1} = \mathfrak{m} \cdot \mathfrak{m}^i = \mathfrak{m} \cdot \mathfrak{m}^{i+1} = \mathfrak{m}^{i+2}$ for all $i \geq 1$, and hence $\mathfrak{m}^i = \langle 0 \rangle$.

Next we show (b). Clearly, for $i = 0, \ldots, r - 1$, the $L$-vector spaces $\mathfrak{m}^i/\mathfrak{m}^{i+1}$ are finite dimensional, since they are residue classes of vector subspaces of a finite dimensional $K$-vector space. For residue classes $\overline{\varphi} \in \mathfrak{m}^i/\mathfrak{m}^{i+1}$ and $\overline{\psi} \in \mathfrak{m}^j/\mathfrak{m}^{j+1}$, the rule $\overline{\varphi} \cdot \overline{\psi} = \overline{\varphi \cdot \psi}$ yields a well-defined element in $\mathfrak{m}^{i+j}/\mathfrak{m}^{i+j+1}$. By extending this rule linearly, we turn $\mathrm{gr}_{\mathfrak{m}}(\mathscr{F})$ into an algebra over the field $L = \mathscr{F}/\mathfrak{m}$. Since this algebra is a finite dimensional $L$-vector space, it is finitely generated.

To prove (c), we observe that the first equality follows immediately from (a). The second equality is a standard result in elementary linear algebra.                         $\square$

Subsets of the monoid of power products $\mathbb{T}^n$ which are closed under taking factors have many special names. We use the following.

**Definition 3.4.9**  A set of terms $\mathscr{O} \subset \mathbb{T}^n$ such that $t \in \mathscr{O}$ and $t' \mid t$ implies $t' \in \mathscr{O}$ is called an **order ideal**.

For a local commuting family, we can construct special $K$-bases as follows.

**Proposition 3.4.10 (Special Bases via Associated Graded Rings)**
*Let $(\mathscr{F}, \mathfrak{m})$ be a local family of commuting endomorphisms of $V$, and let $L = \mathscr{F}/\mathfrak{m}$.*

(a) *There exist endomorphisms $\varphi_1, \ldots, \varphi_k \in \mathscr{F}$ such that the tuple of their residue classes $(\overline{\varphi}_1, \ldots, \overline{\varphi}_k)$ is a $K$-basis of the field $L$, and it is possible to choose $\varphi_1 = \mathrm{id}_V$.*

(b) *There exist endomorphisms $\psi_1, \ldots, \psi_\ell \in \mathfrak{m}$ such that the tuple of their residue classes $(\overline{\psi}_1, \ldots, \overline{\psi}_\ell)$ is an $L$-vector space basis of $\mathfrak{m}/\mathfrak{m}^2$.*

(c) *In the setting of (b), there exists an order ideal $\mathscr{O} = \{t_1, \ldots, t_m\}$ in $L[x_1, \ldots, x_\ell]$ such that the tuple of residue classes $(\overline{\tau}_1, \ldots, \overline{\tau}_m)$, where $\tau_i = t_i(\psi_1, \ldots, \psi_\ell)$, is an $L$-basis of $\mathrm{gr}_{\mathfrak{m}}(\mathscr{F})$.*

(d) *Given (a), (b), and (c), the tuple $B = (\varphi_i \tau_j \mid i = 1, \ldots, k$ and $j = 1, \ldots, m)$ is a $K$-basis of $\mathscr{F}$.*

*Proof* Claims (a) and (b) follow immediately from the preceding lemma. To prove (c), we note that the proof of the preceding lemma also shows that $\{\overline{\psi}_1, \ldots, \overline{\psi}_\ell\}$ is a system of $L$-algebra generators of $\mathrm{gr}_{\mathfrak{m}}(\mathscr{F})$. Thus we obtain an epimorphism of $L$-algebras $\varepsilon : L[x_1, \ldots, x_\ell] \longrightarrow \mathrm{gr}_{\mathfrak{m}}(\mathscr{F})$ defined by $\varepsilon(x_i) = \overline{\psi}_i$. Since $\mathrm{gr}_{\mathfrak{m}}(\mathscr{F})$ is a finite dimensional $K$-vector space, the existence of the desired order ideal $\mathscr{O}$ is guaranteed by Macaulay's Basis Theorem (see [15], Theorem 1.5.7).

Finally we show (d). First we prove that $B$ generates the $K$-vector space $\mathscr{F}$. For every element $\varphi \in \mathscr{F}$, Lemma 3.4.8.a implies that there exists an integer $i \geq 0$ such that $\varphi \in \mathfrak{m}^i \setminus \mathfrak{m}^{i+1}$. Let $(t_{j_1}, \ldots, t_{j_\lambda})$ be a tuple of elements of $\mathscr{O}$ whose images in $\mathrm{gr}_{\mathfrak{m}}(\mathscr{F})$ form an $L$-basis of $\mathrm{gr}_{\mathfrak{m}}(\mathscr{F})_i$. Then there exist elements $c_1, \ldots, c_\lambda \in L$ such that we have $\overline{\varphi} = c_1 \overline{\tau}_{j_1} + \cdots + c_\lambda \overline{\tau}_{j_\lambda}$ in $\mathrm{gr}_{\mathfrak{m}}(\mathscr{F})_i$. Now we use (a) and write

$c_j = a_{j1}\bar{\varphi}_1 + \cdots + a_{jk}\bar{\varphi}_k$ with $a_{j1}, \ldots, a_{jk} \in K$ for $j = 1, \ldots, \lambda$. It follows that

$$\tilde{\varphi} = \varphi - \sum_{\mu=1}^{\lambda} (a_{\mu 1}\varphi_1 + \cdots + a_{\mu k}\varphi_k) \cdot \tau_{j_\mu} \in \mathfrak{m}^{i+1}$$

By applying the same argument to $\tilde{\varphi}$ and continuing in this way, we end up with an element of $\mathfrak{m}^r = \langle 0 \rangle$. This shows that $\varphi$ is a $K$-linear combination of the elements of $B$. To finish the proof, we use Lemma 3.4.8.c to get $\dim_K(\mathcal{F}) = m \cdot k$. Thus $B$ is a system of $K$-vector space generators having exactly $\dim_K(\mathcal{F})$ elements. Hence $B$ is a $K$-basis of $\mathcal{F}$.                                                                          $\square$

Notice that the construction in parts (b) and (c) of the proposition implies that $\{1, x_1, \ldots, x_\ell\}$ is contained in the order ideal $\mathcal{O}$. Let us construct the bases of this proposition in some instances.

**Example 3.4.11** Let $K = \mathbb{Q}$, let $V = K^7$, and let $\varphi_1$ and $\varphi_2$ be the endomorphisms of $V$ defined by

$$A_1 = \begin{pmatrix} 0 & 0 & 0 & 0 & 0 & 0 & 0 \\ 0 & 0 & 0 & 0 & 0 & -12 & 0 \\ 0 & 0 & 0 & 0 & 0 & 0 & -12 \\ 1 & 0 & 0 & 0 & 0 & 0 & 0 \\ 0 & 1 & 0 & 0 & 0 & 7 & 0 \\ 0 & 0 & 0 & 1 & 0 & 0 & 0 \\ 0 & 0 & 0 & 0 & 1 & 3 & 0 \end{pmatrix} \text{ and } A_2 = \begin{pmatrix} 0 & 0 & 0 & 0 & 0 & 0 & 0 \\ 1 & 0 & 0 & 0 & 0 & 0 & 0 \\ 0 & 1 & 0 & 0 & 0 & 0 & 0 \\ 0 & 0 & 0 & 0 & 0 & 0 & 0 \\ 0 & 0 & 0 & 1 & 0 & 0 & 0 \\ 0 & 0 & 0 & 0 & 0 & 0 & 0 \\ 0 & 0 & 0 & 0 & 0 & 1 & 0 \end{pmatrix}$$

respectively. The maps $\varphi_1$ and $\varphi_2$ generate a commuting family $\mathcal{F} = K[\varphi_1, \varphi_2]$. We let $P = K[x, y]$ and $\Phi = (\varphi_1, \varphi_2)$. Using the Buchberger Möller Algorithm for Matrices 2.1.4 with respect to DegRevLex, we calculate

$$\mathrm{Rel}_P(\Phi) = \langle y^3, \, xy^2, \, x^3 - 3x^2 y - 7xy + 12y \rangle$$

It is immediate to check that the radical of this ideal is $\langle x, y \rangle$. Therefore $\mathcal{F}$ is a local family with maximal ideal $\mathfrak{m} = \langle \varphi_1, \varphi_2 \rangle$.

Next we use the Tangent Cone Algorithm (see [16], Corollary 6.5.26) to compute a presentation of the associated graded ring and get $\mathrm{gr}_\mathfrak{m}(\mathcal{F}) = K[x, y]/\langle x^7, y \rangle$. We deduce that the tuple $B = (\mathrm{id}_V, \varphi_1, \varphi_1^2, \varphi_1^3, \varphi_1^4, \varphi_1^5, \varphi_1^6)$ is a $K$-basis of $\mathrm{gr}_\mathfrak{m}(\mathcal{F})$ as in part (c) of the proposition. Since $\mathcal{F}/\mathfrak{m} \cong K$, this tuple $B$ is also the $K$-basis of $\mathcal{F}$ given in part (d) of the proposition.

Let us look at the family $\mathcal{F}$ more closely. The above computation of $\mathrm{Rel}_P(\Phi)$ shows $\mu_{\varphi_2}(z) = z^3$, and thus $\chi_{\varphi_2}(z) = z^7$, since $\dim_\mathbb{Q}(V) = 7$. Consequently, the endomorphism $\varphi_2$ is derogatory. On the other hand, if we calculate the minimal polynomial of $\varphi_1$, we get $\mu_{\varphi_1}(z) = z^7$. Hence the endomorphism $\varphi_1$ is commendable, and consequently it generates the family $\mathcal{F}$ (see Theorem 3.2.4).

This observation suggests that we compute the representation of $\varphi_2$ as a polynomial expression in $\varphi_1$, or equivalently of $A_2$ as a polynomial expression in $A_1$. For instance, we can do this by computing the Gröbner basis of $\mathrm{Rel}_P(\Phi)$ with respect to the lexicographic term ordering $\sigma$ such that $y >_\sigma x$. The result

$$\mathrm{Rel}_P(\Phi) = \left\langle x^7, \ y + \tfrac{847}{20736}x^6 + \tfrac{85}{1728}x^5 + \tfrac{7}{144}x^4 + \tfrac{1}{12}x^3 \right\rangle$$

implies that

$$A_2 = -\tfrac{847}{20736}A_1^6 - \tfrac{85}{1728}A_1^5 - \tfrac{7}{144}A_1^4 - \tfrac{1}{12}A_1^3$$

Looking back at the matrices $A_1$ and $A_2$, who would have guessed this relation?

The following example is another case in point.

**Example 3.4.12** Let $K = \mathbb{Q}$, let $V = K^6$, let the endomorphisms $\varphi_1$ and $\varphi_2$ of $V$ be given by the matrices

$$A_1 = \begin{pmatrix} 0 & 0 & 0 & 0 & 0 & -1 \\ 1 & 0 & 0 & 0 & 0 & 0 \\ 0 & 0 & 0 & 0 & -1 & 0 \\ 0 & 1 & 0 & 0 & 0 & -2 \\ 0 & 0 & 1 & 0 & 0 & 0 \\ 0 & 0 & 0 & 1 & 0 & 0 \end{pmatrix} \quad \text{and} \quad A_2 = \begin{pmatrix} 0 & 0 & 0 & 0 & 0 & 0 \\ 0 & 0 & 0 & 0 & 0 & 0 \\ 1 & 0 & 0 & -1 & 0 & 0 \\ 0 & 0 & 0 & 0 & 0 & 0 \\ 0 & 1 & 0 & 0 & 0 & -1 \\ 0 & 0 & 0 & 0 & 0 & 0 \end{pmatrix}$$

respectively, and let $\mathscr{F} = K[\varphi_1, \varphi_2]$. It is easy to check that $\mathscr{F}$ is a commuting family. We let $P = K[x_1, x_2]$ and $\Phi = (\varphi_1, \varphi_2)$ and use the Buchberger-Möller Algorithm for Matrices 2.1.4 to compute $\mathrm{Rel}_P(\Phi) = \langle (x_1^2 + 1)^2, x_1^2 x_2 + x_2, x_2^2 \rangle$. Thus the tuple $(\mathrm{id}_V, \varphi_1, \varphi_2, \varphi_1^2, \varphi_1\varphi_2, \varphi_1^3)$ is a $K$-basis of $\mathscr{F}$.

By looking at $\mathrm{Rel}_P(\Phi)$, we see that $\mathscr{F}$ is a local family and that its maximal ideal $\mathfrak{m}$ is generated by $\psi_1 = \varphi_1^2 + \mathrm{id}_V$ and $\psi_2 = \varphi_2$. Hence we can choose the images of $\mathrm{id}_V$ and $\varphi_1$ to form a $K$-basis of $\mathscr{F}/\mathfrak{m}$ as in part (a) of the proposition. Using the Tangent Cone Algorithm (see [16], Corollary 6.5.26), we calculate the presentation

$$\mathrm{gr}_{\mathfrak{m}}(\mathscr{F}) \cong (\mathscr{F}/\mathfrak{m})[y_1, y_2]/\langle y_1^2, y_1 y_2, y_2^2 \rangle$$

where the residue class of $y_i$ is the image of $\psi_i$ in $\mathfrak{m}/\mathfrak{m}^2$ for $i = 1, 2$. Consequently, the order ideal $\mathscr{O} = \{1, y_1, y_2\}$ yields the $\mathscr{F}/\mathfrak{m}$-basis $(\overline{\mathrm{id}}_V, \bar{\psi}_1, \bar{\psi}_2)$ of $\mathrm{gr}_{\mathfrak{m}}(\mathscr{F})$ given in part (c) of the proposition. Putting these results together, we arrive at the $K$-basis $B = (\mathrm{id}_V, \varphi_1, \psi_1, \varphi_1\psi_1, \psi_2, \varphi_1\psi_2)$ of $\mathscr{F}$ mentioned in part (d) of the proposition. The matrices which represent $\psi_1$ and $\psi_2$ with respect to $B$ are

$$B_1 = \begin{pmatrix} 0 & 0 & 0 & 0 & 0 & 0 \\ 0 & 0 & 0 & 0 & 0 & 0 \\ 1 & 0 & 0 & 0 & 0 & 0 \\ 0 & 1 & 0 & 0 & 0 & 0 \\ 0 & 0 & 0 & 0 & 0 & 0 \\ 0 & 0 & 0 & 0 & 0 & 0 \end{pmatrix} \quad \text{and} \quad B_2 = \begin{pmatrix} 0 & 0 & 0 & 0 & 0 & 0 \\ 0 & 0 & 0 & 0 & 0 & 0 \\ 0 & 0 & 0 & 0 & 0 & 0 \\ 0 & 0 & 0 & 0 & 0 & 0 \\ 1 & 0 & 0 & 0 & 0 & 0 \\ 0 & 1 & 0 & 0 & 0 & 0 \end{pmatrix}$$

respectively. In particular, we see that $\mathrm{Ker}(\mathfrak{m})$ is 4-dimensional.

## 3.5  Dual Families

*The highest moments in the life of a mathematician*
*are the first few moments after one has proved the result,*
*but before one finds the mistake.*

The main theorem in this section, called the Duality Theorem, required a substantial effort for its proof, and we were quite pleased after finishing it. One of the main reasons why was it so difficult for us is that it talks about dual families on dual vector spaces. Thus it requires a certain kind of *dual thinking*. Let us explain what we mean by this.

Given a $K$-vector space $V$, we have the dual vector space $V^* = \operatorname{Hom}_K(V, K)$ consisting of the linear forms on $V$. Every endomorphism $\varphi$ of $V$ yields a dual endomorphism $\varphi^\vee$ of $V^*$ by letting $\varphi^\vee(\ell) = \ell \circ \varphi$. For a commuting family $\mathscr{F}$ on $V$, we get a dual family $\mathscr{F}^\vee = K[\varphi^\vee \mid \varphi \in \mathscr{F}]$. So far, so good. The first small mind twister occurs when we dualize again. It is well-known that there is a canonical isomorphism between $V$ and $V^{**}$, but no canonical isomorphism between $V$ and $V^*$. However, for a commuting family $\mathscr{F}$, the map $\varphi \mapsto \varphi^\vee$ provides a canonical isomorphism of $K$-algebras between $\mathscr{F}$ and $\mathscr{F}^\vee$.

Further properties of the dual family $\mathscr{F}^\vee$ are straightforward. For instance, it will come to no one's surprise that, given a $K$-basis of $V$, the matrices describing the dual family with respect to the dual basis are the transposes of the matrices describing $\mathscr{F}$. However, the next logical development of the theory leads us on a wrong track. A unigenerated family $\mathscr{F} = K[\varphi]$ is commendable if and only if the dual family $\mathscr{F}^\vee = K[\varphi^\vee]$ is commendable. These conditions are also equivalent to $V$ being a cyclic $\mathscr{F}$-module and to $V^*$ being a cyclic $\mathscr{F}^\vee$-module.

Unfortunately, when we turn to more general commuting families, some of these equivalences break down and the picture becomes more complicated. In fact, the Duality Theorem 3.5.12 says that the dual family $\mathscr{F}^\vee$ is commendable if and only if $V$ is a cyclic $\mathscr{F}$-module. Furthermore, the dual vector space $V^*$ is a cyclic $\mathscr{F}$-module if and only if $\mathscr{F}$ is commendable. A beautiful application of this duality is that is allows us to check algorithmically whether a commuting family is commendable: it suffices to apply the Cyclicity Test 3.1.4 to the dual family $\mathscr{F}^\vee$. If, after all this dualizing, you feel somewhat dizzy, take heart in the fact that we did not fare better. This is not a bug of the theory of dual families, it is a feature.

*Any sufficiently advanced bug*
*is indistinguishable from a feature.*
(Rich Kulawiec)

As usual, let $K$ be a field and $V$ a finite dimensional $K$-vector space of dimension $d$. Recall that the **dual vector space** of $V$ is $V^* = \operatorname{Hom}_K(V, K)$, i.e. the set of all $K$-linear maps $\ell : V \longrightarrow K$. For a map $\varphi \in \operatorname{End}_K(V)$, the **dual endomorphism** of $\varphi$ is the $K$-linear map $\varphi^\vee : V^* \longrightarrow V^*$ given by $\varphi^\vee(\ell) = \ell \circ \varphi$.

**Definition 3.5.1** Let $\mathcal{F}$ be a family of commuting endomorphisms in $\mathrm{End}_K(V)$.

(a) We call $\mathcal{F}^{\vee} = K[\varphi^{\vee} \mid \varphi \in \mathcal{F}]$ the **dual family** of $\mathcal{F}$.
(b) Given an ideal $I$ of $\mathcal{F}$, we call $I^{\vee} = \langle \varphi^{\vee} \mid \varphi \in I \rangle$ the **dual ideal** of $I$.

It is clear that $\mathcal{F}^{\vee}$ is a family of commuting endomorphisms of the dual vector space $V^*$. Moreover, given an ideal $I$ of $\mathcal{F}$, it is easy to check that $I^{\vee}$ is an ideal in the $K$-algebra $\mathcal{F}^{\vee}$. We should be careful not to confuse $\mathcal{F}^{\vee}$ with $\mathcal{F}^* = \mathrm{Hom}_K(\mathcal{F}, K)$ which is the dual vector space of $\mathcal{F}$, i.e. the vector space of linear maps from $\mathcal{F}$ to $K$.

For a finite dimensional $K$-vector space $V$, there is no canonical isomorphism between $V$ and $V^*$, although both vector spaces have the same dimension. However, there is a canonical isomorphism $\Theta : V \longrightarrow V^{**}$ defined by $\Theta(v)(\ell) = \ell(v)$ for $v \in V$ and $\ell \in V^*$. It is called the **canonical isomorphism to the bidual**. Instead of $\Theta(v)$, we sometimes write $v^{**}$. In the case of commuting families, the following proposition holds.

**Proposition 3.5.2 (Properties of the Dual Family)**
*Let $\mathcal{F}$ be a family of commuting endomorphisms of $V$, and let $\Theta : V \longrightarrow V^{**}$ be the canonical isomorphism to the bidual.*

(a) *Let $\delta_{\mathcal{F}} : \mathcal{F} \longrightarrow \mathcal{F}^{\vee}$ be the canonical map defined by $\delta_{\mathcal{F}}(\varphi) = \varphi^{\vee}$. Then $\delta_{\mathcal{F}}$ is an isomorphism of $K$-algebras.*
(b) *In particular, the canonical map $\delta_{\mathcal{F}^{\vee}} : \mathcal{F}^{\vee} \longrightarrow \mathcal{F}^{\sim}$ satisfies $\delta_{\mathcal{F}^{\vee}}(\varphi^{\vee}) = \varphi^{\sim}$ for all $\varphi^{\vee} \in \mathcal{F}^{\vee}$ and is an isomorphism of $K$-algebras.*
(c) *For every $v \in V$ and every $\varphi \in \mathcal{F}$, we have $\varphi^{\sim}(v^{**}) = (\varphi(v))^{**}$.*
(d) *For every $\varphi \in \mathcal{F}$, we have $\varphi^{\sim} = \Theta \circ \varphi \circ \Theta^{-1}$.*

*Proof* First we show (a). The map $\delta$ is clearly $K$-linear. To show that it is injective, we let $\varphi \in \mathcal{F} \setminus \{0\}$. Then there exists a vector $v \in V$ such that $\varphi(v) \neq 0$, and therefore there exists an element $\ell \in V^*$ such that $\ell(\varphi(v)) \neq 0$. Consequently, we have the equality $\varphi^{\vee}(\ell) = \ell \circ \varphi \neq 0$. The fact that the map $\delta$ is surjective follows from the definition of $\mathcal{F}^{\vee}$. Claim (b) follows by applying (a) to $\mathcal{F}^{\vee}$.

To prove (c), we observe that $\varphi^{\sim}(v^{**}) = v^{**} \circ \varphi^{\vee}$ by definition. Consequently, for every $\ell \in V^*$, we have

$$\varphi^{\sim}(v^{**})(\ell) = (v^{**} \circ \varphi^{\vee})(\ell) = v^{**}(\ell \circ \varphi) = (\ell \circ \varphi)(v) = \ell(\varphi(v))$$

On the other hand, we have $(\varphi(v))^{**}(\ell) = \ell(\varphi(v))$ by definition, and the conclusion follows.

Finally, we prove (d). It suffices to show that the two maps agree on every element of $V^{**}$. We know that $\Theta$ is an isomorphism. Hence every element of $V^{**}$ is of the form $v^{**}$ with $v \in V$. Using (c), we know that $\varphi^{\sim}(v^{**}) = (\varphi(v))^{**}$. On the other hand, we have

$$(\Theta \circ \varphi \circ \Theta^{-1})(v^{**}) = (\Theta \circ \varphi)(v) = (\Theta(\varphi(v)) = (\varphi(v))^{**}$$

and the proof is complete.                                                                    $\square$

The vector space $V^*$ has the structure of both an $\mathscr{F}^{\vee}$-module and an $\mathscr{F}$-module as follows.

**Remark 3.5.3**  As $\mathscr{F}^{\vee}$ is a family of commuting endomorphisms of $V^*$, it turns the vector space $V^*$ into an $\mathscr{F}^{\vee}$-module via $\varphi^{\vee} \cdot \ell = \ell \circ \varphi$ for all $\varphi^{\vee} \in \mathscr{F}^{\vee}$ and $\ell \in V^*$. Notice that the first part of the proposition implies that $V^*$ carries also the structure of an $\mathscr{F}$-module via $\varphi \cdot \ell = \ell \circ \varphi$ for $\varphi \in \mathscr{F}$ and $\ell \in V^*$.

The next step is to represent the endomorphisms in the dual family by matrices. For this we recall the following definition.

**Definition 3.5.4**  Let $V$ be a $d$-dimensional $K$-vector space and let $B = (v_1, \dots, v_d)$ be a $K$-basis of $V$.

(a) The **dual basis** of $B$ is the tuple $B^* = (v_1^*, \dots, v_d^*)$ of linear forms in $V^*$ where $v_i^*$ is defined by $v_i^*(v_j) = \delta_{ij}$ for $i, j = 1, \dots, d$. (Here $\delta_{ij}$ is **Kronecker's delta**, i.e., we have $\delta_{ij} = 1$ if $i = j$ and $\delta_{ij} = 0$ if $i \neq j$.)
(b) Given a vector $w = c_1 v_1 + \cdots + c_d v_d \in V$ with $c_i \in K$, the linear map $w^* \in V^*$ defined by $w^*(v_i) = c_i$ for $i = 1, \dots, d$ is called the **dual linear form** of $w$ with respect to $B$.

### Proposition 3.5.5 (Dual Bases and Transposed Matrices)
*Let $B = (v_1, \dots, v_d)$ be a $K$-basis of $V$, let $\varphi \in \mathrm{End}_K(V)$, and let $M_B(\varphi)$ be the matrix which represents $\varphi$ with respect to $B$. Then the dual map $\varphi^{\vee}$ is represented by $M_{B^*}(\varphi^{\vee}) = M_B(\varphi)^{\mathrm{tr}}$ with respect to the dual basis $B^*$.*

*Proof* Let $M_B(\varphi) = (a_{ij})$ with $a_{ij} \in K$. For $i, j = 1, \dots, d$, we have $\varphi^{\vee}(v_j^*)(v_i) = (v_j^* \circ \varphi)(v_i) = v_j^*(\sum_{k=1}^{d} a_{ki} v_k) = a_{ji}$, and therefore $\varphi^{\vee}(v_j^*) = \sum_{i=1}^{d} a_{ji} v_i^*$. This proves the claim.  $\square$

The next proposition says that the ideal of algebraic relations of a system of generators of $\mathscr{F}$ agrees with the ideal of algebraic relations of the dual system of generators of $\mathscr{F}^{\vee}$.

### Proposition 3.5.6 (The Ideal of Relations of the Dual Family)
*Let $\varphi_1, \dots, \varphi_n \in \mathrm{End}_K(V)$ be commuting endomorphisms, let $\Phi = (\varphi_1, \dots, \varphi_n)$, and let $\mathscr{F} = K[\Phi]$.*

(a) *The dual family $\mathscr{F}^{\vee}$ is generated by $\Phi^{\vee} = (\varphi_1^{\vee}, \dots, \varphi_n^{\vee})$.*
(b) *In the polynomial ring $P = K[x_1, \dots, x_n]$, we have $\mathrm{Rel}_P(\Phi) = \mathrm{Rel}_P(\Phi^{\vee})$.*

*Proof* Both claims follow immediately from the isomorphism $\delta : \mathscr{F} \longrightarrow \mathscr{F}^{\vee}$ given in Proposition 3.5.2.a.  $\square$

According to the following remark, this proposition has to be interpreted care-fully.

**Remark 3.5.7** In the setting of the proposition, the identification of $\mathrm{Rel}_P(\Phi)$ with $\mathrm{Rel}_P(\Phi^\vee)$ depends on the fact that we use the same polynomial ring $P$ to repre-sent the family $\mathscr{F}$ as $\mathscr{F} \cong P/\mathrm{Rel}_P(\Phi)$ and the dual family as $\mathscr{F}^\vee \cong P/\mathrm{Rel}_P(\Phi^\vee)$. Let $\psi_1, \ldots, \psi_s$ be generators of a maximal ideal $\mathfrak{m}$ of $\mathscr{F}$. If they are represented by polynomials $f_1, \ldots, f_s \in P$ generating a maximal ideal of $P$, then the residue classes of the same polynomials in $\mathscr{F}^\vee$ are the system of generators $\psi_1^\vee, \ldots, \psi_s^\vee$ of the dual maximal ideal $\mathfrak{m}^\vee$.

However, if we now choose a $K$-basis $B$ of $V$ and represent $\psi_1, \ldots, \psi_s$ by ma-trices $A_1, \ldots, A_s \in \mathrm{Mat}_d(K)$ with respect to $B$, then Proposition 3.5.5 says that $\psi_1^\vee, \ldots, \psi_s^\vee$ are represented by $A_1^{\mathrm{tr}}, \ldots, A_s^{\mathrm{tr}}$ with respect to the dual basis $B^*$ of $V^*$. To compute the joint eigenspace $\mathrm{Ker}(\mathfrak{m})$ using Algorithm 2.2.17, we have to find the kernel of the matrix $\mathrm{Col}(A_1, \ldots, A_s)$. On the other hand, to compute $\mathrm{Ker}(\mathfrak{m}^\vee)$, we have to compute the kernel of $\mathrm{Col}(A_1^{\mathrm{tr}}, \ldots, A_s^{\mathrm{tr}})$ which is, in general, very differ-ent from $\mathrm{Ker}(\mathfrak{m})$ (see for instance Example 3.5.10 below). Thus the identification of the preimages of $\mathfrak{m}$ and $\mathfrak{m}^\vee$ in $P$ does not extend to an identification of the kernels of these ideals. This general phenomenon is partially obscured in the case $s = 1$ in which the two kernels have the same dimension.

The next proposition examines the dual family of a unigenerated family. In this case, not only the conditions to be a commendable family and to define a cyclic module structure on $V$ agree, but these conditions are also equivalent to the same conditions for the dual family.

**Proposition 3.5.8 (Sixth Characterization of Commendable Endomorphisms)**
*For an endomorphism $\varphi$ of $V$, the following conditions are equivalent.*

(a) *The endomorphism $\varphi$ is commendable.*
(b) *The vector space $V$ is a cyclic $K[\varphi]$-module.*
(c) *The family $K[\varphi]$ is commendable.*
(d) *The dual endomorphism $\varphi^\vee$ is commendable.*
(e) *The vector space $V^*$ is a cyclic $K[\varphi^\vee]$-module.*
(f) *The dual family $K[\varphi^\vee]$ is commendable.*

*Proof* The equivalence of (a), (b), and (c) follows from Theorem 1.5.8 and Propo-sition 3.3.5. The same argument shows that (d) is equivalent to (e) and (f). So, it re-mains to be proved that (b) and (e) are equivalent. By Theorem 1.5.8, condition (b) is equivalent to $\mu_\varphi(z) = \chi_\varphi(z)$. Now $\mu_\varphi(z) = \mu_{\varphi^\vee}(z)$ and $\chi_\varphi(z) = \chi_{\varphi^\vee}(z)$ follow from Proposition 3.5.5. Hence condition (b) is equivalent to $\mu_{\varphi^\vee}(z) = \chi_{\varphi^\vee}(z)$, and by applying Theorem 1.5.8 again, we see that this is equivalent to (e).                    □

Next, we show that a similar statement holds for reduced families.

**Proposition 3.5.9 (Dual Families of Reduced Families)**
*Let $\mathscr{F}$ be a reduced commuting family of endomorphisms of $V$. Then the following conditions are equivalent.*

(a) *The vector space $V$ is a cyclic $\mathscr{F}$-module.*
(b) *The family $\mathscr{F}$ is commendable.*
(c) *The vector space $V^*$ is a cyclic $\mathscr{F}^{\vee}$-module.*
(d) *The dual family $\mathscr{F}^{\vee}$ is commendable.*

*Proof* First we prove the equivalence between (a) and (b). By Proposition 3.3.7 and Corollary 3.1.9, we may assume that $\mathscr{F}$ has a unique maximal ideal. Together with the fact that it is reduced, this implies that $\mathscr{F}$ is a field. Its unique maximal ideal is $\mathfrak{m} = \langle 0 \rangle$ and we have $V = \mathrm{Ker}(\mathfrak{m})$. Consequently we obtain the two equalities $\dim_K(\mathscr{F}/\mathfrak{m}) = \dim_K(\mathscr{F})$ and $\dim_K(\mathrm{Ker}(\mathfrak{m})) = \dim_K(V)$.

Now, if $V$ is a cyclic $\mathscr{F}$-module, then we have $\dim_K(\mathscr{F}) = \dim_K(V)$, and hence $\dim_K(\mathscr{F}/\mathfrak{m}) = \dim_K(\mathrm{Ker}(\mathfrak{m}))$. Thus $\mathscr{F}$ is a commendable family. Conversely, if $\mathscr{F}$ is commendable, then $\dim_K(\mathscr{F}/\mathfrak{m}) = \dim_K(\mathrm{Ker}(\mathfrak{m}))$. Hence we have $\dim_K(V) = \dim_K(\mathscr{F})$. This implies $\dim_{\mathscr{F}}(V) = 1$, and thus $V$ is a cyclic $\mathscr{F}$-vector space (see Proposition 3.1.1.d).

From the isomorphism between $\mathscr{F}$ and $\mathscr{F}^{\vee}$ given in Proposition 3.5.2.a we deduce that also $\mathscr{F}^{\vee}$ is reduced. Hence the proof of the equivalence between (a) and (b) yields the equivalence between (c) and (d) as well. Finally, we observe that the conditions $\dim_K(V) = \dim_K(\mathscr{F})$ and $\dim_K(V^*) = \dim_K(\mathscr{F}^{\vee})$ are clearly equivalent. Together with Proposition 3.1.1.d, this implies the equivalence of (a) and (c) and finishes the proof. $\qquad\square$

The two propositions above fail for general families as the following continuation of Example 3.3.4 shows.

**Example 3.5.10** Let $K = \mathbb{Q}$, let $V = \mathbb{Q}^3$, and let $\mathscr{F} = K[\varphi_1, \varphi_2]$ be the derogatory family studied in Example 3.3.4.

Now we examine the dual family $\mathscr{F}^{\vee} = K[\varphi_1^{\vee}, \varphi_2^{\vee}]$. The endomorphisms $\varphi_1^{\vee}$ and $\varphi_2^{\vee}$ are given by the matrices $A_1^{\mathrm{tr}}$ and $A_2^{\mathrm{tr}}$, respectively. Notice that the family $\mathscr{F}^{\vee}$ is commendable, since the joint eigenspace $\langle e_1 \rangle$ is 1-dimensional. However, it does not pass the Cyclicity Test 3.1.4: one has to check whether the determinant of $\begin{pmatrix} z_1 & z_2 & z_3 \\ z_2 & 0 & 0 \\ z_3 & 0 & 0 \end{pmatrix}$ is non-zero and this is not true.

This example leads us to investigate the relation between commendability of a commuting family $\mathscr{F}$ or the $\mathscr{F}$-cyclicity of $V$ and the corresponding properties of their dual objects. Since this will be the main result of the current chapter, we introduce the following preparatory result. Recall that a commuting family having only one maximal ideal is a local ring.

**Lemma 3.5.11** *Let $\mathscr{F}$ be a commendable local family of commuting endomorphisms with maximal ideal $\mathfrak{m}$, let $v \in \mathrm{Ker}(\mathfrak{m}) \setminus \{0\}$, and let $w \in V \setminus \{0\}$. Then there exists an endomorphism $\varphi \in \mathscr{F}$ such that $\varphi(w) = v$.*

*Proof* Let us consider the ideal $\mathrm{Ann}_{\mathscr{F}}(w) :_{\mathscr{F}} \mathfrak{m} = \{\psi \in \mathscr{F} \mid \mathfrak{m} \cdot \psi \subseteq \mathrm{Ann}_{\mathscr{F}}(w)\} = \{\psi \in \mathscr{F} \mid (\eta\psi)(w) = 0 \text{ for all } \eta \in \mathfrak{m}\}$. First we want to show that it contains $\mathrm{Ann}_{\mathscr{F}}(w)$ properly. For a contradiction, suppose that $\mathrm{Ann}_{\mathscr{F}}(w) = \mathrm{Ann}_{\mathscr{F}}(w) : \mathfrak{m}$. Since we have $(\mathrm{Ann}_{\mathscr{F}}(w) : \mathfrak{m}^i) : \mathfrak{m} = \mathrm{Ann}_{\mathscr{F}}(w) : \mathfrak{m}^{i+1}$, induction on $i$ yields $\mathrm{Ann}_{\mathscr{F}}(w) = \mathrm{Ann}_{\mathscr{F}}(w) : \mathfrak{m}^i$ for every $i \geq 1$. By Proposition 2.2.5.b, there is a number $r \geq 1$ such that $\mathfrak{m}^r = \langle 0 \rangle$. Consequently, it follows that $\mathrm{Ann}_{\mathscr{F}}(w) = \mathscr{F}$, in contradiction to $w = \mathrm{id}_V(w) \neq 0$.

Therefore we may now choose an element $\psi$ of $\mathrm{Ann}_{\mathscr{F}}(w) : \mathfrak{m}$ which is not contained in $\mathrm{Ann}_{\mathscr{F}}(w)$. It follows that the map $\psi$ satisfies $\psi(w) \in \mathrm{Ker}(\mathfrak{m}) \setminus \{0\}$. Recall that $\mathrm{Ker}(\mathfrak{m})$ is a vector space over the field $L = \mathscr{F}/\mathfrak{m}$ and that the commendability of the family $\mathscr{F}$ is equivalent to $\dim_L(\mathrm{Ker}(\mathfrak{m})) = 1$. Hence the vector $v$ is an $L$-basis of this vector space and there exists an element $\eta \in \mathscr{F} \setminus \mathfrak{m}$ such that $\psi(w) = \bar{\eta} \cdot v = \eta(v)$. As $L$ is a field, the element $\bar{\eta}$ is invertible, i.e. there exist endomorphisms $\kappa \in \mathscr{F}$ and $\varrho \in \mathfrak{m}$ such that $\kappa\eta - \mathrm{id}_V = \varrho$. Hence, if we put $\varphi = \kappa\psi$, we obtain $\varphi(w) = \kappa\psi(w) = \kappa\eta(v) = v + \varrho(v) = v$, as desired.                                    □

Now we are ready to prove the main theorem of this section. It says that commendability of a commuting family $\mathscr{F}$ and cyclicity of the $\mathscr{F}$-module $V$ are dual properties in the following sense.

**Theorem 3.5.12 (Duality Theorem)**
*Let $\mathscr{F}$ be a family of commuting endomorphisms of $V$.*

(a) *If $V$ is a cyclic $\mathscr{F}$-module, then $\mathscr{F}^\vee$ is commendable.*
(b) *If $\mathscr{F}$ is commendable, then $V^*$ is a cyclic $\mathscr{F}^\vee$-module.*

*Proof* First we prove (a). Let $\mathfrak{m}_1, \ldots, \mathfrak{m}_s$ be the maximal ideals of $\mathscr{F}$, and let $\mathfrak{m}_1^\vee, \ldots, \mathfrak{m}_s^\vee$ be their images in $\mathscr{F}^\vee$ under the isomorphism of Proposition 3.5.2.a. Since $V$ is a cyclic $\mathscr{F}$-module, it is isomorphic to $\mathscr{F}$ by Proposition 3.1.1.c, and we need to show that $\dim_K(\mathrm{Ker}(\mathfrak{m}_i^\vee)) = \dim_K(\mathscr{F}^\vee/\mathfrak{m}_i^\vee)$ for $i = 1, \ldots, s$. Using the relations

$$\mathrm{Ker}(\mathfrak{m}_i^\vee) = \{\ell \in V^* \mid \varphi^\vee(\ell) = \ell \circ \varphi = 0 \text{ for all } \varphi \in \mathfrak{m}_i\}$$

$$\cong \{\ell \in \mathrm{Hom}_K(\mathscr{F}, K) \mid \ell(\mathfrak{m}_i) = 0\} \cong \mathrm{Hom}_K(\mathscr{F}/\mathfrak{m}_i, K)$$

we obtain

$$\dim_K(\mathrm{Ker}(\mathfrak{m}_i^\vee)) = \dim_K(\mathrm{Hom}_K(\mathscr{F}/\mathfrak{m}_i, K)) = \dim_K(\mathscr{F}/\mathfrak{m}_i) = \dim_K(\mathscr{F}^\vee/\mathfrak{m}_i^\vee)$$

for $i = 1, \ldots, s$, as was to be shown.

To prove (b), we use Propositions 3.3.7 and Corollary 3.1.9 to reduce the problem to the case of a local family $\mathscr{F}$ with maximal ideal $\mathfrak{m}$. Let $v \in \mathrm{Ker}(\mathfrak{m}) \setminus \{0\}$. By completing $v$ to a $K$-basis of $V$, we construct a $K$-linear map $v^* : V \longrightarrow K$ which is the projection to $v$. Our goal is to show that $\mathscr{F}^\vee \cdot v^* = V^*$. To prove the claim it suffices to show that the joint kernel $\{w \in V \mid \ell(w) = 0 \text{ for all } \ell \in \mathscr{F}^\vee \cdot v^*\}$ is zero.

In other words, we want to prove that, for every vector $w \in V \setminus \{0\}$, there exists an endomorphism $\varphi^\vee$ in $\mathscr{F}^\vee$ satisfying $(\varphi^\vee \cdot v^*)(w) \neq 0$. Let $\varphi \in \mathscr{F}$ be the map given by Lemma 3.5.11 such that $\varphi(w) = v$. Then we get $(\varphi^\vee \cdot v^*)(w) = v^*(\varphi(w)) = v^*(v) = 1$, and the proof is complete. $\qquad\square$

**Remark 3.5.13** The proof of part (a) shows that the vector space $\mathrm{Hom}_K(\mathscr{F}/\mathfrak{m}_i, K)$ can be identified with $\mathrm{Ker}(\mathfrak{m}_i^\vee)$. According to Theorem 2.4.3, it can therefore be interpreted as a joint eigenspace of $\mathscr{F}^\vee$.

To illustrate the first part of this theorem, we reconsider Example 3.4.12 in the present context.

**Example 3.5.14** Let $V$ be the $\mathbb{Q}$-vector space $V = \mathbb{Q}^6$, and let $\mathscr{F} = \mathbb{Q}[\varphi_1, \varphi_2]$ be the commuting family defined in Example 3.4.12. First we use the Cyclicity Test 3.1.4 to check whether the $\mathscr{F}$-module $V$ is cyclic. The determinant $\Delta$ computed for this test is

$$
\begin{aligned}
z_1^6 &+ 3z_1^4 z_2^2 + 3z_1^2 z_2^4 + z_2^6 - 6z_1^5 z_4 - 12z_1^3 z_2^2 z_4 - 6z_1 z_2^4 z_4 + 15z_1^4 z_4^2 + 18z_1^2 z_2^2 z_4^2 \\
&+ 3z_2^4 z_4^2 - 20z_1^3 z_4^3 - 12z_1 z_2^2 z_4^3 + 15z_1^2 z_4^4 + 3z_2^2 z_4^4 - 6z_1 z_4^5 + z_4^6 - 6z_1^4 z_2 z_6 \\
&- 12z_1^2 z_2^3 z_6 - 6z_2^5 z_6 + 24z_1^3 z_2 z_4 z_6 + 24z_1 z_2^3 z_4 z_6 - 36z_1^2 z_2 z_4^2 z_6 - 12z_2^3 z_4^2 z_6 \\
&+ 24z_1 z_2 z_4^3 z_6 - 6z_2 z_4^4 z_6 + 3z_1^4 z_6^2 + 18z_1^2 z_2^2 z_6^2 + 15z_2^4 z_6^2 - 12z_1^3 z_4 z_6^2 \\
&- 36z_1 z_2^2 z_4 z_6^2 + 18z_1^2 z_4^2 z_6^2 + 18z_2^2 z_4^2 z_6^2 - 12z_1 z_4^3 z_6^2 + 3z_4^4 z_6^2 - 12z_1^2 z_2 z_6^3 \\
&- 20z_2^3 z_6^3 + 24z_1 z_2 z_4 z_6^3 - 12z_2 z_4^2 z_6^3 + 3z_1^2 z_6^4 + 15z_2^2 z_6^4 - 6z_1 z_4 z_6^4 + 3z_4^2 z_6^4 \\
&- 6z_2 z_6^5 + z_6^6
\end{aligned}
$$

Hence $V$ is a cyclic $\mathscr{F}$-module. Notice that using the polynomial $\Delta$ we can decide which vectors generate the $\mathscr{F}$-module $V$.

Next we examine the family $\mathscr{F}$ and the dual family $\mathscr{F}^\vee$ for commendability. From Example 3.4.12 we know that the family $\mathscr{F}$ has only one maximal ideal, namely $\mathfrak{m} = \langle \varphi_1^2 + \mathrm{id}_V, \varphi_2 \rangle$, and that $\dim_{\mathbb{Q}}(\mathrm{Ker}(\mathfrak{m})) = 4$. Since we have $\dim_{\mathbb{Q}}(\mathscr{F}/\mathfrak{m}) = \dim_{\mathbb{Q}}(\mathbb{Q}[x_1, x_2]/\langle x_1^2 + 1, x_2 \rangle) = 2$, we conclude that the family $\mathscr{F}$ is derogatory. On the other hand, using the matrices $B_1$ and $B_2$ given in Example 3.4.12, we see that $\dim_{\mathbb{Q}}(\mathrm{Ker}(\mathfrak{m}^\vee)) = 2$. Hence the family $\mathscr{F}^\vee$ is commendable.

Using the canonical isomorphism to the bidual, we can restate the Duality Theorem as follows.

**Corollary 3.5.15** *Let $\mathscr{F}$ be a family of commuting endomorphisms of $V$.*

(a) *The vector space $V$ is a cyclic $\mathscr{F}$-module if and only if $\mathscr{F}^\vee$ is commendable.*
(b) *The vector space $V^*$ is a cyclic $\mathscr{F}^\vee$-module if and only if $\mathscr{F}$ is commendable.*

*Proof* This follows from the theorem, since $V^{**}$ is canonically isomorphic to $V$ and $\mathscr{F}^{\smile\smile}$ is canonically isomorphic to $\mathscr{F}$ by Proposition 3.5.2. $\qquad\square$

As mentioned in Sect. 3.3, one can check the commendability of a family based on Definition 3.3.1. Using the Duality Theorem, we have an alternative way.

---

**Algorithm 3.5.16 (Checking Commendability of a Commuting Family)**
*Let $\mathscr{F}$ be a commuting family, let $\{\varphi_1, \ldots, \varphi_r\}$ be a set of $K$-algebra genera-tors of $\mathscr{F}$, and let $B$ be a $K$-basis of $V$. Then the following algorithm checks whether the family $\mathscr{F}$ is commendable.*

(1) *For $i = 1, \ldots, r$, represent $\varphi_i$ by a matrix $A_i$ with respect to $B$.*
(2) *Apply the Cyclicity Test 3.1.4 to the family $\mathscr{F}^{\smile} = K[\check{\varphi}_1, \ldots, \check{\varphi}_r]$ where $\check{\varphi}_i$ is represented by the matrix $A_i^{\mathrm{tr}}$ with respect to the basis $B^*$ of $V^*$.*
(3) *If $\mathscr{F}^{\smile}$ is cyclic, return* "Commendable" *and stop. Otherwise, return* "Not Commendable".

---

Let us point out the following consequence of the commendability of a commut-ing family.

**Corollary 3.5.17**   *For a commendable family $\mathscr{F}$, we have $\dim_K(\mathscr{F}) = \dim_K(V)$.*

*Proof* Theorem 3.5.12 implies that $V^*$ is an $\mathscr{F}^{\smile}$-cyclic module. Hence Proposi-tion 3.1.1 yields the equality $\dim_K(V^*) = \dim_K(\mathscr{F}^{\smile})$. On the other hand, we have $\dim_K(V) = \dim_K(V^*)$ and $\dim_K(\mathscr{F}) = \dim_K(\mathscr{F}^{\smile})$ by Proposition 3.5.2.a. The combination of these equalities proves the claim. $\qquad\square$

## 3.6  Extended Families

> *The advantage of growing up with siblings*
> *is that you become very good at fractions.*
> (Robert Brault)

On several occasions we have seen that a large base field leads to good properties of a commuting family. Fortunately, in Linear Algebra it is quite easy to extend the base field of a commuting family. It suffices to tensor everything with an extension field $L$ of the given field $K$, including the vector space $V$, the commuting fam-ily $\mathscr{F}$, and the polynomial ring $P$. The result is a better commuting family $\mathscr{F}_L$. For instance, if $L$ is large enough, it contains a splitting endomorphism (see Sect. 2.5).

In this section we prove that the properties of $\mathscr{F}$ to be a commendable family and of $V$ to be a cyclic module over $\mathscr{F}$ are preserved under base field extensions. Even better, if they hold for the extended family $\mathscr{F}_L$, they must be true for the family $\mathcal{F}$ itself. Notice that for both properties we first show the equivalence under some additional assumption and then get rid of this assumption in a clever way. This is a rather subtle result and its proof needs quite a bit of work, but the possibility of enlarging the base field and the family makes it a worthwhile endeavour.

> *There is no cure for laziness*
> *but a large family helps.*
> (Herbert Prochnow)

Let $K$ be a field and $V$ a finite dimensional $K$-vector space of dimension $d$. We use the following notation. If $K \subset L$ is a field extension, we denote $V \otimes_K L$ by $V_L$. For a family of commuting endomorphisms $\mathcal{F}$ on $V$, we let $\mathcal{F}_L = \mathcal{F} \otimes_K L$ and call $\mathcal{F}_L$ the **extended family** of $\mathcal{F}$ with respect to the field extension $K \subset L$. Similarly, the extension of a polynomial ring $P = K[x_1, \ldots, x_n]$ to the coefficient field $L$ is denoted by $P_L = P \otimes_K L = L[x_1, \ldots, x_n]$.

**Lemma 3.6.1** *Let $K \subset L$ be a field extension.*

(a) *If $V$ is a cyclic $\mathcal{F}$-module then $V_L$ is a cyclic $\mathcal{F}_L$-module.*
(b) *Conversely, if $\operatorname{card}(K) > \dim_K(V)$ and $V_L$ is $\mathcal{F}_L$-cyclic, then $V$ is $\mathcal{F}$-cyclic.*

*Proof* Claim (a) can be shown using the Cyclicity Test 3.1.4. In fact, if in Step (4) there exists a tuple $(c_1, \ldots, c_d) \in K^d$ such that $\Delta(c_1, \ldots, c_d) \neq 0$, then the same tuple, viewed as a tuple in $L^d$, shows that the $\mathcal{F}_L$-module $V_L$ is cyclic, too.

To prove (b) we use the Cyclicity Test again. Let $\Delta = \det(C)$ be the determinant of the matrix $C \in \operatorname{Mat}_d(K[z_1, \ldots, z_d])$ computed in Step (4). By assumption, there exists a tuple $(\ell_1, \ldots, \ell_d) \in L^d$ such that $\Delta(\ell_1, \ldots, \ell_d) \neq 0$. In particular, the polynomial $\Delta$ is not identically zero. Since $C$ is a matrix of linear forms, we have $\deg_{z_i}(\Delta) \leq d$ for $i = 1, \ldots, d$. The assumption about the cardinality of $K$ and an easy induction on $d$ imply that there exists a tuple $(c_1, \ldots, c_d) \in K^d$ such that $\Delta(c_1, \ldots, c_d) \neq 0$. From this the claim follows by the Cyclicity Test. □

The following result provides a first step towards an extension of Corollary 1.5.9.

**Lemma 3.6.2** *Let $K$ be a perfect field, let $K \subset L$ be a field extension, and let $\mathcal{F}$ be a commuting family of endomorphisms of $V$. Then the following conditions are equivalent.*

(a) *The family $\mathcal{F}$ is commendable.*
(b) *The family $\mathcal{F}_L$ is commendable.*

*Proof* By Proposition 3.3.7.b, we may assume that $V = \operatorname{BigKer}(\mathfrak{m})$, where $\mathfrak{m}$ is a maximal ideal of $\mathcal{F}$. The fact that $K$ is perfect implies that we have a decomposition $\mathfrak{m}\mathcal{F}_L = \mathfrak{M}_1 \cap \cdots \cap \mathfrak{M}_t$, where $\mathfrak{M}_1, \ldots, \mathfrak{M}_t$ are maximal ideals in $\mathcal{F}_L$

(see [15], Proposition 3.7.18). Using the Chinese Remainder Theorem (see [15], Lemma 3.7.4), we get the equalities

$$\dim_K(\mathscr{F}/\mathfrak{m}) = \dim_L(\mathscr{F}_L/\mathfrak{m}\mathscr{F}_L) = \dim_L(\mathscr{F}_L/\mathfrak{M}_1) + \cdots + \dim_L(\mathscr{F}_L/\mathfrak{M}_t)$$

Clearly, we have $\dim_K(\mathrm{Ker}(\mathfrak{m})) = \dim_L(\mathrm{Ker}(\mathfrak{m}\mathscr{F}_L))$, and Lemma 2.2.12.a implies that $\dim_L(\mathrm{Ker}(\mathfrak{m}\mathscr{F}_L)) = \sum_{i=1}^{t} \dim_L(\mathrm{Ker}(\mathfrak{M}_i))$. Since we also know from Proposition 2.2.13.c that $\dim_L(\mathrm{Ker}(\mathfrak{M}_i)) \geq \dim_L(\mathscr{F}_L/\mathfrak{M}_i)$ for $i = 1, \ldots, t$, we obtain

$$\dim_K(\mathscr{F}/\mathfrak{m}) = \sum_{i=1}^{t} \dim_L(\mathscr{F}_L/\mathfrak{M}_i) \leq \sum_{i=1}^{t} \dim_L\big(\mathrm{Ker}(\mathfrak{M}_i)\big) = \dim_K\big(\mathrm{Ker}(\mathfrak{m})\big)$$

Now the claimed equivalence follows immediately. □

The following lemma provides two elementary linear algebra results related to base field extensions.

**Lemma 3.6.3** Let $K \subseteq L$ be a field extension.

(a) We have a canonical isomorphism $(\mathscr{F}_L)^{\vee} \cong (\mathscr{F}^{\vee})_L$.
(b) We have a canonical isomorphism $(V_L)^* \cong (V^*)_L$.

*Proof* Let $\varphi_1, \ldots, \varphi_r \in \mathscr{F}$ be such that $\mathscr{F} = K[\varphi_1, \ldots \varphi_r]$. Then both $(\mathscr{F}_L)^{\vee}$ and $(\mathscr{F}^{\vee})_L$ can by identified with $L[\check{\varphi}_1, \ldots, \check{\varphi}_r]$. To prove (b) it suffices to take a $K$-basis $B = (v_1, \ldots, v_d)$ of $V$ and its dual basis $B^* = (v_1^*, \ldots, v_d^*)$ of $V^*$. Then both $(V_L)^*$ and $(V^*)_L$ can be identified with $Lv_1^* + \cdots + Lv_d^*$. □

In view of this lemma, we drop the parentheses and simply write $\mathscr{F}_L^{\vee}$ and $V_L^*$. Now we are ready to state and prove the main result of this section.

**Theorem 3.6.4 (Cyclicity and Commendability Under Field Extensions)**
Let $K \subseteq L$ be a field extension. Then we have the following equivalences.

(a) The $\mathscr{F}$-module $V$ is cyclic if and only if the $\mathscr{F}_L$-module $V_L$ is cyclic.
(b) The family $\mathscr{F}$ is commendable if and only if the family $\mathscr{F}_L$ is commendable.

*Proof* First we prove (a). If $V$ is $\mathscr{F}$-cyclic then $V_L$ is $\mathscr{F}_L$-cyclic by Lemma 3.6.1.a. If $\mathrm{card}(K) > \dim_K(V)$, the converse holds by Lemma 3.6.1.b. Now suppose that $\mathrm{card}(K) \leq \dim_K(V)$ and that $V_L$ is $\mathscr{F}_L$-cyclic. By the Duality Theorem 3.5.12, we know that the family $\mathscr{F}_L^{\vee}$ is commendable. Since $K$ is finite, it is a perfect field. Hence we can apply Lemma 3.6.2 and conclude that $\mathscr{F}^{\vee}$ is commendable. Using the Duality Theorem 3.5.12 again, we deduce that the $K$-vector space $V^{**}$ is $\mathscr{F}^{\vee\vee}$-cyclic. By applying the inverses of the isomorphism $\vartheta : V \longrightarrow V^{**}$ and the isomorphism $\delta^{\vee} \circ \delta : \mathscr{F} \longrightarrow \mathscr{F}^{\vee\vee}$ following from parts (a) and (b) of Proposition 3.5.2, it follows that $V$ is $\mathscr{F}$-cyclic.

Now we show (b). If $K$ is perfect then Lemma 3.6.2 yields the claim. If the field $K$ is not perfect, it is infinite. In this case, the Duality Theorem 3.5.12 shows that we need to prove the equivalence between the dual space $V^*$ being a cyclic $\mathscr{F}$-module and $V_L^*$ being a cyclic $\mathscr{F}_L$-module. Since $K$ is infinite this follows from Lemma 3.6.1.                                                                                             $\square$

This theorem is useful in the study of the behaviour of solutions of polynomial systems under base field extensions. We end our discussion of commuting families with a new look at Example 3.3.3.

**Example 3.6.5** Let $V$ be the $\mathbb{Q}$-vector space $V = \mathbb{Q}^4$, and let $\mathscr{F}$ be the commuting family $\mathscr{F} = \mathbb{Q}[\varphi_1, \varphi_2]$, where $\varphi_1$ and $\varphi_2$ are given by the matrices

$$A_1 = \begin{pmatrix} 0 & 2 & 0 & 0 \\ 1 & 0 & 0 & 0 \\ 0 & 0 & 0 & 2 \\ 0 & 0 & 1 & 0 \end{pmatrix} \quad \text{and} \quad A_2 = \begin{pmatrix} 0 & 0 & 0 & 0 \\ 0 & 0 & 0 & 0 \\ 1 & 0 & 0 & 0 \\ 0 & 1 & 0 & 0 \end{pmatrix}$$

respectively. In Example 3.3.3 we saw that $\mathscr{F}$ is a commendable family and that it has one maximal ideal, namely $\mathfrak{m} = \langle \varphi_1^2 - 2\,\mathrm{id}_V, \varphi_2 \rangle$.

Now we apply the base field extension $\mathbb{Q} \subseteq L = \mathbb{Q}(\sqrt{2})$. The family $\mathscr{F}_L$ has two maximal ideals, namely the ideal $\mathfrak{M}_1 = \langle (\varphi_1)_L - \sqrt{2}\,\mathrm{id}_{V_L}, (\varphi_2)_L \rangle$ and the ideal $\mathfrak{M}_2 = \langle (\varphi_1)_L + \sqrt{2}\,\mathrm{id}_{V_L}, (\varphi_2)_L \rangle$. For the extension of $\mathfrak{m}$, we have the equality $\mathfrak{m}\mathscr{F}_L = \mathfrak{M}_1 \cap \mathfrak{M}_2$.

It follows that $\mathscr{F}_L$ has two joint eigenspaces, namely $\mathrm{Ker}(\mathfrak{M}_1) = \langle (0, 0, \sqrt{2}, 1) \rangle$ and $\mathrm{Ker}(\mathfrak{M}_2) = \langle (0, 0, -\sqrt{2}, 1) \rangle$, and two joint generalized eigenspaces, namely $\mathrm{BigKer}(\mathfrak{M}_1) = \langle (\sqrt{2}, 1, 0, 0), (0, 0, \sqrt{2}, 1) \rangle$ as well as $\mathrm{BigKer}(\mathfrak{M}_2) = \langle (-\sqrt{2}, 1, 0, 0), (0, 0, -\sqrt{2}, 1) \rangle$. In particular, the family $\mathscr{F}_L$ is commendable, because we have $\dim_L(\mathscr{F}_L/\mathfrak{M}_i) = 1 = \dim_L(\mathrm{Ker}(\mathfrak{M}_i))$ for $i = 1, 2$. Notice that the property of being commendable has been preserved under the base field extension $\mathbb{Q} \subseteq L$, but the number of maximal ideals and the dimensions of the corresponding residue class fields have changed.

Having reached the end of the third chapter, we conclude our tour through some parts of Linear Algebra that we consider little known or even new. Was it worth the effort to write everything down in this generality? Clearly, we think so, and the remainder of the book is devoted to harvesting the fruits of this labour by applying the techniques we introduced to Computational Commutative Algebra.

*Advice to writers:*
*Sometimes you just have to stop writing.*
*Even before you begin.*
(Stanislaw J. Lec)

# Chapter 4
# Zero-Dimensional Affine Algebras

*"Look! I solved this puzzle in two days!"*
*"So what?"*
*"On the box it says 3–6 years."*

The first puzzle of this section is its title: how can something be both zero-dimensional and interesting enough for an entire chapter? To solve this puzzle, we start by looking at the meaning of "zero". In ancient India the intuitive concept of zero was called *śūnya* which meant "nothing". The Arabs used *sifr* to describe an "empty" position in the decimal system. The Italian mathematician Fibonacci brought the numeral system to the western world and used the homophone Latin word *zephirum*, resembling the name of the Greek god *Zephyrus* of the mild west wind. Later, the Venetians shortened the name to *zero*. Thus the puzzling question is whether zero-dimensional algebras are nothing, empty, or a mild west wind.

The solution, of course, is that none of these interpretations is correct. In the language of Commutative Algebra, the property "zero-dimensional" refers to the Krull dimension of a ring, and it means that every prime ideal is maximal. In our case, the ring is an affine $K$-algebra, which means, a ring of the form $R = P/I$, where $P = K[x_1, \ldots, x_n]$ is a polynomial ring over a field $K$ and $I$ is an ideal in $P$. For such rings, the condition to have Krull dimension zero is equivalent to being a finite dimensional $K$-vector space (cf. [15], Proposition 3.7.1).

*This Rubik's cube is really difficult!*
*So far I can solve only five faces.*

The second puzzling question you may ask yourself is what do zero-dimensional affine algebras have to do with the commuting families considered in the preced-

**Electronic supplementary material** The online version of this chapter (DOI:10.1007/978-3-319-43601-2_4) contains supplementary material, which is available to authorized users.

M. Kreuzer and L. Robbiano, *Computational Linear and Commutative Algebra*,
DOI 10.1007/978-3-319-43601-2_4

ing chapters? We can solve this even faster than the world record for Rubik's cubes which stands at 4.90 seconds in 2016. Since a zero-dimensional affine $K$-algebra $R$ is a finite dimensional $K$-vector space, it suffices to observe that the set of multiplication maps $\mathscr{F} = \{\vartheta_f \mid f \in R\}$, where $\vartheta_f : R \longrightarrow R$ is given by $\vartheta_f(g) = fg$, form a commuting family of endomorphisms of $R$. Actually, this multiplication family is a particularly nice commuting family: the ring $R$ is a cyclic $\mathscr{F}$-module, and there is a canonical $K$-algebra isomorphism $\eta : \mathscr{F} \longrightarrow R$ given by $\vartheta_f \mapsto f$. Using this isomorphism, we can go back and forth between $\mathscr{F}$ and $R$. Playing this game over and over, we can translate the results of Chaps. 2 and 3 into statements about the ring $R$. After our extensive preparatory work, some puzzles will solve themselves!

*Rubik's cube: more positions than you have brain cells.*

Before plunging into the discussion of multiplication families, let us briefly have a look at the hypotheses underlying the theorems and algorithms in this chapter and the following ones. Superficially, one could get the impression that many very difficult problems can be solved entirely by Linear Algebra methods. This is not the whole story, since our starting assumption is always that we know a vector space basis and that we can represent all elements and endomorphisms effectively in terms of this basis. But how do we get such a basis?

To be given a zero-dimensional affine $K$-algebra $R = P/I$ as above usually means to be given a system of generators of the ideal $I$. Then several well-known algorithms can be used to find a set of polynomials $\mathcal{O}$ whose residue classes form a $K$-basis of $R$, for instance Buchberger's Algorithm to compute a Gröbner basis (see [15], Theorem 2.5.5), the Border Basis Algorithm (see [16], Theorem 6.4.36), or, if $I$ is the vanishing ideal of a set of points, the Buchberger-Möller Algorithm (see [16], Theorem 6.3.10). Notice that the set $\mathcal{O}$ is an order ideal of terms in these cases, i.e., a set of terms such that $t \in \mathcal{O}$ and $\tilde{t} \mid t$ implies $\tilde{t} \in \mathcal{O}$, and that we have efficient methods of representing the residue classes in $R$ in terms of $\mathcal{O}$. Usually, the calculation of $\mathcal{O}$ is the hardest part of solving a problem, and neither our brain cells nor the memory cells of our computer may suffice to do it.

*Our whole life is solving puzzles.*
(Erno Rubik)

After all this yackety-yak, it is time to talk about the actual content of the chapter. Many of the constructions we introduced for commuting families translate to well-known concepts in ring theory when we apply them to the multiplication family $\mathscr{F}$ of a zero-dimensional affine $K$-algebra $R$. The kernel of an ideal $\tilde{\mathfrak{a}}$ in $\mathscr{F}$ becomes the annihilator $\operatorname{Ann}_R(\mathfrak{a})$ of the corresponding ideal in $R$, the big kernel of $\tilde{\mathfrak{a}}$ becomes the big annihilator $\operatorname{BigAnn}_R(\mathfrak{a}) = \bigcup_{i \geq 1} \operatorname{Ann}_R(\mathfrak{a}^i)$, and the ideals $\tilde{\mathfrak{q}}_i = \operatorname{Ann}_{\mathscr{F}}(\operatorname{BigKer}(\tilde{\mathfrak{m}}_i))$ of $\mathscr{F}$ become the primary components $\mathfrak{q}_i$ of the zero ideal in $R$, where $\tilde{\mathfrak{m}}_1, \ldots, \tilde{\mathfrak{m}}_s$ are the maximal ideals in $\mathscr{F}$.

Next we examine the puzzling question of what the ring-theoretic view of the joint generalized eigenspace decomposition $R = \operatorname{BigKer}(\tilde{\mathfrak{m}}_1) \oplus \cdots \oplus \operatorname{BigKer}(\tilde{\mathfrak{m}}_s)$

might be. It turns out that the big kernels $\mathrm{BigKer}(\tilde{\mathfrak{m}}_i)$ are isomorphic to the local factors $R/\mathfrak{q}_i$ of $R$, and the decomposition corresponds to the ring isomorphism $R \cong R/\mathfrak{q}_1 \times \cdots \times R/\mathfrak{q}_s$ given by the Chinese Remainder Theorem.

So, what about individual joint eigenvectors? As we show in Theorem 4.2.11, a joint eigenvector $f \in \mathrm{Ker}(\tilde{\mathfrak{m}}_i)$ is nothing but a *separator* of $R$, i.e., an element whose principal ideal has the smallest possible dimension $\dim_K(R/\mathfrak{m}_i)$. Under the decomposition of $R$ into local factors, these elements correspond to a socle element in $R/\mathfrak{q}_i$ and to zero in all $R/\mathfrak{q}_j$ with $j \neq i$.

Getting better and better at playing this game, we then consider the question of when the multiplication family $\mathcal{F}$ contains a commendable endomorphism. In Sect. 3.2 we learned that this condition has far-reaching consequences for $\mathcal{F}$. In the language of Commutative Algebra, it is related to $R$ being weakly curvilinear. If the maximal ideals of $R$ are linear and $K$ is not too small, it is equivalent to $R$ being curvilinear, i.e., to the local factors being of the form $R/\mathfrak{q}_i = K[z]/\langle z^{m_i} \rangle$ with $m_i \geq 1$.

Of course, this result inspires the next puzzling question. How is the commendability of the family $\mathcal{F}$ reflected in the ring structure of $R$? To solve this, we employ the Duality Theorem 3.5.12 which says that $\mathcal{F}$ is commendable if and only if $\mathcal{F}^{\vee}$ is cyclic. This leads us yet another puzzle: what is the algebraic interpretation of $\mathcal{F}^{\vee}$? Traditionally, the elements of $\mathcal{F}^{\vee}$ have been interpreted using partial derivatives on $P$ which vanish on $I$. Since we are looking for a characteristic-free approach, we use the canonical module $\omega_R = \mathrm{Hom}_R(R, K)$ instead, and note that the elements of $\mathcal{F}^{\vee}$ are nothing but the multiplication maps on $\omega_R$. Hence the family $\mathcal{F}$ is commendable if and only if $\omega_R$ is a cyclic $R$-module, and the latter condition is equivalent to $R$ being a Gorenstein ring. In particular, by applying the Cyclicity Test 3.1.4 to $\mathcal{F}^{\vee}$, we can check whether $R$ is a Gorenstein ring. More generally, to get more information about the module structure of $\omega_R$, we use an idea similar to the construction of a special basis in Proposition 3.4.10 and devise an algorithm to compute the minimal generators of $\omega_R$ (see Algorithm 4.5.14).

> *This video game puzzle has ruined my life.*
> *Now I have only two left.*

Turning our attention to the case of a local ring $R$, or, equivalently, a local multiplication family $\mathcal{F}$, we find that the maximal ideal $\mathfrak{m}$ of $R$ is nilpotent, and a large part of the ring structure is encoded in the filtration of $R$ given by the powers of $\mathfrak{m}$. So, every $f \in R$ has an index of nilpotency $\mathrm{nix}(f) = \min\{i \geq 1 \mid f^i = 0\}$ which is bounded by $\mathrm{nix}(\mathfrak{m}) = \min\{i \geq 1 \mid \mathfrak{m}^i = \langle 0 \rangle\}$. Notice that *Nix* is also one of the moons of Pluto, named after the Greek goddess of darkness, very far away and extremely small. Although Nix certainly does not sustain any life, we will be looking for linear elements $\ell$ of $R$ having a maximal $\mathrm{nix}(\ell)$, i.e., elements trying to keep their life as long as possible. After travelling a long distance, and assuming a suitable base field, we can show that they exist in abundance (see Theorem 4.4.12).

Finally, in the last section of this chapter, we have a brief glance at the Algebraic Geometry point of view. The ring $R$ is the coordinate ring of a zero-dimensional subscheme of an affine space over $K$. Many geometric properties of this scheme are represented by ring-theoretic properties of $R$, and hence by properties of the multiplication family. Besides introducing the affine Hilbert function, we provide algorithms to check for the well-known Cayley-Bacharach property.

Despite all of these successes, there are some puzzles which have no general solution. What is a commendable partner? Should families multiply? Will the interest rate be zero forever? Is this going to destroy the economy?

*The only function of economic forecasting*
*is to make astrology look respectable.*
(John Kenneth Galbraith)

## 4.1 Multiplication Endomorphisms

*Some guy hit my fender and I told him,*
*"Be fruitful and multiply,"*
*but not in those words.*
(Woody Allen)

To multiply by a fixed element in a 0-dimensional affine algebra can be a very fruitful undertaking. Since our rings are commutative, the various multiplications provide us with a commuting family, and we can use the knowledge gathered in the preceding three chapters. This section provides the dictionary between the world of commuting families and the world of commutative algebras. Kernels will be annihilators, big kernels will be big annihilators, and joint generalized eigenspaces will be local factors. And although, as it is generally known, the original is unfaithful to the translation, we will get useful and deep results which lend themselves to further interpretations and extensions in the subsequent sections.

In the following we let $K$ be a field and $R$ a zero-dimensional affine $K$-algebra, i.e., a $K$-algebra of the form $R = P/I$ with a polynomial ring $P = K[x_1, \ldots, x_n]$ and a zero-dimensional ideal $I$ in $P$. Thus $R$ is a finite dimensional $K$-vector space, and we let $d = \dim_K(R)$. The following commuting family will play a central role in this chapter.

**Definition 4.1.1** Let $R$ be a zero-dimensional affine $K$-algebra.

(a) For every $f \in R$, the multiplication by $f$ yields a $K$-linear map $\vartheta_f : R \longrightarrow R$ such that $\vartheta_f(g) = f \cdot g$ for all $g \in R$. It is called the **multiplication endomorphism** by $f$ on $R$.

(b) The family $\mathscr{F} = K[\vartheta_f \mid f \in R]$ is called the **multiplication family** of $R$.

Since $R$ is a commutative ring, the family $\mathscr{F}$ is a family of commuting endomorphisms of $R$. Our next proposition collects some properties of this family.

**Proposition 4.1.2 (Basic Properties of Multiplication Families)**
*Let $R = P/I$ be a zero-dimensional affine $K$-algebra, where $P = K[x_1, \ldots, x_n]$
and $I$ is an ideal in $P$, and let $\mathscr{F}$ be the multiplication family of $R$.*

(a) *We have $\mathscr{F} = K[\vartheta_{x_1}, \ldots, \vartheta_{x_n}]$, where $\vartheta_{x_i}$ denotes the multiplication by $x_i + I$
    on $R$.*
(b) *The map $\iota : R \longrightarrow \mathscr{F}$ given by $f \mapsto \vartheta_f$ is an isomorphism of $K$-algebras. Its
    inverse is the map $\eta : \mathscr{F} \longrightarrow R$ given by $\varphi \mapsto \varphi(1)$.*
(c) *Let $\Phi = (\vartheta_{x_1}, \ldots, \vartheta_{x_n})$. If we represent the family $\mathscr{F}$ as $\mathscr{F} \cong P/\mathrm{Rel}_P(\Phi)$,
    where $\vartheta_{x_i} \mapsto x_i + \mathrm{Rel}_P(\Phi)$, we have $\mathrm{Rel}_P(\Phi) = I$.*
(d) *Via $\eta$, the $K$-vector space $R$ is a cyclic $\mathscr{F}$-module generated by $\{1\}$.*
(e) *The map $\Theta : \mathrm{End}_K(R) \longrightarrow \mathrm{End}_K(\mathscr{F})$ defined by $\Theta(\varphi) = \iota \circ \varphi \circ \eta$ is an isomor-
    phism of $K$-algebras.*

*Proof* To prove (a), it suffices to observe that $\vartheta_{f(x_1, \ldots, x_n)} = f(\vartheta_{x_1}, \ldots, \vartheta_{x_n})$ for
all $f \in P$. Next we show (b). The map $\iota$ is injective, since $\vartheta_f = 0$ implies
$f = \vartheta_f(1) = 0$, and it is surjective by (a). The fact that $\iota$ and $\eta$ are inverse to each
other follows from $(\eta \circ \iota)(f) = \eta(\vartheta_f) = f$ for $f \in R$ and $(\iota \circ \eta)(\vartheta_f) = \iota \circ \vartheta_f(1) =
\iota(f) = \vartheta_f$. Claim (c) follows immediately from (b). Claim (d) follows from (a)
and (b), and claim (e) is a special case of Proposition 3.1.10.                        $\square$

The isomorphism in part (b) of this proposition allows several useful identifica-
tions.

**Remark 4.1.3** Let $R$ be a zero-dimensional affine $K$-algebra, and let $\iota : R \longrightarrow \mathscr{F}$
be the isomorphism of part (b) of the preceding proposition.

(a) The minimal polynomial of an element $f$ of $R$ agrees with the minimal polyno-
    mial of the corresponding multiplication map $\iota(f) = \vartheta_f$.
(b) Let $V$ be a finite dimensional $K$-vector space, and let $\mathscr{F}$ be a family of com-
    muting endomorphisms of $V$. Then $\mathscr{F}$ is a zero-dimensional affine $K$-algebra.
    Let $\widetilde{\mathscr{F}}$ be the multiplication family of $\mathscr{F}$. By (a), the minimal polynomial of
    an element $\varphi$ of $\mathscr{F}$ agrees with the minimal polynomial of the corresponding
    multiplication map $\vartheta_\varphi \in \widetilde{\mathscr{F}}$. Notice that the analogous statement for the char-
    acteristic polynomial does not hold, as Example 1.5.11 shows. There we have
    $\mu_\varphi(z) = \mu_{\vartheta_\varphi}(z) = \chi_{\vartheta_\varphi} = z^2$ and $\chi_\varphi(z) = z^3$.
(c) Using the isomorphism $\iota : \mathscr{F} \longrightarrow \widetilde{\mathscr{F}}$, we can view $\widetilde{\mathscr{F}}$, and hence $\mathscr{F}$, as a cyclic
    $\mathscr{F}$-module.

Given a $K$-basis of $R$, we can represent multiplication maps by matrices which
we denote as follows.

**Definition 4.1.4** Let $\mathscr{F}$ be the multiplication family of a zero-dimensional affine
$K$-algebra $R$, let $B = (t_1, \ldots, t_d)$ be a $K$-basis of $R$, and let $f \in R$. Then the matrix
$M_B(\vartheta_f) \in \mathrm{Mat}_d(K)$ which represents $\vartheta_f$ with respect to the basis $B$ is called the
**multiplication matrix** of $f$ with respect to $B$.

Multiplication matrices are not necessarily triangularizable and can be singular, as the following examples show.

**Example 4.1.5**  Let $K = \mathbb{Q}$, let $R = \mathbb{Q}[x]/\langle x^2 + 1 \rangle$, and let $f = x$. Then the tuple $B = (1, x)$ represents a $\mathbb{Q}$-basis of $R$ and the multiplication matrix of $f$ with respect to this basis is $M_B(\vartheta_f) = \begin{pmatrix} 0 & -1 \\ 1 & 0 \end{pmatrix}$. Since its minimal polynomial is $z^2 + 1$, this multiplication matrix is not triangularizable over $\mathbb{Q}$.

**Example 4.1.6**  Let $K = \mathbb{Q}$, let $R = \mathbb{Q}[x]/\langle x^2 - x \rangle$, and let $f = x$. Then the tuple $B = (1, x)$ represents a $\mathbb{Q}$-basis of $R$. The multiplication matrix of $f$ with respect to this basis is the singular matrix $\begin{pmatrix} 0 & 0 \\ 1 & 1 \end{pmatrix}$.

The following remark points out some possibilities for getting $K$-bases of $R$ and the corresponding multiplication matrices from a description which uses $K$-algebra generators and relations.

**Remark 4.1.7**  Let $R$ be a zero-dimensional affine $K$-algebra, and let $R = P/I$ be a representation of $R$ as a residue class ring of a polynomial ring $P = K[x_1, \ldots, x_n]$ by an explicitly given ideal $I$. We can find a $K$-basis of $R$ in one of the following ways.

(a) Let $\sigma$ be a term ordering on $\mathbb{T}^n$. If we compute a $\sigma$-Gröbner basis of $I$, the leading terms of its elements generate the leading term ideal $\mathrm{LT}_\sigma(I)$. Then Macaulay's Basis Theorem (see [15], Theorem 1.5.7) says that the residue classes of the terms in $\mathcal{O}_\sigma(I) = \mathbb{T}^n \setminus \mathrm{LT}_\sigma(I)$ form a $K$-basis of $I$.

(b) Let $\mathcal{O} = \{t_1, \ldots, t_d\} \subset \mathbb{T}^n$ be an order ideal (see Definition 3.4.9). If the ideal $I$ has an $\mathcal{O}$-border basis (see [16], Sect. 6.4), the elements of the set $\mathcal{O}$ represent a $K$-basis of $R$. For instance, border bases can be found using the Border Basis Algorithm (see [16], Theorem 6.4.36 or [28], Sect. 10.2).

In both cases we have algorithms which allow us to express residue classes in $P/I$ in terms of the basis $\mathcal{O}$. In particular, we can compute the multiplication matrices of $R$ with respect to $\mathcal{O}$.

Let us see an example in which we get different $K$-bases for $R$.

**Example 4.1.8**  Over the field $K = \mathbb{Q}$, let $P = K[x, y]$, let $I$ be the ideal in $P$ defined by $I = \langle xy + \frac{1}{2}y^2 - \frac{1}{2}x, x^2 + x, y^3 - \frac{3}{2}y^2 - \frac{1}{2}x \rangle$, and let $R = P/I$.

(a) Using the term ordering $\sigma = \mathtt{DegRevLex}$, we get $\mathrm{LT}_\sigma(I) = \langle x^2, xy, y^3 \rangle$. Therefore $B = (1, x, y, y^2)$ is a tuple of terms whose residue classes form a $K$-basis of $R$. With respect to this bases we have the multiplication matrices

$$M_B(\vartheta_x) = \begin{pmatrix} 0 & 0 & 0 & 0 \\ 1 & -1 & \frac{1}{2} & 0 \\ 0 & 0 & 0 & 0 \\ 0 & 0 & -\frac{1}{2} & -1 \end{pmatrix} \quad \text{and} \quad M_B(\vartheta_y) = \begin{pmatrix} 0 & 0 & 0 & 0 \\ 0 & \frac{1}{2} & 0 & \frac{1}{2} \\ 1 & 0 & 0 & 0 \\ 0 & -\frac{1}{2} & 1 & \frac{3}{2} \end{pmatrix}.$$

(b) The ideal $I$ has a border basis with respect to the order ideal $\mathcal{O} = (1, x, y, xy)$, namely $(x^2 + x, x^2y + xy, xy^2 + x - 2xy, y^2 - x + 2xy)$. Hence $\mathcal{O}$ yields a $K$-basis of $R$. The corresponding multiplication matrices are

$$M_{\mathcal{O}}(\vartheta_x) = \begin{pmatrix} 0 & 0 & 0 & 0 \\ 1 & -1 & 0 & 0 \\ 0 & 0 & 0 & 0 \\ 0 & 0 & 1 & -1 \end{pmatrix} \quad \text{and} \quad M_{\mathcal{O}}(\vartheta_y) = \begin{pmatrix} 0 & 0 & 0 & 0 \\ 0 & 0 & 1 & -1 \\ 1 & 0 & 0 & 0 \\ 0 & 1 & -2 & 2 \end{pmatrix}$$

In analogy with the definition of the big kernel of an endomorphism, we can define the big annihilator of an ideal as follows.

**Definition 4.1.9** Let $R$ be a ring, let $M$ be an $R$-module, and let $\mathfrak{a}$ be an ideal in $R$. Then the $R$-submodule

$$\text{BigAnn}_M(\mathfrak{a}) = \bigcup_{i=1}^{\infty} \text{Ann}_M(\mathfrak{a}^i) = \{m \in M \mid \mathfrak{a}^i \cdot m = 0 \text{ for some } i \geq 1\}$$

of $M$ is called the **big annihilator** of $\mathfrak{a}$ in $M$.

Given generators of an ideal, its big annihilator can be described and computed in several ways, as the following remark shows.

**Remark 4.1.10** In the setting of the definition, let $f_1, \ldots, f_m \in R$ be such that we have $\mathfrak{a} = \langle f_1, \ldots, f_m \rangle$. It is easy to check that $\text{BigAnn}_M(\mathfrak{a})$ can be obtained in one of the following ways.

(a) We have $\text{BigAnn}_M(\mathfrak{a}) = \bigcup_{i \geq 1} \text{Ann}_M(\langle f_1^i, \ldots, f_m^i \rangle)$.
(b) There exists a number $i \geq 1$ such that $\text{BigAnn}_M(\mathfrak{a}) = \text{Ann}_M(\langle f_1^j, \ldots, f_m^j \rangle)$ for all $j \geq i$.
(c) For $i = 1, \ldots, m$, there exists a number $v_i \geq 1$ such that

$$\text{Ann}_M(f_i) \subset \text{Ann}_M(f_i^2) \subset \cdots \subset \text{Ann}_M(f_i^{v_i}) = \text{Ann}_M(f_i^{v_i+1}) = \cdots$$

and we have $\text{BigAnn}_M(\mathfrak{a}) = \bigcap_{i=1}^{m} \text{Ann}_M(f_i^{v_i})$.

The next proposition uses the isomorphisms $\iota$ and $\eta$ of Proposition 4.1.2.b to translate the concepts of kernels and big kernels into the language of rings.

**Proposition 4.1.11 (Kernels as Annihilators)**
Let $\mathcal{F}$ be the multiplication family of $R$, let $\mathfrak{a}$ be an ideal in $R$, and let $\tilde{\mathfrak{a}} = \iota(\mathfrak{a})$ be the corresponding ideal in $\mathcal{F}$.

(a) For every $f \in R$, we have $\text{Ker}(\vartheta_f) = \text{Ann}_R(f)$ and $\text{BigKer}(\vartheta_f) = \text{BigAnn}_R(f)$.
(b) We have $\text{Ker}(\tilde{\mathfrak{a}}) = \text{Ann}_R(\mathfrak{a})$.
(c) We have $\text{BigKer}(\tilde{\mathfrak{a}}) = \text{BigAnn}_R(\mathfrak{a})$.

*Proof* To prove (a) it suffices to note that $\mathrm{Ker}(\vartheta_f) = \{g \in R \mid fg = 0\} = \mathrm{Ann}_R(f)$.
Next we prove (b). Let $f_1, \ldots, f_m \in R$ be elements such that $\mathfrak{a} = \langle f_1, \ldots, f_m \rangle$.
Then the endomorphisms $\vartheta_{f_1}, \ldots, \vartheta_{f_m}$ generate $\tilde{\mathfrak{a}}$. By Proposition 2.2.10.c and (a),
we get

$$\mathrm{Ker}(\tilde{\mathfrak{a}}) = \bigcap_{i=1}^{m} \mathrm{Ker}(\vartheta_{f_i}) = \bigcap_{i=1}^{m} \mathrm{Ann}_R(f_i) = \mathrm{Ann}_R(\mathfrak{a})$$

Finally, claim (c) follows from $\mathrm{BigKer}(\tilde{\mathfrak{a}}) = \bigcup_{i=1}^{\infty} \mathrm{Ker}(\tilde{\mathfrak{a}}^i)$ and (b).                    □

By Theorem 2.4.3, the ring $R$ has a joint generalized eigenspace decomposition
with respect to the multiplication family $\mathscr{F}$. Next we examine this decomposition
closely and interpret it in the language of commutative algebra.

**Proposition 4.1.12 (Joint Generalized Eigenspaces of a Multiplication Family)**
*Let $R$ be a zero-dimensional affine $K$-algebra with maximal ideals $\mathfrak{m}_1, \ldots, \mathfrak{m}_s$,
and let $\mathscr{F}$ be its multiplication family. For $i = 1, \ldots, s$, we let $\tilde{\mathfrak{m}}_i = \iota(\mathfrak{m}_i)$, we let
$\tilde{\mathfrak{q}}_i = \mathrm{Ann}_{\mathscr{F}}(\mathrm{BigKer}(\tilde{\mathfrak{m}}_i))$, and we define $\mathfrak{q}_i = \eta(\tilde{\mathfrak{q}}_i)$.*

(a) *We have $\mathfrak{q}_1 \cap \cdots \cap \mathfrak{q}_s = \langle 0 \rangle$, and there exist integers $a_i \geq 1$ such that $\mathfrak{q}_i = \mathfrak{m}_i^{a_i}$.
    In particular, we have $\mathrm{Rad}(\mathfrak{q}_i) = \mathfrak{m}_i$ for $i = 1, \ldots, s$.*
(b) *For $i = 1, \ldots, s$, the joint generalized eigenspace $U_i = \mathrm{BigKer}(\tilde{\mathfrak{m}}_i)$ is a princi-
    pal ideal of $R$ generated by an idempotent, and we have $R = U_1 \times \cdots \times U_s$.*
(c) *We have canonical isomorphisms of $K$-algebras*

$$\mathrm{BigKer}(\tilde{\mathfrak{m}}_i) = U_i \cong \mathscr{F}_{U_i} \cong \mathscr{F} / \mathrm{Ann}_{\mathscr{F}}(U_i) = \mathscr{F} / \tilde{\mathfrak{q}}_i \cong R / \mathfrak{q}_i.$$

(d) *If we view $U_i = \mathrm{BigKer}(\tilde{\mathfrak{m}}_i)$ as a ring, it is local with maximal ideal $U_i \cap \mathfrak{m}_i$.
    Moreover, the ring $R/\mathfrak{q}_i$ is local with maximal ideal $\mathfrak{m}_i/\mathfrak{q}_i$.*
(e) *For $i = 1, \ldots, s$, we have $\mathfrak{q}_i = \mathrm{Ann}_R(U_i)$ and $U_i = \mathrm{Ann}_R(\mathfrak{q}_i)$.*

*Proof* Claim (a) follows from parts (b) and (d) of Proposition 2.4.6 by applying the
isomorphism $\eta$.

To prove (b), we note that the vector subspaces $U_i$ are ideals in $R$ by Proposition 4.1.11.c and the fact the annihilators of ideals in $R$ are ideals in $R$. Thus Theorem 2.4.3 provides the desired decomposition, and this is in fact a direct product
of rings. Hence the ideal $U_i$ is generated by the idempotent $(0, \ldots, 0, 1, 0, \ldots, 0)$
where the number 1 occurs in the $i$-th position.

Next we show (c). Corollary 3.1.9 says that $U_i$ is a cyclic $\mathscr{F}_{U_i}$-module. Thus
we have an isomorphism $\iota_{U_i} : U_i \longrightarrow \mathscr{F}_{U_i}$, and the remaining isomorphisms of
claim (c) follow from Corollary 2.1.10 and Proposition 2.4.6.c.

Furthermore, we note that claim (d) is a consequence of (a) and (c). Finally, the
first claim in (e) follows from the definition of $\mathfrak{q}_i$ and the isomorphism $\eta$, and the
second claim in (e) follows from (c).                                               □

In view of the preceding proposition, the joint generalized eigenspace decompo-
sition $R \cong \prod_{i=1}^{s} \mathrm{BigKer}(\tilde{\mathfrak{m}}_i) \cong \prod_{i=1}^{s} R/\mathfrak{q}_i$ deserves a special name.

**Definition 4.1.13** In the setting of Proposition 4.1.12, the decomposition

$$R \cong \prod_{i=1}^{s} \operatorname{BigKer}(\tilde{\mathfrak{m}}_i) \cong \prod_{i=1}^{s} R/\mathfrak{q}_i$$

is called the **decomposition** of $R$ **into local rings**. The rings $R/\mathfrak{q}_i$ are also called the **local factors** of $R$.

Next we see that the ideals $\mathfrak{m}_i$ and $\mathfrak{q}_i$ can be described as annihilator ideals in $R$.

**Corollary 4.1.14** *In the setting of the proposition, let $i \in \{1, \ldots, s\}$.*

(a) *We have $\mathfrak{m}_i = \operatorname{Ann}_R(\operatorname{Ann}_R(\mathfrak{m}_i))$.*
(b) *We have $\mathfrak{q}_i = \operatorname{Ann}_R(\operatorname{BigAnn}_R(\mathfrak{m}_i))$.*

*Proof* The proof of (a) follows from Proposition 2.4.6.a. To prove (b) we observe that Proposition 4.1.11.c yields $\tilde{\mathfrak{q}}_i = \operatorname{Ann}_{\mathscr{F}}(\operatorname{BigAnn}_R(\mathfrak{m}_i))$, and the conclusion follows. □

*A meeting is an event*
*at which minutes are kept*
*and hours are lost.*

## 4.2 Primary Decomposition and Separators

*The primary ideal of the primaries*
*is to choose the political joke*
*with the best chances to be elected.*

In this section we introduce the notion of a primary ideal. This is no joke, but a fundamental construction in Commutative Algebra. For ideals in polynomial rings, the decomposition into primary ideals was first proven by Emanuel Lasker in 1905 (see [17]). Nowadays, a general description can be found in most books on Commutative Algebra (see for instance [15], 5.6.B). In the next two chapters we will study methods for computing primary decompositions and use them to solve polynomial systems. Thus primary ideals may very well be chosen to be the key for understanding polynomial ideals and their zeros.

In the following we let $K$ be a field, $P = K[x_1, \ldots, x_n]$, and $I$ a zero-dimensional ideal in $P$. We start by connecting the decomposition of the zero-dimensional affine $K$-algebra $R = P/I$ into local rings with the primary decomposition of the zero ideal of $R$. For this purpose we introduce the following terminology.

**Definition 4.2.1** Let $S$ be a ring.

(a) An ideal $I$ in $S$ is called **primary** if $fg \in I$ for $f, g \in S$ implies that we have $f \in I$ or $g \in \mathrm{Rad}(I)$.
(b) A representation of an ideal $I$ in $S$ as a finite intersection of primary ideals is called a **primary decomposition** of $I$.

Primary ideals can be characterized as follows.

**Lemma 4.2.2** *Let $S$ be a ring, and let $I$ be an ideal in $S$.*

(a) *If the ideal $I$ is primary then $\mathrm{Rad}(I)$ is a prime ideal.*
(b) *If $\mathfrak{m}$ is a maximal ideal of $S$ and $\mathrm{Rad}(I) = \mathfrak{m}$ then $I$ is primary.*

*Proof* To prove (a), let $f, g \in S$ such that $fg \in \mathrm{Rad}(I)$. Then there exists an integer $a \geq 1$ such that $f^a g^a \in I$. If we assume that $g \notin \mathrm{Rad}(I)$, and thus $g^a \notin \mathrm{Rad}(I)$, then Definition 4.2.1 implies $f^a \in I$, and hence $f \in \mathrm{Rad}(I)$.

To show (b) we observe that if $f, g \in S$ satisfy $fg \in I$ and $g \notin \mathfrak{m} = \mathrm{Rad}(I)$, then no maximal ideal of $S$ contains both $I$ and $g$, as it would contain both $\mathrm{Rad}(I) = \mathfrak{m}$ and $g \notin \mathfrak{m}$. Hence the two ideals $I$ and $\langle g \rangle$ are comaximal, i.e. there exist elements $h \in I$ and $s \in S$ such that $1 = h + sg$. Therefore we get $f = hf + sfg \in I$.      □

In particular, in the case of a zero-dimensional ring $R$, every prime ideal is maximal. Thus, for every primary ideal $I$ in $R$, its radical is a maximal ideal $\mathfrak{m} = \mathrm{Rad}(I)$. In this case we say that $I$ is an $\mathfrak{m}$-**primary** ideal. Now we can interpret and extend Proposition 4.1.12.a as follows.

**Proposition 4.2.3 (Primary Decomposition)**
*Let $R$ be a zero-dimensional affine $K$-algebra with maximal ideals $\mathfrak{m}_1, \ldots, \mathfrak{m}_s$, and let $\mathfrak{q}_i = \mathrm{Ann}_R(\mathrm{BigAnn}_R(\mathfrak{m}_i))$ for $i = 1, \ldots, s$.*

(a) *For $i = 1, \ldots, s$, the ideal $\mathfrak{q}_i$ is $\mathfrak{m}_i$-primary.*
(b) *We have $\langle 0 \rangle = \mathfrak{q}_1 \cap \cdots \cap \mathfrak{q}_s$. This intersection is a primary decomposition of the ideal $\langle 0 \rangle$.*
(c) *We have $\mathrm{Rad}(0) = \mathfrak{m}_1 \cap \cdots \cap \mathfrak{m}_s$.*
(d) *The intersection in (b) is unique in the following sense. If $\mathfrak{q}_1', \ldots, \mathfrak{q}_s'$ are ideals in $R$ such that $\mathfrak{q}_i'$ is $\mathfrak{m}_i$-primary for $i = 1, \ldots, s$, and if $\mathfrak{q}_1' \cap \cdots \cap \mathfrak{q}_s' = \langle 0 \rangle$, then we have $\mathfrak{q}_i' = \mathfrak{q}_i$ for $i = 1, \ldots, s$.*
(e) *For every $i \in \{1, \ldots, s\}$, there exists a number $a_i \geq 1$ such that $\mathfrak{q}_i = (\mathfrak{m}_i)^j$ for every $j \geq a_i$.*

*Proof* In order to show (a), (b) and (c) it suffices to note that $\mathfrak{q}_i$ is the image of $\tilde{\mathfrak{q}}_i = \mathrm{Ann}_{\mathscr{F}}(\mathrm{BigKer}(\mathfrak{m}_i))$ in $R$ and to apply Proposition 4.1.12.a and part (b) of the lemma.

Now we prove claim (d). Using the decomposition of $R$ into local rings of Proposition 4.1.12.c, we see that $\mathfrak{m}_i$ is of the form $\prod_{j=1}^s \mathfrak{m}_{ij}$ with $\mathfrak{m}_{ij} = \langle 1 \rangle$ for $j \neq i$ and the unique maximal ideal $m_{ii}$ of $\mathrm{BigKer}(\tilde{m}_i)$ in the $i$-th component. Since the ideal $q_i'$ is $m_i$-primary, it follows that the decomposition of $q_i'$ is of the form $\prod_{j=1}^s q_{ij}'$ with $q_{ij}' = \langle 1 \rangle$ for $j \neq i$ and an $m_{ii}$-primary ideals $q_{ii}'$ in the $i$-th component. Now the assumption $q_1' \cap \cdots \cap q_s' = \langle 0 \rangle$ yields $q_{ii}' = \langle 0 \rangle$, and thus $q_i' = q_i$ for $i = 1, \ldots, s$.

Finally, claim (e) follows from Proposition 4.1.12.a and the decomposition of $R$ into local rings. $\qquad\square$

Since the primary decomposition of an ideal is an important construction in Commutative Algebra, we introduce the following terminology.

**Definition 4.2.4** Let $R$ be a zero-dimensional affine $K$-algebra, assume that $\langle 0 \rangle = q_1 \cap \cdots \cap q_s$ is the primary decomposition of the zero ideal of $R$, and let $\mathrm{Rad}(0) = m_1 \cap \cdots \cap m_s$ be the corresponding primary decomposition of the zero radical of $R$.

(a) The ideals $q_1, \ldots, q_s$ are called the **primary components** of the zero ideal of $R$. More precisely, for $i = 1, \ldots, s$, the ideal $q_i$ is called the $m_i$-**primary component** of the zero ideal of $R$.
(b) The ideals $m_1, \ldots, m_s$ are called the **maximal components** of the zero ideal of $R$.

In general, the radical ideals of the primary components of an ideal are called its **prime components**. In the zero-dimensional case, all non-zero prime ideals are maximal. Therefore the prime components of the zero ideal of $R$ are also called its maximal components, justifying the definition. In Chap. 5 we discuss methods for computing the primary and maximal components in the zero-dimensional case. The following example illustrates part (e) of the proposition.

**Example 4.2.5** Let $K = \mathbb{Q}$, let $P = K[x, y]$, let $I = \langle x, y^2 \rangle \cap \langle x - 1, y \rangle$, and let $R = P/I$. We have the equality $I = \langle xy, x^2 - x, y^2 \rangle$, the ring $R$ is a zero-dimensional affine $K$-algebra, and the primary decomposition of its zero ideal is $\langle 0 \rangle = q_1 \cap q_2$ with $q_1 = \langle \bar{x}, \bar{y}^2 \rangle$ and $q_2 = \langle \bar{x} - 1, \bar{y} \rangle$.

For the maximal ideal $m_1 = \langle \bar{x}, \bar{y} \rangle$, the $m_1$-primary component of $\langle 0 \rangle$ is the ideal $q_1$. It satisfies $q_1 = (m_1)^j$ for every $j \geq 2$.

Given a representation of $R$ as a residue class ring of a polynomial ring, the primary decomposition of the zero ideal of $R$ can be interpreted in the polynomial ring as follows.

**Remark 4.2.6** Let $P = K[x_1, \ldots, x_n]$ be a polynomial ring over a field $K$, and let $R = P/I$ with a zero-dimensional ideal $I$ of $P$. In this situation we can interpret the primary decomposition of Proposition 4.2.3 in the ring $P$ as follows.

(a) Let $m_1, \ldots, m_s$ be the maximal ideals of $R$, and for $i = 1, \ldots, s$ let $\mathfrak{M}_i$ be the preimage of $m_i$ in $P$. Then the ideals $\mathfrak{M}_i$ are maximal ideals in $P$ and they satisfy $\mathfrak{M}_1 \cap \cdots \cap \mathfrak{M}_s = \mathrm{Rad}(I)$. They are called the **maximal components** of $I$.

(b) For $i = 1, \ldots, s$, let $q_i$ be the $\mathfrak{m}_i$-primary component of $\langle 0 \rangle$ in $R$, and let $\mathfrak{Q}_i$ be the preimage of $q_i$ in $P$. Then the ideal $\mathfrak{Q}_i$ is $\mathfrak{M}_i$-primary for $i = 1, \ldots, s$, and we have $I = \mathfrak{Q}_1 \cap \cdots \cap \mathfrak{Q}_s$. This intersection is called the primary decomposition of $I$. The ideals $\mathfrak{Q}_i$ are called the **primary components** of $I$.

(c) We have $R \cong \prod_{i=1}^{s} R/\bar{q}_i \cong \prod_{i=1}^{s} P/\mathfrak{Q}_i$. For $i = 1, \ldots, s$, the ring $P/\mathfrak{Q}_i$ is a local ring, and its maximal ideal is $\mathfrak{M}_i/\mathfrak{Q}_i$.

For actual computations, we shall use this interpretation and represent all primary ideals and maximal ideals by systems of polynomials generating their preimages in $P$.

By Theorem 2.4.3.b, the joint eigenspaces of the multiplication family of a zero-dimensional affine algebra are the kernels of its maximal ideals. When we use the isomorphism $\eta : \mathscr{F} \longrightarrow R$ to translate this result to a statement about elements of $R$, we recover the following well-known notion from Algebraic Geometry. A local ring $R$ with maximal ideal $\mathfrak{m}$ will be denoted by $(R, \mathfrak{m})$.

**Definition 4.2.7** Let $R$ be a zero-dimensional affine $K$-algebra, let $\mathfrak{m}_1, \ldots, \mathfrak{m}_s$ be its maximal ideals, let $q_1, \ldots, q_s$ be the primary components of its zero ideal, and let $i \in \{1, \ldots, s\}$.

(a) An element $f \in R$ is called a **separator** for $\mathfrak{m}_i$ if we have $\dim_K \langle f \rangle = \dim_K (R/\mathfrak{m}_i)$ and $f \in q_j$ for every $j \neq i$.

(b) If $s = 1$, i.e. if $(R, \mathfrak{m})$ is a local ring, the ideal $\mathrm{Soc}(R) = \mathrm{Ann}_R(\mathfrak{m})$ is called the **socle** of $R$.

The socle of a local zero-dimensional affine algebra can be interpreted and computed as follows.

**Remark 4.2.8** Let $(R, \mathfrak{m})$ be a local zero-dimensional affine $K$-algebra. According to Proposition 4.1.11.b, we have $\mathrm{Ann}_R(\mathfrak{m}) = \mathrm{Ker}(\tilde{\mathfrak{m}})$. Then Theorem 2.4.3.b implies that $\mathrm{Soc}(R)$ is the set of joint eigenvectors of the multiplication family of $R$. The computation of $\mathrm{Soc}(R)$ can be performed in different ways.

(a) If the multiplication matrices of a set of generators of $\mathfrak{m}$ are known with respect to a given $K$-basis $B$ of $R$, i.e., if a set of generators of $\tilde{\mathfrak{m}}$ is known, we can use Algorithm 2.2.17 to compute $\mathrm{Ker}(\tilde{\mathfrak{m}})$, and thus $\mathrm{Soc}(R)$, in terms of $B$.

(b) If we know polynomials representing a set of generators of $\mathfrak{m}$, we can use Gröbner basis techniques to compute $\langle 0 \rangle :_R \mathfrak{m} = \mathrm{Ann}_R(\mathfrak{m})$ (see [15], Sect. 3.2).

In the language of algebraic geometry, the condition to be a separator for a maximal ideal of $R$ can be expressed as follows.

**Remark 4.2.9** Let $i \in \{1, \ldots, s\}$. An element $f \in R$ is a separator for $\mathfrak{m}_i$ if and only if the closed subscheme of $\mathrm{Spec}(R) = \{\mathfrak{m}_1, \ldots, \mathfrak{m}_s\}$ defined by the ideal $\langle f \rangle$ differs from $\mathrm{Spec}(R)$ only at the point $\mathfrak{m}_i$ and its colength has the minimal possible value $\dim_K (R/\mathfrak{m}_i)$. If $R$ is reduced, this is equivalent to requiring that the subscheme defined by $\langle f \rangle$ is obtained from $\mathrm{Spec}(R)$ by removing the point $\mathfrak{m}_i$. If $R$ is reduced and its maximal ideals are $K$-linear, a separator $f$ for $\mathfrak{m}_i$ corresponds under the decomposition into local rings $R \cong K^s$ to a tuple $(0, \ldots, 0, c_i, 0, \ldots, 0)$ where $c_i \neq 0$.

In order to characterize separators using the joint eigenspaces of the multiplication family, we first describe the annihilator of a maximal ideal $\mathfrak{m}_i$ in terms of the decomposition of $R$ into local rings.

**Lemma 4.2.10** *In the setting of Definition* 4.2.7, *the annihilator of the maximal ideal* $\mathfrak{m}_i$ *satisfies* $\mathrm{Ann}_R(\mathfrak{m}_i) = (\mathfrak{q}_i :_R \mathfrak{m}_i) \cdot \prod_{j \neq i} \mathfrak{q}_j$.

*Proof* Using the decomposition $R \cong \prod_{i=1}^{s} R/\mathfrak{q}_i$ of Proposition 4.1.12, we see that the maximal ideal $\mathfrak{m}_i$ corresponds to $\prod_{j=1}^{s} \mathfrak{m}_{ij}$, where $\mathfrak{m}_{ij} = \langle 1 \rangle$ for $j \neq i$, and where $\mathfrak{m}_{ii} = \mathfrak{m}_i/\mathfrak{q}_i$. Clearly, the annihilator of the latter ideal is $\prod_{j=1}^{s} \mathfrak{q}_{ij}$ with $\mathfrak{q}_{ij} = \mathfrak{q}_j$ for $j \neq i$ and $\mathfrak{q}_{ii} = \mathfrak{q}_i :_R \mathfrak{m}_i$. Now an application of the inverse of the above isomorphism finishes the proof.                                        $\square$

Now we are ready to prove that the following conditions characterize separators.

**Theorem 4.2.11 (Characterization of Separators)**
*Let $R$ be a zero-dimensional affine $K$-algebra, let $\mathfrak{m}_1, \ldots, \mathfrak{m}_s$ be its maximal ideals, let $\mathfrak{q}_1, \ldots, \mathfrak{q}_s$ be the primary components of its zero ideal, and let $i \in \{1, \ldots, s\}$. For an element $f \in R$, the following conditions are equivalent.*

(a) *The element $f$ is a separator for $\mathfrak{m}_i$.*
(b) *We have $\mathrm{Ann}_R(f) = \mathfrak{m}_i$.*
(c) *The element $f$ is a non-zero joint eigenvector of the multiplication family $\mathcal{F}$ of $R$ and is contained in the joint eigenspace $\mathrm{Ker}(\tilde{\mathfrak{m}}_i)$, where $\tilde{\mathfrak{m}}_i$ is the image of $\mathfrak{m}_i$ under the isomorphism $\iota : R \longrightarrow \mathcal{F}$.*
(d) *The element $f$ is a non-zero element of $(\mathfrak{q}_i :_R \mathfrak{m}_i) \cdot \prod_{j \neq i} \mathfrak{q}_j$.*
(e) *The image of $f$ is a non-zero element in the socle of the local ring $R/\mathfrak{q}_i$ and, for $j \neq i$, the image of $f$ is zero in $R/\mathfrak{q}_j$.*

*Proof* First we prove that (a) implies (b). The image of $f$ under the decomposition $R \cong \prod_{j=1}^{s} R/\mathfrak{q}_j$ of $R$ into local rings is of the form $(0, \ldots, 0, \bar{f}, 0, \ldots, 0)$ where $\bar{f}$ is the residue class of $f$ in $R/\mathfrak{q}_i$. Since $f$ is a separator, we have $\dim_K(\langle \bar{f} \rangle) = \dim_K(R/\mathfrak{m}_i)$, and therefore $\mathrm{Ann}_{R/\mathfrak{q}_i}(\bar{f}) = \mathfrak{m}_i/\mathfrak{q}_i$. For $j \neq i$, the annihilator of the zero ideal of the ring $R/\mathfrak{q}_j$ is its unit ideal. Reversing the decomposition into local rings then shows that the annihilator of $f$ is the ideal $\mathfrak{m}_i$.

To prove that (c) is a consequence of (b), it suffices to note that $\mathrm{Ann}_R(f) = \mathfrak{m}_i$ implies $\mathfrak{m}_i \cdot f = 0$ and $f \neq 0$ and to apply Proposition 4.1.11.b. Thus the element $f$ is a non-zero vector in $\mathrm{Ann}_R(\mathfrak{m}_i) = \mathrm{Ker}(\tilde{\mathfrak{m}}_i)$.

Since the equivalence between (c) and (d) is shown in the lemma, and since (e) is nothing but a reformulation of (d) using the decomposition of $R$ into local rings, it remains to prove that (c) implies (a). Using the equality $\mathrm{Ann}_R(\mathfrak{m}_i) = \mathrm{Ker}(\tilde{\mathfrak{m}}_i)$, we obtain $\mathfrak{m}_i \cdot f = 0$ with $f \neq 0$. It follows that the principal ideal $\langle f \rangle$ is isomorphic to $R/\mathfrak{m}_i$. By applying the decomposition of $R$ into local rings to $\mathfrak{m}_i \cdot f = 0$, we see that the factors $R/\mathfrak{q}_j$ with $j \neq i$ annihilate the image of $f$, i.e., that $f \in \mathfrak{q}_j$ for $j \neq i$. Therefore $f$ is a separator for $\mathfrak{m}_i$. $\qquad\square$

**Remark 4.2.12 (Existence of Separators)**
Recall that Proposition 2.2.8 says that, for a maximal ideal $\mathfrak{m}_i$ in $R$, there is an element $f \in R \setminus \{0\}$ such that $\mathfrak{m}_i = \mathrm{Ann}_R(f)$. Hence there is a separator for every $\mathfrak{m}_i$.

Let us end this section with some examples of separators.

**Example 4.2.13**  Let $K = \mathbb{Q}$, and let $R = \mathbb{Q}[x, y]/\langle x^2 - x, xy, y^2 - y \rangle$. Then $R$ is reduced and the primary decomposition of its zero ideal is $\langle 0 \rangle = \mathfrak{m}_1 \cap \mathfrak{m}_2 \cap \mathfrak{m}_3$, where $\mathfrak{m}_1 = \langle x, y \rangle$, $\mathfrak{m}_2 = \langle x, y - 1 \rangle$, and $\mathfrak{m}_3 = \langle x - 1, y \rangle$. A $K$-basis of $R$ is given by the residue classes of $(1, x, y)$.

We have $\mathrm{Ann}_R(\mathfrak{m}_1) = \mathfrak{m}_2 \cap \mathfrak{m}_3$ and the separators for $\mathfrak{m}_1$ are $f_1 = x + y - 1$ and its non-zero scalar multiples. Similarly, the separators for $\mathfrak{m}_2$ are the non-zero scalar multiples of $f_2 = y$, and those for $\mathfrak{m}_3$ are the non-zero scalar multiples of $f_3 = x$.

The following example contains a separator in a non-reduced ring.

**Example 4.2.14**  Let $K = \mathbb{Q}$, and let $R = \mathbb{Q}[x]/\langle x^8 + x^6 - 4x^5 - 4x^3 + 4x^2 + 4 \rangle$. Then we have $\langle x^8 + x^6 - 4x^5 - 4x^3 + 4x^2 + 4 \rangle = \mathfrak{q}_1 \cap \mathfrak{q}_2$ with $\mathfrak{q}_1 = \langle (x^3 - 2)^2 \rangle$ and $\mathfrak{q}_2 = \langle x^2 + 1 \rangle$. The corresponding maximal ideals are $\mathfrak{m}_1 = \langle x^3 - 2 \rangle$ and $\mathfrak{m}_2 = \mathfrak{q}_2$.

The polynomial $f_1 = x^5 + x^3 - 2x^2 - 2 = (x^3 - 2)(x^2 + 1)$ is a separator for $\mathfrak{m}_1$, since $f \in \mathrm{Ann}_R(\mathfrak{m}_1) \setminus \{0\}$. The polynomial $f_2 = (x^3 - 2)^2$ is a separator for $\mathfrak{m}_2$.

Recall that the primary component $\mathfrak{q}_i$ is always a power of the maximal ideal $\mathfrak{m}_i$ by Proposition 4.1.12.a. The next example shows that the vector space $\mathrm{Ann}_R(\mathfrak{m}_i)$ containing the separators for $\mathfrak{m}_i$ is not necessarily a power of $\mathfrak{m}_i$.

**Example 4.2.15**  Let $K = \mathbb{Q}$, and let $R = \mathbb{Q}[x, y]/\langle x^2, xy, y^3 \rangle$. Then $R$ is a local ring with maximal ideal $\mathfrak{m} = \langle \bar{x}, \bar{y} \rangle$. We have $\mathrm{Ann}_R(\mathfrak{m}) = \langle \bar{x}, \bar{y}^2 \rangle$, i.e. the separators for $\mathfrak{m}$ are the non-zero elements of the form $a\bar{x} + b\bar{y}^2$ with $a, b \in \mathbb{Q}$. Notice that the ideal $\mathrm{Ann}_R(\mathfrak{m})$ is not a power of $\mathfrak{m}$, because we have $\mathfrak{m}^2 \neq \langle 0 \rangle$ and $\mathfrak{m}^3 = \langle 0 \rangle$, but the element $\bar{x}$ satisfies $\bar{x} \in \mathrm{Ann}_R(\mathfrak{m}) \setminus \mathfrak{m}^2$.

So, primary ideals have maximal powers. Will this joke win the election?

*The people who cast the votes decide nothing.*
*The people who count the votes decide everything.*
(Joseph Stalin)

## 4.3   Commendable and Splitting Multiplication Endomorphisms

> Golden Rule for Math Teachers:
> *You must tell the truth,*
> *and nothing but the truth,*
> *but not the whole truth.*

In Chaps. 2 and 3 we studied the existence of special endomorphisms such as commendable and splitting endomorphisms in commuting families. When the commuting family is the multiplication family $\mathscr{F}$ of a zero-dimensional affine algebra $R$ over a field $K$, these abstract linear algebra results obtain their true algebraic and geometric meaning.

Our starting point is the notion of a curvilinear ring which captures the idea of local factors of the special form $K[z]/\langle z^m \rangle$. The whole truth here would be to speak about a special way how functions pass through the points of the affine scheme $\mathrm{Spec}(R)$, but we leave this to the adepts of Algebraic Geometry. If all maximal ideals are linear and the field $K$ has a large enough characteristic, we prove that the ring $R$ is curvilinear if and only if its multiplication family contains a commendable endomorphism. The latter property has far reaching implications for the structure of $\mathscr{F}$, as we saw in earlier chapters.

Another useful property of a multiplication map is to be a splitting endomorphism. In particular, we would like this to be the case for the multiplication by a generic linear form. Again the whole truth is not totally elementary. If $R$ has linear maximal ideals and $K$ is large enough, then the multiplication map $\vartheta_\ell$ by a generic linear form $\ell$ is a splitting endomorphism. To prove this for more general rings $R$, we need to assume that $K$ is infinite and invest a good deal of cogitation (see Theorem 4.3.11).

And what happens if the base field is finite? Can we still find a map $\vartheta_\ell$ which is a splitting endomorphism? Clearly, this is not always the case. We have to carefully edit the true statement and use a *generically extended* multiplication endomorphism, i.e., a linear combination $y_1 \vartheta_{x_1} + \cdots + y_n \vartheta_{x_n}$, and work over the field $K(y_1, \ldots, y_n)$ (see Theorem 4.3.14).

In view of all these subtleties, why do we want to find a splitting multiplication endomorphism in the first place? The answer is given by Corollary 2.5.3: it allows us to compute the primary components of the zero ideal very easily by calculating the appropriate powers of its eigenfactors. These and further parts of the whole truth will be revealed in Chap. 5 ... in a carefully edited way.

> *The best way to lie is to tell the truth*
> *... carefully edited truth.*

Let $K$ be a field, let $R$ be a zero-dimensional affine $K$-algebra, and let $\mathscr{F}$ be the multiplication family of $R$. In the first part of this section we study the question when $\mathscr{F}$ contains a commendable endomorphism. For general commuting families, Proposition 3.2.14 provides a characterization of the existence of a commendable endomorphism using the property of being locally unigenerated (or weakly curvi-linear). In order to apply this characterization to the multiplication family $\mathscr{F}$, we adapt and specialize Definition 3.2.12 as follows.

**Definition 4.3.1** Let $R$ be a zero-dimensional affine $K$-algebra, and let $q_1, \ldots, q_s$ be the primary components of its zero ideal.

(a) The ring $R$ is called **weakly curvilinear** if there exist monic irreducible poly-nomials $p_i(z) \in K[z]$ and numbers $m_i$ such that $R/q_i \cong K[z]/\langle p_i(z)^{m_i} \rangle$ for $i = 1, \ldots, s$.
(b) The ring $R$ is called **curvilinear** if there exist numbers $m_i$ such that we have $R/q_i \cong K[z]/\langle z^{m_i} \rangle$ for $i = 1, \ldots, s$.

Clearly, a curvilinear ring is weakly curvilinear. Moreover, note that the defini-tion of a curvilinear ring implies that its maximal ideals $m_i = \mathrm{Rad}(q_i)$ are linear. Comparing Definition 4.3.1.a to Definition 3.2.12, we see that the ring $R$ is weakly curvilinear if and only if its multiplication family is weakly curvilinear. Therefore Proposition 3.2.14 yields the following characterization of multiplication families containing a commendable endomorphism.

**Proposition 4.3.2 (Commendable Endomorphisms in Multiplication Families)**
*Let $R$ be a zero-dimensional affine $K$-algebra having $s$ maximal ideals. Then the following conditions are equivalent.*

(a) *There exists a commendable endomorphism in the multiplication family of $R$.*
(b) *The ring $R$ is weakly curvilinear and the polynomials $p_i(z)$ in Definition 4.3.1.a can be chosen pairwise distinct.*

*Moreover, if all maximal ideals of $R$ are linear, these conditions are also equiva-lent to the following one.*

(c) *The ring $R$ is curvilinear and $\mathrm{card}(K) \geq s$.*

*Proof* The equivalence of (a) and (b) follows immediately from Proposition 3.2.14. Now we prove the implication (a) $\Rightarrow$ (c). Let $\varphi$ be a commendable endomorphism in the multiplication family $\mathscr{F}$ of $R$. Then Theorem 3.2.4 implies that $\mathscr{F}$ is unigen-erated by $\varphi$. By Proposition 2.3.8, the eigenfactors of $\varphi$ are linear. Consequently, the ring $\mathscr{F} \cong K[z]/\langle \mu_\varphi(z) \rangle$ is curvilinear. Hence also the ring $R \cong \mathscr{F}$ is curvilinear.

Let $m_1, \ldots, m_s$ be the maximal ideals of $R$. For $i = 1, \ldots, s$, let $c_i \in K$ be such that $z - c_i$ is the $m_i$-eigenfactor of $\varphi$. By Proposition 3.2.10, the map $\varphi$ is a split-ting endomorphism, i.e., its eigenfactors are pairwise distinct. Thus the elements $c_1, \ldots, c_s$ are pairwise distinct, and this implies $\mathrm{card}(K) \geq s$.

Finally, we show that (c) implies (a). Since $\mathrm{card}(K) \geq s$, there exist pairwise distinct elements $c_1, \ldots, c_s \in K$. Hence we can write $R/q_i \cong K[z]/\langle (z - c_i)^{m_i} \rangle$ for $i = 1, \ldots, s$, and the claim follows from the equivalence of (a) and (b).   $\square$

The hypothesis $\text{card}(K) \geq s$ in part (c) of this proposition is for instance satisfied if $\text{card}(K) \geq \dim_K(\mathcal{F})$ (see Proposition 2.2.4). The following example shows that this hypothesis is essential for the implication (c) $\Rightarrow$ (a) to hold.

**Example 4.3.3** Let $K = \mathbb{F}_2$ and $R = K[x, y]/I$ with $I = \langle x^2 + x, xy, y^2 + y \rangle$. In Example 2.5.8 we showed that $I = \langle x, y \rangle \cap \langle x + 1, y \rangle \cap \langle x, y + 1 \rangle$. Therefore the ring

$$R \cong K[x, y]/\langle x, y \rangle \times K[x, y]/\langle x + 1, y \rangle \times K[x, y]/\langle x, y + 1 \rangle$$

is curvilinear. But as we showed in Example 2.5.8, the multiplication family $\mathcal{F}$ of $R$ contains no splitting endomorphism, and consequently no commendable endomorphism (see Corollary 3.2.6).

However, the equivalence of (a) and (b) in Proposition 4.3.2 does not need the assumption $\text{card}(K) \geq s$. The following example illustrates this fact.

**Example 4.3.4** Recall the setting of Example 2.5.6. The multiplication family of the ring $R = K[x, y]/I$, where $K = \mathbb{F}_2$ and

$$I = \langle x^4 + x, y + x^2 + x \rangle = \langle x, y \rangle \cap \langle x + 1, y \rangle \cap \langle x^2 + x + 1, y + 1 \rangle$$

is the family $\mathcal{F}$ given there. The multiplication map $\vartheta_x$ corresponds to the map $\varphi_1$ in $\mathcal{F}$. Since it is a splitting endomorphism and generates $\mathcal{F}$, it is a commendable endomorphism. Notice that the ring $R$ has three maximal ideals, while the field $K$ has only two elements.

In the following we fix a presentation $R \cong K[x_1, \ldots, x_n]/I$ with a zero-dimensional ideal $I$, and for $i = 1, \ldots, n$ we denote the residue class of $x_i$ in $R$ by $\bar{x}_i$. In this setting, a linear maximal ideal of $R$ corresponds to a point in $K^n$ as follows.

**Definition 4.3.5** Given a linear maximal ideal $\mathfrak{m}$ of $R$, the isomorphism $R/\mathfrak{m} \cong K$ shows that there are elements $a_1, \ldots, a_n \in K$ such that $\mathfrak{m} = \langle \bar{x}_1 - a_1, \ldots, \bar{x}_n - a_n \rangle$. Then the tuple $\wp = (a_1, \ldots, a_n) \in K^n$ is called the **associated point** of $\mathfrak{m}$.

In this setting every $f \in R$ has a well-defined **value** $f(\wp) = f(a_1, \ldots, a_n)$ at the point $\wp$. If all maximal ideals of $R$ are linear, a splitting multiplication endomorphism $\vartheta_f$ can be detected by looking at the values of $f$ at the associated points as follows.

**Remark 4.3.6** Assume that the maximal ideals $\mathfrak{m}_1, \ldots, \mathfrak{m}_s$ of $R$ are linear, let $\wp_1, \ldots, \wp_s$ be their associated points, and let $f \in R$.

(a) By Corollary 2.3.9.a, the eigenfactors of the multiplication endomorphism $\vartheta_f$ are given by $z - f(\wp_i)$ for $i = 1, \ldots, s$.

(b) By (a) and Definition 2.5.1, the multiplication endomorphism $\vartheta_f$ is a splitting endomorphism if and only if the values $f(\wp_1), \ldots, f(\wp_s)$ are pairwise distinct.

In addition to having pairwise distinct values, we need a further property of a multiplication map in order to get a commendable endomorphism. The following proposition provides the precise condition.

**Proposition 4.3.7 (Characterization of Commendable Multiplication Maps)**
*Let $R$ be a zero-dimensional affine $K$-algebra with maximal ideals $\mathfrak{m}_1, \ldots, \mathfrak{m}_s$, and assume that these maximal ideals are linear. Let $\wp_1, \ldots, \wp_s \in K^n$ be their associated points, and let $\mathfrak{q}_i$ be the $\mathfrak{m}_i$-primary component of $\langle 0 \rangle$ for $i = 1, \ldots, s$. For $f \in R$, the following conditions are equivalent.*

(a) *The multiplication endomorphism $\vartheta_f$ is commendable.*
(b) *The ring $R$ is curvilinear, the values $f(\wp_1), \ldots, f(\wp_s)$ are pairwise distinct, and for every $i \in \{1, \ldots, s\}$ such that $\mathfrak{q}_i \subset \mathfrak{m}_i$ we have $f - f(\wp_i) \in \mathfrak{m}_i \setminus \mathfrak{m}_i^2$.*

*Proof* First we prove that (a) implies (b). By Proposition 4.3.2, the ring $R$ is curvilinear. By Proposition 3.2.10, the map $\vartheta_f$ is a splitting endomorphism and therefore the values $f(\wp_1), \ldots, f(\wp_s)$ are pairwise distinct.

Now let $i \in \{1, \ldots, s\}$. Let $m_i \geq 2$ be such that $\mathfrak{q}_i = \mathfrak{m}_i^{m_i}$. Since the endomorphism $\vartheta_f$ is commendable, it generates the multiplication family of $R$ by Theorem 3.2.4. Hence the residue class of $f - f(\wp_i)$ generates the $K$-algebra $R/\mathfrak{q}_i$. By Proposition 4.3.2, we have $R/\mathfrak{q}_i \cong K[z]/\langle z^{m_i} \rangle$. Under this isomorphism, the residue class of $f - f(\wp_i) \in \mathfrak{m}_i$ is mapped to a $K$-algebra generator of $K[z]/\langle z^{m_i} \rangle$. This implies $f - f(\wp_i) \notin \mathfrak{m}_i^2$.

Next we prove the implication (b) $\Rightarrow$ (a). By the preceding remark, we know that $\vartheta_f$ is a splitting endomorphism. So, by Proposition 3.2.10, it remains to show that $\vartheta_f$ is locally commendable. Therefore we may assume that $R$ is local with maximal ideal $\mathfrak{m}$ and associated point $\wp$. Since $R$ is curvilinear, we have $\mathcal{F} \cong R \cong K[z]/\langle (z - f(\wp))^m \rangle$ for some $m \geq 1$. If $m = 1$, it is clear that $\vartheta_f$ is commendable. If $m > 1$, the fact that $f - f(\wp) \in \mathfrak{m} \setminus \mathfrak{m}^2$ implies that the minimal polynomial of $\vartheta_f$ is $(z - f(\wp))^m$, and hence $\vartheta_f$ is commendable by Theorem 3.2.4.                                              $\square$

> *Telling the truth to people who misunderstand you*
> *is promoting a falsehood, isn't it?*

In the last part of this section we are telling you the truth about generic linear forms. To this end, we have to consider the case of a large enough base field $K$. We fix a presentation $R = P/I$ with $P = K[x_1, \ldots, x_n]$ and a zero-dimensional ideal $I \subseteq P$. In accordance with Definition 2.5.14, we say that a property $\mathscr{P}$ holds for a **generic linear form** of the presentation $R = P/I$ (or shortly "of $R$" if the presentation is clear from the context) if there exists a non-empty Zariski open subset $U \subseteq K^n$ such that the property holds for elements $\ell = a_1 x_1 + \cdots + a_n x_n + I$ with $(a_1, \ldots, a_n) \in U$. The following proposition says that a generic linear form of $R$

yields a commendable multiplication endomorphism if the maximal ideals of $R$ are linear and the base field $K$ has sufficiently many elements.

**Proposition 4.3.8**  *Let $R = P/I$ zero-dimensional affine $K$-algebra having $s$ linear maximal ideals.*

(a) *If $K$ has more than $\binom{s}{2}$ elements, the multiplication map $\vartheta_\ell$ of a generic linear form $\ell$ of $R$ is a splitting endomorphism.*
(b) *If $K$ is infinite and $R$ is curvilinear then the multiplication map $\vartheta_\ell$ of a generic linear form $\ell$ of $R$ is commendable.*

*Proof*  Let $\wp_1, \ldots, \wp_s$ be the associated points of the maximal ideals $\mathfrak{m}_1, \ldots, \mathfrak{m}_s$ of $R$. From Remark 4.3.6.b and the observation that, for all $i \neq j$, the condition $\ell(\wp_i) = \ell(\wp_j)$ defines a Zariski closed subset of the vector space of linear forms of $R$, we see that the set of all linear forms of $R$ for which the multiplication map is a splitting endomorphism is Zariski open. By the hypothesis on $K$ and an easy extension of Proposition 3.7.22 of [15], this Zariski open set is non-empty. This proves claim (a).

Now we show (b). To prove that $\vartheta_\ell$ is commendable for a generic linear form $\ell$ of $R$, we use Proposition 4.3.7.b. If we show for a fixed $i \in \{1, \ldots, s\}$ with $\mathfrak{q}_i \subset \mathfrak{m}_i$ that the set of all linear forms $\ell$ of $R$ with $\ell - \ell(\wp_i) \in \mathfrak{m}_i \setminus \mathfrak{m}_i^2$ contains a non-empty Zariski open set then we can intersect these Zariski open sets and the set of all tuples of coefficients of $\ell$ for which $\vartheta_\ell$ is a splitting endomorphism. We obtain a non-empty Zariski open set corresponding to linear forms $\ell$ for which $\vartheta_\ell$ is commendable.

So, let $i \in \{1, \ldots, s\}$ with $\mathfrak{q}_i \subset \mathfrak{m}_i$. Using Corollary 1.5.10 and the fact that a constant term cancels in the computation of $\ell - \ell(\wp_i)$, we can assume that $\mathfrak{m}_i = \langle x_1, \ldots, x_n \rangle$. Since $\ell - \ell(\wp_i) \in \mathfrak{m}_i$ is always true, it suffices to prove that there exists a non-empty Zariski open subset $U$ of $K^n$ such that $\ell - \ell(\wp_i) \notin \mathfrak{m}_i^2$ for $\ell = a_1 x_1 + \cdots + a_n x_n + I$ with $(a_1, \ldots, a_n) \in U$.

For a contradiction, assume that $\ell - \ell(\wp_i) \in \mathfrak{m}_i^2$ for all $(a_1, \ldots, a_n)$ outside a proper Zariski closed subset of $K^n$. Since $K$ is infinite, there exist elements $\ell_1, \ldots, \ell_m \in R$ of the given form such that $\mathfrak{m}_i = \langle \ell_1 - \ell_1(\wp_i), \ldots, \ell_m - \ell_m(\wp_i) \rangle$ and $\ell_j - \ell_j(\wp_i) \in \mathfrak{m}_i^2$ for $j = 1, \ldots, m$. Thus we obtain $\mathfrak{m}_i = \mathfrak{m}_i^2$, and therefore $\mathfrak{m}_i = \mathfrak{m}_i^r$ for every $r \geq 1$. This contradicts $\mathfrak{q}_i \subset \mathfrak{m}_i$ and the fact that $\mathfrak{q}_i$ is a power of $\mathfrak{m}_i$ by Proposition 4.2.3.e.                                     $\square$

Our next goal is to prove that part (a) this proposition holds in fact without the hypothesis that the maximal ideals of $R$ are linear. The key ingredient for this generalization is the following result about the behavior of non-empty Zariski open sets with respect to taking preimages under base field extensions.

In the following, given a field $K$ and $f \in K[x_1, \ldots, x_n]$, we denote the set $\{(c_1, \ldots, c_n) \in K^n \mid f(c_1, \ldots, c_n) \neq 0\}$ by $D_f(K)$.

**Proposition 4.3.9 (Preimages of Zariski Open Sets Under Field Extensions)**
*Let $K$ be an infinite field, let $K \subset L$ be a field extension, let $V$ be a finite-dimensional $K$-vector space, let $V_L = V \otimes_K L$ be the extension of $V$, and let $U$*

*be a non-empty Zariski open subset of* $V_L$. *Then* $U \cap V$ *is a non-empty Zariski open subset of* $V$.

*Proof* We let $d = \dim_K(V)$ and choose an isomorphism $V \cong K^d$. It extends to an isomorphism $V_L \cong L^d$, so that we may assume that $V = K^d$ and $V_L = L^d$. Every Zariski open set in $L^d$ is a union of open sets of the form $D_f(L)$ with $f \in L[x_1, \ldots, x_d]$. Therefore it suffices to prove the claim for these open sets.

Let us fix an element $f \in L[x_1, \ldots, x_d] \setminus \{0\}$. Let $B = (b_i \mid i \in \Sigma)$ be a basis of $L$ as a $K$-vector space. We write $f = \sum_{i \in \Sigma} g_i b_i$, where we have $g_i \in K[x_1, \ldots, x_d]$ and where $g_i = 0$ for all but finitely many indices $i \in \Sigma$. Since $D_f(L) \cap K^d = \bigcup_{i \in \Sigma} D_{g_i}(K)$, the set $D_f(L) \cap K^d$ is Zariski open in $K^d$. Moreover, the condition $f \neq 0$ implies that at least one of the polynomials $g_i$ is non-zero. Therefore it suffices to show $D_{g_i}(K) \neq \emptyset$. This follows from the fact that $K$ is infinite and induction on $d$ (see Proposition [16], 5.5.21.a). $\qquad\square$

The following example shows that the assumption that $K$ is infinite is essential in this proposition.

**Example 4.3.10** Let $K = \mathbb{F}_2$, let $V = K$, and let $L = \mathbb{F}_2(x)$ where $x$ is an indeterminate. We have $V_L = \mathbb{F}_2(x)$, and the set $U = \{c \in V_L \mid c^2 + c \neq 0\}$ is a non-empty Zariski open subset of $V_L$. However, we have $U \cap V = \emptyset$.

The next theorem contains the desired generalization of Proposition 4.3.8.a.

**Theorem 4.3.11 (Generic Splitting for Linear Forms)**
*Let* $K$ *be an infinite field. Then the multiplication map* $\vartheta_\ell$ *of a generic linear form* $\ell$ *of* $R$ *is a splitting endomorphism.*

*Proof* Let $K \subseteq L$ be a field extension such that all maximal ideals of $R_L = R \otimes_K L$ are linear. For instance, let $L$ be the algebraic closure of $K$. By Proposition 4.3.8.a, the multiplication map of the generic linear form of $R_L$ is a splitting endomorphism for the multiplication family of $R_L$. Let $U \subseteq L^n$ be the Zariski open subset such that for every element $\ell = a_1 x_1 + \cdots + a_n x_n + I$ with $(a_1, \ldots, a_n) \in U$ the corresponding multiplication map $\vartheta_\ell$ is a splitting endomorphism for the multiplication family of $R_L$.

Now we consider the set $V = U \cap K^n$ which is a non-empty Zariski open set by Proposition 4.3.9. We want to show that, for every linear element $\ell \in R$ whose coordinate tuple is contained in $V$, the multiplication map $\vartheta_\ell$ is a splitting endomorphism.

For a maximal ideal $\mathfrak{m}$ of $R$, the extension $\mathfrak{m}R_L$ is a zero-dimensional ideal and therefore of the form $\mathfrak{m}R_L = \mathfrak{n}_1^{\alpha_1} \cap \cdots \cap \mathfrak{n}_t^{\alpha_t}$ with maximal ideals $\mathfrak{n}_j$ of $R_L$ and with $\alpha_i > 0$ (see Proposition 4.1.12.a). Notice that we have $\mathfrak{m} = \mathfrak{n}_i \cap R$ for $i = 1, \ldots, t$. By definition, the eigenfactor $p_{\mathfrak{m}, \vartheta_\ell}(z)$ generates the kernel of the $L$-algebra homomorphism

$$\Phi_L : L[z] \longrightarrow R_L / \mathfrak{m}R_L \cong R_L / \mathfrak{n}_1^{\alpha_1} \times \cdots \times R_L / \mathfrak{n}_t^{\alpha_t}$$

Hence its prime factorization in $L[z]$ is $p_{\mathfrak{m},\vartheta_\ell}(z) = \mathrm{lcm}(p_{\mathfrak{n}_i,\vartheta_\ell}(z)^{\beta_i} \mid i = 1, \ldots, t)$ with $\beta_i \le \alpha_i$ for $i = 1, \ldots, t$ (see Corollary 2.2.2). Given two maximal ideals $\mathfrak{m}, \mathfrak{m}'$ of $R$, the corresponding maximal ideals $\mathfrak{n}_j, \mathfrak{n}'_k$ of $R_L$ satisfy $p_{\mathfrak{n}_i,\vartheta_\ell}(z) \ne p_{\mathfrak{n}'_j,\vartheta_\ell}(z)$ by the choice of $\ell$. In particular, the polynomials $p_{\mathfrak{m},\vartheta_\ell}(z)$ and $p_{\mathfrak{m}',\vartheta_\ell}(z)$ have different prime factorizations when viewed as elements of $L[z]$. Hence they are distinct. Altogether, it follows that $\vartheta_\ell$ is a splitting endomorphism for the multiplication family of $R$ when the coordinate tuple of $\ell$ is in $V$.    □

The following example indicates a way how to use this theorem.

**Example 4.3.12** Let $K = \mathbb{Q}$ and $R = K[x, y]/I$, where $I$ is the intersection of the ideals $J_1 = \langle (y - x)^2, (x^2 + 1)^2 \rangle$ and $J_2 = \langle (y^2 - 2)^2, (x - 1)(y^2 - 2), (x - 1)^3 \rangle$. Let $\bar{x} = x + I$ and $\bar{y} = y + I$. To find a splitting multiplication endomorphism of $R$, we randomly choose $\ell = 3\bar{x} + 12\bar{y}$. We compute the minimal polynomial of $\ell$ and get

$$\mu_\ell(z) = (z^2 + 225)^3 (z^2 - 6z - 279)^3$$

Thus we have two eigenfactors $p_{\mathfrak{m}_1,\vartheta_\ell}(z) = z^2 + 225$ and $p_{\mathfrak{m}_2,\vartheta_\ell}(z) = z^2 - 6z - 279$, both of which appear with exponent three.

To check that $\vartheta_\ell$ is indeed a splitting multiplication endomorphism, we use Corollary 2.5.3. We compute the principal ideals $\langle p_{\mathfrak{m}_1,\ell}(\ell)^3 \rangle = \langle (\ell^2 + 225)^3 \rangle$ and $\langle p_{\mathfrak{m}_2,\ell}(\ell)^3 \rangle = \langle (\ell^2 - 6\ell - 279)^3 \rangle$, and check that the first ideal is $\mathfrak{q}_1 = J_1/I$ and the second one is $\mathfrak{q}_2 = J_2/I$. Looking at $J_1$ and $J_2$, it is easy to see that the radical of $\mathfrak{q}_1$ is $\mathfrak{m}_1 = \langle \bar{x} - \bar{y}, \bar{y}^2 + 1 \rangle$ and the radical of $\mathfrak{q}_2$ is $\mathfrak{m}_2 = \langle \bar{x} - 1, \bar{y}^2 - 2 \rangle$. Thus the primary decomposition of $\langle 0 \rangle$ in $R$ is $\mathfrak{q}_1 \cap \mathfrak{q}_2$ and $p_{\mathfrak{m}_i,\vartheta_\ell}(z)$ is the $\mathfrak{m}_i$-eigenfactor of $\vartheta_\ell$ for $i = 1, 2$. Hence $\vartheta_\ell$ is a splitting endomorphism.

Now let us try another linear form of $R$, for instance $\ell' = \bar{x} + \bar{y}$. The minimal polynomial of $\ell'$ is $\mu_{\ell'}(z) = (z^2 + 4)^3 (z^2 - 2z - 1)^3$. Using $p_{\mathfrak{m}_1,\vartheta_{\ell'}}(z) = z^2 + 4$ and $p_{\mathfrak{m}_2,\vartheta_{\ell'}}(z) = z^2 - 2z - 1$, we check $\mathfrak{q}_1 = \langle p_{\mathfrak{m}_1,\vartheta_{\ell'}}(\ell')^3 \rangle$ and $\mathfrak{q}_2 = \langle p_{\mathfrak{m}_2,\vartheta_{\ell'}}(\ell')^3 \rangle$. Therefore also $\vartheta_{\ell'}$ is a splitting multiplication endomorphism of $R$.

Next we generalize Definition 2.5.12 to introduce the notion of a generically extended endomorphism with respect to a system of $K$-algebra generators of a commuting family $\mathscr{F}$. Notice that a $K$-basis of $\mathscr{F}$ is a particular system of $K$-algebra generators.

**Definition 4.3.13** Let $\mathscr{F}$ be a commuting family, and let $\Phi = (\varphi_1, \ldots, \varphi_n)$ be a system of $K$-algebra generators of $\mathscr{F}$. We introduce new indeterminates $y_1, \ldots, y_n$ and let $\mathscr{F} \otimes_K K(y_1, \ldots, y_n)$ be the commuting family obtained by the corresponding base field extension. Then the element $\Theta_{y,\Phi} = y_1\varphi_1 + \cdots + y_n\varphi_n$ is called the **generically extended endomorphism** associated to $\Phi$.

We observe that if $\mathscr{F}$ is the multiplication family of a zero-dimensional affine $K$-algebra $R = K[x_1, \ldots, x_n]/I$ and we take $\Xi = (\vartheta_{x_1}, \ldots, \vartheta_{x_n})$ as a system of generators of $\mathscr{F}$, the generically extended endomorphism $\Theta_{y,\Xi} = y_1\vartheta_{x_1} + \cdots + y_n\vartheta_{x_n}$

associated to $\Xi$ is the multiplication map of the element $y_1 \bar{x}_1 + \cdots + y_n \bar{x}_n$ of the $K(y_1, \ldots, y_n)$-algebra $R \otimes_K K(y_1, \ldots, y_n)$. Now we are ready to generalize Proposition 2.5.13 and Theorem 2.5.15 as follows.

### Theorem 4.3.14 (Splitting of Generically Extended Endomorphisms)

*Let $R = P/I$ be a zero-dimensional affine $K$-algebra, where $P = K[x_1, \ldots, x_n]$ and $I$ is an ideal in $P$. Let $\mathscr{F}$ be the multiplication family of $R$, let $y_1, \ldots, y_n$ be indeterminates, and let $\mathscr{F} \otimes_K K(y_1, \ldots, y_n)$ be the commuting family obtained by the corresponding base field extension.*

*Then the generically extended endomorphism associated to $(\vartheta_{x_1}, \ldots, \vartheta_{x_n})$ is a splitting endomorphism for the family $\mathscr{F} \otimes_K K(y_1, \ldots, y_n)$.*

*Proof* Let $y = (y_1, \ldots, y_n)$, let us denote $K(y_1, \ldots, y_n)$ by $K(y)$, and let $s$ be the number of maximal ideals in $\mathscr{F}$. By Proposition 2.5.13.a, the maximal ideals in the $K(y)$-algebra $\mathscr{F} \otimes_K K(y)$ are extensions of maximal ideals of $\mathscr{F}$. As in the proof of that proposition, it follows that $s$ is also the number of maximal ideals in the family $\mathscr{F} \otimes_K K(y)$.

Notice that the tuple $\Xi = (\vartheta_{x_1}, \ldots, \vartheta_{x_n})$ is a system of $K(y)$-algebra generators of $\mathscr{F} \otimes_K K(y)$. We denote the corresponding generically extended endomorphism $y_1 \vartheta_{x_1} + \cdots + y_n \vartheta_{x_n}$ by $\Theta_{y, \Xi}$. Since the field $K(y)$ is infinite, we can use Theorem 4.3.11 to deduce that the multiplication map of a generic linear element of $R \otimes_K K(y) \cong K(y)[x_1, \ldots, x_n]/I K(y)[x_1, \ldots, x_n]$ is a splitting endomorphism for $\mathscr{F} \otimes_K K(y)$.

Let $U \subseteq K(y)^n$ be a non-empty Zariski open subset such that for every tuple $f = (f_1, \ldots, f_n) \in U$, the corresponding multiplication endomorphism $\Theta_{f, \Xi} = \vartheta_{f_1 x_1 + \cdots + f_n x_n} = f_1 \vartheta_{x_1} + \cdots + f_n \vartheta_{x_n}$ is a splitting endomorphism for $\mathscr{F} \otimes_K K(y)$. By definition of the Zariski topology, we may assume that $U$ is the complement in $K(y)^n$ of the zero-set of a non-zero polynomial $p(y, z_1, \ldots, z_n) \in K(y)[z_1, \ldots, z_n]$.

Using induction, we now show that there exists a tuple $(f_1, \ldots, f_n) \in U$ whose entries are algebraically independent over $K$. When we view $p(y, z_1, \ldots, z_n)$ as a univariate polynomial in the indeterminate $z_1$ with coefficients in $K(y, z_2, \ldots, z_n)$, we see that it has only finitely many zeros in $K(y)$. Hence we may choose an element $f_1 \in K(y) \setminus K$ such that $p(y, f_1, z_2, \ldots, z_n) \neq 0$. Since $f_1$ is not in $K$, it is transcendental over $K$.

Now assume that for $1 \leq i < n$ we have found $f_1, \ldots, f_i \in K(y)$ which are algebraically independent over $K$ and $p(y, f_1, \ldots, f_i, z_{i+1}, \ldots, z_n) \neq 0$. We view $p(y, f_1, \ldots, f_i, z_{i+1}, \ldots, z_n)$ as a univariate polynomial in $z_{i+1}$ with coefficients in $K(y, z_{i+2}, \ldots, z_n)$. (Here we read $K(y, z_{i+2}, \ldots, z_n)$ as $K(y)$ if $i = n - 1$.) Since we have $i < n$ and $\operatorname{trdeg}_K(K(y)) = n$, there are infinitely many elements $f_{i+1} \in K(y)$ such that $f_1, \ldots, f_{i+1}$ are algebraically independent over $K$. On the other hand, if we consider $p(y, f_1, \ldots, f_i, z_{i+1}, z_{i+2}, \ldots, z_n) \neq 0$ as a univariate polynomial in $z_{i+1}$, it has only finitely many zeros in $K(y)$. Hence we can find an element $f_{i+1}$ in $K(y)$ such that $f_1, \ldots, f_{i+1}$ are algebraically independent over $K$ and $p(f_1, \ldots, f_{i+1}, z_{i+2}, \ldots, z_n) \neq 0$. We conclude that there exist $f_1, \ldots, f_n \in K(y)$ which are algebraically independent over $K$ and such that $(f_1, \ldots, f_n) \in U$.

Therefore the $K$-algebra homomorphism $\iota : K(y) \longrightarrow K(f_1, \ldots, f_n)$ which sends $y_i$ to $f_i$ for $i = 1, \ldots, n$ is an isomorphism. Using $\iota$, it follows that the minimal polynomials of $\Theta_{f,\Xi}$ and $\Theta_{y,\Xi}$ coincide, and this finishes the proof. $\square$

Comparing this theorem with Theorem 4.3.11, we observe that it does not require $K$ to be infinite.

*Asking someone to name something*
*that is easier done than said*
*is easier said than done.*

## 4.4 Local Multiplication Families

*He who is not satisfied with a little*
*is satisfied with nothing.*
(Epicurus, 341–270 B.C.)

Recall that a zero-dimensional affine algebra $R$ over a field $K$ is called *local* if it has a unique maximal ideal $\mathfrak{m}$. In this case the ideal $\mathfrak{m}$ consists entirely of nilpotent elements and is itself nilpotent in the sense that $\mathfrak{m}^i = \langle 0 \rangle$ for some $i \geq 1$. All other elements of $R$, i.e., the elements outside $\mathfrak{m}$, are units. So, to study such rings, we have to look into the finer structure of their set of nilpotent elements.

The first useful invariants are the index of nilpotency $\mathrm{nix}(f)$ of an element $f \in \mathfrak{m}$ and $\mathrm{nix}(\mathfrak{m})$, the nilpotency index of $\mathfrak{m}$. In German, *nix* is a colloquial form of *nichts*, meaning "nothing". Are we satisfied with "nothing"? Of course, we are not. The main goal in this section is to show that a generic linear form of $\mathfrak{m}$ has the maximal possible nilpotency index $\mathrm{nix}(\mathfrak{m})$. In this statement both the terms "generic" and "linear" have to be interpreted with care. In order to choose a "generic" element with the desired properties, we need sufficiently many elements to choose from, i.e., a large enough base field, and the meaning of the word "linear" depends on the choice of a presentation $R = P/I$ with $P = K[x_1, \ldots, x_n]$ and an ideal $I$ in $P$.

Nevertheless, after overcoming these technical difficulties by showing that a power of a generic linear form is not contained in a given hyperplane (see Proposition 4.4.7), we can prove the existence of linear elements with maximal nilpotency (see Theorem 4.4.12). This result is far from nothing and ready to serve as the basis of substantial applications later on.

*This nothing's more than matter.*
(William Shakespeare)

Let $(R, \mathfrak{m})$ be a local zero-dimensional affine $K$-algebra. According to Proposition 2.2.5.b, every ideal $\mathfrak{a} \subseteq \mathfrak{m}$ satisfies $\mathfrak{a}^t = 0$ for some $t \geq 1$. Moreover, note that for every element $f \in R$, the multiplication map $\vartheta_f$ has only one eigenfactor $p_f(z)$, since there is exactly one maximal ideal in $R$ (see Proposition 2.3.2). Further, recall that $p_f(f) \in \mathfrak{m}$ for every $f \in R$, and that we have $p_f(f) = f$ if $f \in \mathfrak{m}$ (see Proposition 2.3.2.a).

**Definition 4.4.1** Let $(R, \mathfrak{m})$ be a local zero-dimensional affine $K$-algebra.

(a) For $f \in \mathfrak{m}$, the smallest number $t \geq 1$ such that $f^t = 0$ is called the **index of nilpotency** (or the **nilpotency index**) of $f$ and is denoted by $\mathrm{nix}(f)$. More generally, for $f \in R$, the index of nilpotency of $p_f(f)$ is denoted by $\mathrm{nix}(f)$.
(b) The smallest number $t$ such that $\mathfrak{m}^t = 0$ is called the **index of nilpotency** (or the **nilpotency index**) of $\mathfrak{m}$ and is denoted by $\mathrm{nix}(\mathfrak{m})$.
(c) An element $f \in R$ is said to be **of maximal nilpotency** if we have $\mathrm{nix}(f) = \mathrm{nix}(\mathfrak{m})$.

In the following lemma, we compare the index of nilpotency of an element of a local zero-dimensional algebra to the index of nilpotency of the maximal ideal. In particular, part (b) motivates the above definition of maximal nilpotency.

**Lemma 4.4.2** *Let $(R, \mathfrak{m})$ be a local zero-dimensional affine $K$-algebra and $f \in R$.*

(a) *If $\mu_f(z) = p_f(z)^m$ then we have $\mathrm{nix}(f) = m$. In other words, we have the equality $\mathrm{nix}(f) = \mathrm{mmult}(\vartheta_f, p_f(z))$.*
(b) *We have $\mathrm{nix}(f) \leq \mathrm{nix}(\mathfrak{m})$.*

*Proof* Claim (a) follows from the fact that $0 = \mu_f(f) = p_f(f)^m$ and the definition of $\mu_f$. Claim (b) follows from the observation that $p_f(f) \in \mathfrak{m}$.                         $\square$

The next proposition shows the importance of elements of maximal nilpotency.

**Proposition 4.4.3** *Let $(R, \mathfrak{m})$ be a local zero-dimensional affine $K$-algebra, and let $f \in R$ be an element of maximal nilpotency.*

(a) *We have $p_f(f)^{\mathrm{nix}(f)-1} \in \mathrm{Soc}(R) \setminus \{0\}$.*
(b) *We have $\mathfrak{m} = \mathrm{Ann}_R(p_f(f)^{\mathrm{nix}(f)-1})$.*

*Proof* Let $m = \mathrm{nix}(f)$. By assumption, we have $m = \mathrm{nix}(\mathfrak{m})$ and $p_f(f)^{m-1} \neq 0$. Therefore we get $\mathrm{Ann}_{\mathscr{F}}(p_f(f)^{m-1}) \subseteq \mathfrak{m}$.

On the other hand, for every $g \in \mathfrak{m}$ we have $g \cdot p_f(f)^{m-1} \in \mathfrak{m}^m = \langle 0 \rangle$. This shows $p_f(f)^{m-1} \in \mathrm{Soc}(R)$ and $\mathfrak{m} \subseteq \mathrm{Ann}_R(p_f(f)^{m-1})$. Thus both claims are proved.                                                                     $\square$

Notice that the proposition makes sense even when $R$ is a field, because in that case we have $0^0 = 1 \in \mathrm{Soc}(R)$. Furthermore, we note that part (b) of this proposition implies that, in order to compute $\mathfrak{m}$, it suffices to find a single non-zero element of maximal nilpotency. Unfortunately, in positive characteristic, the set of elements of maximal nilpotency can be empty, as our next example shows.

**Example 4.4.4**   Let $K$ be a field of positive characteristic $p$, and let us consider the ring $R = K[x, y]/\langle x^p, y^p \rangle$. The unique maximal ideal of $R$ is $\mathfrak{m} = \langle \bar{x}, \bar{y} \rangle$. It is clear that $\bar{x}^{p-1}\bar{y}^{p-1} \neq 0$, while every monomial $t$ in $K[x, y]$ of degree $\deg(t) \geq 2p - 1$ is such that $\bar{t} = 0$. Therefore we have $\mathrm{nix}(\mathfrak{m}) = 2p - 1$ and $\mathrm{Soc}(R) = \langle \bar{x}^{p-1}\bar{y}^{p-1} \rangle$.

On the other hand, every element $f \in R$ can be written as $f = g + c$ with $g \in \mathfrak{m}$. By Remark 4.4.5 we get $p_f(f) = g$, and it is clear that $g^p = 0$ which implies that $\mathrm{nix}(f) \leq p$. Therefore there is no element of maximal nilpotency in $R$. Notice that the field $K$ may even be infinite here.

In the following part of this section we show that, under a suitable hypothesis on the characteristic of the base field, elements of maximal nilpotency do exist. To this end, we suppose we are given a presentation $R = P/I$ with $P = K[x_1, \ldots, x_n]$ and an ideal $I \subseteq P$. We start with an easy remark.

**Remark 4.4.5**   Let $(R, \mathfrak{m})$ be a local zero-dimensional affine $K$-algebra, let $f \in \mathfrak{m}$, let $g = f + c$ with $c \in K$, and let $p_f(z)$ and $p_g(z)$ be their $\mathfrak{m}$-eigenfactors.

(a)  By Proposition 2.3.2.b, we have $p_g(z) = p_c(z) = z - c$.
(b)  By (a) and the fact that $p_f(z) = z$, we have $p_g(g) = f = p_f(f)$.
(c)  By (b), we have $\mathrm{nix}(f) = \mathrm{nix}(g)$.

For every $r \geq 0$, let $P_r$ be the $K$-vector subspace of $P$ of **forms**, i.e., of homogeneous polynomials of degree $r$.

**Lemma 4.4.6**   *Let $r \geq 1$, assume that the field $K$ has more than $r$ elements, and let $f \in P_r$ be a non-zero form. Then there exists a point $(a_1, \ldots, a_n) \in K^n$ such that $f(a_1, \ldots, a_n) \neq 0$.*

*Proof*   This follows by induction on $n$. For $n = 1$, not all elements of $K$ can be zeros of a form of degree $r$ since $K$ has more than $r$ elements. Now let $n > 1$, and write $f = g_0 x_n^0 + \cdots + g_s x_n^s$ with $g_i \in K[x_1, \ldots, x_{n-1}]_{r-i}$ and $g_s \neq 0$. By induction, there exists a point $(a_1, \ldots, a_{n-1}) \in K^{n-1}$ such that $g_s(a_1, \ldots, a_{n-1}) \neq 0$. Then the degree of $f(a_1, \ldots, a_{n-1}, x_n) \in K[x_n]$ is $s$. Since $s \leq r$, the case $n = 1$ yields an element $a_n \in K$ with $f(a_1, \ldots, a_n) \neq 0$.                                                $\square$

Let $r \geq 0$. The terms of degree $r$ form a $K$-basis of $P_r$, and we have the equality $\dim_K(P_r) = \binom{n+r-1}{r}$. Let us denote this number by $m$, and let $t_1, \ldots, t_m$ be the terms of degree $r$. A **hyperplane** in the vector space $P_r$ is a $K$-vector subspace of dimension $m - 1$. In other words, a hyperplane in $P_r$ is the set of all homogeneous polynomials $\sum_{i=1}^{m} a_i t_i \in P_r$ whose coefficient tuple $(a_1, \ldots, a_m) \in K^m$ is a solution of a fixed linear equation $\sum_{i=1}^{m} c_i y_i = 0$ with $(c_1, \ldots, c_m) \in K^m \setminus \{0\}$. In accordance with Definition 2.5.14, we say that a **generic linear form** has a certain property $\mathscr{P}$ if a generic element of $P_1$ has property $\mathscr{P}$.

The following proposition is the key ingredient for showing that elements of maximal nilpotency exist if the characteristic of the base field is large enough or zero.

**Proposition 4.4.7**   Let $r \geq 1$, let $P = K[x_1, \ldots, x_n]$, and let $H$ be an hyperplane in $P_r$.

(a) If $\mathrm{char}(K) > r$ then there exists a linear form $L$ in $P_1$ such that $L^r \notin H$.
(b) If $\mathrm{char}(K) = 0$ or if $K$ is an infinite field with $\mathrm{char}(K) > r$ then a generic linear form $L$ in $P_1$ satisfies $L^r \notin H$.

*Proof*   Using the notation introduced above, let $H$ be defined by the linear equation $\sum_{i=1}^{m} c_i y_i = 0$ with $(c_1, \ldots, c_m) \in K^m \setminus \{(0, \ldots, 0)\}$. If we are given a linear form $L = a_1 x_1 + \cdots + a_n x_n$ with $a_i \in K$, we calculate

$$L^r = \sum_{i=1}^{m} \binom{r}{\alpha_{i1}, \ldots, \alpha_{in}} a_1^{\alpha_{i1}} \cdots a_n^{\alpha_{in}} x_1^{\alpha_{i1}} \cdots x_n^{\alpha_{in}}$$

Therefore, if we substitute the coordinates of $L^r$ into the linear polynomial $\sum_{i=1}^{m} c_i y_i$ defining $H$, we get

$$\sum_{i=1}^{m} c_i \binom{r}{\alpha_{i1}, \ldots, \alpha_{in}} a_1^{\alpha_{i1}} \cdots a_n^{\alpha_{in}} \tag{$*$}$$

To prove (a), we have to show that this value is non-zero for some linear form $L$. Since we assumed $\mathrm{char}(K) > r$, we have $\binom{r}{\alpha_{i1}, \ldots, \alpha_{in}} \neq 0$. Hence $(*)$ is the value of a non-zero homogeneous polynomial of degree $r$ at the point $(a_1, \ldots, a_n)$. By the lemma, there exits a point $(a_1, \ldots, a_n) \in K^n$ such that this value is non-zero, and thus the corresponding linear form $a_1 x_1 + \cdots + a_n x_n$ satisfies $L^r \notin H$.

Now we prove (b). We have to show that the value of $(*)$ is non-zero for a generic point $(a_1, \ldots, a_n)$ in $K^n$. Since $K$ is infinite, the proof of (a) shows that the polynomial $\sum_{i=1}^{m} c_i \binom{r}{\alpha_{i1}, \ldots, \alpha_{in}} t_i$ is non-zero. Then, by [16], Proposition 5.5.21.a, a generic point $(a_1, \ldots, a_n) \in K^n$ is not contained in the hypersurface defined by this polynomial, and the claim follows.                                                                                       $\square$

Although we are not going to use it further, we note that from this proposition and a straightforward induction we get the following result about vector spaces spanned by powers of generic linear forms.

**Proposition 4.4.8**   Let $P = K[x_1, \ldots, x_n]$, let $r, s > 0$, and let $K$ be a field of characteristic zero or an infinite field with $\mathrm{char}(K) > r$. Then a generic tuple of linear forms $(L_1, \ldots, L_s) \in (P_1)^s$ satisfies

$$\dim_K \left( \langle L_1^r, \ldots, L_s^r \rangle_K \right) = \min \left\{ \binom{n+r-1}{r}, s \right\}$$

i.e., the vector space spanned by $L_1^r, \ldots, L_s^r$ in $P_r$ has maximal dimension.

*Proof*   To prove the claim, we proceed by induction on $s$. For $s = 1$, the equality is obviously true. Now we assume that we have proved the claim for a generic tuple of $s - 1$ linear forms $(L_1, \ldots, L_{s-1})$ and we prove the equality for a generic tuple $(L_1, \ldots, L_s)$. Thus we may assume that the $K$-vector subspace $W$ of $P_r$ spanned by $\{L_1^r, \ldots, L_{s-1}^r\}$ has dimension $\dim_K(W) = \min\{\binom{n+r-1}{r}, s - 1\}$.

If we have $\dim_K(W) = \binom{n+r-1}{r}$ here, we trivially get the claimed equality. Hence we may assume that $\dim_K(W) = s - 1 < \binom{n+r-1}{r}$. Therefore $W$ is contained in a hyperplane of $P_r$. The preceding proposition shows that, for a generic linear form $L_s$, the polynomial $L_s^r$ is not contained in this hyperplane. Hence we have the equality $\dim_K(W + \langle L_s^r \rangle_K) = s$ for a generic linear form $L_s$, and the conclusion follows. □

If the base field has characteristic zero or large enough prime characteristic, our next proposition provides the non-emptiness of the set of elements of maximal nilpotency under the additional hypothesis that the maximal ideal $\mathfrak{m}$ is linear.

**Proposition 4.4.9**  *Let* $(R, \mathfrak{m})$ *be a local zero-dimensional affine $K$-algebra, write* $R = P/I$ *with* $P = K[x_1, \ldots, x_n]$ *and an ideal $I$ in $P$, and assume that $\mathfrak{m}$ is linear.*

(a) *If* $\mathrm{char}(K) = 0$ *or if* $\mathrm{char}(K) \geq \mathrm{nix}(\mathfrak{m})$ *then there exists a linear form $\ell$ in $R$ which is of maximal nilpotency.*

(b) *If* $\mathrm{char}(K) = 0$ *or if $K$ is an infinite field with* $\mathrm{char}(K) \geq \mathrm{nix}(\mathfrak{m})$ *then a generic linear form $\ell$ of $R$ is of maximal nilpotency.*

(c) *If* $\mathrm{char}(K) = 0$ *or if $K$ is an infinite field with* $\mathrm{char}(K) \geq \mathrm{nix}(\mathfrak{m})$ *then a generic form $\ell$ of $R$ is of maximal nilpotency.*

*Proof*  Since the maximal ideal $\mathfrak{m}$ is linear, we may perform a linear change of coordinates and assume that $\mathfrak{m} = \langle \bar{x}_1, \ldots, \bar{x}_n \rangle$. In view of Remark 4.4.5, it still suffices to prove the claims for linear forms in $R$.

To prove (a) and (b), we write $\ell = a_1 \bar{x}_1 + \cdots + a_n \bar{x}_n$ with $a_i \in K$. Now we let $r = \mathrm{nix}(\mathfrak{m}) - 1$, $L = a_1 x_1 + \cdots + a_n x_n \in P_1$ and $\mathfrak{M} = \langle x_1, \ldots, x_n \rangle$. Then we have $\mathfrak{M}^{r+1} \subseteq I$ and $\mathfrak{M}^r \not\subseteq I$. Consequently, the vector space $I \cap P_r$ is a proper vector subspace of $P_r$. Hence there exists a hyperplane $H$ in $P_r$ such that $I \cap P_r \subseteq H$. By part (a) (resp. part (b)) of Proposition 4.4.7, we have $L^r \notin H$ for some $L \in P_1$ (resp. for a generic linear form $L \in P_1$). Then the residue class $\ell$ of $L$ in $R$ has maximal nilpotency.

Finally, to prove (c), we choose a $K$-basis $B = (b_1, \ldots, b_d)$ of $R$ and we write $R = K[y_1, \ldots, y_d]/J$ with an ideal $J$ in the polynomial ring $K[y_1, \ldots, y_d]$ such that $b_i = y_i + J$ for $i = 1, \ldots, d$. Then the claim follows by applying (b) to this presentation of $R$.  □

In the following example we determine an explicit Zariski open subset $U$ of the set of linear forms of $R$ such that all elements in $U$ have maximal nilpotency.

**Example 4.4.10**  Let $K = \mathbb{Q}$, let $V = K^4$, and let $\varphi_1$ and $\varphi_2$ be the endomorphisms of $V$ described in Example 3.3.2. The commuting family $\mathscr{F} = K[\varphi_1, \varphi_2]$ is the multiplication family of $R = P/I$ with $P = K[x, y]$ and $I = \langle x^2, y^2 \rangle$, and the ring $R$ is local with linear maximal ideal $\mathfrak{m} = \langle \bar{x}, \bar{y} \rangle$. It is clear that $\mathrm{nix}(\mathfrak{m}) = 3$. So, let us determine which linear forms $\ell = a_1\bar{x} + a_2\bar{y}$ of $R$ with $a_1, a_2 \in K$ satisfy $\ell^2 \neq 0$ and thus are of maximal nilpotency.

Since $I \cap P_2 = Kx^2 + Ky^2$, we have $L^2 = a_1^2x^2 + 2a_1a_2xy + a_2^2y^2 \notin I$ if and only if $a_1a_2 \neq 0$. Hence the Zariski open set $U = \{(a_1, a_2) \in K^2 \mid a_1a_2 \neq 0\}$ is precisely the set of all pairs $(a_1, a_2)$ such that $\ell = a_1\bar{x} + a_2\bar{x}$ is of maximal nilpotency.

Notice that the situation is completely different if we use a base field $K$ with $\mathrm{char}(K) = 2$ here. Example 4.4.4 shows that there are no elements of maximal nilpotency in this case.

In the remaining part of this section we prove a generalization of Proposition 4.4.9. Our goal is to remove the hypothesis that $\mathfrak{m}$ is linear. First we need some preparatory results.

**Lemma 4.4.11** *Let $(R, \mathfrak{m})$ be a local zero-dimensional affine $K$-algebra, let $\bar{K}$ be the algebraic closure of $K$, let $\bar{R} = R \otimes_K \bar{K}$, let $\delta = \dim_K(R/\mathfrak{m})$, and assume that $K \subseteq R/\mathfrak{m}$ is a separable field extension.*

(a) *The ideal $\mathfrak{m}\bar{R}$ is the intersection of $\delta$ linear maximal ideals $\mathfrak{m}_1, \ldots, \mathfrak{m}_\delta$ of $\bar{R}$.*
(b) *The zero ideal of $\bar{R}$ is the intersection of $\delta$ primary ideals $\mathfrak{q}_1, \ldots, \mathfrak{q}_\delta$, where $\mathfrak{q}_i$ is $\mathfrak{m}_i$-primary for $i = 1, \ldots, \delta$.*
(c) *There is a canonical isomorphism $\varepsilon : \bar{R} \longrightarrow R_1 \times \cdots \times R_\delta$ where $R_i = \bar{R}/\mathfrak{q}_i$ is local and $\bar{\mathfrak{m}}_i = \mathfrak{m}_i/\mathfrak{q}_i$ is the maximal ideal of $R_i$ for $i = 1, \ldots, \delta$.*
(d) *We have $\mathrm{nix}(\mathfrak{m}) = \max\{\mathrm{nix}(\bar{\mathfrak{m}}_1), \ldots, \mathrm{nix}(\bar{\mathfrak{m}}_\delta)\}$.*
(e) *Let $\ell \in R$ and $\varepsilon(\ell) = (\ell_1, \ldots, \ell_\delta)$. Then $p_\ell(z) = \mathrm{lcm}(p_{\ell_1}(z), \ldots, p_{\ell_\delta}(z))$ in $\bar{K}[z]$ and $p_{\ell_i}(z)$ is of the form $p_{\ell_i}(z) = z - c_i$ with $c_i \in \bar{K}$ for $i = 1, \ldots, \delta$.*
(f) *In the setting of (e), we have $\mathrm{nix}(p_\ell(\ell_i)) = \mathrm{nix}(\ell_i - c_i) = \mathrm{nix}(\ell_i)$ in $R_i$ for all $i = 1, \ldots, \delta$.*
(g) *In the setting of (e), we have $\mathrm{nix}(\ell) = \max\{\mathrm{nix}(\ell_1), \ldots, \mathrm{nix}(\ell_\delta)\}$.*

*Proof* First we prove (a). It follows from [31], Chap. VII, Sect. 11, Theorem 37 that $\mathfrak{m}\bar{R}$ is the intersection of distinct maximal ideals of $\bar{R}$. By Hilbert's Nullstellensatz, these maximal ideals are linear (see. [15], Corollary 2.6.9). Assume that the number of the maximal ideals is $s$. The Chinese Remainder Theorem 2.2.1 yields $\eta : \bar{R}/\mathfrak{m}\bar{R} \cong \bar{R}/\mathfrak{m}_1 \times \cdots \times \bar{R}/\mathfrak{m}_s \cong \bar{K}^s$. As we have $\delta = \dim_K(R/\mathfrak{m}) = \dim_{\bar{K}}(\bar{R}/\mathfrak{m}\bar{R})$, it follows that $s = \delta$.

To show (b), we note that, by (a), the primary decomposition of the zero ideal in $\bar{R}$ is of the form $\langle 0 \rangle = \mathfrak{q}_1 \cap \cdots \cap \mathfrak{q}_\delta$, where $\mathfrak{q}_i$ is an $\mathfrak{m}_i$-primary ideal.

Claim (c) is nothing but the decomposition of $\bar{R}$ into local rings. Notice that the canonical isomorphism $\varepsilon$ is induced by the epimorphisms $\bar{R} \twoheadrightarrow \bar{R}/\mathfrak{q}_i$.

To prove (d), we observe that $\mathfrak{m}^i = 0$ if and only if $(\mathfrak{m}\bar{R})^i = 0$ (see [15], Proposition 2.6.12) and $\varepsilon((\mathfrak{m}\bar{R})^i) = \mathfrak{m}_1^i \times \cdots \times \mathfrak{m}_\delta^i$ for every $i \geq 1$. This implies the claim.

Next we show (e). By definition, the polynomial $p_\ell(z)$ is the monic generator of the kernel of the $K$-algebra homomorphism $K[z] \longrightarrow R/\mathfrak{m}$ given by $z \mapsto \ell + \mathfrak{m}$. Hence it is also the monic generator of the kernel of the $\bar{K}$-algebra homomorphism

$\bar{K}[z] \longrightarrow \bar{R}/\mathfrak{m}\bar{R}$ given by $z \mapsto \ell + \mathfrak{m}\bar{R}$. When we combine this map with the isomorphism $\eta$ above, we see that the kernel of the composition $\bar{K}[z] \longrightarrow \prod_{i=1}^{\delta} \bar{R}/\mathfrak{m}_i$ has the monic generator $p_\ell(z)$. By definition of the polynomials $p_{\ell_i}(z)$, the monic generator of the kernel of this composition is also given by $\mathrm{lcm}(p_{\ell_1}(z), \ldots, p_{\ell_\delta}(z))$ (see Corollary 2.2.2), and the first claim follows. The second claim is an immediate consequence of the fact that $\bar{R}/\mathfrak{m}_i \cong \bar{K}$.

Now we prove (f). Using (e), we have $\mathrm{nix}(\ell_i) = \min\{j \geq 1 \mid p_{\ell_i}(\ell_i)^j = 0\} = \min\{j \geq 1 \mid (\ell_i - c_i)^j = 0\}$. Since $\ell_i - c_i \in \bar{\mathfrak{m}}_i$ and since $z - c_i$ is a linear factor of $p_\ell(z) = \mathrm{lcm}(z - c_1, \ldots, z - c_\delta)$, also the element $p_\ell(\ell_i)$ is contained in $\bar{\mathfrak{m}}_i$. Thus we have $\mathrm{nix}(p_\ell(\ell_i)) = \min\{j \geq 1 \mid p_\ell(\ell_i)^j = 0\}$. The polynomial $p_\ell(z)$ is a product of distinct linear factors. When we substitute $z \mapsto \ell_i$ we get $\ell_i - c_i \in \bar{\mathfrak{m}}_i$ and for $c_j \neq c_i$ the element $\ell_i - c_j$ is a unit of $R_i$. Thus we get

$$\mathrm{nix}\big(p_\ell(\ell_i)\big) = \min\{j \geq 1 \mid p_\ell(\ell_i)^j = 0\} = \min\{j \geq 1 \mid (\ell_i - c_i)^j = 0\}$$
$$= \mathrm{nix}(\ell_i - c_i)$$

Finally, we prove (g). By applying $\varepsilon$, we get $\varepsilon(p_\ell(\ell)) = (p_\ell(\ell_1), \ldots, p_\ell(\ell_\delta))$. By definition, we have $\mathrm{nix}(\ell) = \min\{j \geq 1 \mid p_\ell(\ell)^j = 0\}$. On the other hand, from (f) we obtain the equalities $\mathrm{nix}(\ell_i) = \min\{j \geq 1 \mid p_\ell(\ell_i)^j = 0\}$. Therefore the claim follows from the observation that the tuple $(p_\ell(\ell_1)^j, \ldots, p_\ell(\ell_\delta)^j)$ is zero if and only if all of its entries are zero.                                                   $\square$

Now we are ready to prove the desired generalization of Proposition 4.4.9.

**Theorem 4.4.12 (Existence of Linear Elements of Maximal Nilpotency)**
*Let $(R, \mathfrak{m})$ be a local zero-dimensional affine $K$-algebra. We write $R = P/I$ with $P = K[x_1, \ldots, x_n]$ and an ideal $I$ in $P$, and assume that $K \subseteq R/\mathfrak{m}$ is a separable field extension*

(a) *If $\mathrm{char}(K) = 0$ or if $\mathrm{char}(K) \geq \mathrm{nix}(\mathfrak{m})$, there exists a linear form $\ell$ of $R$ which is of maximal nilpotency.*
(b) *If $\mathrm{char}(K) = 0$ or if $K$ is an infinite field with $\mathrm{char}(K) \geq \mathrm{nix}(\mathfrak{m})$ then a generic linear form $\ell$ of $R$ is of maximal nilpotency.*
(c) *If $\mathrm{char}(K) = 0$ or if $K$ is an infinite field with $\mathrm{char}(K) \geq \mathrm{nix}(\mathfrak{m})$ then a generic element $\ell$ of $R$ is of maximal nilpotency.*

*Proof* Let $\bar{K}$ be the algebraic closure of $K$, let $\bar{R}$, $\mathfrak{m}_i$, $\mathfrak{q}_i$, $R_i$, and $\bar{\mathfrak{m}}_i$ be defined as in the lemma, and let $r = \mathrm{nix}(\mathfrak{m}) - 1$.

First we prove claim (a). By the lemma, there exists an index $i \in \{1, \ldots, \delta\}$ such that $\mathrm{nix}(\mathfrak{m}) = \mathrm{nix}(\bar{\mathfrak{m}}_i)$. Since the maximal ideal $\bar{\mathfrak{m}}_i$ of $R_i$ is linear, we can apply Proposition 4.4.9.a and get an element $\ell_i \in R_i$ of maximal nilpotency such that $\ell_i$ is the image of a linear form $L$ in $\bar{P} = \bar{K}[x_1, \ldots, x_n]$ under $R_i \cong (\bar{P}/I\bar{P})/\mathfrak{q}_i$. In fact, the proof of Proposition 4.4.9.a uses Proposition 4.4.7.a. It we reconsider the proof of that proposition, we see that we have to avoid a certain hypersurface $H$ defined by a non-zero polynomial $\sum_{i=1}^{m} c_i \binom{r}{\alpha_{i1}, \ldots, \alpha_{in}} t_i$ of degree $r$ in $\bar{K}^n$. Since we have char$(K) > r$, Lemma 4.4.6 shows that we can actually find a point in $K^n$ avoiding this hypersurface. Consequently, we get a linear form $L$ in $K[x_1, \ldots, x_n]$ whose residue class $\ell_i$ in $R_i$ is of maximal nilpotency. By part (f) of the lemma, it follows that the residue class $\ell$ of $L$ in $R$ has maximal nilpotency.

Next we show claim (b). By the lemma, there exists an index $i \in \{1, \ldots \delta\}$ such that nix$(\mathfrak{m}) = $ nix$(\bar{\mathfrak{m}}_i)$. Since the maximal ideal $\bar{\mathfrak{m}}_i$ of $R_i$ is linear, we can apply Proposition 4.4.9.b and get a non-empty Zariski open subset $U$ of $\bar{K}x_1 + \cdots + \bar{K}x_n$ such that the residue class $\ell \in R$ of every element $L \in U$ has maximal nilpotency. Now we use the hypothesis that $K$ is infinite and Proposition 4.3.9 to conclude that $U' = U \cap (Kx_1 + \cdots + Kx_n)$ is a non-empty Zariski open set. The residue classes of all linear forms in $U'$ are linear forms of $R$ of maximal nilpotency.

Finally, we note that claim (c) follows in the same way from Proposition 4.4.9.c.

$\square$

In Proposition 4.4.3 we saw that an element of maximal nilpotency can be used to compute the maximal ideal $\mathfrak{m}$ of $R$. Thus, to apply the preceding theorem, we have to find conditions under which its hypotheses are satisfied. The following remark provides some important cases.

**Remark 4.4.13** Suppose that we are in the setting of the theorem.

(a) If char$(K) \geq \dim_K(R)$ then the hypothesis of the theorem holds, since we have nix$(\mathfrak{m}) < \dim_K(R)$.
(b) If the field $K$ is perfect, the hypothesis of the theorem (namely that $K \subseteq R/\mathfrak{m}$ is a separable field extension) holds, as already noted in Remark 3.4.6. E.g., this holds if $K$ is finite or if char$(K) = 0$.

Finally, we note that for a local family $(\mathscr{F}, \mathfrak{m})$ of commuting endomorphisms of a finite dimensional $K$-vector space $V$ and for an element $\varphi \in \mathscr{F}$, the eigenfactor of the multiplication map $\vartheta_\varphi : \mathscr{F} \longrightarrow \mathscr{F}$ agrees with the $\mathfrak{m}$-eigenfactor $p_{\mathfrak{m},\varphi}(z)$ of $\varphi$ (see Remark 4.1.3.b). Thus we can define the index of nilpotency of an element $\varphi$ of $\mathscr{F}$ by letting nix$(\varphi) = \min\{j \geq 1 \mid p_{\mathfrak{m},\varphi}(\varphi)^j = 0\}$. Using this definition, the theorem yields the following corollary.

**Corollary 4.4.14** *Let $\mathscr{F}$ be a local family of commuting endomorphisms of a finite dimensional $K$-vector space $V$, let $\mathfrak{m}$ be its maximal ideal, and let $(\varphi_1, \ldots, \varphi_n)$ be a system of $K$-algebra generators of $\mathscr{F}$.*

(a) *If char$(K) = 0$ or if char$(K) \geq$ nix$(\mathfrak{m})$, there exists a $K$-linear combination of $\varphi_1, \ldots, \varphi_n$ which is of maximal nilpotency.*
(b) *If char$(K) = 0$ or if $K$ is an infinite field with char$(K) \geq$ nix$(\mathfrak{m})$ then a generic $K$-linear combination of $\varphi_1, \ldots, \varphi_n$ is of maximal nilpotency.*
(c) *If char$(K) = 0$ or if $K$ is an infinite field with char$(K) \geq$ nix$(\mathfrak{m})$ then a generic element of $\mathscr{F}$ is of maximal nilpotency.*

## 4.5 Dual Multiplication Families

> *The professor told me:*
> *"To fully understand the canonical module,*
> *you have to develop your dual thinking."*
> *But we do not agree with this.*

In Sect. 3.5 we defined and studied the dual family of a commuting family $\mathscr{F}$. The Duality Theorem 3.5.12.a says that the dual family is commendable if the vector space is a cyclic $\mathscr{F}$-module. Wait a second! If $\mathscr{F}$ is the multiplication family of a zero-dimensional affine $K$-algebra $R$ then $R$ is a cyclic $\mathscr{F}$-module. Hence $\mathscr{F}^{\vee}$ is a commendable family, and we have to start developing our dual thinking.

Q: On which vector space does $\mathscr{F}^{\vee}$ operate naturally?

A: Of course, on the dual vector space $\omega_R = \mathrm{Hom}_K(R, K)$.

Q: And is $\mathscr{F}^{\vee}$ a multiplication family again? Is this $\omega_R$ a ring?

A: No, it is not a ring, but an $R$-module via $r \cdot \ell(s) = \ell(rs)$ for $r, s \in R$ and $\ell \in \omega_R$. It is called the *canonical module* of $R$.

Q: So, the dual family $\mathscr{F}$ is a family of endomorphisms of this canonical module and it is commendable. Does it help us to fully understand this module?

A: If you develop your dual thinking, you can use it to compute a minimal system of generators of $\omega_R$.

Q: Isn't this module cyclic, too?

A: In general, it isn't. But we can prove that it is a cyclic $R$-module if and only if $R$ is a *Gorenstein* ring.

Q: And then we can use the Duality Theorem again and conclude that the ring $R$ is Gorenstein if and only if $\mathscr{F}$ is commendable.

A: Wow! This seems to be a cool property. Can we check it algorithmically?

Q: Yes, we can. Just use the characterization via the cyclicity of $\omega_R$ and apply the Cyclicity Test 3.1.4 to $\mathscr{F}^{\vee}$. But wasn't I the one who was supposed to ask the questions?

A: This dual thinking gets us all mixed up.

> *To dualize "Dammit! I'm mad!",*
> *just spell it backwards.*

Let $K$ be a field, let $R$ be a zero-dimensional affine $K$-algebra, and let $\mathscr{F}$ be the multiplication family of $R$. Since $R$ is a cyclic $\mathscr{F}$-module, Theorem 3.5.12.a shows that the dual family $\mathscr{F}^{\vee}$ is commendable. A suitable $R$-module on which the dual family $\mathscr{F}^{\vee}$ operates is defined as follows.

**Definition 4.5.1** Let us consider the operation of the ring $R$ on the dual vector space $R^* = \mathrm{Hom}_K(R, K)$ given by

$$R \times R^* \longrightarrow R^*$$
$$(f, \ell) \longmapsto \big(g \mapsto \ell(fg)\big)$$

It is easy to check that this definition turns $R^*$ into an $R$-module. This module is denoted by $\omega_R$ and called the **canonical module** (or the **dualizing module**) of $R$.

The dual family $\mathscr{F}^\vee$ is a family of endomorphisms of $\omega_R$. They can be interpreted as follows.

**Remark 4.5.2**  Consider the dual family $\mathscr{F}^\vee = \{\vartheta_f^\vee \mid f \in R\}$ of the multiplication family $\mathscr{F}$ of $R$. It consists of endomorphisms of $\omega_R = R^*$. For every element $f \in R$, we can describe the action of $\vartheta_f^\vee$ on an element $\ell \in \omega_R$ by $\vartheta_f^\vee(\ell) = \ell \circ \vartheta_f$. In other words, for every $g \in R$ we have $(\vartheta_f^\vee(\ell))(g) = \ell(fg) = (f \cdot \ell)(g)$. Therefore the action of $\vartheta_f^\vee$ on $\omega_R$ is the multiplication by $f$ on this $R$-module. In other words, the $\mathscr{F}^\vee$-module structure on $\omega_R$ agrees with its $R$-module structure as defined above.

Given a $K$-basis of $R$, we can describe the $R$-module $\omega_R$ and its commuting family $\mathscr{F}^\vee$ as follows.

**Remark 4.5.3**  Let $B = (t_1, \ldots, t_d)$ be a $K$-basis of $R$. According to Definition 3.5.4, the dual basis $B^* = (t_1^*, \ldots, t_d^*)$ is the $K$-basis of $\omega_R$ which satisfies $t_i^*(t_j) = \delta_{ij}$ for $i, j \in \{1, \ldots, d\}$. For an element $f \in R$, the endomorphism $\vartheta_f : R \longrightarrow R$ is represented with respect to $B$ by the multiplication matrix $M_B(\vartheta_f)$. Then Proposition 3.5.5 implies that the matrix which represents the dual map $\vartheta_f^\vee : \omega_R \longrightarrow \omega_R$ with respect to the basis $B^*$ is $M_{B^*}(\vartheta_f^\vee) = M_B(\vartheta_f)^{\mathrm{tr}}$.

Subsequently, we will use the following slightly imprecise notation to simplify the reading of this section. Given a presentation $R = K[x_1, \ldots, x_n]/I$ of a zero-dimensional affine $K$-algebra $R$, we denote the residue class of $x_i$ in $R$ by $x_i$ again. The next two examples illustrate the structure of $\omega_R$ as an $R$-module. The results presented here can be verified directly, but more advanced techniques are explained later (see Algorithm 4.5.16 and Example 4.5.17).

**Example 4.5.4**  Let $K = \mathbb{Q}$, and let $R = K[x, y]/I$ where $I = \langle x^2 - xy, y^2 \rangle$. It is straightforward to see that $R$ is a local ring with maximal ideal $\langle x, y \rangle$ and its socle is $\mathrm{Soc}(R) = \langle xy \rangle$. A $K$-basis of $R$ is given by $B = (1, x, y, xy)$.

Thus the dual basis $B^* = (1^*, x^*, y^*, (xy)^*)$ is a $K$-basis of the canonical module $\omega_R$. Let us examine the $R$-module structure of $\omega_R$ in terms of this basis. From a straightforward verification of the equalities

$$1^* = xy(xy)^*, \quad x^* = y(xy)^*, \quad \text{and} \quad y^* = (x - y)(xy)^*$$

we conclude that $(xy)^*$ generates $\omega_R$ as an $R$-module.

A slightly more involved example containing a non-trivial residue class field extension is given as follows.

**Example 4.5.5** Let $K = \mathbb{Q}$, and let $R = K[x]/I$ where $I = \langle(x^2 + 1)^3\rangle$. Let us denote the residue class of $x^2 + 1$ in $R$ by $a$. Clearly, the ring $R$ is local with maximal ideal $\mathfrak{m} = \langle a \rangle$ and its socle is $\text{Soc}(R) = \langle a^2 \rangle$. Notice that the socle of $R$ is a 2-dimensional $K$-vector space with basis $(a^2, xa^2)$, and at the same time it is a 1-dimensional vector space over the field $K[x]/\langle x^2 + 1\rangle$ with basis $(a^2)$.

The tuple $B = (1, x, a, xa, a^2, xa^2)$ is a $K$-basis of $R$. Consequently, the tuple $B^* = (1^*, x^*, a^*, (xa)^*, (a^2)^*, (xa^2)^*)$ is a $K$-basis of $\omega_R$. To determine the $R$-module structure of $\omega_R$, we calculate

$$1^* = a^2(a^2)^*, \quad x^* = -xa^2(a^2)^*, \quad a^* = a(a^2)^*$$
$$(xa)^* = (-xa - xa^2)(a^2)^*, \quad \text{and} \quad (xa^2)^* = -(x + xa + xa^2)(a^2)^*$$

It follows that $(a^2)^*$ generates $\omega_R$ as an $R$-module. Notice that $-(x + xa + xa^2)$ is the multiplicative inverse of $x$, since we have the following equalities $x(-x - xa - xa^2) = -x^2(1 + a + a^2) = (1 - a)(1 + a + a^2) = 1 - a^3 = 1$.

An important property of $\omega_R$ is that the canonical module of a residue class ring $R/I$ of $R$ corresponds to a submodule of $\omega_R$ as follows.

**Proposition 4.5.6 (The Canonical Module of a Residue Class Ring)**
*Let $I$ be an ideal of $R$, let $\varepsilon : R \longrightarrow R/I$ be the canonical epimorphism, and let $\omega_{R/I} = \text{Hom}_K(R/I, K)$ be the canonical module of $R/I$.*

(a) *The map $\varepsilon^{\vee} : \omega_{R/I} \longrightarrow \omega_R$ given by $\ell \mapsto \ell \circ \varepsilon$ is an injective $R$-module homomorphism.*
(b) *The image of the map $\varepsilon^{\vee}$ is the $R$-submodule $\text{Ann}_{\omega_R}(I) = \{\ell \in \omega_R \mid I \cdot \ell = 0\}$ of $\omega_R$.*

*Proof* To show (a) we note that, for a linear map $\ell \in \omega_{R/I}$, the equality $\ell \circ \varepsilon = 0$ implies $\ell = 0$, because $\varepsilon$ is surjective. Now let us prove (b). For every $\ell \in \omega_{R/I}$, we have $(\ell \circ \varepsilon)(I) = \ell(0) = 0$. Hence the image of $\omega_{R/I}$ under $\varepsilon^{\vee}$ is contained in $\text{Ann}_{\omega_R}(I)$. Conversely, let $\ell' \in \omega_R$ be a linear map such that $\ell'(I) = 0$. By the universal property of the residue class ring, the map $\ell'$ factors in the form $\ell' = \ell \circ \varepsilon$ with $\ell \in \omega_{R/I}$. This observation finishes the proof. $\square$

The main result of Sect. 3.5 is Theorem 3.5.12 which says that the dual family $\mathscr{F}^{\vee}$ is commendable if the vector space is a cyclic $\mathscr{F}$-module. In the current setting this yields the following result.

**Proposition 4.5.7** *Let $R$ be a zero-dimensional affine $K$-algebra, and let $\mathscr{F}$ be the multiplication family of $R$. Then the dual family $\mathscr{F}^{\vee}$ is commendable.*

*Proof* This follows from Theorem 3.5.12.a and the fact that the ring $R$ is a cyclic $\mathscr{F}$-module. $\square$

Our next topic is the explicit construction of a system of $R$-module generators of the canonical module $\omega_R$. We need some preparatory results. The following lemma provides a criterion under which a vector subspace of $\omega_R$ coincides with $\omega_R$.

**Lemma 4.5.8** *For a $K$-vector subspace $U$ of $\omega_R$, the following conditions are equivalent.*

(a) *We have $U = \omega_R$.*
(b) *For every $r \in R \setminus \{0\}$, there exists an element $\ell \in U$ such that $\ell(r) \neq 0$.*

*Proof* To prove the implication (a) $\Rightarrow$ (b), we extend $r$ to a $K$-basis $(b_1, \ldots, b_d)$ of $R$, where $b_1 = r$, and observe that $b_1^*(b_1) = 1$.

To prove the converse, we argue by contradiction. Let $d = \dim_K(R) = \dim_K(\omega_R)$ and suppose that $e = \dim_K(U) < d$. Then there exists a $K$-basis $\ell_1, \ldots, \ell_e$ of $U$ and since $\dim_K(\mathrm{Ker}(\ell_i)) = d - 1$, the intersection $W = \mathrm{Ker}(\ell_1) \cap \cdots \cap \mathrm{Ker}(\ell_e)$ satisfies $\dim_K(W) \geq d - e \geq 1$. Now every $r \in W \setminus \{0\}$ yields a contradiction to (b). $\qquad\square$

The following lemma contains useful properties of the socle of $R$.

**Lemma 4.5.9** *Let $(R, \mathfrak{m})$ be a local zero-dimensional affine $K$-algebra. Moreover, let $r \in R \setminus \{0\}$.*

(a) *There exists an element $r' \in R$ such that $rr' \in \mathrm{Soc}(R) \setminus \{0\}$.*
(b) *Let $f_1, \ldots, f_s \in R$ be elements whose residue classes form an $R/\mathfrak{m}$-basis of $\mathrm{Soc}(R)$. There exists an element $r'' \in R$ such that $rr''$ is an element in $\mathrm{Soc}(R)$ of the form $rr'' = a_1 f_1 + \cdots + a_s f_s$ with $a_i = 1$ for some $i \in \{1, \ldots, s\}$.*

*Proof* First we show (a). Given an element $r \in R \setminus \{0\}$, there are two cases. Either we have $r \in \mathrm{Soc}(R) \setminus \{0\}$, or there exists an element $m_1 \in \mathfrak{m}$ such that $rm_1 \in R \setminus \{0\}$. Continuing in the same way with $rm_1$, we find elements $m_1, m_2, \ldots$ in $\mathfrak{m}$ such that we either get a non-zero element $rm_1 \cdots m_i$ of the socle of $R$ or a non-zero element $rm_1 \cdots m_{i+1}$ of $R$. Since we have $\mathfrak{m}^k = \langle 0 \rangle$ for $k \gg 0$, the first case has to hold eventually.

To prove (b), we use (a) to find $r', a_i' \in R$ with $rr' = a_1' f_1 + \cdots + a_s' f_s \neq 0$. Then there exists an index $i \in \{1, \ldots, s\}$ such that $a_i' \neq 0$. Let $b \in R$ be an element such that $ba_i' = 1 + m$ with $m \in \mathfrak{m}$. Then the element $rr'b$ is of the desired form. $\qquad\square$

The next theorem provides a method for computing a minimal system of generators of $\omega_R$ in the local case.

**Theorem 4.5.10 (A Minimal System of Generators of the Canonical Module)**
*Let $(R, \mathfrak{m})$ be a local zero-dimensional affine $K$-algebra. Let $f_1, \ldots, f_s$ be elements of the ring $R$ whose residue classes form an $R/\mathfrak{m}$-basis of $\mathrm{Soc}(R)$, and let $b_1, \ldots, b_\delta \in R$ be elements whose residue classes form a $K$-basis of $R/\mathfrak{m}$.*

(a) *For $i = 1, \ldots, s$, the tuple $(b_1 f_i, \ldots, b_\delta f_i)$ is a $K$-basis of $Rf_i$. In particular, the elements $b_i f_j$ such that $i \in \{1, \ldots, \delta\}$ and $j \in \{1, \ldots, s\}$ form a $K$-basis of $\mathrm{Soc}(R)$.*

(b) *Let B be a K-basis of R which extends the K-basis of* $\mathrm{Soc}(R)$ *given in (a), and let* $B^*$ *be the dual basis of* $\omega_R$. *Then the elements* $f_1^*, \ldots, f_s^*$ *of* $B^*$ *form a minimal system of R-module generators of* $\omega_R$.

*Proof* To prove (a), we note that the $R$-module $Rf_i$ is annihilated by $\mathfrak{m}$. Hence the tuple $(b_1 f_i, \ldots, b_\delta f_i)$ is a $K$-basis of $Rf_i$ for $i = 1, \ldots, s$. The additional claim follows from the observation that the $\mathrm{Soc}(R)$ is the direct sum of its $R$-submodules $Rf_i$, because the $R$-module structure agrees with the $R/\mathfrak{m}$-vector space structure.

Next we show (b). To prove that the $R$-submodule $Rf_1^* + \cdots + Rf_s^*$ is equal to $\omega_R$, we apply Lemma 4.5.8. Given an element $r \in R \setminus \{0\}$, Lemma 4.5.9.b yields an element $r'' \in R$ and an index $i \in \{1, \ldots, s\}$ such that $rr' = a_1 f_1 + \cdots + a_s f_s$ with $a_1, \ldots, a_s \in R$ and $a_i = 1$. Since we have $a_j f_j \in \langle b_1 f_j, \ldots, b_\delta f_j \rangle_K$ by (a), the construction of $B^*$ implies $f_i^*(a_j f_j) = 0$ for $j \neq i$. This shows $f_i^*(rr') = 1$, and therefore $(r' f_i^*)(r) \neq 0$. Hence the hypothesis of Lemma 4.5.8 is satisfied and we get $Rf_1^* + \cdots + Rf_s^* = \omega_R$.

It remains to show that the system of generators $\{f_1^*, \ldots, f_s^*\}$ is minimal. Assume that there exist an index $i \in \{1, \ldots, s\}$ and elements $a_j \in R$ for $j \neq i$ such that $f_i^* = \sum_{j \neq i} a_j f_j^*$. Then we get $1 = f_i^*(f_i) = \sum_{j \neq i} f_j^*(a_j f_i)$. Since $a_j f_i$ is contained in $\langle b_1 f_i, \ldots, b_\delta f_i \rangle_K$, the right-hand side is zero. Now the contradiction $1 = 0$ shows that our assumption was false, and the proof of the theorem is complete.                                                                                $\square$

Recall the local version of the **Lemma of Nakayama** (see for instance [11], Sect. 7.8): a set of elements of a finitely generated module $M$ of a local ring $(R, \mathfrak{m})$ is a minimal system of $M$ if and only if their residue classes form an $R/\mathfrak{m}$-vector space basis of $M/\mathfrak{m}M$, and consequently, all minimal systems of generators of $M$ have the same cardinality. Hence the theorem shows that every minimal system of generators of $\omega_R$ has $\dim_{R/\mathfrak{m}}(\mathrm{Soc}(R))$ elements.

In the following we strive to turn this theorem into an algorithm for computing a minimal system of generators of $\omega_R$ and for expressing arbitrary elements of $\omega_R$ in terms of this system of generators. Both tasks will be performed with respect to an arbitrary given $K$-basis of $R$. Let us start by introducing the following definition.

**Definition 4.5.11** Let $R$ be a zero-dimensional affine $K$-algebra. Assume that $B = (r_1, \ldots, r_d)$ is a tuple of elements of $R$, and $L = (\ell_1, \ldots, \ell_s)$ is a tuple of elements of $\omega_R$. Then the matrix

$$\mathrm{Eval}(L, B) = \big(\ell_i(r_j)\big) \in \mathrm{Mat}_{s,d}(K)$$

is called the **evaluation matrix** of $L$ at $B$.

The following lemma contains useful properties to control the behaviour of evaluation matrices under base changes.

**Lemma 4.5.12** *Let R be a zero-dimensional affine K-algebra, let* $B = (r_1, \ldots, r_d)$ *be a tuple of elements of R, and let* $L = (\ell_1, \ldots, \ell_s)$ *be a tuple of elements of* $\omega_R$.

*Then, let $t, u$ be positive integers, let $M \in \mathrm{Mat}_{s,t}(K)$, and let $N \in \mathrm{Mat}_{d,u}(K)$.*

(a) *We have* $\mathrm{Eval}(L \cdot M, B) = M^{\mathrm{tr}} \cdot \mathrm{Eval}(L, B)$.
(b) *We have* $\mathrm{Eval}(L, B \cdot N) = \mathrm{Eval}(L, B) \cdot N$.

*Proof* Notice that the matrices on the right and left hand sides of the equality in (a) are in $\mathrm{Mat}_{t,d}(K)$, and the matrices in (b) are in $\mathrm{Mat}_{s,u}(K)$.

To prove (a), we compare the entry in position $(i, j)$ of both sides. The $i$-th entry of $L \cdot M$ is $\sum_{k=1}^{s} m_{ki}\ell_k$. Hence the entry in position $(i, j)$ of the matrix $\mathrm{Eval}(L \cdot M, B)$ is $\sum_{k=1}^{s} m_{ki}\ell_k(b_j)$. On the other hand, the $i$-th row of $M^{\mathrm{tr}}$ is $(m_{1i}, \ldots, m_{si})$. This row has to be multiplied by the $j$-th column of $\mathrm{Eval}(L, B)$. The latter is $(\ell_1(b_j), \ldots, \ell_s(b_j))^{\mathrm{tr}}$. Therefore the entry in position $(i, j)$ of $M^{\mathrm{tr}} \cdot \mathrm{Eval}(L, B)$ is $\sum_{k=1}^{s} m_{ki}\ell_k(b_j)$, and claim (a) is proved. A similar calculation yields claim (b). $\qquad\square$

Next we use this lemma to determine the relation between a change of basis of $R$ and the corresponding change of basis of $\omega_R$.

**Proposition 4.5.13** *Let $R$ be a zero-dimensional affine $K$-algebra. Then let $d = \dim_K(R)$, let $B_1$ and $B_2$ be two $K$-bases of $R$, and let $A \in \mathrm{Mat}_d(K)$ be such that $B_2 = B_1 \cdot A$.*

(a) *We have* $B_2^* = B_1^* \cdot (A^{-1})^{\mathrm{tr}}$.
(b) *We have* $\mathrm{Eval}(B_2^*, B_1) = A^{-1}$.

*Proof* First we show (a). Let $I_d$ be the identity matrix of size $d$. Since it is clear that $\mathrm{Eval}(B_2^*, B_2) = I_d$, it suffices to show the equality $\mathrm{Eval}(B_1^* \cdot (A^{-1})^{\mathrm{tr}}, B_2) = I_d$. By the hypothesis, we have $\mathrm{Eval}(B_1^* \cdot (A^{-1})^{\mathrm{tr}}, B_2) = \mathrm{Eval}(B_1^* \cdot (A^{-1})^{\mathrm{tr}}, B_1 \cdot A)$. From the lemma we get $\mathrm{Eval}(B_1^* \cdot (A^{-1})^{\mathrm{tr}}, B_1 \cdot A) = A^{-1} \cdot \mathrm{Eval}(B_1^*, B_1) \cdot A$, and hence the conclusion follows.

To prove (b) we use the lemma and (a) to get

$$\mathrm{Eval}\big(B_2^*, B_1\big) = \mathrm{Eval}\big(B_1^* \cdot \big(A^{-1}\big)^{\mathrm{tr}}, B_1\big) = A^{-1} \cdot \mathrm{Eval}\big(B_1^*, B_1\big) = A^{-1}$$

Consequently, the proof is complete. $\qquad\square$

Thus we can now construct two algorithms. Given a $K$-basis $B$ of a local zero-dimensional affine $K$-algebra $(R, \mathfrak{m})$, the first algorithm computes a minimal set of generators of the $R$-module $\omega_R$ in terms of the dual basis $B^*$. The non-trivial task of computing $\mathfrak{m}$ from the multiplication family of $R$ will be discussed in Chap. 5.

**Algorithm 4.5.14 (Computing Minimal Generators of $\omega_R$)**
*Let $(R, \mathfrak{m})$ be a local zero-dimensional affine $K$-algebra, let $d = \dim_K(R)$, let $\delta = \dim_K(R/\mathfrak{m})$, and let $B = (r_1, \ldots, r_d)$ be a $K$-basis of $R$. Consider the following sequence of instructions.*

(1) *Compute a tuple $(b_1, \ldots, b_\delta)$ of elements of $R$ such that $b_1 = 1$ and their residue classes form a $K$-basis of $R/\mathfrak{m}$.*

(2) *Using one of the methods explained in Remark 4.2.8, compute a $K$-basis $G = (g_1, \ldots, g_{s\delta})$ of $\mathrm{Soc}(R)$ where $\dim_{R/\mathfrak{m}}(\mathrm{Soc}(R)) = s$.*

(3) *Starting with $f_1 = g_1$, find elements $f_1, \ldots, f_s$ in $G$ such that*

$$f_i \notin \langle b_j f_k \mid j \in \{1, \ldots, \delta\}, k \in \{1, \ldots, i-1\} \rangle_K$$

(4) *Extend the tuple $(b_1 f_1, \ldots, b_\delta f_s)$ to a $K$-basis $B'$ of $R$.*

(5) *Let $f_1^*, \ldots, f_s^*$ be the dual linear forms of $f_1, \ldots, f_s$ with respect to $B'$. Compute the invertible matrix $A$ such that $B' = B \cdot A$ and use Proposition 4.5.13.a to compute representations of $f_1^*, \ldots, f_s^*$ in terms of $B^*$.*

(6) *Return the representations of $f_1^*, \ldots, f_s^*$ in terms of $B^*$ and stop.*

*This is an algorithm which computes a minimal set of generators of the $R$-module $\omega_R$ in terms of the $K$-basis $B^*$.*

*Proof* By construction, the $s\delta$-tuple $(b_1 f_1, \ldots, b_\delta f_s)$ is a $K$-basis of $\mathrm{Soc}(R)$, and the elements $f_1, \ldots, f_s$ are entries of this tuple since $b_1 = 1$. Therefore the tuple $(f_1, \ldots, f_s)$ is an $R/\mathfrak{m}$-basis of $\mathrm{Soc}(R)$. According to Theorem 4.5.10, the set $\{f_1^*, \ldots, f_s^*\}$ is a minimal system of generators of $\omega_R$. Since the elements $f_1^*, \ldots, f_s^*$ are entries of $B^*$, they have obvious representations in terms of this basis. Moreover, from the equality $B' = B \cdot A$ and Proposition 4.5.13 we get $(B')^* = B^* \cdot (A^{-1})^{\mathrm{tr}}$. Hence the correctness of the algorithm follows. $\square$

The next corollary provides the essential ingredient for the algorithm below.

**Corollary 4.5.15** *In the setting of Algorithm 4.5.14, let $L$ be the tuple*

$$L = \left( r_1 f_1^*, r_2 f_1^*, \ldots, r_d f_1^*, \ldots, r_1 f_s^*, r_2 f_s^*, \ldots, r_d f_s^* \right)$$

*Then we have $\mathrm{rank}(\mathrm{Eval}(L, B)) = d$. In particular, there exists a subtuple $L'$ of $L$ consisting of $d$ elements such that the evaluation matrix $\mathrm{Eval}(L', B)$ is invertible.*

*Proof* Since $f_1^*, \ldots, f_s^*$ form a system of $R$-module generators of $\omega_R$, the elements in the tuple $L$ are a system of generators of $\omega_R$ as a $K$-vector space. Hence they contain a $K$-basis of $\omega_R$, and the evaluation matrix of that $K$-basis at the basis $B$ of $R$ is invertible by Proposition 4.5.13.b. $\square$

After computing a minimal system of generators of $\omega_R$, we can express an arbitrary element of $\omega_R$ as an $R$-linear combination of this system of generators as follows.

**Algorithm 4.5.16 (Explicit Membership in $\omega_R$)**

In the setting of Algorithm 4.5.14, let $f_1^*, \ldots, f_s^*$ be the system of $R$-module generators of $\omega_R$ computed by that algorithm. Given an element $\ell = \sum_{i=1}^{d} c_i r_i^*$ of $\omega_R$, where $c_i \in K$ and where $B^* = (r_1^*, \ldots, r_d^*)$, consider the following sequence of instructions.

(1) Let $L = (r_1 f_1^*, r_2 f_1^*, \ldots, r_d f_1^*, \ldots, r_1 f_s^*, r_2 f_s^*, \ldots, r_d f_s^*)$ and compute the matrix $\text{Eval}(L, B)$.
(2) Find $d$ $K$-linearly independent rows of $\text{Eval}(L, B)$ and use them to form an invertible matrix $A \in \text{Mat}_d(K)$.
(3) Using $A^{-1}$, compute the representations of the elements of $B^*$ as linear combinations of elements of $L$.
(4) Using the result of Step (3), compute the representation of $\ell$ as a $K$-linear combination of the elements of $L$.
(5) Return the resulting representation of $\ell$ as an $R$-linear combination of the elements $f_1^*, \ldots, f_s^*$.

This is an algorithm which computes a tuple $(h_1, \ldots, h_s) \in R^s$ such that we have $\ell = \sum_{i=1}^{s} h_i f_i^*$.

*Proof* The fact that there exist $d$ linearly independent rows in $\text{Eval}(L, B)$ was shown in Corollary 4.5.15. The corresponding elements $b_i f_j^*$ are then a $K$-basis of $\omega_R$. Thus it suffices to collect the coefficients as done in Steps (4) and (5) to get a representation of $\ell$ as an $R$-linear combination of $f_1^*, \ldots, f_s^*$.                          $\square$

Let us see these two algorithms at work. As already pointed out, they can be used to perform the computations underlying Examples 4.5.4 and 4.5.5.

**Example 4.5.17**   Let $K = \mathbb{Q}$, let $R = K[x, y]/I$, where $I$ is the ideal $I = \langle (x^2 + 1)^3, (x^2 + 1)y, y^2 \rangle$, and let $a$ denote the image of $x^2 + 1$ in $R$. To ease the reading, we denote the images of $x$ and $y$ in residue class rings of $K[x, y]$ by $x$ and $y$ again. Clearly, the ring $R$ has a unique maximal ideal, namely $\mathfrak{m} = \langle a, y \rangle$, and so $(R, \mathfrak{m})$ is a local $K$-algebra. A $K$-basis of $R$ is given by $B = (1, y, x, xy, x^2, x^3, x^4, x^5)$. We compute the socle of $R$ and a system of generators of $\omega_R$ using Algorithm 4.5.14.

(0) A $K$-basis of $\mathfrak{m}$ is given by $(a, xa, a^2, xa^2, y, xy)$ where we write $x, y$ instead of $\bar{x}, \bar{y}$ to ease the reading. The $K$-basis $B' = (a, xa, a^2, xa^2, y, xy, 1, x)$ of $R$ extends the basis of $\mathfrak{m}$. Here $(1, x)$ is a $K$-basis of $R/\mathfrak{m}$.
(1) We compute $\text{Soc}(R) = \text{Ker}(\tilde{\mathfrak{m}})$ and get the $K$-basis $(a^2, xa^2, y, xy)$.
(2) An $R/\mathfrak{m}$-basis of $\text{Soc}(R)$ is then given by $(a^2, y)$.
(3) Since $a^2$ and $y$ are contained in the basis $B$, their dual linear forms $(a_2)^*$ and $y^*$ form the desired minimal system of generators of $\omega_R$.

If we want to represent the $R$-module generators $(a_2)^*$ and $y^*$ of $\omega_R$ in terms of the $K$-basis $B^*$, we compute the matrix $A$ such that $B' = B \cdot A$ and then use the formula $(B')^* = B^* \cdot (A^{-1})^{\mathrm{tr}}$ to get the equalities $(a_2)^*_{B'} = -2(x^4)^*_B + (x^2)^*_B$ and $y^*_{B'} = y^*_B$. Next we assume that we are given an element

$$\ell = c_1 1^* + c_2 x^* + c_3(a^2)^* + c_4 y^* + c_5(xa^2)^* + c_6(xy)^* + c_7 a^* + c_8(xa)^* \in \omega_R$$

with $c_i \in K$, and we use Corollary 4.5.15 to represent $\ell$ as an $R$-linear combination of $(a^2)^*$ and $y^*$.

(1) We let $L = ((a^2)^*, x(a^2)^*, a^2(a^2)^*, y(a^2)^*, xa^2(a^2)^*, xy(a^2)^*, a(a^2)^*,$ $xa(a^2)^*, y^*, xy^*, a^2y^*, yy^*, xa^2y^*, xyy^*, ay^*, xay^*)$ and compute the matrix $\mathrm{Eval}(L, B)$. The result is

|            | $1$ | $x$ | $a^2$ | $y$ | $xa^2$ | $xy$ | $a$ | $xa$ |
|------------|-----|-----|-------|-----|--------|------|-----|------|
| $(a^2)^*$      | 0 | 0 | 1 | 0 | 0 | 0 | 0 | 0 |
| $x(a^2)^*$     | 0 | 0 | 0 | 0 | -1 | 0 | 0 | 1 |
| $a^2(a^2)^*$   | 1 | 0 | 0 | 0 | 0 | 0 | 0 | 0 |
| $y(a^2)^*$     | 0 | 0 | 0 | 0 | 0 | 0 | 0 | 0 |
| $xa^2(a^2)^*$  | 0 | -1 | 0 | 0 | 0 | 0 | 0 | 0 |
| $xy(a^2)^*$    | 0 | 0 | 0 | 0 | 0 | 0 | 0 | 0 |
| $a(a^2)^*$     | 0 | 0 | 0 | 0 | 0 | 0 | 1 | 0 |
| $xa(a^2)^*$    | 0 | 1 | 0 | 0 | 0 | 0 | 0 | -1 |
| $y^*$          | 0 | 0 | 0 | 1 | 0 | 0 | 0 | 0 |
| $xy^*$         | 0 | 0 | 0 | 0 | 0 | -1 | 0 | 0 |
| $a^2y^*$       | 0 | 0 | 0 | 0 | 0 | 0 | 0 | 0 |
| $yy^*$         | 1 | 0 | 0 | 0 | 0 | 0 | 0 | 0 |
| $xa^2y^*$      | 0 | 0 | 0 | 0 | 0 | 0 | 0 | 0 |
| $xyy^*$        | 0 | -1 | 0 | 0 | 0 | 0 | 0 | 0 |
| $ay^*$         | 0 | 0 | 0 | 0 | 0 | 0 | 0 | 0 |
| $xay^*$        | 0 | 0 | 0 | 0 | 0 | 0 | 0 | 0 |

(2) Using the indicated rows, we extract the following $8 \times 8$ invertible matrix:

|            | $1$ | $x$ | $a^2$ | $y$ | $xa^2$ | $xy$ | $a$ | $xa$ |
|------------|-----|-----|-------|-----|--------|------|-----|------|
| $(a^2)^*$      | 0 | 0 | 1 | 0 | 0 | 0 | 0 | 0 |
| $x(a^2)^*$     | 0 | 0 | 0 | 0 | -1 | 0 | 0 | 1 |
| $a^2(a^2)^*$   | 1 | 0 | 0 | 0 | 0 | 0 | 0 | 0 |
| $xa^2(a^2)^*$  | 0 | -1 | 0 | 0 | 0 | 0 | 0 | 0 |
| $a(a^2)^*$     | 0 | 0 | 0 | 0 | 0 | 0 | 1 | 0 |
| $xa(a^2)^*$    | 0 | 1 | 0 | 0 | 0 | 0 | 0 | -1 |
| $y^*$          | 0 | 0 | 0 | 1 | 0 | 0 | 0 | 0 |
| $xy^*$         | 0 | 0 | 0 | 0 | 0 | -1 | 0 | 0 |

(3) The inverse of this matrix is

$$
\begin{pmatrix}
0 & 0 & 1 & 0 & 0 & 0 & 0 & 0 \\
0 & 0 & 0 & -1 & 0 & 0 & 0 & 0 \\
1 & 0 & 0 & 0 & 0 & 0 & 0 & 0 \\
0 & 0 & 0 & 0 & 0 & 0 & 1 & 0 \\
0 & -1 & 0 & -1 & 0 & -1 & 0 & 0 \\
0 & 0 & 0 & 0 & 0 & 0 & 0 & -1 \\
0 & 0 & 0 & 0 & 1 & 0 & 0 & 0 \\
0 & 0 & 0 & -1 & 0 & -1 & 0 & 0
\end{pmatrix}
$$

Using it, we compute

$$
1^* = a^2(a^2)^*, \quad x^* = -xa^2(a^2)^*, \quad (a^2)^* = (a^2)^*, \quad y^* = y^*,
$$
$$
(xa^2)^* = -x(a^2)^* - xa^2(a^2)^* - xa(a^2)^*, \quad (xy)^* = xy^*,
$$
$$
a^* = a(a^2)^*, \quad \text{and} \quad (xa)^* = -xa^2(a^2)^* - xa(a^2)^*
$$

(4) Finally, we get

$$
\ell = \left(c_1 a^2 - c_2 xa^2 + c_3 - c_5(x + xa^2 + xa) + c_7 a - c_8(xa^2 + xa)\right)(a^2)^*
$$
$$
+ (c_4 + xc_6)y^*
$$

which is a representation of $\ell$ as an $R$-linear combination of $(a^2)^*$ and $y^*$.

As we noted in Remark 4.5.2, the family $\mathscr{F}^{\vee}$ describes the multiplication on $\omega_R$ by elements of $R$. Let us see how the joint eigenspaces and the joint generalized eigenspaces of this family can be interpreted in terms of the canonical module.

**Remark 4.5.18** Let $R$ be a zero-dimensional affine $K$-algebra, let $\mathscr{F}$ be its multiplication family, and let $\omega_R$ be the canonical module of $R$.

(a) Letting $\mathfrak{m}_1, \ldots, \mathfrak{m}_s$ be the maximal ideals of $R$, we have $\mathrm{Rad}(0) = \mathfrak{m}_1 \cap \cdots \cap \mathfrak{m}_s$ and $R/\mathrm{Rad}(0) \cong R/\mathfrak{m}_1 \times \cdots \times R/\mathfrak{m}_s$ (see Proposition 2.2.5.b and Theorem 2.2.1). Together with Proposition 4.5.6 and Proposition 4.1.11.b, this implies

$$
\omega_{R/\mathrm{Rad}(0)} = \mathrm{Ann}_{\omega_R}(\mathfrak{m}_1 \cap \cdots \cap \mathfrak{m}_s) = \bigoplus_{i=1}^{s} \mathrm{Ann}_{\omega_R}(\mathfrak{m}_i)
$$
$$
= \bigoplus_{i=1}^{s} \mathrm{Ker}(\mathfrak{m}_i^{\vee}) \cong \bigoplus_{i=1}^{s} \omega_{R/\mathfrak{m}_i}
$$

Here the vector spaces $\omega_{R/\mathfrak{m}_i} \cong \mathrm{Ker}(\mathfrak{m}_i^{\vee})$ are identified with the joint eigenspaces of the family $\mathscr{F}^{\vee}$ (see Remark 3.5.13).

(b) Letting $\mathfrak{q}_i$ be the $\mathfrak{m}_i$-primary component of the zero ideal of $R$ for $i = 1, \ldots, s$, we have $R \cong R/\mathfrak{q}_1 \times \cdots \times R/\mathfrak{q}_s$. By Proposition 4.5.6 and Proposition 4.1.11.c, we get

$$\omega_R = \bigoplus_{i=1}^{s} \mathrm{Ann}_{\omega_R}(\mathfrak{q}_i) = \bigoplus_{i=1}^{s} \mathrm{BigAnn}_{\omega_R}(\mathfrak{q}_i)$$
$$= \bigoplus_{i=1}^{s} \mathrm{BigKer}(\mathfrak{m}_i^{\vee}) = \omega_{R/\mathfrak{q}_1} \oplus \cdots \oplus \omega_{R/\mathfrak{q}_s}$$

Here the vector spaces $\omega_{R/\mathfrak{q}_i} \cong \mathrm{BigKer}(\mathfrak{m}_i^{\vee})$ are identified with the joint generalized eigenspaces of the family $\mathscr{F}^{\vee}$. Moreover, if we choose bases $B_i$ of the images of the factors $R/\mathfrak{q}_i$ in $R$ then $B = B_1 \cup \cdots \cup B_s$ is a $K$-basis of $R$ and the dual basis $B^* = B_1^* \cup \cdots \cup B_s^*$ is a $K$-basis of $\omega_R$ which is consistent with the direct sum decomposition $\omega_R = \bigoplus_{i=1}^{s} \omega_{R/\mathfrak{q}_i}$.

An important class of zero-dimensional affine $K$-algebras, defined by the condition that all local socles are 1-dimensional vector spaces, is given the following name.

**Definition 4.5.19**   Let $R$ be a zero-dimensional affine $K$-algebra.

(a) Let $(R, \mathfrak{m})$ be a local ring. We say that $R$ is a **Gorenstein local ring** if we have $\dim_{R/\mathfrak{m}}(\mathrm{Soc}(R)) = 1$.
(b) Let $\mathfrak{q}_1, \ldots, \mathfrak{q}_s$ be the primary components of the zero ideal in $R$. We say that $R$ is a **Gorenstein ring** if $R/\mathfrak{q}_i$ is a Gorenstein local ring for $i = 1, \ldots, s$.

The following remark provides a large class of Gorenstein rings.

**Remark 4.5.20**   A field is clearly a Gorenstein ring. Consequently, every reduced zero-dimensional affine $K$-algebra $R$ is a Gorenstein ring, as we can see by applying the isomorphism $R \cong R/\mathfrak{m}_1 \times \cdots \times R/\mathfrak{m}_s$ induced by the primary decomposition of $R$ (see Proposition 2.2.5.c).

The next theorem characterizes Gorenstein rings and provides a link between this property of the ring $R$ and the commendability of its multiplication family $\mathscr{F}$.

**Theorem 4.5.21 (Characterization of Zero-Dimensional Gorenstein Algebras)**
*Let $R$ be a zero-dimensional affine $K$-algebra. The following conditions are equivalent.*

(a) *The ring $R$ is a Gorenstein ring.*
(b) *The multiplication family $\mathscr{F}$ of $R$ is commendable.*
(c) *The canonical module $\omega_R$ is a cyclic $R$-module.*

*Proof* First we prove that conditions (a) and (b) are equivalent. By the definition of a Gorenstein ring and the local nature of commendability (see Proposition 3.3.7), we may assume that $(R, \mathfrak{m})$ is a local ring. Then Proposition 4.1.11.b yields $\mathrm{Soc}(R) = \mathrm{Ker}(\tilde{\mathfrak{m}})$, where $\tilde{\mathfrak{m}} = \iota(\mathfrak{m})$ is the corresponding maximal ideal of $\mathscr{F}$. Hence we have $\dim_{R/\mathfrak{m}}(\mathrm{Soc}(R)) = 1$ if and only if $\dim_{\mathscr{F}/\tilde{\mathfrak{m}}}(\mathrm{Ker}(\tilde{\mathfrak{m}})) = 1$, and this is equivalent to the commendability of $\mathscr{F}$ by Definition 3.3.1.

It remains to prove that conditions (b) and (c) are equivalent. By Remark 4.5.2, the dual family $\mathscr{F}^{\vee}$ describes the multiplication by elements of $R$ on the module $\omega_R$. Therefore we see that the module $\omega_R$ is a cyclic $R$-module if and only if it is a cyclic $\mathscr{F}^{\vee}$-module. Consequently, the claim follows from Theorem 3.5.12.    □

With the help of this theorem, we can write down an algorithm which checks whether $R$ is a Gorenstein ring.

---

**Algorithm 4.5.22 (Checking the Gorenstein Property)**

*Let $R = K[x_1, \ldots, x_n]/I$ be a zero-dimensional affine $K$-algebra, given by a set of generators of $I$. The following instructions define an algorithm checks whether $R$ is a Gorenstein ring or not.*

(1) *Compute a set of polynomials $B = \{t_1, \ldots, t_d\}$ whose residue classes form a $K$-basis of $R$.*

(2) *Compute the multiplication matrices $M_B(\vartheta_{x_1}), \ldots, M_B(\vartheta_{x_n})$.*

(3) *Apply the Cyclicity Test 3.1.4 to $(M_B(\vartheta_{x_1})^{\mathrm{tr}}, \ldots, M_B(\vartheta_{x_n})^{\mathrm{tr}})$.*

(4) *If the test is positive, return "Gorenstein" and stop. Otherwise, return "Not Gorenstein".*

---

*Proof* By the theorem, the ring $R$ is Gorenstein if and only if the $\mathscr{F}^{\vee}$-module $\omega_R$ is cyclic. By Remark 4.5.3, the matrices $M_B(\vartheta_{x_i})^{\mathrm{tr}}$ represent a system of generators of the dual family $\mathscr{F}^{\vee}$. Consequently, the application of the Cyclicity Test 3.1.4 in Step (4) determines whether $R$ is a Gorenstein ring or not.    □

Some ways to perform Steps (1) and (2) of this algorithm effectively were suggested in Remark 4.1.7. Let us see some examples.

**Example 4.5.23**  Let $K = \mathbb{Q}$, let $P = K[x, y, z]$, let

$$I = \langle x^2 - x,\ xz - x,\ yz - x,\ xy - x,\ z^3 - z^2,\ y^3 - y^2 \rangle$$

and let $R = P/I$. We want to check whether $R$ is a Gorenstein ring or not. The given set of generators is the reduced Gröbner basis of $I$ with respect to DegRevLex. Hence a $K$-basis $B$ of $R$ is given by the residue classes of the elements in the tuple $(1, x, y, z, y^2, z^2)$. The transposed multiplication matrices of $x$, $y$, and $z$ with respect to $B$ are

$$M_x = \begin{pmatrix} 0 & 1 & 0 & 0 & 0 & 0 \\ 0 & 1 & 0 & 0 & 0 & 0 \\ 0 & 1 & 0 & 0 & 0 & 0 \\ 0 & 1 & 0 & 0 & 0 & 0 \\ 0 & 1 & 0 & 0 & 0 & 0 \\ 0 & 1 & 0 & 0 & 0 & 0 \end{pmatrix} \quad M_y = \begin{pmatrix} 0 & 0 & 1 & 0 & 0 & 0 \\ 0 & 1 & 0 & 0 & 0 & 0 \\ 0 & 0 & 0 & 0 & 1 & 0 \\ 0 & 1 & 0 & 0 & 0 & 0 \\ 0 & 0 & 0 & 0 & 1 & 0 \\ 0 & 1 & 0 & 0 & 0 & 0 \end{pmatrix} \quad M_z = \begin{pmatrix} 0 & 0 & 0 & 1 & 0 & 0 \\ 0 & 1 & 0 & 0 & 0 & 0 \\ 0 & 1 & 0 & 0 & 0 & 0 \\ 0 & 0 & 0 & 0 & 0 & 1 \\ 0 & 1 & 0 & 0 & 0 & 0 \\ 0 & 0 & 0 & 0 & 0 & 1 \end{pmatrix}$$

To apply the Cyclicity Test 3.1.4, we form the matrix $C \in K[z_1, \ldots, z_6]$ of size $6 \times 6$ whose columns are the products of $I_6, M_x, M_y, M_z, M_y^2, M_z^2$ with $(z_1, \ldots, z_6)^{\text{tr}}$. The matrix is

$$C = \begin{pmatrix} z_1 & z_2 & z_3 & z_4 & z_5 & z_6 \\ z_2 & z_2 & z_2 & z_2 & z_2 & z_2 \\ z_3 & z_2 & z_5 & z_2 & z_5 & z_2 \\ z_4 & z_2 & z_2 & z_6 & z_2 & z_6 \\ z_5 & z_2 & z_5 & z_2 & z_5 & z_2 \\ z_6 & z_2 & z_2 & z_6 & z_2 & z_6 \end{pmatrix}$$

We obtain $\det(C) = 0$ and conclude that $R$ is not a Gorenstein ring.

Next we modify the preceding example slightly.

**Example 4.5.24**   Let $K = \mathbb{Q}$, let $P = K[x, y, z]$, let

$$I = \langle xy - x, \ x^2 - x, \ xz - x, \ yz - x, \ y^2 - y, \ z^4 - z^3 \rangle$$

and let $R = P/I$. We want to check whether $R$ is a Gorenstein ring or not. Again the given generators form the reduced DegRevLex-Gröbner basis of $I$. Thus a $K$-basis $B$ of $R$ is given by the residue classes of the elements in the tuple $(1, x, y, z, z^2, z^3)$. The transposed multiplication matrices of $x$, $y$, and $z$ with respect to $B$ are

$$M_x = \begin{pmatrix} 0 & 1 & 0 & 0 & 0 & 0 \\ 0 & 1 & 0 & 0 & 0 & 0 \\ 0 & 1 & 0 & 0 & 0 & 0 \\ 0 & 1 & 0 & 0 & 0 & 0 \\ 0 & 1 & 0 & 0 & 0 & 0 \\ 0 & 1 & 0 & 0 & 0 & 0 \end{pmatrix} \quad M_y = \begin{pmatrix} 0 & 0 & 1 & 0 & 0 & 0 \\ 0 & 1 & 0 & 0 & 0 & 0 \\ 0 & 0 & 1 & 0 & 0 & 0 \\ 0 & 1 & 0 & 0 & 0 & 0 \\ 0 & 1 & 0 & 0 & 0 & 0 \\ 0 & 1 & 0 & 0 & 0 & 0 \end{pmatrix} \quad M_z = \begin{pmatrix} 0 & 0 & 0 & 1 & 0 & 0 \\ 0 & 1 & 0 & 0 & 0 & 0 \\ 0 & 1 & 0 & 0 & 0 & 0 \\ 0 & 0 & 0 & 0 & 1 & 0 \\ 0 & 0 & 0 & 0 & 0 & 1 \\ 0 & 0 & 0 & 0 & 0 & 1 \end{pmatrix}$$

To apply the Cyclicity Test 3.1.4, we form the matrix $C \in K[z_1, \ldots, z_6]$ of size $6 \times 6$ whose columns are the products of $I_6, M_x, M_y, M_z, M_y^2, M_z^2$ with $(z_1, \ldots, z_6)^{\text{tr}}$. The matrix is

$$C = \begin{pmatrix} z_1 & z_2 & z_3 & z_4 & z_5 & z_6 \\ z_2 & z_2 & z_2 & z_2 & z_2 & z_2 \\ z_3 & z_2 & z_3 & z_2 & z_2 & z_2 \\ z_4 & z_2 & z_2 & z_5 & z_6 & z_6 \\ z_5 & z_2 & z_2 & z_6 & z_6 & z_6 \\ z_6 & z_2 & z_2 & z_6 & z_6 & z_6 \end{pmatrix}$$

Since $\det(C) = -z_2^3 z_5^3 + z_2^2 z_3 z_5^3 + 3z_2^3 z_5^2 z_6 - 3z_2^2 z_3 z_5^2 z_6 + z_2^2 z_5^3 z_6 - z_2 z_3 z_5^3 z_6 - 3z_2^3 z_5 z_6^2 + 3z_2^2 z_3 z_5 z_6^2 - 3z_2^2 z_5^2 z_6^2 + 3z_2 z_3 z_5^2 z_6^2 + z_2^3 z_6^3 - z_2^2 z_3 z_6^3 + 3z_2^2 z_5 z_6^3 - 3z_2 z_3 z_5 z_6^3 - z_2^2 z_6^4 + z_2 z_3 z_6^4$ shows $\det(C) \neq 0$, we conclude that $R$ is a Gorenstein ring.

If we know that $R$ is a Gorenstein ring, and thus that $\omega_R$ is a free $R$-module of rank one, we can compute an $R$-basis element of $\omega_R$. The first step of this computation is the calculation of the primary components of the zero ideal in $R$. Various methods for performing this step are discussed in Chap. 5.

---

**Algorithm 4.5.25 (Computing the $R$-Basis of $\omega_R$ for a Gorenstein Ring)**
*Let $R = P/I$ be a zero-dimensional affine $K$-algebra with $P = K[x_1, \ldots, x_n]$ and an ideal $I$ in $P$, let $\mathfrak{Q}_1, \ldots, \mathfrak{Q}_s$ be the primary components of $I$, let $\mathfrak{q}_1, \ldots, \mathfrak{q}_s$ be their images in $R$, let $\mathfrak{m}_1, \ldots, \mathfrak{m}_s$ be the corresponding maximal ideals of $R$, and assume that $R$ is a Gorenstein ring. Then the following instructions define an algorithm which computes a $K$-basis $B^*$ of $\omega_R$ and an element $\ell^* \in \omega_R$ which is an $R$-basis of $\omega_R$.*

(1) *For $i = 1, \ldots, s$, use the method explained in Remark 4.2.8 to compute a polynomial $\tilde{f}_i \in P$ such that $\mathrm{Soc}(P/\mathfrak{Q}_i) = \mathrm{Soc}(R/\mathfrak{q}_i)$ is generated by the residue class of $\tilde{f}_i$ as an $R/\mathfrak{m}_i$-vector space.*
(2) *For $i = 1, \ldots, s$, compute a tuple of polynomials $\tilde{B}_i$ whose residue classes are a $K$-basis of $P/\mathfrak{Q}_i$ such that $B_i$ contains the polynomial $\tilde{f}_i$.*
(3) *Using the Chinese Remainder Theorem 2.2.1, compute tuples of polynomials $B_1, \ldots, B_s$ such that $B = B_1 \cup \cdots \cup B_s$ is a $K$-basis of $R$ and such that the tuple $B_i$ corresponds to the tuple $\tilde{B}_i$ under the decomposition $R = R/\mathfrak{q}_1 \times \cdots \times R/\mathfrak{q}_s$.*
(4) *For $i = 1, \ldots, s$, let $f_i$ be the element of $B_i$ corresponding to $\tilde{f}_i$. Return the dual basis $B^*$ and the element $\ell^* = f_1^* + \cdots + f_s^*$ of $\omega_R$.*

---

*Proof* For $i = 1, \ldots, s$, we know from Remark 4.5.18.b that $\omega_R$ is the direct sum of the submodules $\omega_{R/\mathfrak{q}_i}$. By Theorem 4.5.10, the element $\tilde{f}_i^*$ is an $R/\mathfrak{q}_i$-basis of the module $\omega_{R/\mathfrak{q}_i}$. Hence we can use Proposition 3.1.7.b to conclude that $f_1^* + \cdots + f_s^*$ is an $R$-basis of $\omega_R$.                                                    $\square$

In the next example we apply the preceding algorithm to compute an $R$-basis of $\omega_R$ for certain zero-dimensional affine $K$-algebra which is a Gorenstein ring.

**Example 4.5.26** Let $K$ be a field, let $P = K[x, y]$, let $I = \langle x^2, y^2 - y \rangle$, and let $R = P/I$. It is straightforward to check that the primary decomposition of $I$ is given by $I = \mathfrak{Q}_1 \cap \mathfrak{Q}_2$ with $\mathfrak{Q}_1 = \langle x^2, y \rangle$ and $\mathfrak{Q}_2 = \langle x^2, y - 1 \rangle$. Therefore we have a canonical isomorphism $R \cong R/\mathfrak{q}_1 \times R/\mathfrak{q}_2$, where $\mathfrak{q}_i$ denotes the image of $\mathfrak{Q}_i$ in $R$.

In the following, to ease the notation, we denote the images of $x$ resp. $y$ in residue class rings of $P$ by $x$ resp. $y$ again. Let us follow the steps of Algorithm 4.5.25.

(1) We have $R/\mathfrak{m}_i \cong K$ and the residue class of $x$ in $R/\mathfrak{q}_i$ is a $K$-basis of $\mathrm{Soc}(R/\mathfrak{q}_i)$ for $i = 1, 2$.
(2) For $i = 1, 2$, the tuple $(1, x)$ is a $K$-basis of $R/\mathfrak{q}_i$.
(3) By lifting the bases of $R/\mathfrak{q}_1$ and $R/\mathfrak{q}_2$ to $R$ via the Chinese Remainder Theorem 2.2.1, we get the $K$-basis $B = (1 - y, (1 - y)x, y, yx)$ of $R$. Here $(1 - y)x$ corresponds to $(x, 0)$ and $yx$ to $(0, x)$.
(4) The algorithm returns the $K$-basis $((1 - y)^*, ((1 - y)x)^*, y^*, (yx)^*)$ as well as the $R$-basis $((1 - y)x)^* + (yx)^*$ of $\omega_R$.

Finally, we collect some observations about the minimal number of generators of $\omega_R$ in the general case. The claims in the following remark can be shown using arguments similar to the proof of Algorithm 4.5.25.

**Remark 4.5.27** Let $(R, \mathfrak{m})$ be a local affine zero-dimensional $K$-algebra, let the primary components of $\langle 0 \rangle$ in $R$ be denoted by $\mathfrak{q}_1, \ldots, \mathfrak{q}_s$, and let $\mathfrak{m}_1, \ldots, \mathfrak{m}_s$ be the corresponding maximal ideals of $R$.

(a) The minimal number of generators of $\omega_R$ is the maximum of the socle dimensions $\dim_{R/\mathfrak{m}_i}(\mathrm{Soc}(R/\mathfrak{q}_i))$ for $i = 1, \ldots, s$.
(b) For $i = 1, \ldots, s$, let $f_{i1}, \ldots, f_{iv_i} \in R$ be elements in $R/\mathfrak{q}_i$ which form an $R/\mathfrak{m}_i$-basis of the socle of the ring $R/\mathfrak{q}_i$. Then we use the isomorphism $\iota : R \cong R/\mathfrak{q}_1 \times \cdots \times R/\mathfrak{q}_s$ and $t = \max\{v_1, \ldots, v_s\}$ to find elements $g_1, \ldots, g_t$ with $\iota(g_i) = (f_{1i}, \ldots, f_{si})$, where we let $f_{ij} = 0$ for $j > v_i$. Then the elements $g_1, \ldots, g_t$ form a minimal system of generators of the $R$-module $\omega_R$.
(c) Using the base change technique of Lemma 4.5.12 and Proposition 4.5.13, we can express the minimal system of generators given in (b) with respect to any basis $B^*$ of $\omega_R$ which is dual to a given $K$-basis $B$ of $R$.

> *Why is it called common sense*
> *if so many people don't have it?*

## 4.6 Hilbert Functions and the Cayley-Bacharach Property

> *Ubi materia, ibi geometria.*
> [Where there is matter, there is geometry.]
> (Johannes Kepler)

In fact, as an old professor used to say, without geometry life would be *pointless*. Although this book deals mainly with computational algebra, zero-dimensional affine $K$-algebras have an intrinsic geometric meaning. Consequently, we decided to provide the reader with a glimpse of the geometric nature of these special rings.

To simplify the treatment, we limit ourselves to the case of zero-dimensional affine $K$-algebras whose maximal ideals are linear. Geometrically speaking, we consider the case of zero-dimensional schemes supported at $K$-rational points.

Let $K$ be a field, and let us fix a presentation $R = P/I$ with a polynomial ring $P = K[x_1, \ldots, x_n]$ and an ideal $I$ in $P$. In the language of algebraic geometry, we consider a fixed embedding of the scheme $\mathrm{Spec}(R)$ into an affine space $\mathbb{A}_K^n$. More down to earth, if we assume that $R$ is reduced and its maximal ideals are linear, we look at a finite set of *simple points* where the set of points is the set of zeros $\mathscr{Z}(I) = \{\wp \in K^n \mid f(\wp) = 0 \text{ for all } f \in I\}$ of $I$ (see Definition 4.3.5). Since the points correspond to maximal ideals of $R$, we can apply the linear algebra methods developed earlier to study *geometric* properties of a zero-dimensional affine $K$-algebra $R$. What is the geometry of these points? What can one say about the hypersurfaces passing through them?

As we shall see, a lot of geometric information is encoded in the (affine) Hilbert function of $R$. The definition of this function is based on the following observations. A basic information which is available is the *degree* of a polynomial. More generally, we can filter the elements of $R$ according to their degrees as follows (see also [16], Sect. 6.5).

**Definition 4.6.1** Let $P = K[x_1, \ldots, x_n]$, let $I$ be an ideal in $P$, and let $R = P/I$.

(a) For every $i \in \mathbb{Z}$, we let $F_i P = \{f \in P \setminus \{0\} \mid \deg(f) \le i\} \cup \{0\}$. Then the family $(F_i P)_{i \in \mathbb{Z}}$ is called the **degree filtration** on $P$.
(b) For every $i \in \mathbb{Z}$, let $F_i I = F_i P \cap I$, and let $F_i R = F_i P / F_i I$. Then the family $(F_i R)_{i \in \mathbb{Z}}$ is called the **degree filtration** on $R$.

The degree filtrations on $P$ and $R$ have the following properties.

**Remark 4.6.2** The degree filtration on $P$ is a $\mathbb{Z}$-filtration in the sense of [16], Definition 6.5.1, i.e. it has the following properties:

(a) For $i \le j$ we have $F_i P \subseteq F_j P$ and $F_i I \subseteq F_j I$.
(b) We have $\bigcup_{i \in \mathbb{Z}} F_i P = P$ and $\bigcup_{i \in \mathbb{Z}} F_i I = I$. Hence we also have $\bigcup_{i \in \mathbb{Z}} F_i R = R$.
(c) For $i, j \in \mathbb{Z}$, we have $F_i P \cdot F_j P \subseteq F_{i+j} P$.
(d) We have $F_0 P = K$ and $F_i P = \{0\}$ for $i < 0$.

The family $(F_i I)_{i \in \mathbb{Z}}$ is the **induced filtration** on $I$ in the sense of [16], Proposition 6.5.9, and the family $(F_i R)_{i \in \mathbb{Z}}$ is a $\mathbb{Z}$-filtration on $R$ by part (c) of the same proposition.

Notice that property (b) in this remark and the fact that $R$ is zero-dimensional imply that we have $F_i R = R$ for $i \gg 0$. In particular, the degree filtration on $R$ has only finitely many different parts. In order to examine the degree filtration on $R$ more closely, we need a $K$-basis of $R$ which is compatible with the degree filtration in the following sense.

**Definition 4.6.3** A $K$-basis $B = \{t_1, \ldots, t_d\}$ of $R$ is called a **degree-filtered basis** of $R$ if $F_i B = B \cap F_i R$ is a $K$-basis of $F_i R$ for every $i \in \mathbb{Z}$.

One element of a degree filtered basis of $R$ is a non-zero constant. Hence we may assume $t_1 = 1$ in the following. Our next remark provides some ways of computing a degree filtered basis of $R$.

**Remark 4.6.4** Let $R = P/I$ be a zero-dimensional affine $K$-algebra.

(a) Let $\sigma$ be a degree compatible term ordering on $\mathbb{T}^n$. We calculate a $\sigma$-Gröbner basis of $I$ and let $\mathcal{O} = \mathbb{T}^n \setminus \mathrm{LT}_\sigma(I)$. Then the residue classes of the terms in $\mathcal{O}$ form a degree filtered $K$-basis of $R$.

(b) Suppose that we know a Macaulay basis $\{f_1, \ldots, f_r\}$ of $I$, i.e., a set of polynomials such that the degree form ideal $\mathrm{DF}(I)$ is generated by the degree forms $\mathrm{DF}(f_1), \ldots, \mathrm{DF}(f_r)$ (see [16], Sect. 4.2). For $i = 0, 1, 2, \ldots$ we then compute a tuple of polynomials $B_i$ whose residue classes form a $K$-basis of the $i$-th homogeneous component of $P/\mathrm{DF}(I)$ until this homogeneous component is zero. Then the tuple $B = B_0 \cup B_1 \cup \cdots$ is a degree filtered basis of $R$.

Let us have a look at an easy example.

**Example 4.6.5** Let $K = \mathbb{Q}$, let $P = K[x, y]$, let $I$ be the ideal of $P$ generated by $\{x - y^2, y^3\}$, and let $R = P/I$. The pair $(x - y^2, y^3)$ is the reduced Gröbner basis of $I$ with respect to the lexicographic term ordering. Hence the residue classes of the elements of $(1, y, y^2)$ form a $K$-basis of $R$. However, this is not a degree-filtered basis, since we have $\bar{y}^2 = \bar{x}$ in $R$. Instead, the reduced Gröbner basis of $I$ with respect to DegRevLex is $(y^2 - x, xy, x^2)$. Since this term ordering is degree compatible, the residue classes of the elements of $(1, y, x)$ form a degree-filtered $K$-basis of $R$.

The vector space dimensions of the filtered components $F_i R$ are important invariants of the presentation $R = P/I$ known by the following name.

**Definition 4.6.6** Let $R = P/I$ be a presentation of an affine $K$-algebra, and let $(F_i R)_{i \in \mathbb{Z}}$ be the degree filtration of $R$.

(a) The map $\mathrm{HF}_R^a : \mathbb{Z} \longrightarrow \mathbb{Z}$ given by $\mathrm{HF}_R^a(i) = \dim_K(F_i R)$ for $i \in \mathbb{Z}$ is called the **(affine) Hilbert function** of $R$.

(b) The map $\Delta\,\mathrm{HF}_R^a : \mathbb{Z} \longrightarrow \mathbb{Z}$ defined by $\Delta\,\mathrm{HF}_R^a(i) = \mathrm{HF}_R^a(i) - \mathrm{HF}_R^a(i-1)$ for $i \in \mathbb{Z}$ is called the **Hilbert difference function** or the **Castelnuovo function** of $R$.

In [16], Sect. 5.6.A, the affine Hilbert function of $R$ is defined and a number of general properties are shown. For a zero-dimensional affine $K$-algebra, we have the following additional properties.

**Proposition 4.6.7** *Let* $R = P/I$ *be a zero-dimensional affine $K$-algebra, and let* $d = \dim_K(R)$.

(a) *We have* $\mathrm{HF}_R^a(0) = 1$ *and* $\mathrm{HF}_R^a(i) = 0$ *for* $i < 0$.
(b) *There exists a number* $r = \mathrm{ri}(R)$, *called the* **regularity index** *of $R$, such that* $\mathrm{HF}_R^a(i) = d$ *for* $i \geq r$ *and such that $r$ is the smallest integer with this property.*
(c) *The Castelnuovo function of $R$ satisfies* $\triangle\,\mathrm{HF}_R^a(i) = 0$ *for* $i < 0$ *and* $i > r$, *and* $\triangle\,\mathrm{HF}_R^a(i) > 0$ *for* $0 \leq i \leq r$. *In particular, the Hilbert function of $R$ is non-decreasing, and it is strictly increasing for* $0 \leq i \leq r$.

*Proof* Claim (a) follows from Remark 4.6.2.d and claim (b) is a consequence of parts (a) and (b) of this remark and the fact that $d = \dim_K(R)$ is finite. Hence, to prove (c), it remains to show that $\mathrm{HF}_R^a$ increases strictly until it reaches the value $d = \dim_K(R)$. Suppose that $\mathrm{HF}_R^a(i) = \mathrm{HF}_R^a(i+1)$ for some $i \geq 0$. Then the equality $F_{i+1}R = F_i R$ implies

$$F_{i+2}R = x_1 F_{i+1}R + \cdots + x_n F_{i+1}R = x_1 F_i R + \cdots + x_n F_i R = F_{i+1}R$$

Continuing in this way, we see that $\mathrm{HF}_R^a(j) = \mathrm{HF}_R^a(i)$ for all $j \geq i$, and therefore $\mathrm{HF}_R^a(i) = d$.                                                                           $\square$

Next we discuss various ways to compute the Hilbert function of $R = P/I$.

**Remark 4.6.8** Let $R = P/I$ be a zero-dimensional affine $K$-algebra.

(a) If we are given a degree-filtered $K$-basis $B$ of $R$, it obviously suffices to set $\mathrm{HF}_R^a(i) = \#(B \cap F_i R)$ for $i \in \mathbb{Z}$.
(b) If we are given a system of generators of the ideal $I$, we can compute a Gröbner basis of $I$ with respect to a degree compatible term ordering $\sigma$ and use the formula $\mathrm{HF}_R^a(i) = \sum_{j=0}^{i} \dim_K(P_i/\mathrm{LT}_\sigma(I)_i)$ (see [16], Proposition 5.6.3.a) together with the algorithms in [16], Sect. 5.3, for the computation of $\dim_K(P_i/\mathrm{LT}_\sigma(I)_i)$.

The Hilbert function of $R = P/I$ encodes important geometric information about the set of zeros (or, more generally, the **zero scheme**) of the ideal $I$. In [16], Sect. 6.3 we have already provided a number of hints in this direction. Let us point out a few elementary observations.

**Remark 4.6.9** Let $R = P/I$ be a reduced zero-dimensional affine $K$-algebra, and assume that all maximal ideals of $R$ are linear. Geometrically, this means that the set $\mathbb{X} = \mathscr{Z}(I)$ consists of $K$-rational points.

(a) The set $\mathbb{X}$ is contained in a hyperplane of $\mathbb{A}_K^n$ if and only if $\mathrm{HF}_R^a(1) \leq n$. This follows from the fact that $\mathrm{HF}_R^a(1) \leq n$ if and only if the ideal $I$ contains a non-zero linear polynomial.
(b) The set $\mathbb{X}$ contains at most $\mathrm{ri}(R)$ points on a line. This is a consequence of the following considerations. First, we observe that the equality $F_{\mathrm{ri}(R)}R = R$

implies that the ideal $I$ is generated by polynomials of degree $\leq \mathrm{ri}(R)$. If the set $\mathbb{X}$ contains $\mathrm{ri}(R)+1$ points on a line then every polynomial of degree $\leq \mathrm{ri}(R)$ in $I$ must be a multiple of the linear polynomial which defines the line. This contradicts the fact the $I$ is zero-dimensional.

The notion of a separator was discussed in the last part of Sect. 4.2. In the following it turns out that the degree of a separator contains useful geometric information. More precisely, we need the notion of the order of an element of $R$ which is defined as follows (see [16], Definition 6.5.10).

**Definition 4.6.10** Let $R = P/I$ be a zero-dimensional affine algebra, let $\mathfrak{m}_1, \ldots, \mathfrak{m}_s$ be the maximal ideals of $R$, and let $(F_i R)_{i \in \mathbb{Z}}$ be the degree filtration of $R$.

(a) For a non-zero element $f \in R$, the number $\mathrm{ord}(f) = \min\{i \in \mathbb{Z} \mid f \in F_i R\}$ is called the **order** of $f$ with respect to the degree filtration of $R$.

(b) Assume that the maximal ideals of $R$ are linear, and let $i \in \{1, \ldots, s\}$. The number $\mathrm{sepdeg}(\mathfrak{m}_i) = \min\{\mathrm{ord}(f) \mid f$ is a separator for $\mathfrak{m}_i\}$ is called the **separator degree** of $\mathfrak{m}_i$ in $R$.

Let us collect a few observations about these concepts.

**Remark 4.6.11** Assume that we are in the setting of the preceding definition.

(a) For every $f, g \in R \setminus \{0\}$ such that $f + g \neq 0$ we have the inequality $\mathrm{ord}(f+g) \leq \max\{\mathrm{ord}(f), \mathrm{ord}(g)\}$. Moreover, if $fg \neq 0$ then $\mathrm{ord}(fg) \leq \mathrm{ord}(f)\,\mathrm{ord}(g)$.

(b) Given a degree filtered basis $B = (b_1, \ldots, b_d)$ such that $b_i$ is the residue class of a polynomial of degree $\mathrm{ord}(b_i)$, the order of a non-zero element $f \in R$ can be computed in the following way: write $f = a_1 b_1 + \cdots + a_d b_d$ with $a_i \in K$ and set $\mathrm{ord}(f) = \max\{\mathrm{ord}(b_i) \mid a_i \neq 0\}$.

(c) Suppose that $R$ is reduced and has linear maximal ideals. Then the separator degree of $\mathfrak{m}_i$ is the least degree of a hypersurface $\mathscr{Z}(f)$ which contains all points of $\mathscr{Z}(I)$ except the point associated to $\mathfrak{m}_i$.

Since we have $F_{\mathrm{ri}(R)} R = R$, every element $f \in R$ satisfies $\mathrm{ord}(f) \leq \mathrm{ri}(R)$. In particular, the separator degree satisfies $\mathrm{sepdeg}(\mathfrak{m}_i) \leq \mathrm{ri}(R)$ for $i = 1, \ldots, s$. Thus the following definition provides a natural property of $R$.

**Definition 4.6.12** Let $R = P/I$ be a zero-dimensional affine $K$-algebra with linear maximal ideals $\mathfrak{m}_1, \ldots, \mathfrak{m}_s$. We say that $R$ has the **Cayley-Bacharach property** if the separator degree of $\mathfrak{m}_i$ satisfies $\mathrm{sepdeg}(\mathfrak{m}_i) = \mathrm{ri}(R)$ for $i = 1, \ldots, s$.

If $I$ is a radical ideal, this definition agrees with the one given for $\mathbb{X} = \mathscr{Z}(I)$ in [16], Tutorial 88: the set $\mathbb{X}$ is said to have the Cayley-Bacharach property if all rings $P/J$, where $J$ is the vanishing ideal of a subset of $\mathbb{X}$ consisting of all points but one, have the same Hilbert function. The name "Cayley-Bacharach property" derives from the mathematicians A. Cayley and I. Bacharach who proved that complete intersection rings $R$ have this property.

The following algorithm allows us to check whether $R$ has the Cayley-Bacharach property. A different algorithm based on the canonical module of $R$ will be presented at the end of this section (see Algorithm 4.6.21).

---

**Algorithm 4.6.13 (Checking the Cayley-Bacharach Property, I)**
*Let $R = P/I$ be a zero-dimensional affine $K$-algebra having linear maximal ideals $\mathfrak{m}_1, \ldots, \mathfrak{m}_s$. Consider the following sequence of instructions.*

(1) *Calculate the primary components $\mathfrak{q}_1, \ldots, \mathfrak{q}_s$ of the zero ideal of $R$. (For instance, use one of the methods presented in Chap. 5.)*
(2) *For $i = 1, \ldots, s$, compute a $K$-basis $S_i$ of the socle of $R/\mathfrak{q}_i$.*
(3) *For $i = 1, \ldots, s$, use Theorem 4.2.11.e and the Chinese Remainder Theorem 2.2.1 to compute a $K$-basis $B_i$ of the space of separators for $\mathfrak{m}_i$.*
(5) *Compute the degree filtration $(F_i R)_{i \in \mathbb{Z}}$ and the regularity index $\mathrm{ri}(R)$ of $R$.*
(4) *For $i = 1, \ldots, s$, calculate $\langle B_i \rangle \cap F_{\mathrm{ri}(R)-1} R$. If this intersection is $\{0\}$, set $\mathrm{sepdeg}(\mathfrak{m}_i) = \mathrm{ri}(R)$.*
(6) *Return TRUE if $\mathrm{sepdeg}(\mathfrak{m}_1) = \cdots = \mathrm{sepdeg}(\mathfrak{m}_s) = \mathrm{ri}(R)$. Otherwise, return FALSE*

*This is an algorithm which checks whether $R$ has the Cayley-Bacharach property and returns the corresponding boolean value.*

---

*Proof* By the definition of a separator for $\mathfrak{m}_i$, the set of all separators for $\mathfrak{m}_i$ is given by $\langle B_i \rangle \setminus \{0\}$. By the definition of the separator degree of $\mathfrak{m}_i$, Step (4) is correct. Finally, the definition of the Cayley-Bacharach property implies that Step (5) yields the correct answer. $\qquad\square$

Recall that the canonical module of $R$ is given by $\omega_R = \mathrm{Hom}_K(R, K)$ (see Sect. 4.5). Our next goal is to define a suitable Hilbert function of $\omega_R$ and to relate it to the Hilbert function of $R$. For this purpose we consider the filtration on $\omega_R$ which is dual to the degree filtration of $R$.

**Definition 4.6.14** Let $R$ be a zero-dimensional affine $K$-algebra with canonical module $\omega_R = \mathrm{Hom}_K(R, K)$.

(a) For every $i \in \mathbb{Z}$, let $G_i \omega_R = \{\ell \in \omega_R \mid \ell(F_{-i}) = 0\}$. Then the family $(G_i \omega_R)_{i \in \mathbb{Z}}$ is called the **degree filtration** of $\omega_R$.
(b) The map $\mathrm{HF}_{\omega_R}^a : \mathbb{Z} \longrightarrow \mathbb{Z}$ defined by $\mathrm{HF}_{\omega_R}^a(i) = \dim_K(G_i \omega_R)$ for $i \in \mathbb{Z}$ is called the **(affine) Hilbert function** of $\omega_R$.

Let us check that part (a) of this definition turns $\omega_R$ into a filtered module over the degree filtered ring $R$ in the sense of [16], Sect. 6.5.

**Remark 4.6.15** The family $(G_i \omega_R)_{i \in \mathbb{Z}}$ is a $\mathbb{Z}$-filtration on $\omega_R$ which is compatible with the degree filtration on $R$. In particular, the following properties hold:

(a) For $i \leq j$, we have $G_i \omega_R \subseteq G_j \omega_R$, since $F_{-j} R \subseteq F_{-i} R$ implies that we have $\ell(F_{-j} R) \subseteq \ell(F_{-i} R) = 0$ for every $\ell \in G_i \omega_R$.
(b) We have $\bigcup_{i \in \mathbb{Z}} G_i \omega_R = \omega_R$, since $G_1 \omega_R = \{\ell \in \omega_R \mid \ell(0) = 0\} = \omega_R$.
(c) For $i, j \in \mathbb{Z}$, we have $(F_i R) \cdot (G_j \omega_R) \subseteq G_{i+j} \omega_R$, since $f \in F_i R$ and $\ell \in G_j \omega_R$ imply $(f\ell)(g) = \ell(fg) \in \ell(F_{-j} R) = 0$ for all $g \in F_{-i-j} R$.

The affine Hilbert function of $\omega_R$ can be determined from the affine Hilbert function of $R$ in the following way.

**Proposition 4.6.16** *Let $\omega_R$ be the canonical module of a zero-dimensional affine algebra $R = P/I$. For every $i \in \mathbb{Z}$, we have*

$$\mathrm{HF}^a_{\omega_R}(i) = \dim_K(R) - \mathrm{HF}^a_R(-i)$$

*Proof* Let $i \in \mathbb{Z}$ and $j = \mathrm{HF}^a_R(-i) = \dim_K(F_{-i} R)$. The requirement to vanish on the $j$-dimensional vector subspace $F_{-i} R$ of $R$ defines a $j$-codimensional vector subspace of the dual vector space $\omega_R = \mathrm{Hom}_K(R, K)$. This observation implies the claim. $\qquad\square$

In particular, the Hilbert function of $\omega_R$ has the following properties.

**Remark 4.6.17** Let $R$ be a zero-dimensional affine algebra.

(a) For $i \leq -\mathrm{ri}_R$, we have $\mathrm{HF}^a_{\omega_R}(i) = 0$.
(b) We have $\mathrm{HF}^a_{\omega_R}(-\mathrm{ri}(R) + 1) \neq 0$.
(c) We have $\mathrm{HF}^a_{\omega_R}(0) = \dim_K(R) - 1$.
(d) For $i \geq 1$, we have $\mathrm{HF}^a_{\omega_R}(i) = \dim_K(R)$.

Given a degree filtered $K$-basis $B = (b_1, \ldots, b_d)$ of $R$, it is clear that the dual basis $B^* = (b_d^*, b_{d-1}^*, \ldots, b_1^*)$ is a degree filtered $K$-basis of $\omega_R$. Moreover, we can define the order of linear forms in $\omega_R$ as follows.

**Definition 4.6.18** Let $(G_i \omega_R)_{i \in \mathbb{Z}}$ be the degree filtration on $\omega_R$. Given an element $\ell \in \omega_R$, the number $\mathrm{ord}(\ell) = \min\{i \in \mathbb{Z} \mid \ell \in G_i \omega_R\}$ is called the **order** of $\ell$ with respect to the degree filtration of $\omega_R$.

In the following proposition we assume that all maximal ideals of our zero-dimensional $K$-algebra $R$ are linear and that $R$ is Gorenstein. This last assumption is for instance satisfied if $R$ is reduced (see Remark 4.5.20). Therefore the claims of the proposition apply to the case of a $K$-algebra associated to a set of simple points. Using the elements of lowest order $-\mathrm{ri}(R) + 1$ in $\omega_R$, we can characterize rings having the Cayley-Bacharach property as follows.

**Theorem 4.6.19 (Characterization of the Cayley-Bacharach Property)**

*Let $K$ be an infinite field, and let $R = P/I$ be a zero-dimensional $K$-algebra. Assume that the maximal ideals $\mathfrak{m}_1, \ldots, \mathfrak{m}_s$ of $R$ are linear and that $R$ is a Gorenstein ring. Then the following conditions are equivalent.*

(a) *The ring $R$ has the Cayley-Bacharach property.*

(b) *For every $i \in \{1, \ldots, s\}$, there is a separator $f_i$ for $\mathfrak{m}_i$ such that $\mathrm{ord}(f_i) = \mathrm{ri}(R)$.*

(c) *There exists an element $\ell \in \omega_R$ such that $\mathrm{ord}(\ell) = -\mathrm{ri}(R) + 1$ and $\mathrm{Ann}_R(\ell) = 0$.*

*Proof* By Theorem 4.2.11.e, the space $B_i$ of separators for $\mathfrak{m}_i$ is isomorphic to the socle of $R/\mathfrak{m}_i$ for $i = 1, \ldots, s$. Since $R$ is a Gorenstein ring, we obtain $\dim_K(B_i) = 1$. Therefore, up to scalar multiples, there is a unique separator $f_i$ for $\mathfrak{m}_i$ for every $i \in \{1, \ldots, s\}$. In particular, we have $\mathrm{ord}(f_i) = \mathrm{sepdeg}(\mathfrak{m}_i)$ for $i = 1, \ldots, s$. Thus conditions (a) and (b) are equivalent.

Next we prove that (b) implies (c). By assumption, the separator $f_i$ satisfies $\mathrm{ord}(f_i) = \mathrm{ri}(R)$ for $i = 1, \ldots, s$. Now consider the space $G_{-\mathrm{ri}(R)+1}\omega_R$, i.e., the space of all $K$-linear maps $\ell : R \longrightarrow K$ with $\ell(F_{\mathrm{ri}(R)-1}R) = 0$. Every map $\ell$ uniquely corresponds to an induced $K$-linear form $\bar{\ell}$ on $F_{\mathrm{ri}(R)}R/F_{\mathrm{ri}(R)-1}R$. Since the residue classes $\bar{f}_1, \ldots, \bar{f}_s$ are non-zero in this vector space, and since $K$ is infinite, we can choose a 1-codimensional vector subspace which avoids all these residue classes. If we take it as the kernel of $\bar{\ell}$, the corresponding linear map $\ell$ satisfies $\ell(f_i) \neq 0$ for $i = 1, \ldots, s$.

Now we claim that this element $\ell$ satisfies $\mathrm{Ann}_R(\ell) = \langle 0 \rangle$. For a contradiction, assume that $g\ell = 0$ for some non-zero element $g \in R$. Using the isomorphism $\iota : R \cong R/\mathfrak{q}_1 \times \cdots \times R/\mathfrak{q}_s$ of Proposition 4.1.11, where $\mathfrak{q}_1, \ldots, \mathfrak{q}_s$ are the primary components of the zero ideal of $R$, we write $\iota(g) = (\bar{g}_1, \ldots, \bar{g}_s)$. Since $g \neq 0$, there exists an index $i \in \{1, \ldots, s\}$ such that $\bar{g}_i \neq 0$. The fact that $R/\mathfrak{q}_i$ is a local Gorenstein ring implies by Lemma 4.5.9.a that there is a non-zero element $\bar{h}_i \in R/\mathfrak{q}_i$ such that $\bar{g}_i \bar{h}_i \in \mathrm{Soc}(R/\mathfrak{q}_i)$. Hence there exists an element $h \in R$ such that we have $\iota(gh) = (0, \ldots, 0, \bar{g}_i \bar{h}_i, 0, \ldots, 0)$, and henceforth we have $gh = cf_i$ for some $c_i \in K \setminus \{0\}$. Now $0 = (g\ell)(h) = \ell(gh) = c\ell(f_i) \neq 0$ yields a contradiction.

Finally, we prove the implication (c) $\Rightarrow$ (b). Let $R \cong R/\mathfrak{q}_1 \times \cdots \times R/\mathfrak{q}_s$ be the decomposition of $R$ into local rings, and let $i \in \{1, \ldots, s\}$. Since $R$ is a Gorenstein ring, the socle of $R/\mathfrak{q}_i$ is a 1-dimensional $K$-vector space. Let $f_i \in R$ be the corresponding separator which is unique up to a scalar multiple. Notice that $\langle f_i \rangle = Kf_i$. Then we have $(f_i\ell)(r) \in K\ell(f_i)$ for every $r \in R$, and $f_i\ell \neq 0$ implies $\ell(f_i) \neq 0$. In particular, since $\ell \in G_{-\mathrm{ri}(R)+1}\omega_R$, it follows from $\ell(F_{\mathrm{ri}(R)-1}R) = 0$ that $\mathrm{ord}(f_i) = \mathrm{ri}(R)$, as we wanted to show.                                      $\square$

Using this theorem, we can derive a second algorithm for checking the Cayley-Bacharach property. We need the following auxiliary result about the annihilator of an element of the canonical module.

**Lemma 4.6.20** *Let $R = P/I$ be a zero-dimensional affine $K$-algebra which has a $K$-basis $B = (b_1, \ldots, b_d)$, and let $\ell \in \omega_R$. We write $\ell = c_1 b_1^* + \cdots + c_d b_d^*$ with $c_1, \ldots, c_d \in K$.*

(a) *For $f \in R$, we have $f\ell = 0$ in $\omega_R$ if and only if $(c_1, \ldots, c_d) \cdot M_B(\vartheta_f) = 0$.*
(b) *Let $\Lambda_c \in \mathrm{Mat}_d(K)$ be the matrix whose $i$-th row is $(c_1, \ldots, c_d) \cdot M_B(\vartheta_{b_i})$ for every $i \in \{1, \ldots, d\}$. Then we have $\mathrm{Ann}_R(\ell) = \langle 0 \rangle$ if and only if $\det(\Lambda_c) \neq 0$.*

*Proof* To prove (a) we note that $(f\ell)(b_i) = \ell(fb_i) = (c_1 b_1^* + \cdots + c_d b_d^*)(fb_i) = (c_1, \ldots, c_d) \cdot M_B(\vartheta_f)_i$, where $M_B(\vartheta_f)_i$ denotes the $i$-th column of the matrix $M_B(\vartheta_f)$.

Claim (b) follows from (a), because $\mathrm{Ann}_R(\ell) = \langle 0 \rangle$ is equivalent to the condition that the products $b_1 \ell, \ldots, b_d \ell$ are $K$-linearly independent.     $\square$

Now we are ready to formulate the desired algorithm.

**Algorithm 4.6.21 (Checking the Cayley-Bacharach Property, II)**
*Let $K$ be an infinite field, and let $R = P/I$ be a zero-dimensional affine $K$-algebra which is a Gorenstein ring and whose maximal ideals $\mathfrak{m}_1, \ldots, \mathfrak{m}_s$ are linear. Consider the following sequence of instructions.*

(1) *Compute a degree filtered $K$-basis $B = (b_1, \ldots, b_d)$ of the ring $R$. Let $b_{d-\Delta+1}, b_{d-\Delta+2}, \ldots, b_d$ be those elements in $B$ whose order is $\mathrm{ri}(R)$.*
(2) *Introduce new indeterminates $z_1, \ldots, z_\Delta$ and then calculate the square matrix $C \in \mathrm{Mat}_d(K[y_1, \ldots, y_\Delta])$ whose $i$-th row is given by $(0, \ldots, 0, y_1, \ldots, y_\Delta) \cdot M_B(\vartheta_{b_i})$ for $i = 1, \ldots, d$.*
(3) *If $\det(C) \neq 0$, return TRUE. Otherwise, return FALSE.*

*This is an algorithm which checks whether $R$ has the Cayley-Bacharach property and returns the corresponding boolean value.*

*Proof* Since this algorithm is clearly finite, it suffices to prove correctness. By the preceding theorem, we have to check whether there exists an element $\ell \in \omega_R$ of order $\mathrm{ord}(\ell) = -\mathrm{ri}(R) - 1$ such that $\mathrm{Ann}_R(\ell) = \langle 0 \rangle$. The definition of $\Delta$ implies that $b_{d-\Delta+1}^*, \ldots, b_d^*$ are precisely the elements of order $-\mathrm{ri}(R) + 1$ in $B^*$. Hence they form a $K$-basis of $G_{-\mathrm{ri}(R)+1}\omega_R$. Therefore we are looking for an element $\ell = c_1 b_{d-\Delta+1}^* + \cdots + c_\Delta b_d^*$ with $c_1, \ldots, c_\Delta \in K$ such that $\mathrm{Ann}_R(\ell) = \langle 0 \rangle$. By the lemma, this equivalent to $\det(\Lambda_c) \neq 0$ for the corresponding matrix $C_c$. Now the claim follows from the facts that $C_c$ results from the substitution $c_i \mapsto y_i$ in $C_y$ and that the field $K$ is infinite.     $\square$

Let us use this algorithm to check some rings for the Cayley-Bacharach property.

**Example 4.6.22** Let $K = \mathbb{Q}$, let $P = K[x, y]$, and let $I = \langle xy - x, x^2 - x, y^3 - y \rangle$. It turns out that the primary decomposition of $I$ is $I = \mathfrak{M}_1 \cap \mathfrak{M}_2 \cap \mathfrak{M}_3 \cap \mathfrak{M}_4$, where $\mathfrak{M}_1 = \langle x, y \rangle$, $\mathfrak{M}_2 = \langle x - 1, y \rangle$, $\mathfrak{M}_3 = \langle x - 1, y - 1 \rangle$, and $\mathfrak{M}_4 = \langle x - 1, y + 1 \rangle$ (see for instance the methods explained in Chap. 5).

Hence $R = P/I$ is a Gorenstein ring and we can apply the algorithm. The reduced Gröbner basis of $I$ with respect to DegRevLex is $(xy - x, x^2 - x, y^3 - y)$, and a degree-filtered $K$-basis of $R$ is given by the residue classes of the elements in the tuple $(1, x, y, y^2)$. We have $\Delta = 1$, and the corresponding matrix $C$ is

$$\begin{pmatrix} 0 & 0 & 0 & z \\ 0 & 0 & 0 & z \\ 0 & 0 & z & 0 \\ z & z & 0 & z \end{pmatrix}$$

Since we have $\det(C) = 0$, we conclude that $R$ does not have the Cayley-Bacharach property. The geometric explanation is that $I$ is the ideal of four points such that three are on a line and one is outside the line. Therefore one separator has degree 1, while the other separators have degree 2.

The following example is related to the Cayley-Bacharach Theorem. More precisely, it is a special case of the fact that if one removes a point from a reduced complete intersection, the remaining set of points still has the Cayley-Bacharach property (see [7], Corollary 2.17).

**Example 4.6.23** Let $K = \mathbb{Q}$, and let $I = \langle y^3 - y, x^3 - x, x^2y^2 - x^2 - y^2 + 1 \rangle$ in $P = K[x, y]$. Using the algorithms in Chap. 5 we can calculate the primary decomposition of $I$ and get $I = \mathfrak{M}_1 \cap \cdots \cap \mathfrak{M}_8$, where

$$\mathfrak{M}_1 = \langle x + 1, y + 1 \rangle \quad \mathfrak{M}_2 = \langle x + 1, y \rangle \quad \mathfrak{M}_3 = \langle x + 1, y - 1 \rangle \quad \mathfrak{M}_4 = \langle x, y + 1 \rangle$$
$$\mathfrak{M}_5 = \langle x, y - 1 \rangle \qquad \mathfrak{M}_6 = \langle x - 1, y \rangle \quad \mathfrak{M}_7 = \langle x - 1, y - 1 \rangle$$
$$\mathfrak{M}_8 = \langle x - 1, y + 1 \rangle$$

Therefore we see that $R = P/I$ is a Gorenstein ring. Let us apply the algorithm to check whether $R$ has the Cayley-Bacharach property. The reduced Gröbner basis of $I$ with respect to DegRevLex is $(y^3 - y, x^3 - x, x^2y^2 - x^2 - y^2 + 1)$, and a degree-filtered $K$-basis of $R$ is given by the residue classes of the elements in the tuple $(1, x, y, x^2, xy, y^2, x^2y, xy^2)$. Hence we have $\Delta = 2$, and the corresponding matrix $C$ is

$$C = \begin{pmatrix} 0 & 0 & 0 & 0 & 0 & 0 & z_1 & z_2 \\ 0 & 0 & 0 & 0 & z_1 & z_2 & 0 & 0 \\ 0 & 0 & 0 & z_1 & z_2 & 0 & 0 & 0 \\ 0 & 0 & z_1 & 0 & 0 & 0 & z_1 & z_2 \\ 0 & z_1 & z_2 & 0 & 0 & 0 & z_2 & z_1 \\ 0 & z_2 & 0 & 0 & 0 & 0 & z_1 & z_2 \\ z_1 & 0 & 0 & z_1 & z_2 & z_1 & 0 & 0 \\ z_2 & 0 & 0 & z_2 & z_1 & z_2 & 0 & 0 \end{pmatrix}$$

Since we have $\det(C) = z_1^6 z_2^2 - 2z_1^4 z_2^4 + z_1^2 z_2^6 \neq 0$, we conclude that $R = P/I$ has the Cayley-Bacharach property.

# Chapter 5
# Computing Primary and Maximal Components

*The single most important component of a camera
is the twelve inches behind it.*

(Ansel Adams)

Every week, a **M**athematician and a **P**hotographer take a walk. Some months ago, they had the following conversation.

P: Why are you looking so sad?

M: I have this ideal. I tried to find the primary components, but I couldn't.

P: Well, that is easy. The primary components of a photograph are $(x, y, z)$, colour and content.

M: The linear maximal ideal is o.k., but colour and content? Are they radical ideals?

P: Colour and content are inseparable. We see in colour.

M: And I compute in slow motion. I think that photography is easier than finding primary and maximal components, but I can't prove it.

P: Did you use a generic approach?

M: No, not yet. But that is an excellent idea! Let me try it immediately.

In the following days the mathematician worked out Sect. 5.1.A. Given a zero-dimensional ideal $I$ in a polynomial ring $P = K[x_1, \ldots, x_n]$ over a field $K$, one can introduce new indeterminates $y_1, \ldots, y_n$, form the generically extended linear form $L = y_1 x_1 + \cdots + y_n x_n$, and compute the primary components of $I$ from the factorization of the minimal polynomial $\mu_{L+I}(z)$ in $K(y_1, \ldots, y_n)[z]$ (see Algorithm 5.1.5). By calculating suitable colon ideals, we also get the maximal components of $I$ (see Algorithm 5.1.7). A week later the two friends met again.

M: The generic approach to computing primary and maximal ideals worked, but barely. It is so slow that I am afraid I need a better computer soon.

**Electronic supplementary material** The online version of this chapter (DOI:10.1007/978-3-319-43601-2_5) contains supplementary material, which is available to authorized users.

M. Kreuzer and L. Robbiano, *Computational Linear and Commutative Algebra*, DOI 10.1007/978-3-319-43601-2_5

P: A good picture does not need a good camera: it needs a good photographer! If you want to take good pictures, you need to follow a few basic guidelines such as the rule of thirds and the golden ratio. What are your primary ideals?

M: I just want to enjoy my commuting families and apply them to some non-trivial problems.

P: Maybe some random choices will help you find out the good settings.

M: That is an excellent idea! I should try random linear elements!

After returning home, the mathematician immediately started drafting Sect. 5.1.B. He was able to prove that the fundamental idempotents of $P/I$ correspond to the primary components of $I$, and that a randomly chosen linear element is able to split the primary components with overwhelmingly high probability (see Algorithm 5.1.15). Consequently, he succeeded in calculating the joint generalized eigenspaces of a commuting family as the big kernels of the primary components (see Algorithm 5.1.21). He found a way of getting the joint eigenspaces, too (see Algorithm 5.1.20). Their next walk started in a much better mood.

P: Did you know that photography means "writing with light"? Of course, to do so you must know the alphabet: **a**perture, white **b**alance, **c**olour saturation, **d**epth of field, **e**xposure time, **f**ocal length, and so on. Did you find some good parameters for your calculations?

M: I did, but they required a bit of luck. I still do not know what to do if I have a very small field.

P: This is a common problem. After all, a photograph merely involves the projection of a 3-dimensional scene into a small 2-dimensional frame. That is why you do not *take* a picture, you *make* it. You have to apply your mental powers!

M: Powers! That's it! I have to use the powers of Frobenius! Thank you so much for this hint.

Once again, the mathematician could not wait to get home and start working on Sect. 5.2. He discovered that, for a zero-dimensional algebra $R = P/I$ over a finite base field $K = \mathbb{F}_q$, the Frobenius space $\mathrm{Frob}_q(R) = \{f \in R \mid f^q - f = 0\}$ holds the keys to computing the primary decomposition of $I$: the fundamental idempotents are a $K$-basis of $\mathrm{Frob}_q(R)$, its dimension is the number of primary components, and the minimal polynomials of its non-zero elements can be used to split the primary components. On their next encounter, the mathematician was enthusiastic.

M: Your suggestion worked beautifully! The Frobenius space knows everything.

P: Everybody has a photographic memory. Some just don't have film.

M: I wonder whether you have also a suggestion for how to compute the maximal components of a zero-dimensional ideal directly.

P: The maximal components of photography are light and time. They are the factors that rule every picture.

M: So, which factors determine the maximal components? Maybe I have an idea. Maybe we can factor minimal polynomials over extension fields. Let's head home. I urgently want to try some examples.

This time around, the effort needed to make his idea work was much greater than the mathematician had bargained for. Several preparations were required to get Sect. 5.3 off the ground. First the behaviour of minimal polynomials under field extensions had to be studied carefully. Then an algorithm for factoring polynomials over extension fields had to be devised and then improved carefully (see Algorithm 5.3.8). Finally, he constructed a recursive algorithm to compute the maximal components of a zero-dimensional ideal via repeated factorizations of minimal polynomials over extension fields (see Algorithm 5.3.12). As a nice byproduct, every maximal component was automatically given by a triangular system of generators. Thus the next walk was delayed by a fortnight.

P: Where have you been?
M: I got bogged down by this factoring business. There must be better ways to achieve my primary goals. I think I need a different point of view.
P: There are two people in every photograph: the photographer and the viewer. Maybe it is time for you to change perspective and look for a more radical solution.
M: Yes, that's it! I can go via the radical. The maximal components of the ideal and the maximal components of the radical are the same thing.

A few hours later, Sect. 5.4 was born. The first success was a nice improvement of the algorithm of [15], Corollary 3.7.16, to compute the radical of a zero-dimensional ideal. It is based in the observation that if the squarefree part of the $i$-th elimination polynomial has the maximal degree $\dim_K(P/I)$, one can stop the calculation (see Algorithm 5.4.2). Then it suffices to factor the first $i$ elimination polynomials to get the maximal components of $\mathrm{Rad}(I)$ (see Algorithm 5.4.7). As an extra benefit, one obtains nice ways to check whether a given zero-dimensional ideal is primary or maximal (see Corollary 5.4.10). Finally, several methods for getting the primary components from the maximal ones are available, the recommended one being Algorithm 5.4.14. Simultaneously exhausted and elated, the mathematician joined their next excursion.

M: Now I have so many algorithms for computing primary and maximal components that it has become a knotty problem to unravel their interdependencies.
P: Indeed, you can have too much of a good thing. Nowadays, there is hyperinflation of photographs. Even a smart phone can take a dumb picture.
M: Just like photography, mathematics is a true language, but there may be illiterate people. And like a living creature, it is constantly evolving.
P: Do you really think mathematicians will continue to discover new algorithms?
M: Do you think photographers will continue to discover new motifs?

The mathematician had barely reached home when he found a new preprint about the separable subalgebra of a zero-dimensional ring. It became the inspiration of Sect. 5.5 and led to a nice algorithm for computing the separable and the nilpotent part of an element (see Algorithm 5.5.9), as well as a description of the set of all separable elements (see Proposition 5.5.11). After penning thousands of words, he had finished another chapter of a neverending story.

*A picture is worth a thousand words,*
*but it takes longer to load.*

## 5.1  Computing Primary Decompositions

*The problem with internet quotes is*
*that you can't always depend on their accuracy.*
(Abraham Lincoln, 1864)

As the title of this section indicates, we now start discussing methods for computing the primary decomposition of a zero-dimensional ideal. The basic observation underlying these algorithms is that every zero-dimensional ideal $I$ in a polynomial ring $P = K[x_1, \dots, x_n]$ over a field $K$ is the ideal of algebraic relations of a commuting family by Proposition 4.1.2.c. So we can quote many results from the earlier chapters. Let us hope that they are accurate!

A first, very general approach to computing the primary components of a zero-dimensional polynomial ideal is based on Sect. 4.3 where we showed that a generically extended endomorphism is a splitting endomorphism for the generically extended family. Then we quote some results from Sect. 2.5 which tell us how to use a splitting endomorphism for computing the primary components. With a slight modification, we can also calculate the maximal components in the same way. This approach is not only very general, it is also very slow.

In the second subsection we therefore try to make sense of the common practice of substituting "random" numbers for the generic coefficients used in the first approach. In order to turn this heuristic into a Las Vegas algorithm, we need a reasonably efficient way to check whether the computed ideals are actually primary ideals. To construct such a primary ideal test, we combine some theory of the idempotents of the ring $R = P/I$ with a quote from the future: in Sects. 6.1 and 6.2 we shall describe fast ways of finding the $K$-rational zeros of a zero-dimensional polynomial ideal.

Finally, in the third part of this section, we bring the story full circle. We know that the ideal of algebraic relations of a commuting family is zero-dimensional. Thus, by computing the primary and maximal components of this ideal, and quoting the basic insights gained in Sect. 2.4, we get algorithms for computing the joint eigenspaces and joint generalized eigenspaces of a commuting family.

*What's the use of a good quotation*
*if you can't change it?*
(Doctor Who)

In the following we let $K$ be a field, let $P = K[x_1, \ldots, x_n]$ be a polynomial ring over $K$, and let $I$ be a zero-dimensional ideal in $P$. Then the residue class ring $R = P/I$ is a zero-dimensional $K$-algebra. It has finitely many maximal ideals $\mathfrak{m}_1, \ldots, \mathfrak{m}_s$ (see Proposition 2.2.4). The primary decomposition of the zero ideal in $R$ is of the form $\langle 0 \rangle = \mathfrak{q}_1 \cap \cdots \cap \mathfrak{q}_s$ where $\mathfrak{q}_i$ is $\mathfrak{m}_i$-primary for $i = 1, \ldots, s$. The preimages of the ideals $\mathfrak{q}_i$ and $\mathfrak{m}_i$ in $P$ are denoted by $\mathfrak{Q}_i$ and $\mathfrak{M}_i$, respectively. Then the primary decomposition of the ideal $I$ in $P$ is given by $I = \mathfrak{Q}_1 \cap \cdots \cap \mathfrak{Q}_s$, and the ideal $\mathfrak{Q}_i$ is $\mathfrak{M}_i$-primary for $i = 1, \ldots, s$. The main goal in this section is to compute the ideals $\mathfrak{Q}_i$ from a given system of generators of $I$.

## 5.1.A  Using the Generically Extended Linear Form

> *No one ever says "It's only a game,"*
> *when their team is winning.*

To start the game of computing primary decompositions of zero-dimensional polynomial ideals, we try a very general method. It can be obtained by combining the fact that a generically extended endomorphism splits (see Theorem 4.3.14) with the computation of the primary components of a commuting family from a splitting endomorphism (see Corollary 2.5.3). Is this a winning strategy? Unfortunately, the algorithms in this subsection are not efficient enough to treat larger examples. Thus they are mainly of theoretical value. However, we can still claim that it was only a game until we find a more victorious approach in the next subsection.

> *If all is not lost, where is it?*

In the following we frequently need to compute the minimal polynomial over $K$ of an element of a zero-dimensional $K$-algebra. For the convenience of the reader we recall the following algorithm from [15], Corollary 3.6.4.b which solves this task using elimination.

---

**Algorithm 5.1.1 (The Minimal Polynomial of an Element, I)**
*Let $R = P/I$ be a zero-dimensional affine $K$-algebra, let $f \in P$, and let $\bar{f}$ be its image in $P/I$. The following instructions compute the minimal polynomial of $\bar{f}$.*

(1) *In $P[z]$ form the ideal $J = \langle z - f \rangle + I \cdot P[z]$ and compute $J \cap K[z]$.*
(2) *Return the monic generator of $J \cap K[z]$.*

---

If a $K$-basis $B$ of $R = P/I$ is known and if we have an effective method for representing elements of $R$ in terms of this basis, we can also use the following algorithm to compute the minimal polynomial of an element of $R$. Notice that these hypotheses are satisfied if we know a Gröbner basis or a border basis of $I$.

**Algorithm 5.1.2 (The Minimal Polynomial of an Element, II)**
*Let $R = P/I$ be a zero-dimensional affine $K$-algebra, let $f \in P$, and let $\bar{f}$ be its image in $P/I$. Suppose that we know a $K$-basis $B$ of $R$ and an effective method $\mathrm{NF}_B : P \longrightarrow K^d$ which maps an element of $P$ to the coefficient tuple of its residue class in $R$ with respect to the basis $B$. Then the following instructions compute the minimal polynomial of $\bar{f}$.*

(1) *Let $L = (1)$.*
(2) *For $i = 1, 2, \ldots$, compute $\mathrm{NF}_B(f^i)$ and check whether it is $K$-linearly dependent on the elements in $L$. If this is not the case, append $\mathrm{NF}_B(f^i)$ to the tuple $L$ and continue with the next number $i$.*
(3) *If there exist $c_0, \ldots, c_{i-1} \in K$ such that $\mathrm{NF}_B(f^i) = \sum_{k=0}^{i-1} c_k \mathrm{NF}_B(f^k)$ then return the polynomial $\mu_{\bar{f}}(z) = z^i - \sum_{k=0}^{i-1} c_k z^k$ and stop.*

*Proof* Since at most $d$ tuples in $K^d$ can be linearly independent, it is clear that the loop in Step (2) is finite. Moreover, the computed polynomial $\mu_{\bar{f}}(z)$ is clearly monic. The tuple $\mathrm{NF}_B(f^i) - \sum_{k=0}^{i-1} c_k \mathrm{NF}_B(f^k)$ is the coordinate tuple of $f^i - \sum_{k=0}^{i-1} c_k f_k$ with respect to $B$. Since it is zero, it follows that $\mu_{\bar{f}}(\bar{f}) = 0$. No polynomial of lower degree can vanish at $\bar{f}$, because the corresponding linear combination of tuples $\mathrm{NF}_B(f^0), \ldots, \mathrm{NF}_B(f^j)$ would have been found at an earlier iteration of Step (2). $\qquad\square$

A third method for computing the minimal polynomial is given by the following remark.

**Remark 5.1.3** In the setting of the algorithms above, let $\bar{f} \in R$, and suppose that we know both a $K$-basis $B$ of $R$ and the multiplication matrix of $\bar{f}$ with respect to $B$. Then we can use Algorithm 1.1.8 and the observation that we have $\mu_{\bar{f}}(z) = \mu_{\vartheta_f}(z)$ to compute the minimal polynomial of $\bar{f}$.

The next easy proposition provides a useful property of minimal polynomials of elements of affine algebras.

**Proposition 5.1.4** *Let $R$ and $S$ be zero-dimensional affine $K$-algebras, let $r \in R$, and let $\alpha : R \longrightarrow S$ be a $K$-algebra homomorphism.*

(a) *The minimal polynomial of $\alpha(r)$ is a divisor of the minimal polynomial of $r$.*
(b) *If $\alpha$ is injective, the two minimal polynomials coincide.*

*Proof* Claim (a) follows from the observation that $0 = \alpha(\mu_r(r)) = \mu_r(\alpha(r))$. To prove (b), we note that $0 = \mu_{\alpha(r)}(\alpha(r)) = \alpha(\mu_{\alpha(r)}(r))$ implies $\mu_{\alpha(r)}(r) = 0$, since $\alpha$ is injective. Thus $\mu_r(z)$ divides $\mu_{\alpha(r)}(z)$ is this case. $\qquad\square$

Now we are ready to compute the primary decomposition of a zero-dimensional ideal via the generically extended linear form.

---

**Algorithm 5.1.5 (Primary Decomposition via the Generically Extended Linear Form)**
Let $I$ be a zero-dimensional ideal in $P = K[x_1, \ldots, x_n]$, and let $R = P/I$. Consider the following sequence of instructions.

(1) Let $y_1, \ldots, y_n$ be new indeterminates, let $L = y_1 x_1 + \cdots + y_n x_n$ be the generically extended linear form, and let $\ell = L + I^e$ where $I^e$ is the extension of $I$ to $R \otimes_K K(y_1, \ldots, y_n)$. Compute the minimal polynomial $\mu_\ell(z) \in K(y_1, \ldots, y_n)[z]$.

(2) Factor $\mu_\ell(z)$ in the form $\mu_\ell(z) = p_1(z)^{m_1} \cdots p_s(z)^{m_s}$, where $p_i(z)$ is an irreducible polynomial in $K(y_1, \ldots, y_n)[z]$ for $i = 1, \ldots, s$.

(3) For $i = 1, \ldots, s$, let $\widetilde{\mathfrak{Q}}_i = I + \langle p_i(L)^{m_i} \rangle \subset P \otimes_K K(y_1, \ldots, y_n)$. Choose a term ordering $\sigma$ on $\mathbb{T}(x_1, \ldots, x_n)$ and compute the reduced $\sigma$-Gröbner basis $G_i$ of $\widetilde{\mathfrak{Q}}_i$.

(4) For $i = 1, \ldots, s$, let $\mathfrak{Q}_i = \langle G_i \rangle \subset P$. Return the tuple $(\mathfrak{Q}_1, \ldots, \mathfrak{Q}_s)$ and stop.

This is an algorithm which computes the primary components $\mathfrak{Q}_1, \ldots, \mathfrak{Q}_s$ of $I$.

---

*Proof* Let $\mathscr{F}$ be the multiplication family of $R$. Recall that we have an isomorphism $\mathscr{F} \cong R = P/I$ (see Proposition 4.1.2.b). We use it to identify the primary components of the zero ideal in $R$ with those of the zero ideal in $\mathscr{F}$. Next we recall Proposition 2.5.13.a, where it was shown that the map $\iota : \mathscr{F} \longrightarrow \mathscr{F} \otimes_K K(y_1, \ldots, y_n)$ induces a 1–1 correspondence between the maximal ideals of $\mathscr{F}$ and the maximal ideals of the generically extended family $\mathscr{F} \otimes_K K(y_1, \ldots, y_n)$. More precisely, the maximal ideals of the generically extended family are the extensions $\mathfrak{m}_i \otimes_K K(y_1, \ldots, y_n)$ of the maximal ideals $\mathfrak{m}_i$ of $\mathscr{F}$.

By Proposition 4.1.12, we have the decomposition $\mathscr{F} \cong \prod_{i=1}^{s} \mathscr{F}/\mathfrak{q}_i$ of $\mathscr{F}$ into local rings. Hence the generically extended family satisfies

$$\mathscr{F} \otimes_K K(y_1, \ldots, y_n) = \prod_{i=1}^{s} \left( \mathscr{F} \otimes_K K(y_1, \ldots, y_n) \right) / \left( \mathfrak{q}_i \otimes_K K(y_1, \ldots, y_n) \right)$$

It follows that the primary components of the generically extended family are the ideals $\mathfrak{q}_i \otimes_K K(y_1, \ldots, y_n)$, where $i \in \{1, \ldots, s\}$. The generically extended endomorphism $\Theta_{y,\varPhi}$ associated to $\varPhi = (\vartheta_{x_1}, \ldots, \vartheta_{x_n})$ is the multiplication endomorphism by $\ell$. By Theorem 4.3.14, it is a splitting endomorphism for the family $\mathscr{F} \otimes_K K(y_1, \ldots, y_n)$. Hence Corollary 2.5.3 shows that the primary components of the zero ideal in the generically extended family $\mathscr{F} \otimes_K K(y_1, \ldots, y_n)$ are given by

$$\mathfrak{q}_i \otimes_K K(y_1, \ldots, y_n) = \langle p_i(\vartheta_\ell)^{m_i} \rangle$$

Here we use the fact that the minimal polynomials of $\ell$, of $\vartheta_\ell$ and of the extension of $\vartheta_\ell$ to $\mathscr{F} \otimes_K K(y_1, \ldots, y_n)$ are identical. Therefore the ideals $\tilde{\mathfrak{Q}}_i$ computed in Step (3) of the algorithm are the primary components of $I \otimes_K K(y_1, \ldots, y_n)$.

Since the ideals $\mathfrak{q}_i \otimes K(y_1, \ldots, y_n)$ are the extensions of the ideals $\mathfrak{q}_i$ in $\mathscr{F} \cong R$, the ideals $\tilde{\mathfrak{Q}}_i$ are the extensions of the primary components $\mathfrak{Q}_i$ of $I$ and hence defined over $K$. Thus the reduced Gröbner basis of $\tilde{\mathfrak{Q}}_i$ is contained in $P$ (see [15], Theorem 2.4.17), and Steps (3) and (4) of the algorithm compute $\mathfrak{Q}_i$. Altogether, it follows that the ideals $\mathfrak{Q}_1, \ldots, \mathfrak{Q}_s$ computed by the algorithms are the primary components of $I$, as claimed.      $\square$

To show how this algorithm works, we reconsider Example 2.5.8.

**Example 5.1.6** As we did in Example 2.5.8, we let $K = \mathbb{F}_2$, $P = \mathbb{F}_2[x_1, x_2]$, and $R = P/I$ where $I = \langle x_1^2 + x_1, x_1 x_2, x_2^2 + x_2 \rangle$. To compute the primary decomposition of $I$, we follow the steps of Algorithm 5.1.5.

(1) Let $y_1, y_2$ be new indeterminates, and let $L = y_1 x_1 + y_2 x_2$. We compute the minimal polynomial of the residue class $\ell$ of $L$ in the quotient ring $K(y_1, y_2)[x_1, x_2]/I \cdot K(y_1, y_2)[x_1, x_2]$ and get $\mu_\ell(z) = z^3 + (y_1 + y_2)z^2 + (y_1 y_2)z$.

(2) The factorization of $\mu_\ell(z)$ is $\mu_\ell(z) = z(z + y_1)(z + y_2)$.

(3) Next we substitute $z$ by $L$ and get the three primary ideals

$$\tilde{\mathfrak{Q}}_1 = \langle x_1^2 + x_1, x_1 x_2, x_2^2 + x_2, y_1 x_1 + y_2 x_2 \rangle$$

$$\tilde{\mathfrak{Q}}_2 = \langle x_1^2 + x_1, x_1 x_2, x_2^2 + x_2, y_1 x_1 + y_2 x_2 + y_1 \rangle$$

$$\tilde{\mathfrak{Q}}_3 = \langle x_1^2 + x_1, x_1 x_2, x_2^2 + x_2, y_1 x_1 + y_2 x_2 + y_2 \rangle$$

(4) The reduced Gröbner bases of the ideals $\tilde{\mathfrak{Q}}_i$ generate the ideals $\mathfrak{Q}_1 = \langle x_1, x_2 \rangle$, $\mathfrak{Q}_2 = \langle x_1 + 1, x_2 \rangle$, and $\mathfrak{Q}_3 = \langle x_1, x_2 + 1 \rangle$, respectively. The algorithm returns the triple $(\mathfrak{Q}_1, \mathfrak{Q}_2, \mathfrak{Q}_3)$.

This result agrees with the primary decomposition mentioned in Example 2.5.8.

After calculating the primary components of a zero-dimensional polynomial ideal, we can also find its minimal primes, i.e., the maximal ideals containing it. For this purpose, we can use the description of the maximal ideal of a zero-dimensional local ring given in Proposition 4.4.3.b and the fact that, under a suitable assumption on the base field, the generically extended linear form is of maximal nilpotency by Theorem 4.4.12.b.

**Algorithm 5.1.7 (Computing Maximal Components via the Generically Extended Linear Form)**
*Let $I$ be a zero-dimensional ideal in $P$ and $R = P/I$. Assume that $\mathrm{char}(K) = 0$ or that $K$ is an infinite field with $\mathrm{char}(K) \geq \dim_K(R)$. Consider the following sequence of instructions.*

(1) *Perform Algorithm 5.1.5. In particular, let $L = y_1 x_1 + \cdots + y_n x_n$ be the generically extended linear form, and let $\mathfrak{Q}_1, \ldots, \mathfrak{Q}_s$ be the primary components of $I$.*

(2) *For $i = 1, \ldots, s$, we let $\widetilde{Q}_i = \mathfrak{Q}_i \otimes_K K(y_1, \ldots, y_n)$ and compute the ideal $\widetilde{\mathfrak{M}}_i = \widetilde{\mathfrak{Q}}_i :_P \langle p_i(L)^{m_i - 1} \rangle$.*

(3) *For $i = 1, \ldots, s$, calculate a reduced Gröbner basis $G_i$ of $\widetilde{\mathfrak{M}}_i$. (Notice that $G \subset P$.)*

(4) *For $i = 1, \ldots, s$, let $\mathfrak{M}_i = \langle G_i \rangle$. Return the tuple $(\mathfrak{M}_1, \ldots, \mathfrak{M}_s)$.*

*This is an algorithm which computes the maximal components $\mathfrak{M}_1, \ldots, \mathfrak{M}_s$ of $I$. In particular, it computes $\mathrm{Rad}(I) = \mathfrak{M}_1 \cap \cdots \cap \mathfrak{M}_s$, the primary decomposition of the radical of $I$.*

*Proof* After performing Algorithm 5.1.5, we know the primary components $\mathfrak{Q}_i$ of $I$. Hence the rings $P/\mathfrak{Q}_i$ are zero-dimensional and local. Then Remark 4.4.13 shows that the hypotheses of Theorem 4.4.12.b are satisfied, and hence the residue class $\ell_i$ of $L$ in $(P/\mathfrak{Q}_i) \otimes_K K(y_1, \ldots, y_n)$ is of maximal nilpotency. Now an application of Proposition 4.4.3.b shows that we have $\widetilde{\mathfrak{M}}_i = \widetilde{\mathfrak{Q}}_i :_P \langle p_i(L)^{\mathrm{nix}(\ell_i) - 1} \rangle$. Furthermore, we note that Lemma 4.4.2.a yields $\mathrm{nix}(\ell_i) = m_i$. Finally, as in the proof of Algorithm 5.1.5, it follows that $G_i \subset P$ and that $\mathfrak{M}_1, \ldots, \mathfrak{M}_s$ are the maximal components of $I$. $\qquad\square$

In this algorithm, the assumption that the base field $K$ is not of small characteristic is essential, since Example 4.4.4 shows that, even if $K$ is infinite, it is possible that an element of maximal nilpotency does not exist. More precisely, Theorem 4.4.12.b holds also if $K$ is an infinite field and $\mathrm{char}(K) \geq \mathrm{nix}(\mathfrak{m}_i)$, but since we usually do not know the nilpotency index of the maximal ideal $\mathfrak{m}_i$ of $P/\mathfrak{Q}_i$ *a priori*, we have to resort to the estimate given in Remark 4.4.13 to make sure that the characteristic of the base field is not too small.

Let us see an example which illustrates the above algorithm.

**Example 5.1.8** Let $K = \mathbb{Q}$, $P = K[x_1, x_2, x_3]$, and $I$ the ideal generated by $\{x_1^2,\ x_1 x_2 - x_3^2,\ x_2^2 - x_2 - 1\}$. To compute the maximal ideals containing $I$ via Algorithm 5.1.7, we proceed as follows.

(1) To perform Algorithm 5.1.5, we consider the generically extended linear form $L = y_1 x_1 + y_2 x_2 + y_3 x_3$ and its residue class $\ell$ in $(P/I) \otimes_K K(y_1, y_2, y_3)$. We compute the minimal polynomial of $\ell$ and get $\mu_\ell(z) = (z^2 - y_2 z - y_2^2)^6$. This implies that we have $(L^2 - y_2 L - y_2^2)^6 \in I \cdot K[y_1, y_2, y_3, x_1, x_2, x_3]$ and that there is only one primary component of $I$, namely $I$ itself.

(2) Now, to compute the maximal ideal $\mathfrak{M} = \mathrm{Rad}(I)$, we use Algorithm 5.1.7 and get $\widetilde{\mathfrak{M}} = (I \otimes_K K(y_1, y_2, y_3)) :_P \langle (L^2 - y_2 L - y_2^2)^5 \rangle = \langle x_1, x_2^2 - x_2 - 1, x_3 \rangle$.

(4) Hence the algorithm returns the maximal ideal $\mathfrak{M} = \langle x_1, x_2^2 - x_2 - 1, x_3 \rangle$.

In the following example we compute the maximal ideals containing a zero-dimensional ideal $I$ in a slightly different way, namely by first computing the radical of $I$ and then applying Algorithm 5.1.5.

**Example 5.1.9**   Let $K = \mathbb{Q}$, let $P = K[x_1, x_2, x_3]$, and let $I = \langle x_1x_3 - x_2x_3 - x_3, x_2^2 - x_3^2, x_3^3 - x_2x_3 - x_3, x_2x_3^2 - \frac{1}{3}x_1^2 - x_3^2 + \frac{2}{3}, x_1^2x_2 - 3x_3^2 - 2x_2, x_1^3 - x_1^2 - 3x_3^2 - 2x_1 + 2\rangle$.

First we compute the radical ideal of $I$ using [15], Corollary 3.7.16 and get $\mathrm{Rad}(I) = \langle x_1x_3 - x_2x_3 - x_3, x_2^2 - x_3^2, x_1x_2 - x_3^2 - x_2, x_1^2 - 3x_2 - 2, x_3^3 - x_2x_3 - x_3, x_2x_3^2 - x_3^2 - x_2\rangle$. Next we calculate the primary decomposition of $\mathrm{Rad}(I)$ using Algorithm 5.1.5.

(1) Let $y_1, y_2, y_3$ be new indeterminates, and let $L = y_1x_1 + y_2x_2 + y_3x_3$. We compute the minimal polynomial $\mu_\ell(z)$ of the residue class $\ell$ of $L$ in the $K(y_1, y_2, y_3)$-algebra $(P/\mathrm{Rad}(I)) \otimes_K K(y_1, y_2, y_3)$.
(2) The polynomial $\mu_\ell(z)$ factors in $K(y_1, y_2, y_3)[z]$ as

$$\mu_\ell(z) = \left(z^2 - (3y_1 + y_2 + y_3)z + y_1^2 - y_1y_2 - y_2^2 - y_1y_3 - 2y_2y_3 - y_3^2\right)$$
$$\cdot \left(z^2 - (3y_1 + y_2 - y_3)z + y_1^2 - y_1y_2 - y_2^2 + y_1y_3 + 2y_2y_3 - y_3^2\right)$$
$$\cdot \left(z^2 - 2y_1^2\right)$$

(3) We add the three polynomials $F_1 = L^2 - (3y_1 + y_2 + y_3)L + y_1^2 - y_1y_2 - y_2^2 - y_1y_3 - 2y_2y_3 - y_3^2$, $F_2 = L^2 - (3y_1 + y_2 - y_3)L + y_1^2 - y_1y_2 - y_2^2 + y_1y_3 + 2y_2y_3 - y_3^2$, and $F_3 = L^2 - 2y_1^2$ to the ideal $J$ which is the extension of $\mathrm{Rad}(I)$ to $P \otimes_K K(y_1, y_2, y_3)$ and get the three ideals $\widetilde{M}_1 = J + \langle F_1 \rangle$, $\widetilde{M}_2 = J + \langle F_2 \rangle$, and $\widetilde{M}_3 = J + \langle F_3 \rangle$.
(4) The ideals generated by the reduced Gröbner bases of $\widetilde{\mathfrak{M}}_1$, $\widetilde{\mathfrak{M}}_2$, and $\widetilde{\mathfrak{M}}_3$ are $\mathfrak{M}_1 = \langle x_1 + x_3 - 1, x_2 + x_3, x_3^2 + x_3 - 1 \rangle$, $\mathfrak{M}_2 = \langle x_1 - x_3 - 1, x_2 - x_3, x_3^2 - x_3 - 1 \rangle$, and $\mathfrak{M}_3 = \langle x_1^2 - 2, x_2, x_3 \rangle$, respectively.

Therefore there are three maximal ideals containing $I$, namely $\mathfrak{M}_1$, $\mathfrak{M}_2$, and $\mathfrak{M}_3$.

## 5.1.B  Using Linear Forms and Idempotents

> *Probable impossibilities are to be preferred*
> *to improbable possibilities.*
> (Aristoteles)

A pointed out before, the algorithms in the preceding subsection are not efficient enough to treat larger examples. The reason is that they introduce many new indeterminates. When the base field is sufficiently large, we can try to simplify these methods by replacing the indeterminates $y_1, \ldots, y_n$ with "random" coefficients. Many authors claim that in this way one gets a probabilistic algorithm and that the probability to get the correct result is "very high". Other authors say that "the probability of success is 1". In our opinion these claims are not correct, since the resulting procedures typically do not even satisfy the basic requirements of a **Monte Carlo algorithm**, due to the difficulty of providing a bound for the error probability. So, they should correctly be called **heuristic algorithms**.

The good news is that, in practice, when using elements of the base field in place of the new indeterminates, the result is indeed almost always correct. The bad news is that, if we want to certify the result and construct a **Las Vegas algorithm** for primary decomposition, we have to use methods which can be computationally expensive. The heuristic algorithm based on the choice of "random" coefficients was first reported by C. Monico in [23].

In the following we work in the same setting as in the first subsection. Thus we let $K$ be a field, let $P = K[x_1, \ldots, x_n]$, let $I$ be a zero-dimensional ideal in $P$, and let $R = P/I$. Our goal is to find a Las Vegas algorithm for computing the primary decomposition of $I$. It is based on the idea of using idempotents to check whether a given ideal is primary and whether a given set of ideals is the set of primary components of $I$. Recall that an **idempotent** of $R$ is an element $f \in R$ such that $f^2 = f$. Clearly, the elements 0 and 1 are idempotents. Furthermore, idempotents can be obtained from the decomposition of $R$ into local rings (see Definition 4.1.13). The following proposition provides some basic information.

**Proposition 5.1.10** *In the above setting, let $\langle 0 \rangle = \mathfrak{q}_1 \cap \cdots \cap \mathfrak{q}_s$ be the primary decomposition of the zero ideal of $R$. For $i = 1, \ldots, s$, we let $J_i = \cap_{j \neq i} \mathfrak{q}_j$.*

(a) *For $i = 1, \ldots, s$, there exist uniquely determined elements $e_i \in J_i$ and $f_i \in \mathfrak{q}_i$ such that $e_i + f_i = 1$.*

(b) *Under the isomorphism $\pi : R \cong \prod_{i=1}^{s} R/\mathfrak{q}_j$ we have the equality $\pi(e_i) = (0, \ldots, 0, 1, 0, \ldots, 0)$ with 1 in $i$-th position. In particular, the elements $e_1, \ldots, e_s$ are idempotents of $R$.*

(c) *For $i = 1, \ldots, s$, we have $Re_i \cong R/\mathfrak{q}_i$. Consequently, there exists an isomorphism $R \cong \prod_{i=1}^{s} Re_i$.*

(d) *The ring $R$ is local if and only if its only idempotents are 0 and 1.*

(e) *The set $\{\sum_{i=1}^{s} c_i e_i \mid c_i \in \{0, 1\}\}$ is the set of all idempotents of $R$. It consists of $2^s$ elements.*

*Proof* First we show (a). From the proof of the Chinese Remainder Theorem 2.2.1 we know that the ideals $J_i$ and $\mathfrak{q}_i$ are comaximal for every $i \in \{1, \ldots, s\}$. Hence there exist elements $e_i \in J_i$ and $f_i \in \mathfrak{q}_i$ such that $e_i + f_i = 1$. Now let us assume that $e_i' + f_i' = 1$ with $e_i' \in J_i$ and $f_i' \in \mathfrak{q}_i$. Then $f_i - f_i' = e_i - e_i' \in \cap_{j=1}^{s} \mathfrak{q}_j = \langle 0 \rangle$, and thus $e_i = e_i'$ and $f_i = f_i'$.

Claim (b) follows by applying the isomorphism $\pi$ to the equality $e_i + f_i = 1$. Next we prove (c). It suffices to show that $\operatorname{Ann}_R(e_i) = \mathfrak{q}_i$ for $i = 1, \ldots, s$. To do this, we note that if $g \in \operatorname{Ann}_R(e_i)$ then $ge_i = g - gf_i = 0 \in \mathfrak{q}_i$, and hence $g \in \mathfrak{q}_i$. Conversely, if $g \in \mathfrak{q}_i$ then $ge_i \in \cap_{i=1}^{s} \mathfrak{q}_i = \langle 0 \rangle$.

To prove one implication in (d), we assume $R$ to be local, let $\mathfrak{m}$ be its maximal ideal, and let $e \in R$ be an idempotent. It follows that $e(e-1) = e^2 - e = 0 \in \mathfrak{m}$. Since not both $e$ and $e - 1$ can be contained in $\mathfrak{m}$, one of these two elements is invertible. Then the other one is zero. The converse implication follows from (b).

To prove (e) we first note that the elements in $\{\sum_{i=1}^{s} c_i e_i \mid c_i \in \{0, 1\}\}$ are clearly idempotents. Conversely, if $f \in R$ is idempotent then the element $\pi(f)$ is also idempotent. Hence its components in each local ring $R/\mathfrak{q}_i$ are idempotent, and the claim follows from (d).                                                                                      $\square$

This proposition motivates the following definition.

**Definition 5.1.11**   Let $R = P/I$ be a zero-dimensional affine $K$-algebra as above. The elements $e_1, \ldots, e_s \in R$ constructed in part (a) of the preceding proposition are called the **primitive idempotents** (or the **fundamental idempotents**) of $R$.

The following algorithm can be used to check whether a given polynomial ideal is primary. It requires the computation of the $K$-rational zeros of a zero-dimensional ideal. This task will be examined in Sect. 6.2.

---

**Algorithm 5.1.12 (Primary Ideal Test)**
*Let $I$ be a zero-dimensional ideal in $P = K[x_1, \ldots, x_n]$. The following in-structions define an algorithm which checks whether $I$ is a primary ideal and outputs the corresponding boolean value.*

(1) *Using one of the methods explained in Remark 4.1.7, compute a tuple $B = (t_1, \ldots, t_d)$ of terms in $\mathbb{T}^n$ such that $t_1 = 1$ and such that the residue classes of the entries of $B$ form a $K$-basis of $P/I$.*

(2) *Let $y_1, \ldots, y_d$ be new indeterminates, let $f = \sum_{i=1}^{d} y_i t_i$, and use the expressions of all the products $t_i t_j$ in terms of the basis $B$ to write $f^2 = \sum_{i=1}^{d} g_i t_i + h$ with $g_i \in K[y_1, \ldots, y_d]$ and $h \in I \cdot P[y_1, \ldots, y_d]$.*

(3) *In the polynomial ring $K[y_1, \ldots, y_d]$, let $J = \langle g_1 - y_1, \ldots, g_d - y_d \rangle$. Using one of the methods given in Sect. 6.2, find the $K$-rational zeros of $J$.*

(4) *If the only $K$-rational zeros of $J$ are $(0, 0, \ldots, 0)$ and $(1, 0, \ldots, 0)$ then return TRUE. Otherwise, return FALSE.*

---

*Proof* A $K$-rational zero $(a_1, \ldots, a_d) \in K^d$ of $J$ corresponds 1–1 to a polynomial $\tilde{f} = a_1 t_1 + \cdots a_d t_d$ in $P$ such that $\tilde{f}^2 + I = \tilde{f} + I$, i.e. to an idempotent of $R = P/I$. Therefore the claim follows from Proposition 5.1.10.d.                                    $\square$

Let us invoke Algorithm 5.1.12 in a non-trivial case.

**Example 5.1.13**   Let $K = \mathbb{Q}$, let $P = K[x_1, x_2, x_3]$, and let $I = \langle f_1, f_2, f_3, f_4, f_5 \rangle$, where $f_1 = x_3^2 - x_2 + 1$, $f_2 = x_1 x_3 - x_2 x_3 - x_1 + 2x_2 - 3x_3$, $f_3 = x_2^2 + x_2 x_3 - x_1 -$

$x_3 + 2$, $f_4 = x_1x_2 + 2x_2x_3 - 3x_1 - 4x_3 + 5$, and $f_5 = x_1^2 + 2x_2x_3 - 11x_1 + 3x_2 - 11x_3 + 17$. We follow the steps of the algorithm.

(1) A $K$-basis of $P/I$ is given by the residue classes of the components of the tuple $B = (1, x_1, x_2, x_3, x_2x_3)$.

(2) Let $y_1, \ldots, y_5$ be new indeterminates, and let $f = y_1 + y_2x_1 + y_3x_2 + y_4x_3 + y_5x_2x_3$. We calculate the representation of $f^2$ in the basis $B$ and get $f^2 = (y_1^2 - 17y_2^2 - 10y_2y_3 - 2y_3^2 - y_4^2 + 2y_3y_5 - 4y_4y_5 - 4y_5^2) + (2y_1y_2 + 11y_2^2 + 6y_2y_3 + y_3^2 + 2y_2y_4 + 2y_2y_5 + 2y_4y_5 + 2y_5^2)x_1 + (-3y_2^2 + 2y_1y_3 - 4y_2y_4 + y_4^2 - 2y_4y_5 - 2y_5^2)x_2 + (11y_2^2 + 8y_2y_3 + y_3^2 + 2y_1y_4 + 6y_2y_4 + 4y_2y_5 + 2y_4y_5 + 3y_5^2)x_3 + (-2y_2^2 - 4y_2y_3 - y_3^2 + 2y_2y_4 + 2y_3y_4 + 2y_1y_5 + 10y_2y_5 + 4y_3y_5 - 2y_4y_5 - 2y_5^2)x_2x_3 + h$ with $h \in IK[y_1, \ldots, y_5, x_1, x_2, x_3]$.

(3) In the ring $K[y_1, \ldots, y_5]$, we obtain the ideal $J = \langle y_1^2 - 17y_2^2 - 10y_2y_3 - 2y_3^2 - y_4^2 + 2y_3y_5 - 4y_4y_5 - 4y_5^2 - y_1, 2y_1y_2 + 11y_2^2 + 6y_2y_3 + y_3^2 + 2y_2y_4 + 2y_2y_5 + 2y_4y_5 + 2y_5^2 - y_2, -3y_2^2 + 2y_1y_3 - 4y_2y_4 + y_4^2 - 2y_4y_5 - 2y_5^2 - y_3, 11y_2^2 + 8y_2y_3 + y_3^2 + 2y_1y_4 + 6y_2y_4 + 4y_2y_5 + 2y_4y_5 + 3y_5^2 - y_4, -2y_2^2 - 4y_2y_3 - y_3^2 + 2y_2y_4 + 2y_3y_4 + 2y_1y_5 + 10y_2y_5 + 4y_3y_5 - 2y_4y_5 - 2y_5^2 - y_5 \rangle$. Using one of the methods given in Sect. 6.2, we find that the only $K$-rational zeros of $J$ are $(0, 0, 0, 0, 0)$ and $(1, 0, 0, 0, 0)$. Therefore we conclude that the ideal $I$ is primary. If we compute the radical ideal of $I$, it turns out that $I$ is in fact a maximal ideal of $P$.

The following example shows that the result of Algorithm 5.1.12 sometimes depends on the choice of the base field.

**Example 5.1.14** Let $K = \mathbb{Q}$, let $P = K[x_1, x_2]$, and let $I$ be the ideal in $P$ generated by $\{x_1x_2 - \frac{1}{2}x_2^2 - \frac{9}{2}x_1 + \frac{9}{2}x_2 - \frac{1}{2}, x_1^2 + x_1 - x_2, x_2^3 - \frac{31}{2}x_2^2 - \frac{77}{2}x_1 + \frac{83}{2}x_2 - \frac{5}{2}\}$. Let us apply Algorithm 5.1.12 to check whether $I$ is a primary ideal.

(1) A $K$-basis of $P/I$ is given by the residue classes of $B = (1, x_1, x_2, x_2^2)$.

(2) Using new indeterminates $y_1, y_2, y_3, y_4$, we let $F = y_1 + y_2x_1 + y_3x_2 + y_4x_2^2$. We compute the representation of $f^2$ in the basis $B$ and get $f^2 = (y_1^2 + 5y_2y_3 + 58y_3^2 + y_2y_4 + 7y_3y_4) + (77y_2y_3 + 770y_3^2 + 2y_1y_4 + 9y_2y_4 + 79y_3y_4 - y_4^2)x_1 + (2y_1y_2 - 83y_2y_3 - 814y_3^2 - 9y_2y_4 - 81y_3y_4 + y_4^2)x_2 + (y_3^2 + 2y_1y_3 + 31y_2y_3 + 218y_3^2 + y_2y_4 + 11y_3y_4)x_2^2 + h$ with $h \in IK[y_1, \ldots, y_4, x_1, x_2]$.

(3) In $K[y_1, y_2, y_3, y_4]$ we get the ideal $J = \langle y_1^2 + 5y_2y_3 + 58y_3^2 + y_2y_4 + 7y_3y_4 - y_1, 77y_2y_3 + 770y_3^2 + 2y_1y_4 + 9y_2y_4 + 79y_3y_4 - y_4^2 - y_2, 2y_1y_2 - 83y_2y_3 - 814y_3^2 - 9y_2y_4 - 81y_3y_4 + y_4^2 - y_3, y_3^2 + 2y_1y_3 + 31y_2y_3 + 218y_3^2 + y_2y_4 + 11y_3y_4 - y_4 \rangle$. The only $K$-rational zeros of $J$ are $(0, 0, 0, 0)$ and $(1, 0, 0, 0)$. Hence $I$ is a primary ideal.

It turns out that the reduced Gröbner basis of $I$ with respect to the lexicographic term ordering satisfying $x_2 > x_1$ is $(x_1^4 - 10x_1^2 + 1, x_2 - x_1^2 - x_1)$. This implies that $I$ is a maximal ideal of $P$. The polynomial $x_1^4 - 10x_1^2 + 1$ is irreducible over the rational numbers, but it is reducible modulo every prime (see [21], p. 13). Therefore things change if we replace $K = \mathbb{Q}$ by a field of prime characteristic. For instance, if we let $K = \mathbb{F}_7$, then the ideal $I$ splits into two primary components. The reader is invited to check what happens over base fields $K = \mathbb{F}_p$ for various other prime numbers $p$.

*It is not certain that everything is uncertain.*
(Blaise Pascal)

Now we can modify Algorithm 5.1.5 as follows: we replace the generically extended linear form by an arbitrary (e.g., randomly chosen) linear form in $P$, perform the analogous algorithm, and use Algorithm 5.1.12 to check whether the computed ideals are indeed primary.

---

**Algorithm 5.1.15 (Primary Decomposition via a Linear Form)**
*Let $P = K[x_1, \ldots, x_n]$, let $I$ be a zero-dimensional ideal in $P$, let $R = P/I$, let $L = c_1 x_1 + \cdots + c_n x_n$ be a linear form in $P$, where $c_1, \ldots, c_n \in K$, and let $\ell = L + I$ be the residue class of $L$ in $R = P/I$.*

(1) *Compute the minimal polynomial $\mu_\ell(z)$ and factor it in the form $\mu_\ell(z) = p_1(z)^{m_1} \cdots p_s(z)^{m_s}$, with pairwise distinct, monic, irreducible polynomials $p_1(z), \ldots, p_s(z) \in K[z]$.*

(2) *For $i = 1, \ldots, s$, let $\mathfrak{Q}_i = I + \langle p_i(L)^{m_i} \rangle$ and use Algorithm 5.1.12 (or any other suitable method) to check whether $\mathfrak{Q}_i$ is a primary ideal.*

(3) *Return the tuple $(\mathfrak{Q}_1, \ldots, \mathfrak{Q}_s)$. If all the tests in Step (2) yielded the answer TRUE, output the string* Primary Decomposition. *Otherwise, output the string* Partial Decomposition.

*This is an algorithm which computes a tuple of ideals $(\mathfrak{Q}_1, \ldots, \mathfrak{Q}_s)$ in $P$ such that we have $I = \mathfrak{Q}_1 \cap \cdots \cap \mathfrak{Q}_s$. Moreover, it checks whether this intersection is the primary decomposition of $I$ or merely a partial decomposition and outputs the corresponding string.*

---

*Proof* By Theorem 2.4.12.a, the ideals $\langle p_i(\ell)^{m_i} \rangle$ in $R$ are intersections of some of the primary components of the zero ideal and pairwise coprime. Therefore we have $\bigcap_{i=1}^{s} \langle p_i(\ell)^{m_i} \rangle = \langle \mu_\ell(\ell) \rangle = \langle 0 \rangle$. This proves the first claim. By combining Theorem 2.4.12.b with Algorithm 5.1.12, we obtain the second claim.                    $\square$

A linear form $L$ leading to a decomposition $I = \mathfrak{Q}_1 \cap \cdots \cap \mathfrak{Q}_s$ for which not all ideals $\mathfrak{Q}_i$ are equal and not all are primary corresponds to a multiplication endomorphism $\vartheta_\ell$ of $R$ which is a partially splitting endomorphism in the sense of Definition 2.5.1.a. Likewise, a linear form $L$ for which the decomposition $I = \mathfrak{Q}_1 \cap \cdots \cap \mathfrak{Q}_s$ is the primary decomposition of $I$ corresponds to a multiplication endomorphism $\vartheta_\ell$ of $R$ which is a splitting endomorphism. This suggests the following observation.

**Remark 5.1.16** In the setting of the algorithm, suppose that we know the number $s'$ of primary components of $I$ a priori. Then we can skip the applications of Algorithm 5.1.12 in Step (2) and just check whether $s = s'$, since we know that each of the computed ideals $\mathfrak{Q}_i$ is an intersection of some of the primary components of $I$.

To simplify the terminology, we complement Definition 2.5.1 as follows.

**Definition 5.1.17** Let $R$ be a zero-dimensional affine $K$-algebra, let $r \in R$, and let $\vartheta_r$ be the multiplication endomorphism by $r$ in $R$. Then the element $r$ is called a **splitting element** (resp. a **partially splitting element**) if the multiplication map $\vartheta_r$ is a splitting endomorphism (resp. a partially splitting endomorphism) for the multiplication family of $R$.

Recall that in [16], Tutorial 79, we also introduced the notion of a **splitting element for** $I$. We note that notion is weaker and corresponds to a partially splitting multiplication endomorphism here. In the remaining part of this subsection we try out Algorithm 5.1.15 in a number of different cases. Let us begin by reexamining Example 2.3.10.

**Example 5.1.18** As in Examples 2.3.10 and 2.5.9, we let $K = \mathbb{Q}$, $P = K[x, y]$, and $I = \langle x^2 - 2, y^2 - 8 \rangle$. To compute the primary decomposition of $I$, we choose the linear form $L = 7x + 12y$ and follow the steps of Algorithm 5.1.15.

(1) First we compute the minimal polynomial $\mu_\ell(z) = z^4 - 2500z^2 + 1110916$ and factor it as $\mu_\ell(z) = (z^2 - 578)(z^2 - 1922)$.
(2) Then we form the two ideals $\mathfrak{Q}_1 = \langle L^2 - 578 \rangle + I = \langle x + \frac{1}{2}y, y^2 - 8 \rangle$ and $\mathfrak{Q}_2 = \langle L^2 - 1922 \rangle + I = \langle x - \frac{1}{2}y, y^2 - 8 \rangle$. At this point we can skip the application of Algorithm 5.1.12, since it is clear that $\mathfrak{Q}_1$ and $\mathfrak{Q}_2$ are maximal ideals of $P$.
(3) We conclude that $I = \mathfrak{Q}_1 \cap \mathfrak{Q}_2$ is the primary decomposition of $I$ and that $I$ is a radical ideal.

In the following example we encounter a case where a certain choice of $L$ yields a partially splitting element, while another choice yields a splitting element and allows the computation of the primary decomposition of $I$.

**Example 5.1.19** Let $K = \mathbb{Q}$, $P = K[x, y, z]$, and $I = \langle x^2 - y - z, y^2 - xz, z^2 - x - y \rangle$.

(a) If we choose $L = x - 2y + z$, we get the minimal polynomial $\mu_\ell(z) = z^3(z^2 - 5)$. Its factors yield the two ideals $\mathfrak{Q}_1 = \langle L^3 \rangle + I = \langle x - z, y - z^2 + z, z^4 - 2z^3 \rangle$ and $\mathfrak{Q}_2 = \langle L^2 - 5 \rangle + I = \langle x + z + 1, y - z^2 - z - 1, z^4 + 2z^3 + 4z^2 + 3z + 1 \rangle$. Since the ideal $\mathfrak{Q}_1$ turns out not to be primary, the residue class $L + I$ in $P/I$ is a partially splitting element.
(b) Now we pick the linear form $\tilde{L} = 3x + 5y - 12z$ and calculate the minimal polynomial $\mu_{\tilde{\ell}}(z) = (z + 8)(z^4 - 8z^3 + 524z^2 - 7657z + 47641)z^3$. This yields

the ideals

$$\tilde{\mathfrak{Q}}_1 = I + \langle \tilde{L} + 8 \rangle = \langle x - 2, y - 2, z - 2 \rangle$$

$$\tilde{\mathfrak{Q}}_2 = I + \langle \tilde{L}^4 - 8\tilde{L}^3 + 524\tilde{L}^2 - 7657\tilde{L} + 47641 \rangle$$

$$= \langle x + z + 1, z^2 - y + z + 1, y^2 + y - 1 \rangle$$

$$\tilde{\mathfrak{Q}}_3 = I + \langle \tilde{L}^3 \rangle = \langle x - z, z^2 - y - z, yz + y + z, y^2 - y - z \rangle$$

The ideals $\tilde{\mathfrak{Q}}_1$ and $\tilde{\mathfrak{Q}}_2$ are maximal ideals of $P$, while $\tilde{\mathfrak{Q}}_3$ is a primary ideal whose radical is $\mathfrak{M}_3 = \langle x, y, z \rangle$. In conclusion, the primary decomposition of $I$ is $I = \tilde{\mathfrak{Q}}_1 \cap \tilde{\mathfrak{Q}}_2 \cap \tilde{\mathfrak{Q}}_3$, and we have $\mathrm{Rad}(I) = \tilde{\mathfrak{Q}}_1 \cap \tilde{\mathfrak{Q}}_2 \cap \mathfrak{M}_3$. Thus we see that the residue class $\tilde{L} + I$ in $P/I$ is a splitting element.

Q: *How many consultants does it take to change a light bulb?*
A: *How many can you afford?*

### 5.1.C  Computing Joint Eigenspaces

*We do not put the cart in front of the horse,*
*but we leave the horse behind the cart.*
(Giovanni Trapattoni)

When we introduced joint eigenspaces and generalized joint eigenspaces of a commuting family in Chap. 2, we did not mention how to compute them, but just noted that they are the kernels of the maximal and primary ideals, respectively (see Theorem 2.4.3 and Proposition 2.4.6). Thus we do not put linear algebra methods in front of the computation of joint eigenspaces and joint generalized eigenspaces, but we leave this computation behind the computation of maximal and primary components for which we introduced linear algebra methods in the preceding two subsections. So, let us see how to pull off the computation of joint eigenspaces in this order.

Returning to the setting in earlier chapters, we let $K$ be a field, let $V$ be a finite dimensional $K$-vector space, let $d = \dim_K(V)$, and let $\mathscr{F} \subseteq \mathrm{End}_K(V)$ be a family of commuting endomorphisms of $V$. As already pointed out, knowledge of the generators of the maximal ideals of a family, and hence of their powers, allows us to compute the joint eigenspaces and the joint generalized eigenspaces via Algorithm 2.2.17. Since that result is important for this chapter, let us spell it out explicitly.

---

**Algorithm 5.1.20 (Computing Joint Eigenspaces)**
*In the above setting, let $\Phi = (\varphi_1, \ldots, \varphi_n)$ be a system of $K$-algebra generators of $\mathscr{F}$ and let $P = K[x_1, \ldots, x_n]$. Consider the following sequence of instructions.*

(1) *Using the Buchberger-Möller Algorithm for Matrices 2.1.4, compute the ideal $I = \mathrm{Rel}_P(\Phi)$.*

(2) *Using Algorithm 5.1.7, compute the maximal ideals $\mathfrak{M}_1, \ldots, \mathfrak{M}_s$ of $P$ which contain $I$.*
(3) *For $i = 1, \ldots, s$, let $\mathfrak{m}_i$ be the image of $\mathfrak{M}_i$ in $\mathscr{F}$ via the isomorphism $P/I \cong \mathscr{F}$ given by $x_j + I \mapsto \varphi_j$. Using Algorithm 2.2.17, calculate $K$-bases of $\mathrm{Ker}(\mathfrak{m}_1), \ldots, \mathrm{Ker}(\mathfrak{m}_s)$ and return the result.*

*This is an algorithm which computes the joint eigenspaces of $\mathscr{F}$.*

*Proof* The correctness of this algorithm follows from the fact that the kernels $\mathrm{Ker}(\mathfrak{m}_1), \ldots, \mathrm{Ker}(\mathfrak{m}_s)$ are the joint eigenspaces of $\mathscr{F}$ by Theorem 2.4.3. $\qquad\square$

Of course, in Step (2) of this algorithm we can also use other methods to compute the maximal ideals containing $I$, for instance the ones introduced in Algorithm 5.3.12 and Algorithm 5.4.7. To compute the joint generalized eigenspaces of $\mathscr{F}$ we can use at least two different strategies. The first one is based on the computation of the primary decomposition of $\mathrm{Rel}_P(\Phi)$.

**Algorithm 5.1.21 (Computing Joint Generalized Eigenspaces, I)**
*In the above setting, let $\Phi = (\varphi_1, \ldots, \varphi_n)$ be a system of $K$-algebra generators of $\mathscr{F}$, and let $P = K[x_1, \ldots, x_n]$. Consider the following sequence of instructions.*

(1) *Using the Buchberger-Möller Algorithm for Matrices 2.1.4, compute the ideal $I = \mathrm{Rel}_P(\Phi)$.*
(2) *Using any of the algorithms presented in the preceding subsections, compute the primary decomposition $\mathfrak{Q}_1 \cap \cdots \cap \mathfrak{Q}_s$ of $I$.*
(3) *For $i = 1, \ldots, s$, let $\mathfrak{q}_i$ be the image of $\mathfrak{Q}_i$ via the isomorphism $P/I \cong \mathscr{F}$ which is given by $x_j + I \mapsto \varphi_j$. Using Algorithm 2.2.17, calculate $K$-bases of $\mathrm{Ker}(\mathfrak{q}_1), \ldots, \mathrm{Ker}(\mathfrak{q}_s)$ and return the result.*

*This is an algorithm which computes the joint generalized eigenspaces of $\mathscr{F}$.*

*Proof* The correctness of this algorithm follows from the fact that the kernels $\mathrm{Ker}(\mathfrak{q}_1), \ldots, \mathrm{Ker}(\mathfrak{q}_s)$ are the joint generalized eigenspaces of $\mathscr{F}$ by Proposition 2.4.6.d. $\qquad\square$

A second way to calculate the joint generalized eigenspaces of $\mathscr{F}$ uses the strategy to first compute the maximal ideals containing the relation ideals and then to calculate their big kernels.

**Algorithm 5.1.22 (Computing Joint Generalized Eigenspaces, II)**
*In the above setting, let $\Phi = (\varphi_1, \ldots, \varphi_n)$ be a system of $K$-algebra generators of $\mathscr{F}$, and let $P = K[x_1, \ldots, x_n]$. Consider the following instructions.*

(1) *Using the Buchberger-Möller Algorithm for Matrices 2.1.4, compute the ideal $I = \mathrm{Rel}_P(\Phi)$.*

(2) *Using Algorithm 5.1.7, compute the maximal ideals $\mathfrak{M}_1, \ldots, \mathfrak{M}_s$ of $P$ which contain $I$. For $i = 1, \ldots, s$, let $\mathfrak{M}_i = \langle f_{i1}, \ldots, f_{iv_i} \rangle$ with $f_{ij} \in P$.*

(3) *For $i = 1, \ldots, s$ and $j = 1, \ldots, v_i$ let $\psi_{ij} = f_{ij}(\varphi_1, \ldots, \varphi_n) \in \mathrm{End}_K(V)$. Using Algorithm 1.1.3, compute the big kernel $\mathrm{BigKer}(\psi_{ij})$.*

(4) *For $i = 1, \ldots, s$, calculate $U_i = \bigcap_{j=1}^{v_i} \mathrm{BigKer}(\psi_{ij})$ and return the resulting $K$-bases.*

*This is an algorithm which computes $K$-bases of the joint generalized eigenspaces $\mathrm{BigKer}(\mathfrak{m}_i) = U_i$ of $\mathscr{F}$, where $\mathfrak{m}_i$ is the image of $\mathfrak{M}_i$ in $P/I$ for $i \in \{1, \ldots, s\}$.*

*Proof* The algorithm is obviously finite. Using Proposition 2.2.10, we see that we have $U_i = \mathrm{BigKer}(\mathfrak{m}_i)$, and by Theorem 2.4.3, these are the joint generalized eigenspaces of the family $\mathscr{F}$.                                          $\square$

In the following example we apply the preceding three algorithms to compute the joint eigenspaces and the joint generalized eigenspaces of a commuting family.

**Example 5.1.23** Let $K = \mathbb{Q}$, let $V = K^{10}$, and let $\varphi_1, \varphi_2$ be the endomorphisms of $V$ which are given by the matrices

$$
A_1 = \begin{pmatrix}
0 & 0 & 0 & 0 & 0 & 0 & -9 & 0 & 0 & 0 \\
1 & 0 & 0 & 0 & 0 & 0 & 0 & 0 & 0 & 0 \\
0 & 0 & 0 & 0 & 0 & 0 & 0 & 0 & 0 & -9 \\
0 & 1 & 0 & 0 & 0 & 0 & -6 & 0 & 0 & 0 \\
0 & 0 & 1 & 0 & 0 & 0 & 0 & 0 & 0 & 0 \\
0 & 0 & 0 & 0 & 0 & 1 & 8 & 0 & 0 & 0 \\
0 & 0 & 0 & 1 & 0 & 0 & 0 & 0 & 0 & 0 \\
0 & 0 & 0 & 0 & 1 & 0 & 0 & 0 & 0 & -6 \\
0 & 0 & 0 & 0 & 0 & 0 & 0 & 0 & 1 & 8 \\
0 & 0 & 0 & 0 & 0 & 0 & 0 & 1 & 0 & 0
\end{pmatrix}
\qquad
A_2 = \begin{pmatrix}
0 & 0 & 0 & 0 & 0 & 0 & 0 & 0 & 0 & 0 \\
0 & 0 & 0 & 0 & 0 & 0 & 0 & 0 & 0 & 0 \\
1 & 0 & 0 & 0 & 0 & 0 & 0 & 0 & 0 & 0 \\
0 & 0 & 0 & 0 & 0 & 0 & 0 & 0 & 0 & 0 \\
0 & 1 & 0 & 0 & 0 & 0 & 0 & 0 & 0 & 0 \\
0 & 0 & 1 & 0 & 1 & 0 & 0 & 1 & 2 & 1 \\
0 & 0 & 0 & 0 & 0 & 0 & 0 & 0 & 0 & 0 \\
0 & 0 & 0 & 1 & 0 & 0 & 0 & 0 & 0 & 0 \\
0 & 0 & 0 & 0 & 0 & 1 & 0 & 0 & 0 & 0 \\
0 & 0 & 0 & 0 & 0 & 0 & 1 & 0 & 0 & 0
\end{pmatrix}
$$

respectively. One may check that $A_1$ and $A_2$ commute. Therefore $\mathscr{F} = K[\varphi_1, \varphi_2]$ is a commuting family of endomorphisms of $V$. Let us begin by computing the joint eigenspaces of $\mathscr{F}$ using Algorithm 5.1.20.

By applying the Buchberger-Möller Algorithm for Matrices 2.1.4 to the pair $\Phi = (\varphi_1, \varphi_2)$, we compute $\mathrm{Rel}_P(\Phi) = \langle x_1 x_2^2 - x_2^2, x_2^4 - 2x_2^2, x_1^4 + 6x_1^2 - 8x_2^2 + 9 \rangle$. Then we calculate the maximal components of this ideal and get $\mathfrak{M}_1 = \langle x_1^2 + 3, x_2 \rangle$ and $\mathfrak{M}_2 = \langle x_1 - 1, x_2^2 - 2 \rangle$. Their images in the family $\mathscr{F}$ are the maximal ideals $\mathfrak{m}_1 = \langle \varphi_1^2 + 3\,\mathrm{id}_V, \varphi_2 \rangle$ and $\mathfrak{m}_2 = \langle \varphi_1 - \mathrm{id}_V, \varphi_2^2 - 2\,\mathrm{id}_V \rangle$. The last step is to calculate the kernels of these ideals using Algorithm 2.2.17. It yields the joint eigenspaces

$$\mathrm{Ker}(\mathfrak{m}_1) = \big\langle (0, 0, -3, 0, 0, 0, 0, -1, 2, 0), (0, 0, -3, 0, 3, 0, 0, -1, 0, 1) \big\rangle \quad \text{and}$$

$$\mathrm{Ker}(\mathfrak{m}_2) = \big\langle (0, 0, 0, 0, 0, 1, 0, 0, 0, 0), (0, 0, 0, 0, 0, 0, 0, 0, 1, 0) \big\rangle$$

Next we calculate the joint generalized eigenspaces of $\mathscr{F}$ using Algorithm 5.1.21. The first step is again to determine the ideal $\mathrm{Rel}_P(\Phi)$. Then we compute its primary decomposition and get $\mathrm{Rel}_P(\Phi) = \mathfrak{Q}_1 \cap \mathfrak{Q}_2$, where $\mathfrak{Q}_1 = \langle (x_1^2 + 3)^2, x_2^2 \rangle$ and $\mathfrak{Q}_2 = \langle x_1 - 1, x_2^2 - 2 \rangle = \mathfrak{M}_2$. The images of these ideals in $\mathscr{F}$ are $\mathfrak{q}_1 = \langle (\varphi_1^2 + 3\,\mathrm{id}_V)^2, \varphi_2^2 \rangle$ and $\mathfrak{q}_2 = \mathfrak{m}_2$. Then Algorithm 2.2.17 produces the joint generalized eigenspaces

$$\mathrm{Ker}(\mathfrak{q}_1) = \big\langle (-1,1,0,0,0,0,0,0,0,0), (-1,0,0,1,0,0,0,0,0,0),$$
$$(0,0,-1,0,1,0,0,0,0,0), (-2,0,0,0,0,1,0,0,0,0),$$
$$(-1,0,0,0,0,0,1,0,0,0), (0,0,-1,0,0,0,0,1,0,0),$$
$$(0,0,-2,0,0,0,0,0,1,0), (0,0,-1,0,0,0,0,0,0,1) \big\rangle$$

and $\mathrm{Ker}(\mathfrak{q}_2) = \mathrm{Ker}(\mathfrak{m}_2)$.

Finally, we apply Algorithm 5.1.22 to calculate the joint generalized eigenspaces in a different way. For the two maximal ideals $\mathfrak{m}_1$ and $\mathfrak{m}_2$, we have to compute the big kernels. It turns out that

$$\mathrm{Ker}\big(\varphi_1^2 + 3\,\mathrm{id}_V\big) \subset \mathrm{Ker}\big((\varphi_1^2 + 3\,\mathrm{id}_V)^2\big) = \mathrm{Ker}\big((\varphi_1^2 + 3\,\mathrm{id}_V)^3\big) \quad \text{and}$$
$$\mathrm{Ker}(\varphi_2) \subset \mathrm{Ker}\big(\varphi_2^2\big) = \mathrm{Ker}\big(\varphi_2^3\big)$$

This yields $\mathrm{BigKer}(\mathfrak{m}_1) = \mathrm{Ker}((\varphi_1^2 + 3\,\mathrm{id}_V)^2) \cap \mathrm{Ker}(\varphi_2^2) = \mathrm{Ker}(\mathfrak{q}_1)$. Furthermore, since $\mathrm{Ker}(\varphi_1 - \mathrm{id}_V) = \mathrm{Ker}((\varphi_1 - \mathrm{id}_V)^2)$ and $\mathrm{Ker}(\varphi_2^2 - 2\,\mathrm{id}_V) = \mathrm{Ker}((\varphi_2^2 - 2\,\mathrm{id}_V)^2)$, we have $\mathrm{BigKer}(\mathfrak{m}_2) = \mathrm{Ker}(\mathfrak{m}_2)$.

## 5.2 Primary Decomposition over Finite Fields

*The way to get good ideas is*
*to get lots of ideas*
*and throw the bad ones away.*
(Linus Pauling)

In Sect. 5.1.A we saw general methods for computing the primary and maximal components of a zero-dimensional ideal via a transcendental extension of the base field. To overcome the computational inefficiency of this approach we tried to substitute the additional indeterminates by "random" elements of the base field in Sect. 5.1.B. But what can we do if the base field is finite? In this case it may very well happen that for *all* possible ways of substituting the additional indeterminates by field elements, the algorithms will fail to produce the correct result, in particular if the base field is small.

In this section we produce new, good ideas for working over a finite field $\mathbb{F}_q$. Our main inspiration for getting these ideas is the paper [6] by S. Gao, D. Wan, and M. Wang. Let $R$ be a zero-dimensional affine $\mathbb{F}_q$-algebra. The most powerful additional tool we have available in this setting is the $q$-Frobenius endomorphism $\phi_q : R \longrightarrow R$ which is given by $\phi_q(r) = r^q$. What does it have to do with the primary components of the zero ideal in $R$? The key idea is to look at the eigenspace of $\phi_q$ associated to the eigenfactor $z - 1$, i.e., to consider the $q$-Frobenius space $\mathrm{Frob}_q(R) = \{r \in R \mid r^q = r\}$. It turns out that the primitive idempotents of $R$ form an $\mathbb{F}_q$-basis of this vector space, and that we can therefore compute the number of primary components of the zero ideal even before we compute the components themselves. Carefully exploiting this excellent idea by splitting the problem until we reach the individual primary components yields a really good algorithm (see Algorithm 5.2.11).

So, is there a fly in the ointment? If the number of elements $q$ of the base field is finite, but large, the innocent-looking tasks of forming the matrix of $\phi_q$ and computing the kernel of $\phi_q - \mathrm{id}_R$ become nasty problems. In this case the very linear algebra methods we cherish so much are to be blamed for the failure of these otherwise wonderful ideas involving the Frobenius space.

> *I didn't say it was your fault.*
> *I said I was blaming you.*

In the following we let $K$ be a finite field. Then the characteristic of $K$ is a prime number $p$ and the number of its elements is of the form $q = p^e$ with $e > 0$. Furthermore, it is known that all fields having $q$ elements are isomorphic (see for instance [21]). In the following we say that a field $\mathbb{F}_p[x_1, \ldots, x_n]/\mathfrak{m}$ with $p^e$ elements **represents** $\mathbb{F}_q$. Later we will show how to describe the isomorphisms between two fields which represent the same $\mathbb{F}_q$ (see Sect. 6.2.A).

As in the previous sections, we let $P = K[x_1, \ldots, x_n]$ be a polynomial ring, we let $I$ be a zero-dimensional ideal in $P$, and we let $R$ denote the quotient ring $P/I$. By $\mathscr{F} = K[\vartheta_{x_1}, \ldots, \vartheta_{x_n}]$ we denote the multiplication family of $R$ (see Proposition 4.1.2). In this setting we have the following endomorphisms.

**Definition 5.2.1** Let $p$ be a prime number, and let $R$ be a ring of characteristic $p$.

(a) The map $\phi_p : R \longrightarrow R$ defined by $a \mapsto a^p$ is an $\mathbb{F}_p$-linear ring endomorphism of $R$. It is called the **Frobenius endomorphism** of $R$.
(b) Suppose that $R$ is an algebra over the field $\mathbb{F}_q$. Then the map $\phi_q : R \longrightarrow R$ defined by $a \mapsto a^q$ is an $\mathbb{F}_q$-linear ring endomorphism of $R$. It is called the $q$-**Frobenius endomorphism** of $R$.

The following proposition contains some basic properties of the Frobenius endomorphism.

**Proposition 5.2.2** *Let $R$ be a zero-dimensional $\mathbb{F}_q$-algebra.*

(a) *For every $a \in \mathbb{F}_q$, we have $\phi_q(a) = a$.*
(b) *The polynomial $z - 1$ is an eigenfactor of $\phi_q$.*
(c) *For every $f \in R$, we have $\phi_q(f) - f = f^q - f = \prod_{a \in \mathbb{F}_q}(f - a)$.*

(d) *Given a primary ideal $\mathfrak{q}$ in $R$ and $f \in R$, we have $\phi_q(f) - f = f^q - f \in \mathfrak{q}$ if and only if $f - a \in \mathfrak{q}$ for some $a \in \mathbb{F}_q$.*
(e) *If $R$ is a field, then the map $\phi_q$ is an $\mathbb{F}_q$-automorphism of $R$.*

*Proof* From the fact that $z^q - z = \prod_{a \in \mathbb{F}_q}(z - a)$ (see [10], Sect. 4.13) it follows that $a^q - a = 0$ for every $a \in \mathbb{F}_q$. This proves (a). To show (b), we note that we have $\mathrm{Ker}(\phi_q - \mathrm{id}_R) \neq \{0\}$ since this kernel contains $\mathbb{F}_q$. Hence the linear polynomial $z - 1$ is an eigenfactor of $\phi_q$ by Proposition 1.2.6.

Claim (c) follows by substituting $f$ for $z$ in the equality $z^q - z = \prod_{a \in \mathbb{F}_q}(z - a)$ mentioned above. Clearly, this implies the implication "$\Leftarrow$" in (d). So, let us prove the reverse implication "$\Rightarrow$". Since $\mathfrak{q}$ is a primary ideal, its radical is a maximal ideal $\mathfrak{m} = \mathrm{Rad}(\mathfrak{q})$ of $R$. Using (c) and the fact that $\mathfrak{m}$ is a prime ideal, we see that $f - a \in \mathfrak{m}$ for some $a \in \mathbb{F}_q$. Consequently, we have $f - b \notin \mathfrak{m}$ for all $b \in \mathbb{F}_q \setminus \{a\}$. Thus we have $\prod_{b \neq a}(f - b) \notin \mathfrak{m}$ and the hypothesis that $\mathfrak{q}$ is a primary ideal yields $f - a \in \mathfrak{q}$.

Finally, we show (e). Clearly, the map $\phi_q$ is injective, since a field has no non-trivial nilpotent elements. Then the fact that $R$ is finite implies that $\phi_q$ is bijective. $\square$

The eigenspace of the $q$-Frobenius endomorphism corresponding to the eigen-factor $z - 1$ will turn out to be a crucial tool for computing primary decompositions in the current setting. Thus it deserves a name.

**Definition 5.2.3**   In the above setting the $K$-vector subspace

$$\mathrm{Frob}_q(R) = \mathrm{Eig}(\phi_q, z - 1) = \{f \in R \mid f^q - f = 0\}$$

of $R$, i.e., the fixed-point space of $R$ with respect to $\phi_q$, is called the $q$-**Frobenius space** of $R$.

Our goal is to use the $q$-Frobenius space of $R$ in order to predict the number of primary components of the zero ideal in $R$. For the proof of the next theorem we use some properties of idempotents presented in Proposition 5.1.10. It contains fundamental information about the $q$-Frobenius space of $R$.

**Theorem 5.2.4 (Properties of the $q$-Frobenius Space)**
*Let $R$ be a zero-dimensional affine $\mathbb{F}_q$-algebra, and let $s$ be the number of primary components of the zero ideal of $R$, or equivalently, the number of local factors of $R$ (see Definition 4.1.13).*

(a) *We have $\mathrm{Frob}_q(R) = \{\sum_{i=1}^{s} c_i e_i \mid c_1, \ldots c_s \in \mathbb{F}_q\}$ where $e_1, \ldots, e_s$ are the primitive idempotents of $R$.*
(b) *We have $\dim_{\mathbb{F}_q}(\mathrm{Frob}_q(R)) = s$.*
(c) *We have $\mathrm{Frob}_q(R) = R$ if and only if the primary decomposition of the zero ideal of $R$ is $\langle 0 \rangle = \mathfrak{m}_1 \cap \cdots \cap \mathfrak{m}_s$ where the ideals $\mathfrak{m}_1, \ldots, \mathfrak{m}_s$ are linear maximal ideals.*
(d) *For $f \in R$, we have $f \in \mathrm{Frob}_q(R)$ if and only if $\mu_f(z)$ splits into distinct linear factors.*

*Proof* To prove (a), we first show that $\mathrm{Frob}_q(R) \subseteq \{\sum_{i=1}^{s} c_i e_i \mid c_1, \ldots c_s \in \mathbb{F}_q\}$. Let $\pi : R \cong \prod_{i=1}^{s} R/\mathfrak{q}_i$ be the decomposition of $R$ into local rings. Given $f \in R$, we write $\pi(f) = (f_1, \ldots, f_s)$ and get

$$\mathrm{Frob}_q(R) = \{f \in R \mid f_i^q - f_i = 0 \text{ for } i = 1, \ldots, s\}.$$

In the rings $R/\mathfrak{q}_i$ the zero-ideals are primary. For $f \in \mathrm{Frob}_q(R)$, Proposition 5.2.2.d implies that there exist $c_1, \ldots, c_s \in \mathbb{F}_q$ such that $f_i = c_i$ for $i = 1, \ldots, s$. Consequently, we have the equality $f = \sum_{i=1}^{s} c_i e_i$.

Next we prove the other inclusion. Every fundamental idempotent $e_i$ satisfies $e_i^2 = e_i$, and hence $e_i^q = e_i$. Moreover, by Proposition 5.2.2.a, we have $c_i^q = c_i$ for $i = 1, \ldots, s$. Therefore we get $(\sum_{i=1}^{s} c_i e_i)^q = \sum_{i=1}^{s} c_i e_i$ which concludes the proof.

Notice that (b) is a direct consequence of (a). So, let us prove (c). From the isomorphism $\pi : R \cong \prod_{i=1}^{s} R/\mathfrak{q}_i$ we get the equality $\dim_{\mathbb{F}_q}(R) = \sum_{i=1}^{s} \dim_{\mathbb{F}_q}(R/\mathfrak{q}_i)$. Now, if $\mathrm{Frob}_q(R) = R$, it follows from (b) that $\dim_{\mathbb{F}_q}(R) = s$. Consequently, we have $\dim_{\mathbb{F}_q}(R/\mathfrak{q}_i) = 1$ for $i = 1, \ldots, s$. Hence $\mathfrak{q}_i$ is a linear maximal ideal (see Definition 2.3.7). Conversely, if the ideals $\mathfrak{q}_1, \ldots, \mathfrak{q}_s$ are linear maximal ideals, then we have $\dim_{\mathbb{F}_q}(R/\mathfrak{q}_i) = 1$ for $i = 1, \ldots, s$, and therefore $\dim_{\mathbb{F}_q}(R) = s$. Since $s$ is also the dimension of the vector subspace $\mathrm{Frob}_q(R)$ of $R$, the claim follows.

Finally, to prove (d), we recall that $z^q - z = \prod_{a \in \mathbb{F}_q}(z - a)$, as seen in Proposition 5.2.2.a. We are going to use this relation in both directions. Every element $f \in \mathrm{Frob}_q(R)$ satisfies $f^q - f = 0$. Hence the polynomial $\mu_f(z)$ divides $z^q - z = \prod_{c \in \mathbb{F}_q}(z - c)$. In particular, it splits into distinct linear factors. Conversely, if $\mu_f(z)$ splits into distinct linear factors then it divides $\prod_{c \in \mathbb{F}_q}(z - c)$. Hence it divides $z^q - z$, and therefore we have $f^q - f = 0$.                                                      $\square$

The final part of this proposition shows that many elements of the $q$-Frobenius space are partially splitting multiplication endomorphisms in $\mathscr{F}$.

**Corollary 5.2.5** *Let $R$ be a zero-dimensional affine $\mathbb{F}_q$-algebra whose zero ideal has $s$ primary components, and let $d = \dim_K(R)$.*

(a) *Let $f \in \mathrm{Frob}_q(R)$, and let $r$ be the number of eigenfactors of $\vartheta_f$. Then we have $r = \deg(\mu_{\vartheta_f}(z)) \le \min\{q, s\}$.*
(b) *There exists an element $f \in \mathrm{Frob}_q(R)$ such that $\vartheta_f$ has $\min\{q, s\}$ eigenfactors.*
(c) *If $s > 1$, every element $f \in \mathrm{Frob}_q(R) \setminus \mathbb{F}_q$ yields a partially splitting endomorphism $\vartheta_f$.*
(d) *Given a $K$-basis $B$ of $R$, we have $s = d - \mathrm{rank}(M_B(\phi_q) - I_d)$.*

*Proof* First we show (a). The equality $r = \deg(\mu_{\vartheta_f}(z))$ follows immediately from part (c) of the theorem. To prove the inequality $r \le \min\{q, s\}$, we note that we have $\deg(\mu_{\vartheta_f}(z)) \le q$ by 5.2.2.c. Now we can use part (a) of the theorem and write $f = \sum_{i=1}^{s} c_i e_i$ with $c_i \in \mathbb{F}_q$ and the primitive idempotents $e_1, \ldots, e_s$ of $R$. Using the decomposition of $R$ into local rings and Lemma 1.1.11, it follows that we have $\mu_{\vartheta_f}(z) = \mathrm{lcm}(\mu_{\vartheta_{c_1}}(z), \ldots, \mu_{\vartheta_{c_s}}(z))$. Clearly, we have $\mu_{\vartheta_{c_i}}(z) = z - c_i$ for $i = 1, \ldots, s$, and therefore $\deg(\mu_{\vartheta_f}(z)) \le s$.

To prove claim (b) we construct an element $f \in \mathrm{Frob}_q(R)$ via its representation $f = \sum_{i=1}^{s} c_i e_i$ with $c_i \in K$. If $s \leq q$ we choose pairwise distinct elements $c_1, \ldots, c_s \in \mathbb{F}_q$. Instead, if $s > q$, we let $c_1, \ldots, c_q$ be the elements of $\mathbb{F}_q$ and use $f = \sum_{i=1}^{q} c_i e_i + \sum_{i=q+1}^{s} c_q e_q$. In both cases, the computation of $\mu_{\vartheta_f}(z)$ in the proof of (a) shows that we have the desired equality.

Next we prove (c). The only elements $f \in \mathrm{Frob}_q(R)$ such that $\mu_{\vartheta_f}(z)$ is irreducible are the elements $f = \sum_{i=1}^{s} c_i e_i$ with $c_1 = \cdots = c_s$. Clearly, these are the elements of $\mathbb{F}_q$. Claim (d) follows from part (b) of the theorem.     □

Using part (d) of this corollary, we can check whether a given zero-dimensional ideal of $P$ is maximal.

---

**Algorithm 5.2.6 (Checking Maximal Ideals over Finite Fields)**
*Let $I$ be a zero-dimensional ideal in $P$, let $R = P/I$, let $d = \dim_K(R)$, and let $B$ be a $K$-basis of $R$. Consider the following sequence of instructions.*

(1) *Calculate the radical ideal of $I$. If $\mathrm{Rad}(I) \neq I$ then return* FALSE *and stop.*
(2) *Compute $s = d - \mathrm{rank}(M_B(\phi_q) - I_d)$. If $s > 1$ then return* FALSE *and stop.*
(3) *Return* TRUE.

*This is an algorithm which checks whether $I$ is a maximal ideal and returns the corresponding Boolean value.*

---

To perform step (1) of this algorithm, we can use [15], Corollary 3.7.16 or the improved version given in Algorithm 5.4.2. Now we are interested in an $\mathbb{F}_q$-basis of the $q$-Frobenius space, since its non-constant elements yield partially splitting multiplication endomorphisms by part (c) of the corollary. The following algorithm assumes knowledge of an $\mathbb{F}_q$-basis of $R$ which can be calculated as described in Remark 4.1.7.

---

**Algorithm 5.2.7 (Computing the $q$-Frobenius Space)**
*Let $R$ be a zero-dimensional affine $\mathbb{F}_q$-algebra, and let $B = (b_1, \ldots, b_d)$ be an $\mathbb{F}_q$-basis of $R$, where $d = \dim_K(R)$. Consider the following sequence of instructions.*

(1) *Compute the matrix $M_B(\phi_q)$.*
(2) *Compute an $\mathbb{F}_q$-basis $C$ of the kernel of $M_B(\phi_q) - I_d$ where $I_d$ is the identity matrix of size $d$.*
(3) *Return the list of all elements $c_1 b_1 + \cdots + c_d b_d \in R$ such that $(c_1, \ldots, c_d)$ is an element of $C$.*

*This is an algorithm which computes an $\mathbb{F}_q$-basis of the $q$-Frobenius space of $R$.*

*Proof* By definition, the $q$-Frobenius space of $R$ is the kernel of $\phi_q - \mathrm{id}_R$.	□

By combining this algorithm with the fact that the minimal polynomial of an element of $\mathrm{Frob}_q(R)$ splits into distinct linear factors, we can find the primary decomposition of the zero ideal of $R$ in many cases. Notice that, in order to find an element in $\mathrm{Frob}_q(R) \setminus \mathbb{F}_q$, we can choose the basis $B = (b_1, \ldots, b_d)$ of $R$ such that $b_1 = 1$, and choose non-constant elements in $\mathrm{Frob}_q(R)$ to get a partially splitting multiplication endomorphism. The following example shows that for $s = 2$ this straightforward approach is already sufficient to calculate the desired primary decomposition.

**Example 5.2.8** Let $K = \mathbb{F}_2$, let $P = K[x, y]$, let $I = \langle x^2 + x + 1, y^2 + y + 1 \rangle$, and let $R = P/I$. No matter which term ordering $\sigma$ we choose, the reduced $\sigma$-Gröbner basis of $I$ is $\{x^2 + x + 1, y^2 + y + 1\}$. Therefore the tuple $B = (\bar{1}, \bar{x}, \bar{y}, \bar{x}\bar{y})$ is a $K$-basis of $R$. Now we apply Algorithm 5.2.7 and follow its steps.

(1) For every entry $b$ of $B$, compute the representation of $b^2$ in terms of the basis $B$. Using the coefficient tuples of these representations as columns, we get the matrix
$$M_B(\phi_2) = \begin{pmatrix} 1 & 1 & 1 & 1 \\ 0 & 1 & 0 & 1 \\ 0 & 0 & 1 & 1 \\ 0 & 0 & 0 & 1 \end{pmatrix}$$

(2) The kernel of the matrix $M_B(\phi_2) - I_4$ has the basis $C = ((1, 0, 0, 0), (0, 1, 1, 0))$. Therefore we have $\mathrm{Frob}_2(R) = \langle 1, \bar{x} + \bar{y} \rangle_K$.

Since the zero ideal of the ring $R$ has only two primary components, the element $\ell = \bar{x} + \bar{y}$ of $\mathrm{Frob}_2(R)$ yields not only a partially splitting, but a splitting multiplication endomorphism. The minimal polynomial of $\ell$ is obviously $\mu_\ell(z) = z(z + 1)$.

Now we let $L = x + y$ and add $L$ (resp. $L + 1$) to $I$. We get the two primary components $\mathfrak{Q}_1 = \langle x + y, y^2 + y + 1 \rangle$ and $\mathfrak{Q}_2 = \langle x + y + 1, y^2 + y + 1 \rangle$ of $I$. Clearly, the ideals $\mathfrak{Q}_1$ and $\mathfrak{Q}_2$ are maximal, and hence $I$ is a radical ideal.

Note that we could not have achieved this result by using elements such as $\bar{x}$ or $\bar{y}$ which are not contained in $\mathrm{Frob}_q(R)$. For instance, the minimal polynomial of $\bar{x}$ and $\bar{y}$ is $\mu_{\bar{x}}(z) = \mu_{\bar{y}}(z) = z^2 + z + 1$. Since it is irreducible, it does not help us to find the two primary components.

Also in the following more complicated example we find the primary decomposition of the zero ideal of $R$ by choosing a suitable element of the $q$-Frobenius space.

**Example 5.2.9** Let $K = \mathbb{F}_7$, let $P = K[x, y]$, let $I = \langle f_1, f_2, f_3 \rangle$ be the ideal generated by $f_1 = y^3 - x^2 + 2xy$, $f_2 = x^4 - x^2 - x$, and $f_3 = x^3 y^2 - xy^2 - y^2$, and let $R = P/I$. Then a $K$-basis $B$ of $R$ is given by $B = (1, \bar{x}, \bar{y}, \bar{x}^2, \bar{x}\bar{y}, \bar{y}^2, \bar{x}^3, \bar{x}^2\bar{y}, \bar{x}\bar{y}^2,$

$\bar{x}^3\bar{y}, \bar{x}^2\bar{y}^2$). Now we compute $M_B(\phi_7)$ and get

$$M_B(\phi_7) = \begin{pmatrix}
1 & 0 & 0 & 0 & 0 & 0 & 0 & 0 & 0 & 0 & 0 \\
0 & 1 & -3 & 0 & -2 & 1 & 0 & 0 & 1 & 2 & -3 \\
0 & 0 & 0 & 0 & 0 & 0 & 0 & 0 & 0 & 0 & 0 \\
0 & 2 & -3 & -2 & -1 & 3 & 2 & 1 & -1 & 3 & 2 \\
0 & 0 & 1 & 0 & 1 & 2 & 0 & -2 & 3 & 2 & 3 \\
0 & 0 & 3 & 0 & 2 & 0 & 0 & 0 & -1 & -2 & 3 \\
0 & 1 & 0 & 2 & 2 & -3 & 2 & 0 & -1 & 2 & -3 \\
0 & 0 & 1 & 0 & 1 & -3 & 0 & 2 & 2 & 2 & -3 \\
0 & 0 & 3 & 0 & 1 & -1 & 0 & -1 & -1 & -3 & 0 \\
0 & 0 & -1 & 0 & 1 & 0 & 0 & 0 & -2 & 0 & -3 \\
0 & 0 & 0 & 0 & -2 & -3 & 0 & 0 & 3 & -2 & -2
\end{pmatrix}$$

Next we need to calculate the kernel of $M_B(\phi_7) - I_{11}$. The result is the vector space $\mathrm{Ker}(M_B(\phi_7) - I_{11}) = \langle v_1, v_2, v_3, v_4, v_5 \rangle$, where

$$v_1 = (1, 0, 0, 0, 0, 0, 0, 0, 0, 0, 0)$$
$$v_2 = (0, 2, 0, -1, 0, 0, 0, 0, 0, 0, 0)$$
$$v_3 = (0, 1, 0, 0, 0, 0, -1, 0, 0, 0, 0)$$
$$v_4 = (0, 2, 0, 0, 3, 3, 0, 2, 1, 1, 0)$$
$$v_5 = (0, 1, 0, 0, 0, 1, 0, 2, 2, 0, 1)$$

Thus we know that $I$ has five primary components. Clearly, the vector $v_1$ corresponds to the constant elements of $\mathrm{Frob}_7(R)$. If we multiply the vector $v_2$ with the basis $B$, we get the residue class $\bar{f} = 2\bar{x} - \bar{x}^2$. The minimal polynomial of $\bar{f}$ is $\mu_{\bar{f}}(z) = z(z+1)(z-3)$. Obviously, the corresponding multiplication endomorphism $\vartheta_{\bar{f}}$ is partially splitting, but not splitting. Instead, if we multiply $v_4$ with the basis, we get $\bar{g} = 2\bar{x} + 3\bar{x}\bar{y} + 3\bar{y}^2 + 2\bar{x}^2\bar{y} + \bar{x}\bar{y}^2 + \bar{x}^3\bar{y}$. The minimal polynomial of this element is $\mu_{\bar{g}}(z) = z(z-1)(z-2)(z+1)(z+2)$. Since it has the required number of linear factors, the map $\vartheta_{\bar{g}}$ is a splitting endomorphism and we can use its eigenfactors to compute the primary decomposition of $I$. The result is

$$\mathfrak{Q}_1 = I + \langle g \rangle = \langle x, y^2 \rangle$$

$$\mathfrak{Q}_2 = I + \langle g - 1 \rangle = \langle x + 2, y^2 + y - 3 \rangle$$

$$\mathfrak{Q}_3 = I + \langle g - 2 \rangle = \langle xy - 3y^2 + x - y + 3, x^2 - 2x + 3, y^3 - y^2 + 3x + 2y - 3 \rangle$$

$$\mathfrak{Q}_4 = I + \langle g + 1 \rangle = \langle x + 3y - 1, y^2 + 1 \rangle$$

$$\mathfrak{Q}_5 = I + \langle g + 2 \rangle = \langle y - 1, x + 2 \rangle$$

Notice that $\mathfrak{Q}_2, \mathfrak{Q}_4$, and $\mathfrak{Q}_5$ are maximal ideals, and that we have $\mathrm{Rad}(\mathfrak{Q}_1) = \langle x, y \rangle$. Let us also check whether $\mathfrak{Q}_3$ is maximal or not. The minimal polynomial of $\bar{x}$ in $P/\mathfrak{Q}_3$ is $z^2 - 2z + 3$ which is irreducible. The minimal polynomial of $\bar{y}$ in $P/\mathfrak{Q}_3$ is $z^4 + 3z^2 + 2z + 2$ which is also irreducible. Therefore Seidenberg's Lemma (see [15], Proposition 3.7.15) implies that $\mathfrak{Q}_3$ is a maximal ideal of $P$. (This fact can also be checked using the algorithms given later in this chapter.) Altogether, we obtain the primary decompositions

$$I = \mathfrak{Q}_1 \cap \mathfrak{Q}_2 \cap \mathfrak{Q}_3 \cap \mathfrak{Q}_4 \cap \mathfrak{Q}_5 \quad \text{and} \quad \mathrm{Rad}(I) = \langle x, y \rangle \cap \mathfrak{Q}_2 \cap \mathfrak{Q}_3 \cap \mathfrak{Q}_4 \cap \mathfrak{Q}_5$$

Finally, we note that the cardinality of $\mathrm{Frob}_7(R)$ is $7^5$, while the number of its elements which yield a splitting endomorphism is $7 \cdot 6 \cdot 5 \cdot 4 \cdot 3$. Consequently, the probability to get a splitting endomorphism associated to a randomly chosen element of $\mathrm{Frob}_7(R)$ is about 15 %.

If the base field is too small, a splitting endomorphism may not exist in $\mathrm{Frob}_q(R)$, as the following remark shows.

**Remark 5.2.10** Assume that the base field is $K = \mathbb{F}_2$ and that the ring $R$ is not local. Then we have the equality $\min\{\#K, \dim_K(\mathrm{Frob}_q(R))\} = 2$. If $R$ has more than 2 components, part (d) of the theorem implies that there is no element $f \in \mathrm{Frob}_q(R)$ such that $\vartheta_f$ is a splitting endomorphism. Nevertheless, it is possible that there is an element $f \in R \setminus \mathrm{Frob}_2(R)$ such that $\vartheta_f$ is a splitting endomorphism, as we have seen in Example 4.3.4.

Notwithstanding this remark, it is a fact that all elements in $\mathrm{Frob}_q(R)$ share the nice property that their minimal polynomial splits into distinct linear factors. This observation leads us to the idea of using elements in $\mathrm{Frob}_q(R)$ to get partial splittings, and then continuing to split the ideal in a recursive manner. The following algorithm turns this idea into an effective method.

**Algorithm 5.2.11 (Primary Decomposition over a Finite Base Field)**
*Let $P = \mathbb{F}_q[x_1, \ldots, x_n]$, let $I$ be a zero-dimensional ideal in $P$, and then let $R = P/I$. Consider the following sequence of instructions.*

(1) *Using Corollary 5.2.5, compute the number $s$ of primary components of the zero ideal of $R$. If $s = 1$ then return $(I)$ and stop.*
(2) *Let $L$ be the list consisting of the pair $(I, s)$. Repeat the following steps until the second component of all pairs in $L$ is 1. Then return the tuple consisting of all first components of the pairs in $L$ and stop.*
(3) *Choose the first pair $(J, t)$ in $L$ for which $t > 1$ and remove it from $L$.*
(4) *Using Algorithm 5.2.7, compute the $q$-Frobenius space of $P/J$. Choose a non-constant element $\bar{f}$ in it. Let $f \in P$ be a representative of $\bar{f}$.*
(5) *Calculate the minimal polynomial of the element $\bar{f}$ and factor it in the form $\mu_{\bar{f}}(z) = (z - c_1) \cdots (z - c_u)$ with $c_1, \ldots, c_u \in \mathbb{F}_q$.*
(6) *For $i = 1, \ldots, u$, let $J_i = J + \langle f - c_j \rangle$. Compute the dimension $d_i$ of $\mathrm{Frob}_q(P/J_i)$ and append the pair $(J_i, d_i)$ to $L$.*

*This is an algorithm which calculates the list of primary components of $I$.*

*Proof* Whenever a pair $(J, t)$ is treated in Steps (4)–(6), the element $\bar{f}$ yields a partially splitting multiplication endomorphism $\vartheta_{\bar{f}}$ of $P/J$ by Corollary 5.2.5. Therefore the ideals $J_i$ are intersections of some of the primary components of $J$ in such a way that each primary component contains exactly one of the ideals $J_i$. By Theorem 5.2.4.b, the corresponding number $d_i$ is the number of primary components of $J$ containing $J_i$. In particular, the number $t$ is the number of primary components of $J$ and the sum $d_1 + \cdots + d_u$ equals $t$ throughout the course of the algorithm. This shows that the loop is finite and that the first components of the pairs in $L$ are precisely the primary components of $I$ when the algorithm stops.                □

Let us apply this algorithm to a non-trivial case.

**Example 5.2.12**   Let $K = \mathbb{F}_2$, let $P = K[x, y]$, let $I = \langle f_1, f_2, f_3, f_4, f_5 \rangle$, where $f_1 = x^4 + x^3y + x^3$, $f_2 = x^3y^2 + xy^4 + xy^2$, $f_3 = x^2y^3 + xy^4 + x^3y + x^2y + xy^2$, $f_4 = xy^5 + y^5 + x^3y + xy^3 + x^3 + y^3$, and $f_5 = y^6 + x^3y + y^4 + x^3$, and let $R = P/I$. We follow the steps of Algorithm 5.2.11 to calculate the primary decomposition of $I$.

(1) A $K$-basis $B$ of $R$ is given by

$$B = \left(1, \bar{y}, \bar{y}^2, \bar{y}^3, \bar{y}^4, \bar{y}^5, \bar{x}, \bar{x}\bar{y}, \bar{x}\bar{y}^2, \bar{x}\bar{y}^3, \bar{x}\bar{y}^4, \bar{x}^2, \bar{x}^2\bar{y}, \bar{x}^2\bar{y}^2, \bar{x}^3, \bar{x}^3\bar{y}\right)$$

We compute a $K$-basis of $\mathrm{Ker}(M_B(\phi_2) - I_{16})$ and get $(v_1, v_2, v_3)$, where

$$v_1 = (1, 0, 0, 0, 0, 0, 0, 0, 0, 0, 0, 0, 0, 0, 0, 0)$$
$$v_2 = (0, 0, 0, 0, 0, 0, 1, 0, 0, 0, 1, 0, 0, 0, 1, 0)$$
$$v_3 = (0, 0, 0, 0, 0, 0, 0, 0, 1, 1, 0, 0, 0, 1, 0, 1)$$

Hence there are $s = 3$ primary components of $I$.

(2) Let $L = ((I, 3))$.

(3) Choose the pair $(I, 3)$ and let $L = ()$.

(4) We pick the element $\bar{f} \in \mathrm{Frob}_2(R)$ which corresponds to $v_2$. It is the residue class of $f = x^3y + y^4 + x^3$.

(5) The minimal polynomial of $\bar{f}$ is $\mu_{\bar{f}}(z) = z(z + 1)$.

(6) We let $J_1 = I + \langle f \rangle = \langle x^3, xy^2, x^2y, y^3 \rangle$ and $J_2 = I + \langle f + 1 \rangle = \langle xy^2 + y^3 + y^2 + x + y + 1, x^3 + y^3 + y^2 + y + 1, y^6 + 1 \rangle$. Algorithm 5.2.7 yields the dimensions $\dim_{\mathbb{F}_2}(\mathrm{Frob}_2(P/J_1)) = 1$ and $\dim_{\mathbb{F}_2}(\mathrm{Frob}_2(P/J_2)) = 2$. We append the pairs $(J_1, 1)$ and $(J_2, 2)$ to $L$.

(3) Choose the pair $(J_2, 2)$ and let $L = ((J_1, 1))$.

(4) During the execution of Step (6) above we have already calculated $\mathrm{Frob}_2(P/J_2)$. Its basis is $(1, \bar{g})$, where $g = y^4 + y^2$.

(5) The minimal polynomial of $\bar{g}$ is $\mu_{\bar{g}}(z) = z(z + 1)$.

(6) Let $J_{21} = J_2 + \langle g \rangle = \langle x^3, y^2 + 1 \rangle$ and $J_{22} = J_2 + \langle g + 1 \rangle = \langle x + y + 1, y^4 + y^2 + 1 \rangle$. The calculation of the dimension of the Frobenius spaces of $P/J_{21}$ and $P/J_{22}$ yields $d_{21} = d_{22} = 1$. Hence we append the pairs $(J_{21}, 1)$ and $(J_{22}, 1)$ to $L$.

(2) The algorithm returns the list $(J_1, J_{21}, J_{22})$ and stops.

Altogether, it follows that the primary decomposition of $I$ is

$$I = J_1 \cap J_{21} \cap J_{22} = \langle x, y \rangle^3 \cap \langle x^3, (y+1)^2 \rangle \cap \langle x + y + 1, (y^2 + y + 1)^2 \rangle$$

The last example in this section shows that, in general, the primary decomposition changes if we enlarge the base field.

**Example 5.2.13**  In the setting of Example 5.2.8, assume that the base field is given by $K = \mathbb{F}_2[t]/\langle t^2 + t + 1 \rangle$, i.e., a field with four elements. As in this example, we let $P = K[x, y]$, $I = \langle x^2 + x + 1, y^2 + y + 1 \rangle$, and $R = P/I$. Again it is easy to check that $\{x^2 + x + 1, y^2 + y + 1\}$ is the reduced Gröbner basis of $I$ with respect to every term ordering, and therefore $B = (1, \bar{x}, \bar{y}, \bar{x}\bar{y})$ is a $K$-basis of $R$.

Now we start the calculation of the primary decomposition of $I$. Since we get $M_B(\phi_4) - I_4 = 0$, we have $\mathrm{Frob}_4(R) = R$ and we conclude that $I$ has four primary components. Moreover, since $K$ has four elements, it follows that there exists an element $\bar{f} \in \mathrm{Frob}_q(R)$ such that $\vartheta_{\bar{f}}$ is a splitting endomorphism. The elements of $B$ yield only partially splitting multiplication endomorphisms. We try $\bar{f} = \bar{x} + \bar{t}\bar{y}$ and obtain the minimal polynomial $\mu_{\bar{f}}(z) = z^4 + z = z(z+1)(z+\bar{t})(z+\bar{t}+1)$. Therefore the primary decomposition of $I$ is given by $I = \mathfrak{Q}_1 \cap \mathfrak{Q}_2 \cap \mathfrak{Q}_3 \cap \mathfrak{Q}_4$, where

$$\mathfrak{Q}_1 = I + \langle f \rangle = \langle x + \bar{t}, y + \bar{t} \rangle$$
$$\mathfrak{Q}_2 = I + \langle f + 1 \rangle = \langle x + \bar{t}, y + \bar{t} + 1 \rangle$$
$$\mathfrak{Q}_3 = I + \langle f + \bar{t} \rangle = \langle x + \bar{t} + 1, y + \bar{t} \rangle$$
$$\mathfrak{Q}_4 = I + \langle f + \bar{t} + 1 \rangle = \langle x + \bar{t} + 1, y + \bar{t} + 1 \rangle$$

When we compare this result to Example 5.2.8 we see that the enlargement of the base field has further split the primary components of $I$.

Notice that we can also view $R$ as a zero-dimensional affine $\mathbb{F}_2$-algebra. In this case its presentation is $R = \mathbb{F}_2[t, x, y]/\langle t^2 + t + 1, x^2 + x + 1, y^2 + y + 1 \rangle$. If we use the above algorithm to calculate the primary decomposition of the zero ideal of $R$ via this presentation, we get four primary components containing $\bar{t}^2 + \bar{t} + 1$ which then can be identified with the four components found above in the obvious way.

Finally, let us indicate a connection between the primary decomposition of $I$ and Galois theory. It is known that $\mathrm{Gal}(K/\mathbb{F}_2) \cong \mathbb{Z}/2\mathbb{Z}$ is generated by $\tau = \phi_2$. We may readily check that $\tau(\mathfrak{Q}_1) = \mathfrak{Q}_4$ and $\tau(\mathfrak{Q}_2) = \mathfrak{Q}_3$. Moreover, we have $\mathfrak{Q}_1 \cap \mathfrak{Q}_4 = \langle x + y, y^2 + y + 1 \rangle$ and $\mathfrak{Q}_2 \cap \mathfrak{Q}_3 = \langle x + y + 1, y^2 + y + 1 \rangle$. In this way we get the primary decomposition computed in Example 5.2.8.

## 5.3   Computing Maximal Components via Factorization

> *My opinions may have changed,*
> *but not the fact that I am right.*
> (Ashleigh Brilliant)

Given a zero-dimensional ideal $I$ in a polynomial ring $P = K[x_1, \ldots, x_n]$ over a field $K$, we can find its maximal components using Algorithm 5.1.7. Another possibility is to use Algorithm 5.1.15 and follow it by a calculation of the radicals of the primary components. Are these the best algorithms?

In this section and the next, we change our opinion and present algorithms to find the maximal components of $I$ more directly. The approach followed in this section is based on repeatedly factoring minimal polynomials over algebraic extension fields of $K$. It requires two main ingredients: the computation of the minimal polynomial of an element over an algebraic extension field of $K$, and the factorization of polynomials over extension fields into irreducible factors. These two tasks are examined in the first two subsections. The desired method for computing the maximal components of $I$ is then achieved in Algorithm 5.3.12 in the third subsection.

This method cannot avoid some computation of radicals. In general, if $K$ is perfect, it can be done using [15], Corollary 3.7.16. In Sect. 5.4 we present a variant of this technique. If $K$ is not perfect, one can use the method explained in [13]. However, it is our opinion that the method presented in this section requires the computation of very special radical ideals for which even better methods may be found.

Notice that the entire section is formulated in terms of computations for zero-dimensional polynomial ideals and zero-dimensional affine $K$-algebras. Given a family of commuting endomorphisms $\mathscr{F}$ of a finite dimensional $K$-vector space, we can choose a system $\Phi = (\varphi_1, \ldots, \varphi_n)$ of $K$-algebra generators of $\mathscr{F}$ and use the presentation $\mathscr{F} = P/\operatorname{Rel}_P(\Phi)$. Therefore the algorithms in this section and the next allow us to compute the maximal ideals of a commuting family $\mathscr{F}$ as well.

Now, which of these algorithms is the best to compute those maximal components? We find this hard to decide, and our opinion changes frequently. It really appears to depend on the concrete setting and the actual example.

> *I used to be indecisive. Now I'm not so sure.*

### 5.3.A   Minimal Polynomials and Finite Field Extensions

> *Opera in English is, in the main,*
> *just about as sensible as baseball in Italian.*
> (Henry L. Mencken)

In mathematical English there is a subtle difference between an extension field and a field extension. The first consists of one field, the second of two. And if we are given a finite field extension $K \subseteq L$, the extension field $L$ does not have to be finite at all. Rather, it is a finitely generated vector space over $K$. A further subtlety occurs when we extend a zero-dimensional affine $K$-algebra to an $L$-algebra using a finite field extension $K \subseteq L$: the minimal polynomial of an element of the extension algebra $R_L$ may change when we view $R_L$ as a $K$-algebra or as an $L$-algebra. In this subsection we get this problem under control by computing a Gröbner basis with respect to a very carefully selected term ordering. So, the subtleties of the algorithms mirror the subtleties of the English language.

*Why do we have noses that run and feet that smell?*

In the following we let $K$ be a field and $L$ a finite algebraic field extension of $K$. Then we can represent $L$ in the form $L = K[y_1, \ldots, y_m]/\mathfrak{M}$ with indeterminates $y_1, \ldots, y_m$ and a maximal ideal $\mathfrak{M}$ in $K[y_1, \ldots, y_m]$.

Given a zero-dimensional affine $K$-algebra $R = P/I$ with a polynomial ring $P = K[x_1, \ldots, x_n]$ and an ideal $I$ in $P$, the ring $R_L = R \otimes_K L$ obtained by **base field extension** has a representation

$$R_L = L[x_1, \ldots, x_n]/IL[x_1, \ldots, x_n] = K[y_1, \ldots, y_m, x_1, \ldots, x_n]/J$$

where $J = IK[y_1, \ldots, y_m, x_1, \ldots, x_n] + \mathfrak{M}K[y_1, \ldots, y_m, x_1, \ldots, x_n]$ is a zero-dimensional ideal. Notice that we can view $R_L$ both as an $L$-algebra and as a $K$-algebra.

For an element $r \in R_L$, we can therefore form its minimal polynomial either over $L$ or over $K$. To distinguish these minimal polynomials, we shall write $\mu_{r,L}(z)$ for the minimal polynomial of $r$ in $L[z]$ and $\mu_{r,K}(z)$ for the minimal polynomial in $K[z]$. We note that the polynomial $\mu_{r,K}(z)$ is a multiple of $\mu_{r,L}(z)$ in $L[z]$, since it is contained in the kernel of the composed map

$$K[z] \hookrightarrow L[z] \longrightarrow R_L$$

where the second map is given by $z \mapsto r$. In the following we strive to compute the minimal polynomial $\mu_{r,L}(z)$ of an element $r \in R_L$ by performing calculations over the base field $K$.

*Why is the man who invests all your money called a broker?*
(George Carlin)

Let us start with an auxiliary algorithm which shows how we can compute the monic generator of a non-zero ideal in $L[z]$.

**Algorithm 5.3.1 (Monic Generators over Extension Fields)**
*In the above setting, let $z$ be a new indeterminate, let $J$ be a zero-dimensional ideal in $K[y_1, \ldots, y_m, z]$ which contains $\mathfrak{M}K[y_1, \ldots, y_m, z]$, and let $\bar{J}$ be its image in $L[z]$. Consider the following sequence of instructions.*

(1) *Choose a term ordering $\sigma$ on $\mathbb{T}(y_1, \ldots, y_m, z)$ which is an elimination ordering for $z$ and compute the reduced $\sigma$-Gröbner basis $G$ of $J$.*

(2) *Let $g \in K[y_1, \ldots, y_m][z]$ be the unique polynomial in $G$ which is monic in $z$.*

(3) *Return the residue class of $g$ in $L[z]$.*

*This is an algorithm which computes the monic generator of $\bar{J}$.*

*Proof* Let $\sigma'$ be the term ordering on $\mathbb{T}(y_1, \ldots, y_m)$ induced by the restriction of $\sigma$, and let $G'$ be the reduced $\sigma'$-Gröbner basis of $\mathfrak{M}$. Using Buchberger's Criterion [15], Corollary 2.5.3, it easy to check that $G'$ is also the reduced $\sigma$-Gröbner basis of $\mathfrak{M} K[y_1, \ldots, y_m, z]$. Since ideal $J$ is zero-dimensional in $K[y_1, \ldots, y_m, z]$, it contains $\mathfrak{M} K[y_1, \ldots, y_m, z]$ properly. Consequently, the ideal $\bar{J}$ is a non-zero principal ideal. Let $g \in K[y_1, \ldots, y_m, z] \setminus \{0\}$ be a polynomial such that $\bar{g}$ generates $\bar{J}$. We write $g$ as a polynomial in $z$ with coefficients in $K[y_1, \ldots, y_m]$. By assumption, we have $g \notin \mathfrak{M} K[y_1, \ldots, y_m, z]$. Consequently, we may assume that all coefficients of $g$ in $K[y_1, \ldots, y_m]$ are non-zero modulo $\mathfrak{M}$, and thus invertible modulo $\mathfrak{M}$.

Now we let $a \in K[y_1, \ldots, y_m] \setminus \mathfrak{M}$ be such that $a \cdot z^s$ is the leading monomial of $g$ in $K[y_1, \ldots, y_m][z]$. Since $a$ is invertible modulo $\mathfrak{M}$, there exist polynomials $b \in K[y_1, \ldots, y_m] \setminus \mathfrak{M}$ and $c \in \mathfrak{M}$ such that $ab = 1 + c$. Notice that we can replace $g$ by $bg$. Hence we may assume that the leading monomial of $g$ is $z^s$. Then $G' \cup \{g\}$ generates $J$, and in fact is a $\sigma$-Gröbner basis of $J$. In order to replace $g$ by its normal form $h = \mathrm{NF}_{\sigma, \mathfrak{M} K[y_1, \ldots, y_m, z]}(g)$, we have to divide $g$ by the Gröbner basis $G'$ which is contained in $K[y_1, \ldots, y_m]$. It follows that $G' \cup \{h\}$ is the reduced $\sigma$-Gröbner basis of $J$ and $h$ is its unique element which is monic in $z$. Hence Step (2) of the algorithm correctly identifies the desired element whose residue class generates $\bar{J}$.                                                        $\square$

Let us apply this algorithm in a concrete case.

**Example 5.3.2** Let $K = \mathbb{Q}$, let $\mathfrak{M}$ be the maximal ideal $\mathfrak{M} = \langle y_1^4 + 1, y_2^2 + y_1^2 - 2 \rangle$ in $K[y_1, y_2]$, and let $J$ be the ideal $J = \mathfrak{M} K[y_1, y_2, z] + \langle y_2 z^2 - y_1 z - 5 \rangle$ in $K[y_1, y_2, z]$. The ideal $J$ is clearly zero-dimensional. Our goal is to compute the monic generator of the residue class ideal $\bar{J}$ of $J$ in $L[z]$, where $L$ is the field $L = K[y_1, y_2]/\mathfrak{M}$. We follow the steps of the algorithm.

(1) To define an elimination ordering for $z$ on $\mathbb{T}(y_1, y_2, z)$, we let $V = \begin{pmatrix} 0 & 0 & 1 \\ 1 & 1 & 0 \\ 0 & -1 & 0 \end{pmatrix}$
    and use $\sigma = \mathrm{Ord}(V)$ (see [15], Sect. 1.3). The reduced $\sigma$-Gröbner basis of the ideal $J$ is $\{y_1^2 + y_2^2 - 2, y_2^4 - 4y_2^2 + 5, z^2 + \frac{1}{5}y_1 y_2^3 z - \frac{4}{5}y_1 y_2 z + y_2^3 - 4y_2\}$.

(2) The element of $G$ which is monic in $z$ is $g = z^2 + \frac{1}{5}y_1 y_2^3 z - \frac{4}{5}y_1 y_2 z + y_2^3 - 4y_2$.

(3) Thus the algorithm returns the element $\bar{g} = z^2 + \frac{1}{5}\bar{y}_1 \bar{y}_2^3 z - \frac{4}{5}\bar{y}_1 \bar{y}_2 z + \bar{y}_2^3 - 4\bar{y}_2$
    as the monic generator of the ideal $\bar{J}$ in $L[z]$.

Notice that, in this case, extending the field does not change the minimal polynomial (see Proposition 5.1.4). In other words, given an element $r \in R$ and its image $r_L = r \otimes 1$ in $R_L = R \otimes_K L$, we have $\mu_{r,K}(z) = \mu_{r_L,L}(z)$. However, given an arbitrary element of $R_L$, its minimal polynomials over $K$ and over $L$ may very well differ, and it is not clear how the knowledge of the minimal polynomial over $K$ could be used to determine which of its divisors in $L[z]$ is the minimal polynomial over $L$. Instead, we now extend the method of Algorithm 5.1.1 and compute the minimal polynomial over $L$ directly.

---

**Algorithm 5.3.3 (Minimal Polynomials and Finite Field Extensions)**
*Let $K$ be a field, and let $L = K[y_1, \ldots, y_m]/\mathfrak{M}$ be an extension field of $K$, where $\mathfrak{M}$ is maximal in $K[y_1, \ldots, y_m]$. In $Q = K[y_1, \ldots, y_m, x_1, \ldots, x_n]$, we let $J$ be a zero-dimensional ideal which contains the ideal $\mathfrak{M}Q$. Moreover, assume that we are given an element $f \in Q$. We let $\overline{Q} = Q/J = L[x_1, \ldots, x_n]/\overline{J}$ and denote the residue class of $f$ in $\overline{Q}$ by $\bar{f}$. Consider the following instructions.*

*(1) Choose a term ordering $\sigma''$ on $\mathbb{T}(y_1, \ldots, y_m)$. Extend it to a term ordering $\sigma'$ on $\mathbb{T}(y_1, \ldots, y_m, z)$ which is an elimination ordering for $z$. Then extend $\sigma'$ to a term ordering $\sigma$ on $\mathbb{T}(y_1, \ldots, y_m, z, x_1, \ldots, x_n)$ which is an elimination ordering for $(x_1, \ldots, x_n)$. Compute the reduced $\sigma$-Gröbner basis $G$ of the ideal $J + \langle z - f \rangle$.*

*(2) Let $g$ be the unique polynomial in $G$ which is contained in the ring $K[y_1, \ldots, y_m][z]$ and monic in $z$.*

*(3) Return the residue class $\bar{g}$ of $g$ in $L[z]$.*

*This is an algorithm which computes the minimal polynomial $\mu_{\bar{f},L}(z)$ of $\bar{f}$ in $L[z]$.*

---

*Proof* Consider the ideal $J' = (J + \langle z - f \rangle) \cap K[y_1, \ldots, y_m, z]$. By the construction of the term ordering $\sigma$, the elements of the set $G' = G \cap K[y_1, \ldots, y_m, z]$ form the reduced $\sigma'$-Gröbner basis of the ideal $J'$. Since $J$ is a zero-dimensional ideal in $Q$, the ideal $J + \langle z - f \rangle$ is a zero-dimensional ideal in $Q[z]$. From the fact that $J'$ is the kernel of the composition $K[y_1, \ldots, y_m, z] \hookrightarrow Q[z] \longrightarrow Q[z]/(J + \langle z - f \rangle)$, it follows then that also $K[y_1, \ldots, y_m, z]/J'$ is a finite dimensional $K$-vector space, i.e., that $J'$ is a zero-dimensional ideal in $K[y_1, \ldots, y_m, z]$.

Clearly, as $J$ contains $\mathfrak{M}Q$, the ideal $J'$ contains $\mathfrak{M}K[y_1, \ldots, y_m, z]$. Now Algorithm 5.3.1 shows that the polynomial $\bar{g}$ generates the residue class ideal of $J'$ in $L[z]$. Indeed, by construction, the ideal $J'$ consists of all polynomials in $L[z]$ for which the substitution $z \mapsto \bar{f}$ yields zero. Hence we have $\bar{g} = \mu_{\bar{f},L}(z)$, as claimed.  $\square$

Notice that the elimination of all indeterminates $y_1, \ldots, y_m, x_1, \ldots, x_n$ from the ideal $J + \langle z - f \rangle$ would have resulted in the minimal polynomial $\mu_{\bar{f},K}(z)$ of $\bar{f}$ which is a multiple of $\mu_{\bar{f},L}(z)$, as we noted above. This observation and the correctness of the algorithm are illustrated by the following example.

**Example 5.3.4** Let $K = \mathbb{Q}$, let $L = K[y_1, y_2]/\mathfrak{M}$, where $\mathfrak{M} = \langle y_1^2 - 2, y_2^3 - 5\rangle$, let $Q = K[y_1, y_2, x_1, x_2]$, and let $J = \mathfrak{M}Q + \langle y_1 x_1 - 2y_2 x_2, x_2^2 - x_1 - 3\rangle$. The ideal $J$ is a zero-dimensional ideal of $Q$ which contains the ideal $\mathfrak{M}Q$.

To compute the minimal polynomial $\mu_{\bar{f}, L}(z)$ of the residue class $\bar{f}$ of the polynomial $f = y_1 x_2^2 - 3y_2 x_1 - x_2 + y_1^5$ in the ring $\overline{Q} = Q/J = L[x_1, x_2]/\bar{J}$, we follow the steps of Algorithm 5.3.3.

(1) Let $\sigma = \mathrm{Ord}(V)$ be the term ordering on $\mathbb{T}(y_1, y_2, z, x_1, x_2)$ defined by the matrix

$$V = \begin{pmatrix} 0 & 0 & 0 & 1 & 1 \\ 0 & 0 & 1 & 0 & 0 \\ 1 & 1 & 0 & 0 & 0 \\ 0 & 0 & 0 & 0 & -1 \\ 0 & -1 & 0 & 0 & 0 \end{pmatrix}$$

Notice that $\sigma'' = \mathrm{DegRevLex}$, that $\sigma'$ is an elimination ordering for $z$, and that $\sigma$ is an elimination ordering for $(x_1, x_2)$. Now we compute the reduced $\sigma$-Gröbner basis $G$ of $J + \langle z - f\rangle$ and get

$$G = \{y_1^2 - 2, y_2^3 - 5,$$

$$z^2 - 2y_1 y_2^2 z + y_1 y_2 z - 14y_1 z + 30z - 18y_1 y_2^2 + 16y_2^2 - 30y_1 - 272y_2 + 95,$$

$$x_2 + \tfrac{627}{460631} y_1 y_2^2 z + \tfrac{14720}{460631} y_1 y_2 z - \tfrac{2828}{1381893} y_2^2 z + \tfrac{290}{460631} y_1 z$$
$$+ \tfrac{1196}{1381893} y_2 z + \tfrac{28813}{1381893} z + \tfrac{19796}{1381893} y_1 y_2^2 - \tfrac{8372}{1381893} y_1 y_2$$
$$- \tfrac{8778}{460631} y_2^2 - \tfrac{201691}{1381893} y_1 - \tfrac{206080}{460631} y_2 - \tfrac{4060}{460631},$$

$$x_1 + \tfrac{1196}{1381893} y_1 y_2^2 z + \tfrac{28813}{1381893} y_1 y_2 z + \tfrac{29440}{460631} y_2^2 z - \tfrac{14140}{1381893} y_1 z$$
$$+ \tfrac{580}{460631} y_2 z + \tfrac{6270}{460631} z - \tfrac{206080}{460631} y_1 y_2^2 - \tfrac{4060}{460631} y_1 y_2 - \tfrac{16744}{1381893} y_2^2$$
$$- \tfrac{43890}{460631} y_1 + -\tfrac{403382}{1381893} y_2 + \tfrac{197960}{1381893}\}$$

(2) The unique element in $G \cap K[y_1, y_2, z]$ which is monic in $z$ is the polynomial
$$g = z^2 - (2y_1 y_2^2 - y_1 y_2 + 14y_1 - 30)z - (18y_1 y_2^2 - 16y_2^2 + 30y_1 + 272y_2 - 95).$$
(3) The algorithm returns $\mu_{\bar{f}, L}(z) = \bar{g}$, where $\bar{g}$ is the residue class of $g$ in $L[z]$.

If we eliminate $(y_1, y_2, x_1, x_2)$ from the ideal $J + \langle z - f\rangle$, the result is the principal ideal generated by

$$\mu_{\bar{f}, K}(z) = z^{12} + 180z^{11} + 13014z^{10} + 494820z^9 + 10422807z^8 + 89609400z^7$$

$$- 1090284388z^6 - 44774512920z^5 - 1080832025025z^4$$

$$- 18422725682700z^3 - 176120391920250z^2 - 1251376301141100z$$

$$+ 7483863077173225$$

It is possible to check that this polynomial is a multiple of $\mu_{\bar{f},L}(z)$ in $L[z]$, as implied by the observation above.

Finally, let us point out several further observations regarding the properties of Algorithm 5.3.3.

**Remark 5.3.5**  In the setting of Algorithm 5.3.3, the following statements hold.

(a) Algorithm 5.1.1 is a special case of Algorithm 5.3.3, namely the case where $m = 0$ and $\mathfrak{M} = \langle 0 \rangle$.

(b) If there exists an index $i \in \{1, \ldots, n\}$ such that $f = x_i$, then there is no need to introduce the indeterminate $z$ and the polynomial $z - f$. In this case there is no intermediate term ordering $\sigma'$, the ideal $J + \langle z - f \rangle$ is just $J$, the term ordering $\sigma$ has to be chosen to be an elimination ordering for $(x_1, \ldots, \hat{x}_i, \ldots, x_n)$, and the resulting $\sigma$-Gröbner basis of $J$ contains a unique polynomial from $K[y_1, \ldots, y_m][x_i]$ which is monic in $x_i$. This polynomial is $\mu_{\bar{x}_i,L}(x_i)$.

(c) Algorithm 5.3.3 cannot be modified by skipping the construction of $\sigma'$ and choosing a term ordering $\sigma$ which is an elimination ordering for $(z, x_1, \ldots, x_n)$. For instance, suppose that in Example 5.3.4 we pick $\sigma = \mathrm{Ord}(V)$ where the first row of $V$ is $(0, 0, 1, 1, 1)$. Although $\sigma$ eliminates $(z, x_1, x_2)$, it does not compute the required polynomial and cannot be used for our purposes.

## 5.3.B  Factorizing over Extension Fields

> *If at first you don't succeed,*
> *look in the trash for the instructions.*
> (Sean Keogh)

In this subsection we try to succeed by following the general strategy outlined in the introduction of this section. Our next goal is to find an effective method for factorizing univariate polynomials over finite field extensions. This turns out to be an interesting application of the methods for computing primary decompositions given in the first two sections of this chapter. Since Algorithm 5.3.6 uses the calculation of the maximal components of some zero-dimensional ideal, it may appear that our general intention to use it for exactly this calculation is in jeopardy. But if you check the instructions of the improved version given in Algorithm 5.3.8 carefully, you will see that we can succeed by calculating only the primary components of one zero-dimensional ideal if the base field is perfect.

Continuing to use the same setting as in the preceding subsection, we let $K$ be a field and $L = K[y_1, \ldots, y_m]/\mathfrak{M}$ an extension field of $K$, where $\mathfrak{M}$ is a maximal ideal of $K[y_1, \ldots, y_m]$. The following algorithm for computing the factorization of a polynomial in $L[z]$ is inspired by the methods described by Y. Sun and D. Wang in [29].

**Algorithm 5.3.6 (Factorizing Polynomials over Extension Fields)**
*In the above setting, let $z$ be a further indeterminate, let $f \in K[y_1, \ldots, y_m][z]$ be a polynomial which is not contained in $K[y_1, \ldots, y_m]$ and monic in $z$. Moreover, let $\bar{f}$ be the image of $f$ in $L[z]$. Consider the following instructions.*

(1) *Form the ideal $I = \mathfrak{M}K[y_1, \ldots, y_m][z] + \langle f \rangle$ in $K[y_1, \ldots, y_m][z]$. Compute the primary decompositions $I = \mathfrak{Q}_1 \cap \cdots \cap \mathfrak{Q}_s$ and $\text{Rad}(I) = \mathfrak{M}_1 \cap \cdots \cap \mathfrak{M}_s$. Here $\mathfrak{Q}_1, \ldots, \mathfrak{Q}_s$ are primary ideals and $\mathfrak{M}_1, \ldots, \mathfrak{M}_s$ are maximal ideals in $K[y_1, \ldots, y_m][z]$.*
(2) *Let $\sigma$ be a term ordering on $\mathbb{T}(y_1, \ldots, y_m, z)$ which is an elimination ordering for $z$. For $i = 1, \ldots, s$, compute the reduced $\sigma$-Gröbner basis $G_i$ of $\mathfrak{Q}_i$ and the reduced $\sigma$-Gröbner basis $H_i$ of $\mathfrak{M}_i$.*
(3) *For $i = 1, \ldots, s$, let $g_i \in K[y_1, \ldots, y_m][z]$ be the unique element of $G_i$ which is monic in $z$, and let $h_i \in K[y_1, \ldots, y_m][z]$ be the unique polynomial in $H_i$ which is monic in $z$.*
(4) *For $i = 1, \ldots, s$, let $d_i = \deg_z(h_i)$, and let $m_i = \deg_z(g_i)/d_i$.*
(5) *Return the list of pairs $((\bar{h}_1, m_1), \ldots, (\bar{h}_s, m_s))$ where $\bar{h}_i$ is the image of $h_i$ in $L[z]$.*

*This is an algorithm which computes the irreducible factors and their multiplicities in the factorization of $\bar{f}$ in $L[z]$. More precisely, we have the equality $\bar{f} = \bar{h}_1^{m_1} \cdots \bar{h}_s^{m_s}$ where the polynomials $\bar{h}_i$ are the distinct irreducible factors of $\bar{f}$ in $L[z]$.*

*Proof* First of all, we note that $I$ is a zero-dimensional ideal in $K[y_1, \ldots, y_m][z]$. Hence the computation of the primary decompositions of $I$ and $\text{Rad}(I)$ can be performed with the algorithms explained in the preceding sections. Suppose that $\bar{f}$ factorizes as $\bar{f} = \prod_{i=1}^{s} \bar{f}_i^{m_i}$ where $m_i \geq 1$ and $\bar{f}_i \in L[z]$ is a monic irreducible polynomial for $i = 1, \ldots, s$. Then the image of $\mathfrak{Q}_i$ in $L[z]$ is generated by $\bar{f}_i^{m_i}$ and the image of $\mathfrak{M}_i$ in $L[z]$ is generated by $\bar{f}_i$ for $i = 1, \ldots, s$. Using Algorithm 5.3.1, it therefore follows that we have $\bar{g}_i = \bar{f}_i^{m_i}$ and $\bar{h}_i = \bar{f}_i$ for $i = 1, \ldots, s$. Hence the algorithm computes the irreducible factors of $\bar{f}$ correctly. Since $\deg(\bar{g}_i) = m_i \deg(\bar{f}_i) = m_i \deg(\bar{h}_i)$, Step (5) determines the numbers $m_i$ correctly, and the proof is complete. $\qquad\square$

Let us see this algorithm at work in an easy case.

**Example 5.3.7** Let $K = \mathbb{Q}$, and let $L = K[y]/\mathfrak{M}$, where $\mathfrak{M} = \langle y^2 - 2y + 2 \rangle$ is a maximal ideal, since $y^2 - 2y + 2$ is irreducible over $K$.

The polynomial $f = z^2 + 1$ is irreducible in $K[z]$. But what happens if we try to factorize its image $\bar{f}$ in $L[z]$? Let us follow the steps of Algorithm 5.3.6.

(1) Let $I$ be the ideal $I = \mathfrak{M} K[y, z] + \langle z^2 + 1 \rangle$ in $K[y, z]$. We compute the primary decomposition of $I$ and get $I = \mathfrak{M}_1 \cap \mathfrak{M}_2$, where $\mathfrak{M}_1$ and $\mathfrak{M}_2$ are maximal ideals.

(2) The reduced Gröbner bases of $\mathfrak{M}_1$ and $\mathfrak{M}_2$ with respect to an elimination ordering for $z$ are given by $\{y^2 - 2y + 2, z - y + 1\}$ and $\{y^2 - 2y + 2, z + y - 1\}$, respectively.

(3) Hence we have $g_1 = h_1 = z - y + 1$ and $g_2 = h_2 = z + y - 1$.

(4) We calculate $m_1 = m_2 = 1$.

(5) The algorithm returns the pair $((z - \bar{y} + 1, 1), (z + \bar{y} - 1, 1))$.

Consequently, we have the factorization $\bar{f} = (z - \bar{y} + 1)(z + \bar{y} - 1)$ in $L[z]$.

The preceding algorithm can be simplified substantially if the field $K$ is perfect. In this case it suffices to compute the primary decomposition $I = \mathfrak{Q}_1 \cap \cdots \cap \mathfrak{Q}_s$ of the ideal $I$ is Step (1). Then the polynomials $\bar{h}_i$ can be calculated as the squarefree parts of the polynomials $\bar{g}_i$ in $L[z]$. Notice that [15], Sect. 3.7, provides efficient ways to compute the squarefree part of a univariate polynomial over a perfect field. Let us spell out this improved algorithm in detail.

**Algorithm 5.3.8 (Improved Factorization over Extension Fields)**
*In the above setting, assume that $K$ is a perfect field. Let $z$ be a further inde-
terminate, let $f \in K[y_1, \ldots, y_m][z] \setminus K[y_1, \ldots, y_m]$ be a polynomial which
is monic in $z$, and let $\bar{f}$ be the image of $f$ in $L[z]$. Consider the following
instructions.*

(1) *Form the ideal $I = \mathfrak{M} K[y_1, \ldots, y_m][z] + \langle f \rangle$ in $K[y_1, \ldots, y_m][z]$.
Compute the primary decomposition $I = \mathfrak{Q}_1 \cap \cdots \cap \mathfrak{Q}_s$ of $I$, where
$\mathfrak{Q}_1, \ldots, \mathfrak{Q}_s$ are primary ideals in $K[y_1, \ldots, y_m][z]$.*

(2) *Let $\sigma$ be a term ordering on $\mathbb{T}(y_1, \ldots, y_m, z)$ which is an elimination
ordering for $z$. For $i = 1, \ldots, s$, compute the reduced $\sigma$-Gröbner basis $G_i$
of $\mathfrak{Q}_i$.*

(3) *For $i = 1, \ldots, s$, let $g_i \in K[y_1, \ldots, y_m][z]$ be the unique element of $G_i$
which is monic in $z$.*

(4) *For $i = 1, \ldots, s$, let $\bar{g}_i$ be the image of $g_i$ in $L[z]$. Using [15], Proposition 3.7.9 or 3.7.12, compute the squarefree part $\bar{h}_i = \mathrm{sqfree}(\bar{g}_i)$ of $\bar{g}_i$
in $L[z]$.*

(5) *For $i = 1, \ldots, s$, let $m_i = \deg(\bar{g}_i)/\deg(\bar{h}_i)$. Then return the list of pairs
$((\bar{h}_1, m_1), \ldots, (\bar{h}_s, m_s))$.*

*This is an algorithm which computes the irreducible factors and their multi-
plicities in the factorization of $\bar{f}$ in $L[z]$. More precisely, we have the equality
$\bar{f} = \bar{h}_1^{m_1} \cdots \bar{h}_s^{m_s}$ where the polynomials $\bar{h}_i$ are the distinct irreducible factors
of $\bar{f}$ in $L[z]$.*

*Proof* As in the proof of Algorithm 5.3.6, it follows that $\bar{g}_i$ is a power of an irreducible factor of $\bar{f}$ in $L[z]$. Since $K$ is a perfect field, also $L$ is perfect, and we can use the methods described in [15], Proposition 3.7.9 and 3.7.12 to compute the irreducible factors $\bar{h}_i = \mathrm{sqfree}(\bar{g}_i)$ of $\bar{f}$ in $L[z]$.                                                    $\square$

In the following example we apply Algorithm 5.3.8 to a non-trivial case. We also see that its output depends on the base field $K$.

**Example 5.3.9** Let $K = \mathbb{Q}$, and let $L = K[y_1, y_2, y_3]/\mathfrak{M}$, where $\mathfrak{M}$ is the maximal ideal

$$\mathfrak{M} = \langle y_1 + 2y_2 - 2, \, y_3^2 - y_2 - 1, \, y_2 y_3 - 4, \, y_2^2 + y_2 - 4y_3 \rangle$$

of $K[y_1, y_2, y_3]$. Using Algorithm 5.3.8, we want to find the factorization of the image $\bar{f}$ of the polynomial

$$\begin{aligned}
f = {}& z^4 + (y_1^2 - 1)z^3 - (y_1^4 - 2y_1^2 + y_2^2 - 1)z^2 \\
& - (y_1^6 - 3y_1^4 + 2y_1^2 y_2^2 - y_1^2 - 2y_2^2 + 3)z \\
& - y_1^4 y_2^2 + 2y_1^4 + 2y_1^2 y_2^2 - 4y_1^2 - y_2^2 + 2
\end{aligned}$$

in $L[z]$. The primary decomposition of $I = \mathfrak{M} + \langle f \rangle$ is $I = \mathfrak{Q}_1 \cap \mathfrak{Q}_2$, where we have

$$\mathfrak{Q}_1 = \mathfrak{M} + \langle z^2 + 12y_2 z - 16y_3 z - 3z + y_2 - 4y_3 + 2 \rangle$$

$$\mathfrak{Q}_2 = \mathfrak{M} + \langle z^2 - 24y_2 z + 32y_3 z + 6z + 40y_2 + 672y_3 - 1271 \rangle$$

This yields the polynomials $g_1 = z^2 + 12y_2 z - 16y_3 z - 3z + y_2 - 4y_3 + 2$ as well as $g_2 = z^2 - 24y_2 z + 32y_3 z + 6z + 40y_2 + 672y_3 - 1271$.

Since we have $\mathrm{char}(L) = 0$, we can use [15], Proposition 3.7.9.b to compute

$$\bar{h}_1 = \bar{g}_1/\gcd(\bar{g}_1, \bar{g}_1') = z^2 + 12\bar{y}_2 z - 16\bar{y}_3 z - 3z + \bar{y}_2 - 4\bar{y}_3 + 2$$

$$\bar{h}_2 = \bar{g}_2/\gcd(\bar{g}_2, \bar{g}_2') = z - 12\bar{y}_2 + 16\bar{y}_3 + 3$$

Altogether, in $L[z]$ we get the factorization

$$\bar{f} = (z^2 + 12\bar{y}_2 z - 16\bar{y}_3 z - 3z + \bar{y}_2 - 4\bar{y}_3 + 2)(z - 12\bar{y}_2 + 16\bar{y}_3 + 3)^2$$

Let us point out several aspects of this computation. In order to calculate the polynomials $g_i$, it is important that we use an elimination ordering $\sigma$ for $z$. For instance, if we use $\tau = \mathtt{DegRevLex}$ instead, we get $\mathrm{LT}_\tau(\mathfrak{Q}_1) = \langle y_1, y_3^2, y_2 y_3, y_2^2, y_2 z, y_3 z^2, z^3 \rangle$ and the algorithm does not find the correct polynomial $g_1$, because we obtain $\mathrm{LT}_\tau(z^2 + 12y_2 z - 16y_3 z - 3z + y_2 - 4y_3 + 2) = y_2 z$, and not $z^2$ as it is if $\sigma$ is an elimination ordering for $z$.

Another observation is that the same example, viewed over the base field $K = \mathbb{F}_3$, results in a very different factorization, namely $\bar{f} = (z + \bar{y}_2)(z - \bar{y}_2 - \bar{y}_3)(z + \bar{y}_3)^2$ in $\tilde{L}[z]$, where $\tilde{L} = \mathbb{F}_3[y_1, y_2, y_3]/\langle y_1 + 2y_2 - 2, \, y_3^2 - y_2 - 1, \, y_2 y_3 - 4, \, y_2^2 + y_2 - 4y_3 \rangle$.

In the last part of this section we show that one can use the preceding two algorithms to find all elements in a finite field extension which have the same minimal polynomial as a given one. For a related algorithm over a finite field $K$, we refer the reader to Algorithm 6.3.10.

**Proposition 5.3.10 (Elements with the Same Minimal Polynomial)**
*Let $L = K[y_1, \ldots, y_m]/\mathfrak{M}$ be an extension field of $K$, where $\mathfrak{M}$ is a maximal ideal of $K[y_1, \ldots, y_m]$, and let $a, b \in L$. Then the following conditions are equivalent.*

(a) *We have $\mu_a(z) = \mu_b(z)$ in $K[z]$.*
(b) *The polynomial $z - b$ is a factor of $\mu_a(z)$ in $L[z]$.*

*In particular, using Algorithms 5.3.6 and 5.3.8, we can compute all elements of $L$ which have a given minimal polynomial over $K$.*

*Proof* To prove the implication $(a) \Rightarrow (b)$, we note that the assumption implies $\mu_a(b) = 0$ in $L$. Hence the polynomial $z - b$ divides $\mu_a(z)$ in $L[z]$.

Conversely, if (b) holds, then we have $\mu_a(b) = 0$ in $L$. Therefore the polynomial $\mu_a(z)$ is a multiple of $\mu_b(z)$. Since both polynomials are irreducible and monic in $K[z]$, they are equal.                                                                          □

Let us apply this proposition to compute some elements with the same minimal polynomial.

**Example 5.3.11** Let $K = \mathbb{Q}$, and let $L = K[y_1, y_2]/\mathfrak{M}$, where

$$\mathfrak{M} = \langle y_1^2 + 3y_1 y_2 - y_2^2 - y_1 - 1, \, y_2^2 - y_2 - 1 \rangle$$

It is easy to check that $\mathfrak{M}$ is a maximal ideal and that $L$ is an extension field of $K$ with $\dim_K(L) = 4$. The minimal polynomial of $\bar{y}_1$ is $\mu_{\bar{y}_1}(z) = z^4 + z^3 - 16z^2 + 5z + 5$. In particular, we see that $\bar{y}_1$ is a primitive element of $L/K$.

Are there other elements in $L$ which have the same minimal polynomial over $K$ as $y_1$? To answer this question, we apply Algorithm 5.3.8 to compute the linear factors of $\mu_{\bar{y}_1}(z)$ in $L[z]$. First we calculate the primary decomposition of the ideal $I = \mathfrak{M} + \langle z^4 + z^3 - 16z^2 + 5z + 5 \rangle$ in $K[y_1, y_2, z]$ and get $I = \mathfrak{Q}_1 \cap \mathfrak{Q}_2 \cap \mathfrak{Q}_3$, where

$$\mathfrak{Q}_1 = \mathfrak{M} + \langle z - y_1 \rangle$$

$$\mathfrak{Q}_2 = \mathfrak{M} + \langle z + y_1 + 3y_2 - 1 \rangle$$

$$\mathfrak{Q}_3 = \mathfrak{M} + \langle z^2 - 3y_2 z + 2z + y_2 - 3 \rangle$$

Thus we deduce that there is only one more element with the same minimal polynomial as $\bar{y}_1$, namely the residue class $\bar{f}$ of the polynomial $f = -y_1 - 3y_2 + 1$.

Next we check whether there are other elements with the same minimal polynomial as $\bar{y}_2$, i.e. with the minimal polynomial $\mu_{\bar{y}_2}(z) = z^2 - z - 1$. Using the same technique, we find that there is again only one other element of $L$ with the same minimal polynomial as $\bar{y}_2$, namely the residue class $\bar{g}$ of $g = -y_2 + 1$.

Finally, in the same way, we can check that there is no element of $L$ whose minimal polynomial over $K$ is $z^2 + 3z - 1$.

## 5.3.C  Using Factorizations to Compute Maximal Components

> *Why do people spend half their money buying food,*
> *and the other half trying to lose weight?*

Now it is time to combine the ingredients and to spend our investments wisely. We are ready to describe an algorithm which computes the maximal components of a zero-dimensional polynomial ideal $I$ by working one indeterminate at a time and factoring its minimal polynomial over the extension field defined by the partial maximal ideal calculated up to that point. In this way we are trying to reduce the weight of a single computation, but we have to pay by performing many iterations. It is interesting to observe that the same pattern will be used again in Chap. 6 (see Algorithm 6.2.7). The results of the computation are nice, slim, triangular sets of generators of the maximal components of $I$. Therefore it can also be applied to bring a given system of generators of a maximal ideal into triangular shape.

---

**Algorithm 5.3.12 (Computing Maximal Components via Factorizations)**
*Let $K$ be a field, let $P = K[x_1, \ldots, x_n]$, let $I$ be a zero-dimensional ideal in $P$, and let $R = P/I$. Consider the following sequence of instructions.*

(1) *Compute the minimal polynomial $\mu_{\bar{x}_1}(z)$ of the residue class $\bar{x}_1$ of $x_1$ in $R$.*

(2) *Calculate the factorization $\mu_{\bar{x}_1}(z) = p_1(z)^{\alpha_1} \cdots p_s(z)^{\alpha_s}$ with $\alpha_i > 0$ and pairwise distinct irreducible polynomials $p_i(z)$ in $K[z]$. Then form the tuple of ideals $T_1 = (\langle p_1(x_1) \rangle, \ldots, \langle p_s(x_1) \rangle)$.*

(3) *For $i = 2, \ldots, n$, repeat the following steps. Then return the final tuple $T_n$.*

(4) *Let $T_i$ be the empty tuple. For every ideal $\langle f_1, \ldots, f_{i-1} \rangle$ in $T_{i-1}$ perform the following steps.*

(5) *In $P_{i-1} = K[x_1, \ldots, x_{i-1}]$ form the maximal ideal $\mathfrak{M} = \langle f_1, \ldots, f_{i-1} \rangle$ and let $L = P_{i-1}/\mathfrak{M}$. Using Algorithm 5.3.4, compute the minimal polynomial $\mu_{\bar{x}_i, L}(z)$ of the residue class of $\bar{x}_i$ in $R_L = L[x_i, \ldots, x_n]/\bar{I}$.*

(6) *Using Algorithm 5.3.6, compute polynomials $h_1, \ldots, h_t$ in the ring $K[x_1, \ldots, x_{i-1}, z]$ whose residue classes $\bar{h}_j \in L[z]$ are the distinct irreducible factors of the polynomial $\mu_{\bar{x}_i, L}(z)$.*

(7) *For $j = 1, \ldots, t$, append the ideal $\langle f_1, \ldots, f_{i-1}, h_j(x_1, \ldots, x_{i-1}, x_i) \rangle$ to $T_i$.*

*This is an algorithm which computes a tuple $T_n$ whose entries are ideals $\langle f_1, \ldots, f_n \rangle$ with $f_i \in K[x_1, \ldots, x_i]$ for $i = 1, \ldots, n$, and these ideals are precisely the maximal components of $I$.*

*Proof* Finiteness of this algorithm is obvious. Let $\langle f_1, \ldots, f_n \rangle$ be one of the ideals in the final output $T_n$. The residue class of each $f_i$ is an irreducible polynomial in the ring $L[x_i]$, where $L = K[x_1, \ldots, x_{i-1}]/\langle f_1, \ldots, f_{i-1} \rangle$. Hence $\langle \bar{f_i} \rangle$ is a maximal ideal in this ring, and consequently the ideal $\langle f_1, \ldots, f_i \rangle$ is maximal in $K[x_1, \ldots, x_i]$. In particular, the computed ideal $\langle f_1, \ldots, f_n \rangle$ is maximal in $P$.

Furthermore, we note that the residue class ideal $\bar{I}$ of $I$ in the ring $L[x_n]$, where $L = K[x_1, \ldots, x_{n-1}]/\langle f_1, \ldots, f_{n-1} \rangle$, is a principal ideal generated by the minimal polynomial of $\bar{x}_n$. Hence the residue class of $h_j(x_1, \ldots, x_n)$ in the ring $L[x_n]$ generates a maximal ideal which contains $\bar{I}$. Therefore the ideal $\langle f_1, \ldots, f_n \rangle$ contains $I$ and thus is a maximal component of $I$.

It remains to show that every maximal component of $I$ is one of the ideals in the output $T_n$. Let $\mathfrak{M}$ be a maximal component of $I$. Since the element $\mu_{\bar{x}_i, L}(\bar{x}_i)$ is contained in the residue class ideal $\overline{\mathfrak{M}}$ of $\mathfrak{M}$ in $L[x_i, \ldots, x_n]$, one of its factors $\bar{h}_j(\bar{x}_1, \ldots, \bar{x}_i)$ has to be contained in $\overline{\mathfrak{M}}$. Hence one of the ideals $\langle f_1, \ldots, f_{i-1}, h_j \rangle$ appended in Step (7) is contained in $\mathfrak{M}$. Inductively, it follows that one of the ideals $\langle f_1, \ldots, f_n \rangle$ in the output $T_n$ is contained in $\mathfrak{M}$. Altogether, we have shown that these ideals $\langle f_1, \ldots, f_n \rangle$ are precisely the maximal components of $I$.                                      □

The output of this algorithm suggests the following definition.

**Definition 5.3.13** Let $P = K[x_1, \ldots, x_n]$, and let $f_1, \ldots, f_n$ be polynomials in $P$. The set $\{f_1, \ldots, f_n\}$ is called a **triangular set** if the following conditions are satisfied for $i = 1, \ldots, n$.

(a)  We have $f_i \in K[x_1, \ldots, x_i] \setminus K[x_1, \ldots, x_{i-1}]$.
(b)  When viewed as an element of $K[x_1, \ldots, x_{i-1}][x_i]$, the polynomial $f_i$ is monic in $x_i$.

Notice that it follows immediately from this definition that a triangular set is a Gröbner basis of the ideal it generates with respect to the lexicographic term ordering $\sigma$ such that $x_n >_\sigma \cdots >_\sigma x_1$. As a consequence of the preceding algorithm, we get the following result.

**Corollary 5.3.14 (Maximal Ideals and Triangular Sets)**
*Every maximal ideal in $K[x_1, \ldots, x_n]$ can be generated by a triangular set.*

Let us follow the steps of the algorithm in a concrete case.

**Example 5.3.15** Let $K = \mathbb{Q}$, let $P = K[x_1, x_2, x_3]$, and let $I = \langle x_1^2 - 2, x_2^2 - 8, x_3^2 - 32 \rangle$. (Notice that the multiplication matrices of the ring $R = P/I$ with respect to the $K$-basis $(1, x_1, x_2, x_3, x_1x_2, x_1x_3, x_2x_3, x_1x_2x_3)$ are similar to the multiplication matrices in Example 2.3.10.) To compute the maximal components of $I$, we apply Algorithm 5.3.12.

(1) Calculate $\mu_{\bar{x}_1}(z) = z^2 - 2$.
(2) Since $\mu_{\bar{x}_1}(z)$ is irreducible, we form the tuple $T_1 = ((x_1^2 - 2))$.
(4) Let $i = 2$ and $\langle f_1 \rangle = \langle x_1^2 - 2 \rangle$.
(5) Let $L = K[x_1]/\langle x_1^2 - 2 \rangle$. Using Algorithm 5.3.4, we calculate the minimal polynomial $\mu_{\bar{x}_2, L}(z) = z^2 - 8$ of $\bar{x}_2$ in $L[x_2, x_3]/\bar{I}$.
(6) Using Algorithm 5.3.6, we factorize the polynomial $\mu_{\bar{x}_2, L}(z)$ in $L[z]$ and get
$$\mu_{\bar{x}_2, L}(z) = (z - 2\bar{x}_1)(z + 2\bar{x}_1).$$
(7) We let $T_2 = ((x_1^2 - 2, x_2 - 2x_1), (x_1^2 - 2, x_2 + 2x_1))$.
(4) We let $i = 3$ and work on the ideal $\langle x_1^2 - 2, x_2 - 2x_1 \rangle$.
(5) We let $L = K[x_1, x_2]/\langle x_1^2 - 2, x_2 - 2x_1 \rangle$ and get $\mu_{\bar{x}_3, L}(z) = z^2 - 32$ in the ring $L[x_3]/\bar{I} = L[x_3]/\langle x_3^2 - 32 \rangle$.
(6) The factorization of $\mu_{\bar{x}_3, L}(z)$ is $(z - 4\bar{x}_1)(z + 4\bar{x}_1)$.
(7) We let $T_3 = ((x_1^2 - 2, x_2 - 2x_1, x_3 - 4x_1), (x_1^2 - 2, x_2 - 2x_1, x_3 + 4x_1))$.
(4) Now we work on the ideal $\langle x_1^2 - 2, x_2 + 2x_1 \rangle$ and obtain analogous results in Steps (5) and (6).
(7) The output tuple is $T_3 = ((x_1^2 - 2, x_2 - 2x_1, x_3 - 4x_1), (x_1^2 - 2, x_2 - 2x_1, x_3 + 4x_1), (x_1^2 - 2, x_2 + 2x_1, x_3 - 4x_1), (x_1^2 - 2, x_2 + 2x_1, x_3 + 4x_1))$.

Altogether, the algorithm correctly finds four maximal components of $I$, namely the maximal ideals in $T_3$.

Notice that the order of the indeterminates $x_1, \dots, x_n$ plays an important role for the output of Algorithm 5.3.12. If we permute the indeterminates before applying this algorithm, we get different triangular sets for the same maximal ideal. For instance, if we interchange $x_1$ and $x_2$ in the preceding example, the algorithm computes the same four maximal components, but their triangular systems of generators look very different. E.g., the maximal ideal $\langle x_1 - \frac{1}{2}x_2, x_2^2 - 8, x_3 - 4x_1 \rangle$ is equal to the first maximal ideal computed above, although this may not be apparent.

Finally, we follow the steps of Algorithm 5.3.12 in another example.

**Example 5.3.16**  Let $K = \mathbb{Q}$, let $P = K[x_1, x_2, x_3]$, and let

$$I = \langle x_1^3 - x_2 - 2, x_2^2 + x_3^2 - 2x_2 - 3, x_3^3 - 2x_3^2 - 2x_3 + 4, x_2x_3^2 - x_3^2 - 2x_2 + 2 \rangle$$

To compute the maximal components of $I$, we use the above algorithm and note some of its main steps.

(1) We have $\mu_{\bar{x}_1}(z) = z^9 - 9z^6 + 25x^3 - 21 = (z^3 - 3)(z^6 - 6z^3 + 7)$. Thus we let $T_1 = ((x_1^3 - 3), (x_1^6 - 6x_1^3 + 7))$.
(5) Over $L = K[x_1]/\langle x_1^3 - 3 \rangle$, we get $\mu_{\bar{x}_2, L}(z) = z - 1$.
(7) We let $T_2 = ((x_1^3 - 3, x_2 - 1))$.
(5) Over $L = K[x_1]/\langle x_1^6 - 6x_1^3 + 7 \rangle$, we get $\mu_{\bar{x}_2, L}(z) = z - \bar{x}_1^3 + 2$.

(7)  We let $T_2 = (\langle x_1^3 - 3, x_2 - 1 \rangle, \langle x_1^6 - 6x_1^3 + 7, x_2 - x_1^3 + 2 \rangle)$ and pass to $i = 3$.

(5)  Over the field $L = K[x_1, x_2]/\langle x_1^3 - 3, x_2 - 1 \rangle$, we get $\mu_{\bar{x}_3, L}(z) = z - 2$.

(7)  Thus we set $T_3 = (\langle x_1^3 - 3, x_2 - 1, x_3 - 2 \rangle)$.

(5)  Over $L = K[x_1, x_2]/\langle x_1^6 - 6x_1^3 + 7, x_2 - x_1^3 + 2 \rangle$, we calculate $\mu_{\bar{x}_3, L}(z) = z^2 - 2$.

(6)  In $L[z]$, this polynomial factorizes as $\mu_{\bar{x}_3, L}(z) = (z - \bar{x}_1^3 + 3)(z + \bar{x}_1^3 - 3)$.

(7)  The output of the algorithm is the triple of ideals $T_3 = (\langle x_1^3 - 3, x_2 - 1,$ $x_3 - 2 \rangle, \langle x_1^6 - 6x_1^3 + 7, x_2 - x_1^3 + 2, x_3 - x_1^3 + 3 \rangle, \langle x_1^6 - 6x_1^3 + 7, x_2 - x_1^3 + 2,$ $x_3 + x_1^3 - 3 \rangle)$.

Thus the ideal $I$ has three maximal components which are the ideals given in the output tuple $T_3$.


## 5.4  Primary Decompositions Using Radical Ideals

> *Before you criticize someone, walk a mile in their shoes.*
> *That way, when you do criticize them,*
> *you are a mile away and have their shoes.*

A classical approach to computing the primary decomposition of a zero-dimensional ideal is to first compute its radical, then the maximal components of the radical, and finally the primary components from the maximal components. Since we described a number of direct methods for computing the primary decomposition in Sects. 5.1 and 5.2, one could criticize this approach as unnecessarily complicated. But let us first walk a mile in these shoes and see what happens.

Let $K$ be a field, let $P = K[x_1, \ldots, x_n]$, and let $I$ be a zero-dimensional ideal in $P$. The first step of the planned walk is to compute the radical of $I$. If $K$ is a perfect field, one way to go is to compute the monic polynomials generating the elimination ideals $I \cap K[x_i]$, then to determine their squarefree parts, and finally to add these squarefree parts to the ideal. Methods for computing the squarefree part of a univariate polynomial were explained in [15], Propositions 3.7.9 and 3.7.12. If $K$ is not perfect, the computation of the radical of $I$ can be achieved using the method explained in [13].

But let us not stray from our path and assume that $K$ is perfect. Then the first subsection contains a possible speed up of the above approach which is based on the observation that the monic generator of $I \cap K[x_i]$ can be viewed as the minimal polynomial of $x_i + I$ in $P/I$. This leads to improved algorithms for computing the radical of $I$ (see Algorithm 5.4.2) and for computing the maximal components of $I$ (see Algorithm 5.4.7). Finally, we go the distance and use the Chinese Reminder Theorem to reconstruct the primary components from the maximal components of $I$ (see Algorithm 5.4.14).

> *Until you walk a mile in another man's moccasins*
> *you can't imagine the smell.*
> (Robert Byrne)

Combining these algorithms, we see that a general criticism of the approach to compute primary decompositions using radical ideals smells fishy, and that every method presented in the preceding sections has its merits and its weak points.

## 5.4.A   Maximal Components via Radical Ideals

> *Time flies like an arrow.*
> *Fruit flies like bananas.*

Since time is money and flies like an arrow, our main goal in this subsection is to save time by computing the radical of a zero-dimensional ideal $I$ faster. Since we also get a time-saving way to determine the maximal components of $I$, we have been trying for quite a while now to figure out how to spend all this saved time.

In the following we always work over a perfect field $K$. Let $P = K[x_1, \ldots, x_n]$, and let $I$ be a zero-dimensional ideal in $P$. The standard method for computing the radical of $I$ requires the calculation of many elimination ideals, and thus of many Gröbner bases (see [15], Corollary 3.7.16). The generators of these elimination ideals are the minimal polynomials of the residue classes $x_i + I$ in $P/I$. Therefore we can also use Algorithm 2.1.11 or Algorithm 1.1.8 to find these elimination ideals. In view of these remarks, we now develop a method for computing $\mathrm{Rad}(I)$ faster. It is based on the following observation.

**Lemma 5.4.1**   *In the above setting, let $i \in \{1, \ldots, n\}$, let $g(z) = \mathrm{sqfree}(\mu_{x_i+I}(z))$, and let $J = I + \langle g(x_i) \rangle$. Then we have $g(z) = \mu_{x_i+J}(z)$.*

*Proof* Let $\mu_{x_i+I}(z) = p_1(z)^{\alpha_1} \cdots p_t(z)^{\alpha_t}$ be the decomposition of the minimal polynomial of $x_i + I \in P/I$ into irreducible factors $p_j(z)$. Then we have the factorization $g(z) = p_1(z) \cdots p_t(z)$. Clearly, it follows that $g(x_i + J) = g(x_i) + J = 0$ in $P/J$.

Now suppose that there exists a proper divisor $h(z)$ of $g(z)$ such that $h(x_i + J) = 0$ in $P/J$. Then we can represent $h(x_i)$ in the form $h(x_i) = q_1 + q_2 g(x_i)$ with $q_1 \in I$ and $q_2 \in P$. Since we have $g(x_i)^k \in I$ for large enough $k \geq 1$, it follows that $h(x_i)^k \in I$. Therefore $h(z)^k$ is a multiple of $\mu_{x_i+I}(z)$, in contradiction to the fact that $h(z)^k$ misses some irreducible factor of $\mu_{x_i+I}(z)$.                                        $\square$

Now we are ready to improve the calculation of the radical of a zero-dimensional polynomial ideal as follows.

**Algorithm 5.4.2 (Computing the Radical of a Zero-Dimensional Ideal)**
*Let $K$ be a perfect field, let $P = K[x_1, \ldots, x_n]$, and let $I$ be a zero-dimensional ideal in $P$. Consider the following sequence of instructions.*

*(1) Let $J = I$.*
*(2) For $i = 1, \ldots, n$, perform the following Steps (3)–(6).*

(3) *Compute the minimal polynomial* $\mu_{x_i+J}(z)$ *of* $x_i + J$ *in* $P/J$.

(4) *Calculate* $g_i(z) = \text{sqfree}(\mu_{x_i+J}(z))$.

(5) *Replace* $J$ *with* $J + \langle g_i(x_i) \rangle$.

(6) *If* $\deg(g_i(x_i)) = \dim_K(P/J)$, *return the ideal* $J$ *and stop.*

(7) *Return the ideal* $J$.

   *This is an algorithm which computes the ideal* $\text{Rad}(I)$.

*Proof* If no index $i$ is found such that the condition in Step (6) is satisfied, then the claim follows from the lemma and from [15], Corollary 3.7.16. So, let us assume that an index $i \in \{1, \ldots, n\}$ is found such that

$$\deg(g_i(x_i)) = \dim_K(P/J) = \dim_K(P/(I + \langle g_1(x_1), \ldots, g_i(x_i) \rangle))$$

By the lemma, the canonical homomorphism $K[x_i]/\langle g_i(x_i) \rangle \longrightarrow P/J$ is injective. Since the $K$-vector space dimensions agree, it is therefore an isomorphism. Hence the ring $P/J$ is reduced, which means that $J$ is a radical ideal. From the inclusions $I \subseteq J \subseteq \text{Rad}(I)$ we deduce that $J = \text{Rad}(I)$ which concludes the proof.   □

   Let us see two examples which demonstrate some advantages of this algorithm.

**Example 5.4.3** Let $K = \mathbb{Q}$, let $P = K[x_1, x_2, x_3]$, and let $I = \langle f_1, f_2, f_3 \rangle$ be the ideal generated by the polynomials $f_1 = x_1^4 - x_2 - x_3$, $f_2 = x_2^4 + x_3^4 - 2x_2^3 + 3x_2^2 - 3x_2 + 1$, and $f_3 = x_3^4 - x_2$ in $P$. We apply Algorithm 5.4.2 and follow the steps.

(1) Let $J = I$.

(3) We compute $\mu_{x_1+I}(z)$ and get $\mu_{x_1}(z) = g_1(z)^2$, where

$$g_1(z) = z^{32} - 4z^{28} + 10z^{24} - 16z^{20} + 18z^{16} - 8z^{12} + 4z^8 - 8z^4 + 4$$

(4) The squarefree part of $\mu_{x_1+I}(z)$ is $g_1(z)$.

(5) Let $J = I + \langle g_1(x_1) \rangle$.

(6) We compute $\dim_K(P/J) = 32$ and conclude that $J = \text{Rad}(I)$.

Based on this result, the reduced `DegRevLex`-Gröbner basis of $\text{Rad}(I)$ turns out to be $\{x_1^4 - x_2 - x_3, x_2^2 - x_2 + 1, x_3^4 - x_2\}$.

   Also in the following example the application of Algorithm 5.4.2 saves some work.

**Example 5.4.4** Let $K = \mathbb{Q}$, let $P = K[x_1, x_2, x_3]$, and let $I = \langle f_1, f_2, f_3, f_4 \rangle \subsetneq P$ be generated by $f_1 = x_1 x_2^2 + x_1^2 - x_1 - 1$, $f_2 = x_1^3 + x_1^2 x_3 + \frac{1}{4} x_1 x_3^2 - x_1^2 + \frac{1}{4} x_2^2 - \frac{1}{2} x_1 x_3 - \frac{1}{2} x_1 - \frac{1}{2} x_3 + \frac{1}{4}$, $f_3 = x_2^4 - 2x_2^2 x_3 + 3x_1^2 + 2x_1 x_3 + x_3^2 - 2x_1 - 2$, and $f_4 = x_3^5 - x_1 x_3 - 1$. Again we apply Algorithm 5.4.2 and follow the main steps.

(1) Let $J = I$.

(3) When we compute $\mu_{x_1+I}(z)$, we get $\mu_{x_1+I}(z) = g_1(z)^2$, where

$$g_1(z) = z^{10} - \tfrac{5}{2}z^9 + \tfrac{59}{16}z^7 - \tfrac{29}{32}z^6 - \tfrac{79}{32}z^5 + \tfrac{15}{32}z^4 + \tfrac{15}{16}z^3 - \tfrac{5}{32}z - \tfrac{1}{32}$$

is irreducible.

(5) Let $J = I + \langle g_1(x_1)\rangle$.

(6) We find $\dim_K(P/J) = 20 > \deg(g_1(x_1))$.

(3) Next we compute $\mu_{x_2+J}(z)$ and get the irreducible polynomial

$$\mu_{x_2+J}(z) = z^{20} - \tfrac{5}{2}z^{18} + \tfrac{61}{16}z^{14} - \tfrac{33}{32}z^{12} - \tfrac{19}{4}z^{10} + \tfrac{149}{32}z^8 - \tfrac{159}{32}z^6 + \tfrac{65}{16}z^4$$
$$- \tfrac{47}{32}z^2 + \tfrac{5}{32}$$

(4) Let $g_2(z) = \mu_{x_2+J}(z)$.

(5) Let $J = I + \langle g_1(x_1), g_2(x_2)\rangle$.

(6) Since we now have $\dim_K(P/J) = 20 = \deg(g_2(x_2))$, we conclude that $J$ is the radical of $I$.

Altogether, the reduced DegRevLex-Gröbner basis of $J = \mathrm{Rad}(I)$ turns out to be $\{x_1^2 + \tfrac{1}{2}x_1x_3 - \tfrac{1}{2}x_1 - \tfrac{1}{2}, x_2^2 - x_1 - x_3, x_3^5 - x_1x_3 - 1\}$.

Let us point out some observations about the preceding algorithm.

**Remark 5.4.5**  Suppose we are in the setting of Algorithm 5.4.2.

(a) Since $K$ is a perfect field, we can use [15], Propositions 3.7.9.b and 3.7.12 to perform Step (4) of the algorithm.
(b) Given linear forms $\ell_1, \ldots, \ell_n \in P_1$ which form a $K$-basis of $P_1$, we can use these linear forms instead of $x_1, \ldots, x_n$ in Steps (3)–(6) of the algorithm. In particular, if the field $K$ has sufficiently many elements, we can start with a randomly chosen linear form $\ell_1$ and hope that the algorithm stops after the first iteration.

  However, this is not a sure bet. Given the ideal $I = \langle x_1, x_2\rangle^2$ in the ring $P = K[x_1, x_2]$, no matter which linear form $\ell_1 \in P_1$ we choose, its minimal polynomial is given by $\mu_{\ell_1+I}(z) = z^2$, while $\dim(P/I) = 3$.

The following corollary exhibits a nice property of Algorithm 5.4.2.

**Corollary 5.4.6**  *Assume that we are in the setting of Algorithm 5.4.2.*

(a) *If the polynomial $\mu_{x_1+I}(z)$ in Step (3) of the algorithm is squarefree and has degree $\dim_K(P/I)$, then the algorithm stops and shows that $I$ is a radical ideal.*
(b) *If there exists an linear form $\ell \in P_1$ such that $\mu_{\ell+I}(z)$ is squarefree of degree $\dim_K(P/I)$, then $I$ is a radical ideal.*

Algorithm 5.4.2 suggests an alternative approach to the computation of the maximal components of a zero-dimensional ideal. In fact, as our next algorithms shows, they have essentially been computed if Algorithm 5.4.2 finishes before reaching Step (7).

---

**Algorithm 5.4.7 (Computing Maximal Components via the Radical)**
*Let $K$ be a perfect field, let $P = K[x_1, \ldots, x_n]$, and let $I$ be a zero-dimensional ideal in $P$. Let $i \in \{1, \ldots, n\}$ be such that Algorithm 5.4.2 stops after the $i$-th iteration of the loop in Steps (3)–(6). Consider the following sequence of instructions.*

(1) *Let $J = I + \langle g_1(x_1), \ldots, g_i(x_i) \rangle$ be the ideal returned by Algorithm 5.4.2.*
(2) *Compute the factorization $g_i(x_i) = p_1(x_i) \cdots p_t(x_i)$ with pairwise distinct irreducible factors $p_j(x_i)$ in $K[x_i]$.*
(3) *Return the ideals $\mathfrak{M}_j = I + \langle g_1(x_1), \ldots, g_{i-1}(x_{i-1}), p_j(x_i) \rangle$ for every $j = 1, \ldots, t$.*

*This is an algorithm which computes the maximal components $\mathfrak{M}_1, \ldots, \mathfrak{M}_t$ of the ideal $I$.*

---

*Proof* Since the polynomials $p_1(x_i), \ldots, p_t(x_i)$ are pairwise coprime, the ideals $\mathfrak{M}_1, \ldots, \mathfrak{M}_t$ are pairwise comaximal. By the correctness of Algorithm 5.4.2, we have $I + \langle g_1(x_1), \ldots, g_i(x_i) \rangle = \mathrm{Rad}(I)$. Hence we get $\mathrm{Rad}(I) \subseteq \mathfrak{M}_1 \cap \cdots \cap \mathfrak{M}_t$. Combining this observation with the isomorphism in the Chinese Remainder Theorem 2.2.1, we have the inequality

$$\dim_K\big(P/\mathrm{Rad}(I)\big) \geq \dim_K\big(P/(\mathfrak{M}_1 \cap \cdots \cap \mathfrak{M}_t)\big) = \sum_{j=1}^{t} \dim_K(P/\mathfrak{M}_j) \quad (5.1)$$

On the other hand, for every $j \in \{1, \ldots, t\}$, the polynomial $p_j(x_i) \in \mathfrak{M}_j$ is irreducible. This implies $p_j(z) = \mu_{x_i + \mathfrak{M}_j}(z)$, i.e., the polynomial $p_j(z)$ is the minimal polynomial of the residue class of $x_i$ in $P/\mathfrak{M}_j$. Consequently, the canonical map $\alpha_j : K[x_i]/\langle p_j(x_i) \rangle \longrightarrow P/\mathfrak{M}_j$ is injective, and therefore we have the inequality $\deg(p_j(x_i)) \leq \dim_K(P/\mathfrak{M}_j)$. By the hypothesis, we also have $\deg(g_i(x_i)) = \dim_K(P/\mathrm{Rad}(I))$. Combining these observations, we get the inequality

$$\dim_K\big(P/\mathrm{Rad}(I)\big) = \deg\big(g_i(x_i)\big) = \sum_{j=1}^{t} \deg\big(p_j(x_i)\big) \leq \sum_{j=1}^{t} \dim_K(P/\mathfrak{M}_j) \quad (5.2)$$

Putting together (5.1) and (5.2), we obtain the equality $\mathrm{Rad}(I) = \mathfrak{M}_1 \cap \cdots \cap \mathfrak{M}_t$ and the fact that the maps $\alpha_1, \ldots, \alpha_t$ are isomorphisms. This implies that the ideals $\mathfrak{M}_1, \ldots, \mathfrak{M}_t$ are maximal, and the proof is complete. $\qquad\square$

In the following example Algorithm 5.4.2 finishes in Step (6) after the second iteration, so that the hypothesis of the above algorithm is satisfied.

**Example 5.4.8** Let $K = \mathbb{Q}$, let $P = K[x_1, x_2, x_3]$, and let $I = \langle f_1, f_2, f_3 \rangle$, where we let $f_1 = x_1^2 - x_2^3$, $f_2 = x_1^4 - 2x_1^3 + x_1^2$, and $f_3 = x_1 x_2 - x_1 x_3 - x_3$. We follow the steps of the above algorithms. First we apply Algorithm 5.4.2.

(1)–(4) Let $i = 1$. We compute $\mu_{x_1+I}(z) = z^4 - 2z^3 + z^2 = z^2(z-1)^2 = g_1(z)^2$ with $g_1(z) = z(z-1)$.
(5), (6) For $J = I + \langle g_1(x_1) \rangle$ we have $\dim_K(P/J) = 6 > \deg(g_1(x_1))$.
(2)–(4) Let $i = 2$. We compute $\mu_{x_2+J}(z) = z^6 - z^3 = z^3(z-1)(z^2+z+1)$ and get $g_2(z) = z(z-1)(z^2+z+1)$.
(5), (6) For the ideal $J = I + \langle g_1(x_1), g_2(x_2) \rangle$, we have $\dim_K(P/J) = 4 = \deg(g_2(x_2))$. Hence the algorithm stops and returns $J = \mathrm{Rad}(I)$.

Now we continue with Algorithm 5.4.7. Since $g_2(x_2) = x_2(x_2 - 1)(x_2^2 + x_2 + 1)$, the algorithm returns the maximal components

$$\mathfrak{M}_1 = J + \langle x_2 \rangle = \langle x_1, x_2, x_3 \rangle$$

$$\mathfrak{M}_2 = J + \langle x_2 - 1 \rangle = \left\langle x_1 - 1, x_2 - 1, x_3 - \tfrac{1}{2} \right\rangle$$

$$\mathfrak{M}_3 = J + \left\langle x_2^2 + x_2 + 1 \right\rangle = \left\langle x_1 - 1, x_2 - 2x_3, x_2^2 + \tfrac{1}{2}x_3 + \tfrac{1}{4} \right\rangle$$

What happens if the hypothesis of the preceding algorithm is not satisfied, i.e., if Algorithm 5.4.2 finishes only after Step (7)? In this case we have to content ourselves with the fact that we have found the radical ideal of $I$. Then we can use any of the methods presented in Sects. 5.1 and 5.2 to find the primary components of $\mathrm{Rad}(I)$ which are exactly the maximal components of $I$. Our next example shows that we cannot do better in this case.

**Example 5.4.9** Let $K = \mathbb{Q}$, and let $P = K[x_1, x_2]$.

(a) For the ideal $I = \langle x_1^2 - 2, x_2^2 - 3 \rangle$, Algorithm 5.4.2 yields the irreducible polynomials $g_1(x_1) = x_1^2 - 2$ and $g_2(x_2) = x_2^2 - 3$, both of which have a degree smaller than $\dim_K(P/I) = 4$. Thus the algorithm stops in Step (7) and shows that $I$ is a radical ideal. It is not able to determine that $I$ is maximal.
(b) For the ideal $I = \langle x_1^2 - 2, x_2^2 - 8 \rangle$, Algorithm 5.4.2 again proves that it is a radical ideal, but it does not detect that $I = \mathfrak{M}_1 \cap \mathfrak{M}_2$ is the intersection of two maximal ideals $\mathfrak{M}_1 = \langle x_1^2 - 2, x_2 - 2x_1 \rangle$ and $\mathfrak{M}_2 = \langle x_1^2 - 2, x_2 + 2x_1 \rangle$.

If Algorithm 5.4.2 finishes in Step (6) with an irreducible polynomial $g_i(x_i)$, we have detected a primary ideal $I$ and found its maximal component, as the following corollary shows.

**Corollary 5.4.10 (Checking Primary and Maximal Ideals)**
*Assume that we are in the setting of Algorithm 5.4.7.*

(a) *If the polynomial $g_i(x_i)$ is irreducible then $\mathfrak{M} = I + \langle g_1(x_1), \dots, g_i(x_i) \rangle$ is a maximal ideal and $I$ is $\mathfrak{M}$-primary.*

(b) *If there exists a linear form* $\ell \in P_1$ *such that* $\mu_{\ell+I}(z)$ *is irreducible of degree* $\dim_K(P/I)$ *then* $I$ *is a maximal ideal.*

*Proof* To prove (a) we observe that Algorithm 5.4.7 shows that the ideal $\mathfrak{M}$ is maximal and that we have $\mathfrak{M} = \mathrm{Rad}(I)$. This means that $I$ is an $\mathfrak{M}$-primary ideal.

Next we show (b). If the minimal polynomial $\mu_{x_1+I}(z)$ is irreducible then the ideal $\mathfrak{M} = I + \langle \mu_{x_1+I}(x_1) \rangle = I$ is maximal by part (a). Using this observation and the second part of Remark 5.4.5.c, the conclusion follows.                              $\square$

The following two examples provide cases in which the algorithm stops for $i = 1$ in Step (6) with an irreducible polynomial $\mu_{x_1+I}(z)$, so that we can conclude that $I$ is a maximal ideal.

**Example 5.4.11**  Let $K = \mathbb{Q}$, let $P = K[x_1, x_2]$, and let $I = \langle x_1^3 + 3x_1x_2^2 - x_2^3 - x_1 - 1, x_2^5 - x_2 - 1 \rangle$. We calculate the minimal polynomial of $x_1 + I$ in $P/I$ and get

$$\mu_{x_1+I}(z) = z^{15} - 5z^{13} - 5z^{12} - 8z^{11} + 35z^{10} + 54z^9 - 21z^8 - 31z^7 - 83z^6z^5$$
$$+ 266 + 27z^4 + 3z^3 - 61z^2 - 16z - 4$$

This polynomial is irreducible and its degree is $15 = \dim_K(P/I)$. Therefore the corollary shows that $I$ is a maximal ideal.

In the foreword we mentioned a certain message in a bottle that we received from the Mare Tenebrarum. Although we had trouble dating it, the following text, found on the back of the letter, helped greatly.

> *To find the year, find the field.*
> $xxxxxx + yyyyy + yyyy + yyy + yy$
> $yyyyyy + xxxxx + yy, xy + x + 1$

**Example 5.4.12 (In the Year 2048)**
Let $K = \mathbb{F}_2$, let $P = K[x, y]$, and let $I$ be the ideal $I = \langle x^6 + y^5 + y^4 + y^3 + y^2, y^6 + x^5 + y^2, xy + x + 1 \rangle$ in $P$. It turns out that the given generators of $I$ form the reduced DegRevLex-Gröbner basis of $I$. Consequently, the residue classes of the terms in $(1, y, y^2, y^3, y^5, x, x^2, x^3, x^4, x^5)$ form a $K$-basis of $P/I$. We compute the minimal polynomial of $x + I$ and get $\mu_{x+I}(z) = z^{11} + z^2 + 1$ which is irreducible of degree 11. Therefore the corollary shows that $I$ is a maximal ideal. Hence $P/I$ is a field with 2048 elements.

The following example continues Example 5.4.4. It contains a case in which the algorithm stops for $i = 2$.

**Example 5.4.13**  In the setting of Example 5.4.4, Algorithm 5.4.2 stops for $i = 2$ in Step (6) with an irreducible polynomial $g_2(x_2) = \mu_{x_2+J}(x_2)$. Again it follows that the ideal $\mathfrak{M} = \mathrm{Rad}(I) = I + \langle g_1(x_1), g_2(x_2) \rangle$ is maximal and $I$ is $\mathfrak{M}$-primary.

### 5.4.B  Primary Components from Maximal Components

> *Undetected errors are handled as if no error occurred.*
> (IBM Manual)

In the first part of this subsection we address the following problem. Let $K$ be a field, let $P = K[x_1, \ldots, x_n]$, and let $I$ be a zero-dimensional ideal in $P$. Suppose that we know the maximal components $\mathfrak{M}_1, \ldots, \mathfrak{M}_s$ of $I$. How can we compute the corresponding primary components $\mathfrak{Q}_1, \ldots, \mathfrak{Q}_s$? In the unlikely event that our computation of the maximal components of $I$ contains an undetected error, we proceed as if no error occurred.

---

**Algorithm 5.4.14 (Primary Components via All Maximal Components)**
*In the above setting, let $\mathfrak{M}_1, \ldots, \mathfrak{M}_s$ be the maximal components of $I$, and let $i \in \{1, \ldots, s\}$. Consider the following sequence of instructions.*

(1) *Use the isomorphism $P/\operatorname{Rad}(I) \cong P/\mathfrak{M}_1 \times \cdots \times P/\mathfrak{M}_s$ and the Chinese Remainder Theorem 2.2.1 to compute a polynomial $f_i \in P$ such that $f_i \in \mathfrak{M}_i$ and $f_i + \mathfrak{M}_j = 1 + \mathfrak{M}_j$ for $j \neq i$.*
(2) *Calculate the minimal polynomial $\mu_{f_i+I}(z)$ of the residue class of $f_i$ in $P/I$ and its factorization $\mu_{f_i+I}(z) = z^{\alpha_i}(z-1)^{\beta_i}$ with $\alpha_i, \beta_i \in \mathbb{N}$.*
(3) *Return the ideal $\mathfrak{Q}_i = I + \langle f_i^{\alpha_i} \rangle$.*

*This is an algorithm which computes the $\mathfrak{M}_i$-primary component $\mathfrak{Q}_i$ of $I$.*

---

*Proof* For $j = 1, \ldots, s$, we denote the image of $\mathfrak{M}_j$ in $R = P/I$ by $\mathfrak{m}_j$. Then we apply the isomorphism $\imath : R \longrightarrow \mathscr{F}$ given by $r \mapsto \vartheta_r$ and denote the image of $\mathfrak{m}_j$ in $\mathscr{F}$ by $\tilde{\mathfrak{m}}_j$. Furthermore, we denote the image of the residue class $f_i + I \in R$ in $\mathscr{F}$ by $\varphi_i$. Then we have $\varphi_i \in \tilde{\mathfrak{m}}_i$ and $\varphi_i \notin \tilde{\mathfrak{m}}_j$ for $j \neq i$.

Now we apply Theorem 2.4.12.a. Since $\tilde{\mathfrak{m}}_i$ is the only maximal ideal of $\mathscr{F}$ containing $\varphi_i$, the eigenfactor $z$ of $\varphi_i$ equals $p_{\tilde{\mathfrak{m}}_i, \varphi_i}(z)$ and the cited theorem shows that the $\tilde{\mathfrak{m}}_i$-primary component $\tilde{\mathfrak{q}}_i$ of the zero ideal in $\mathscr{F}$ is given by $\tilde{\mathfrak{q}}_i = \langle \varphi_i^{\alpha_i} \rangle$, where $\alpha_i$ is the multiplicity of the eigenfactor $z$ in $\mu_{\varphi_i}(z)$. Hence it suffices to apply $\imath^{-1}$ and to form the preimage of $\mathfrak{q}_i = \imath^{-1}(\tilde{\mathfrak{q}}_i)$ in $P$ to get $\mathfrak{Q}_i = I + \langle f_i^{\alpha_i} \rangle$, as claimed.     $\square$

Notice that the construction of the element $f_i$ agrees with the method used in Proposition 2.5.4.a. For an application of this algorithm, we continue Example 5.4.8, where we calculated all maximal components of a zero-dimensional ideal.

**Example 5.4.15** Continuing Example 5.4.8, we let $K = \mathbb{Q}$, $P = K[x_1, x_2, x_3]$, and $I = \langle f_1, f_2, f_3 \rangle$ be the ideal in $P$ generated by $f_1 = x_1^2 - x_2^3$, $f_2 = x_1^4 - 2x_1^3 + x_2^2$, and $f_3 = x_1 x_2 - x_1 x_3 - x_3$. We have already seen that the maximal components of $I$ are $\mathfrak{M}_1 = \langle x_1, x_2, x_3 \rangle$, $\mathfrak{M}_2 = \langle x_1 - 1, x_2 - 1, x_3 - \frac{1}{2} \rangle$, and $\mathfrak{M}_3 = \langle x_1 - 1, x_2 - 2x_3, x_3^2 + \frac{1}{2}x_3 + \frac{1}{4} \rangle$. Now let us follow the steps of the preceding algorithm to compute the $\mathfrak{M}_1$-primary component of $I$.

(1) An application of the Chinese Remainder Theorem 2.2.1 yields the polynomial $f_1 = x_1$ which satisfies $f_1 \in \mathfrak{M}_1$ and the equalities $f_1 + \mathfrak{M}_j = 1 + \mathfrak{M}_j$ for $j = 2, 3$.

(2) The minimal polynomial of $f_1 + I$ in $P/I$ is $\mu_{f_1+I}(z) = z^2(z-1)^2$.

(3) Thus the $\mathfrak{M}_1$-primary component of $I$ is the ideal

$$\mathfrak{Q}_1 = I + \langle x_1^2 \rangle = \langle x_1^2, x_1x_2 - x_3, x_1x_3, x_2^3, x_2^2x_3, x_3^2 \rangle$$

Similarly, we can compute the $\mathfrak{M}_2$-primary component of $I$ using the polynomial $f_2 = -\frac{4}{3}x_3^2 - \frac{1}{3}x_1 - \frac{2}{3}x_3 + 1$ and $\mu_{f_2+I}(z) = z^2(z-1)^2$. We get

$$\mathfrak{Q}_2 = I + \langle f_2^2 \rangle = \langle x_1 - \tfrac{12}{7}x_3 - \tfrac{1}{7}, x_2 - \tfrac{8}{7}x_3 - \tfrac{3}{7}, x_3^2 - x_3 + \tfrac{1}{4} \rangle$$

Finally, when we compute the $\mathfrak{M}_3$-primary component of $I$, we get the polynomial $f_3 = \frac{4}{3}x_3^2 - \frac{2}{3}x_1 + \frac{2}{3}x_3 + 1$ and $\mu_{f_3+I}(z) = z^2(z-1)^2$, so that

$$\mathfrak{Q}_3 = I + \langle f_3^2 \rangle = \langle x_1^2 - 2x_1 + 1, x_1x_2 + x_2 - 4x_3, x_1x_3 + x_2 - 3x_3,$$

$$x_2^2 + \tfrac{4}{3}x_1 - \tfrac{1}{3}x_2 + \tfrac{8}{3}x_3 - \tfrac{1}{3}, x_2x_3 + \tfrac{11}{12}x_1 - \tfrac{2}{3}x_2 + \tfrac{7}{3}x_3 - \tfrac{5}{12},$$

$$x_3^2 + \tfrac{7}{12}x_1 - \tfrac{7}{12}x_2 + \tfrac{5}{3}x_3 - \tfrac{1}{3} \rangle$$

> *Eternity is really long,*
> *especially near the end.*
> (Woody Allen)

In the second part of this subsection we present three algorithms which answer the following question: given a single maximal component of $I$, how can we compute its corresponding primary component? This turns out to be a hard computational problem and none of these algorithms is particularly efficient. Thus you may have to wait really long for the result, especially near the end. Nevertheless, these algorithms may be useful, for instance, if we find a single linear maximal ideal containing $I$ via the methods in Sect. 6.1 and 6.2 and want to know the corresponding primary component.

---

**Algorithm 5.4.16 (Primary Components via Powers of Maximal Ideals)**
*In the above setting, let $\mathfrak{M}$ be a maximal component of $I$. Consider the following sequence of instructions.*

(1) *Let $J_0 = I$.*

(2) *For $i = 1, 2, \ldots,$ compute the ideal $J_i = I + \mathfrak{M}^i$ until we have the equality $J_i = J_{i-1}$.*

(3) *Return the ideal $\mathfrak{Q} = J_i$.*

*This is an algorithm which computes the $\mathfrak{M}$-primary component $\mathfrak{Q}$ of $I$.*

*Proof* Let $R = P/I$, and let $\mathfrak{m}$ be the image of $\mathfrak{M}$ in $R$. Then $\mathfrak{m}$ is a maximal ideal of $R$. Now let $\mathscr{F}$ be the multiplication family of $R$, and let $\tilde{\mathfrak{m}}$ be the image of $\mathfrak{m}$ under the isomorphism $\iota : R \longrightarrow \mathscr{F}$ given in Proposition 4.1.2.b. Then $\tilde{\mathfrak{m}}$ is a maximal ideal of $\mathscr{F}$. If we have $\tilde{\mathfrak{m}}^a = \tilde{\mathfrak{m}}^{a+1}$ for some $a \geq 1$ then it follows that $\tilde{\mathfrak{m}}^b = \tilde{\mathfrak{m}}^a$ for all $b \geq a$. In particular, by Proposition 2.4.6.e, it follows that $\tilde{\mathfrak{q}} = \tilde{\mathfrak{m}}^a$ is the $\tilde{\mathfrak{m}}$-primary component of the zero ideal in the family $\mathscr{F}$. Consequently, the ideal $\mathfrak{q} = \iota^{-1}(\tilde{\mathfrak{q}}) = \mathfrak{m}^a$ is the $\mathfrak{m}$-primary component of the zero ideal in $R$, and the ideal $\mathfrak{Q} = \mathfrak{M}^a + I$ is the $\mathfrak{M}$-primary component of $I$. The correctness of the algorithm follows from these observations.                                                                  □

Let us illustrate this algorithm with the following example.

**Example 5.4.17**   Let $K = \mathbb{Q}$, let $P = K[x, y, z]$, and let $I = \langle y^2 - z^2, z^3 - xz,$ $xyz^2 - 2x^3 - x^2y - 4z^2 + 4x + 2y + 2z, x^2z^2 - x^3 - 2z^2 + 2x, x^3z - yz - z^2,$ $x^3y + 2yz^2 - 2xy - yz - z^2, x^4 + 2xz^2 - yz^2 - 2x^2 - xz\rangle$. The ideal $I$ is zero-dimensional, and it is clear that the ideal $\mathfrak{M} = \langle x, y, z\rangle$ is a maximal ideal containing $I$. To compute the $\mathfrak{M}$-primary component of $I$ we use the above algorithm. It turns out that we have $\mathfrak{M} + I \subset \mathfrak{M}^2 + I$, while $\mathfrak{M}^2 + I = \mathfrak{M}^3 + I = \langle x, y^2, z\rangle$. Hence $\mathfrak{Q} = \langle x, y^2, z\rangle$ is the $\mathfrak{M}$-primary component of $I$.

Our second algorithm uses again the isomorphism $\iota : R \to \mathscr{F}$ described in Proposition 4.1.2.b. When we translate the problem to a problem for the multiplication family $\mathscr{F}$ of the ring $R = P/I$, it becomes essentially the problem of computing big kernels.

---

**Algorithm 5.4.18 (Primary Components via Big Kernels)**

*In the above setting, let $\mathfrak{M}$ be a maximal component of $I$, let $\mathfrak{m}$ be its image in $R = P/I$, let $\mathscr{F}$ be the multiplication family of $R$, let $\iota : R \longrightarrow \mathscr{F}$ be the isomorphism given by $r \mapsto \vartheta_r$, and let $\tilde{\mathfrak{m}}$ be the image of $\mathfrak{m}$ in $\mathscr{F}$. Consider the following sequence of instructions.*

*(1) Compute $\mathrm{BigKer}(\tilde{\mathfrak{m}})$.*
*(2) For $i = 1, \dots, n$, compute the restriction $\varrho_i$ of $\vartheta_{x_i}$ to $\mathrm{BigKer}(\tilde{\mathfrak{m}})$.*
*(3) Compute the ideal $\mathfrak{Q} = \mathrm{Rel}_P(\varrho_1, \dots, \varrho_n)$ and return it.*

*This is an algorithm which computes the $\mathfrak{M}$-primary component $\mathfrak{Q}$ of $I$.*

---

*Proof* Let $\mathfrak{Q}$ be the $\mathfrak{M}$-primary component of $I$, and let $\mathfrak{q}$ be its image in $R$. Then the ideal $\mathfrak{q}$ is the $\mathfrak{m}$-primary component of the zero ideal in $R$. By Proposition 4.1.12, we have $\mathfrak{q} = \mathrm{Ann}_R(\mathrm{BigKer}(\tilde{\mathfrak{m}}))$. Therefore an application of Algorithm 2.1.7 finishes the proof.                                                                  □

Let us see an example which illustrates this algorithm.

**Example 5.4.19**  Let $K = \mathbb{Q}$, let $P = K[x_1, x_2]$, and let $I = \langle g_1, g_2, g_3, g_4 \rangle$ be the ideal in $P$ generated by $g_1 = x_1^3 - 5x_1^2 + 8x_1 - 4$, $g_2 = x_1^2 x_2 + 2x_1^2 - 12x_1 - 4x_2 + 16$, $g_3 = x_1 x_2 + x_1^2 - 5x_1 - 2x_2 + 6$, and $g_4 = x_2^2 - x_1^2 + 4x_1 - 2x_2 - 3$. Our goal is to find the primary components of $I$.

To check that the set $G = \{g_1, g_2, g_3, g_4\}$ is an $\mathcal{O}$-border basis of $I$ with respect to $\mathcal{O} = \{1, x_1, x_2, x_1^2\}$, it suffices to verify that the formal multiplication matrices

$$
A_{x_1} = \begin{pmatrix} 0 & 0 & -6 & 4 \\ 1 & 0 & 5 & -8 \\ 0 & 0 & 2 & 0 \\ 0 & 1 & -1 & 5 \end{pmatrix} \quad \text{and} \quad A_{x_2} = \begin{pmatrix} 0 & -6 & 3 & -16 \\ 0 & 5 & -4 & 12 \\ 1 & 2 & 2 & 4 \\ 0 & -1 & 1 & -2 \end{pmatrix}
$$

commute. Hence the residue classes of the elements of $\mathcal{O}$ form a $K$-basis of the ring $R = P/I$. The maximal components of $I$ are $\mathfrak{M}_1 = \langle x_1 - 1, x_2 - 2 \rangle$ and $\mathfrak{M}_2 = \langle x_1 - 2, x_2 - 1 \rangle$.

Now let us follow the steps of the algorithm and compute the $\mathfrak{M}_2$-primary component $\mathfrak{Q}_2$ of $I$.

(1) The ideal $\tilde{\mathfrak{m}}_2$ is generated by $\vartheta_{x_1} - 2\,\mathrm{id}_R$ and $\vartheta_{x_2} - \mathrm{id}_R$. Its big kernel is generated by the vectors $v_1 = (1, -1, 0, 0)$, $v_2 = (2, 0, -1, 0)$, and $v_3 = (1, 0, 0, -1)$ with respect to the basis $\mathcal{O}$.

(2) By applying the $\vartheta_{x_i}$ to the vectors $v_j$, we see that the restrictions $\varrho_i$ of $\vartheta_{x_i}$ to $\mathrm{BigKer}(\tilde{\mathfrak{m}}_2)$ are given by the matrices

$$
M_C(\varrho_1) = \begin{pmatrix} -1 & 3 & -9 \\ 0 & 2 & 0 \\ 1 & -1 & 5 \end{pmatrix} \quad \text{and} \quad M_C(\varrho_2) = \begin{pmatrix} 5 & -5 & 12 \\ 1 & 0 & 3 \\ -1 & 1 & -2 \end{pmatrix}
$$

with respect to the basis $C = (v_1, v_2, v_3)$.

(3) Now we apply the Buchberger-Möller Algorithm for Matrices 2.1.4 to this pair of matrices and get $\mathfrak{Q}_2 = \langle x_1^2 - 4x_1 + 4, x_1 x_2 - x_1 - 2x_2 + 2, x_2^2 - 2x_2 + 1 \rangle$.

A similar application of the algorithm to $\mathfrak{M}_1$ yields $\mathfrak{Q}_1 = \mathfrak{M}_1$.

Following the indications given in [15], Tutorial 43, we can use the description of the primary components of the zero ideal in $R$ as annihilators given in Corollary 4.1.14 to compute the $\mathfrak{M}$-primary component of $I$ as follows.

---

**Algorithm 5.4.20 (Primary Components via Saturation)**
*In the above setting, let $\mathfrak{M}$ be a maximal component of $I$. Consider the following sequence of instructions.*

(1) *Compute the saturation $\mathfrak{N} = I :_P \mathfrak{M}^\infty$. (For instance, use one of the methods given in [15], Sect. 3.5.B.)*

> (2) *Compute the colon ideal* $\mathfrak{Q} = I :_P \mathfrak{N}$. *(For instance, use one of the methods given in* [15], *Sect. 3.2.B or Sect. 3.4.)*
> (3) *Return the ideal* $\mathfrak{Q}$.
>
> *This is an algorithm which computes the* $\mathfrak{M}$-*primary component* $\mathfrak{Q}$ *of I.*

*Proof* Let $R = P/I$, and let $\mathfrak{m}$ be the image of $\mathfrak{M}$ in $R$. Then Corollary 4.1.14.b says that the image $\mathfrak{q}$ of the $\mathfrak{M}$-primary component $\mathfrak{Q}$ of $I$ in $R$ is given by the ideal $\mathfrak{q} = \mathrm{Ann}_R(\mathrm{BigAnn}_R(\mathfrak{m}))$. The preimage of $\mathrm{BigAnn}_R(\mathfrak{m}) = \bigcup_{j \geq 1} \mathrm{Ann}_R(\mathfrak{m}^j)$ in $P$ is the ideal $\mathfrak{N} = I :_P \mathfrak{M}^\infty = \bigcup_{j \geq 1} I :_P \mathfrak{M}^j$. The preimage of the annihilator of $\mathrm{BigAnn}_R(\mathfrak{m})$ in $P$ is $I :_P \mathfrak{N}$. The claim follows from these observations. $\square$

Let us apply this algorithm in an explicit case.

**Example 5.4.21** Let $K = \mathbb{Q}$, let $P = K[x_1, x_2, x_3]$, and let $I = \langle f_1, f_2, f_3 \rangle$ be the ideal generated by $f_1 = x_2 x_3 - 2x_3^2 - x_1 - 4x_3$, $f_2 = x_3^3 - x_2 + 2x_3 + 1$, and $f_3 = x_1^2 - 2x_1 x_2 + 34x_2^2 - 56x_1 x_3 + 1369x_3^2 + 256x_1 - 794x_2 - 448x_3 + 2091$.

Let us check that $\mathfrak{M} = \langle x_1 + 2, x_2 - 4, x_3 - 1 \rangle$ is a maximal component of $I$. We compute the minimal polynomial of the residue class $x_1 + I$ in $R = P/I$ and get $\mu_{x_1 + I}(z) = (z + 2)^2 (z^2 - 242z + 14740)^3$. Then we add $\langle x_1 + 2 \rangle$ to $I$ and get $\mathfrak{M}$.

Now we want to compute the $\mathfrak{M}$-primary component of $I$ using the preceding algorithm. First we calculate $\mathfrak{N} = I :_P \mathfrak{M}^\infty$ and obtain $\mathfrak{N} = \langle x_2 x_3 - 2x_3^2 - x_1 - 4x_3, x_2^2 + 359x_3^2 + 29x_1 - 2x_2 + 87x_3 + 1332, x_1 x_2 + 31x_1 x_3 + 99x_3^2 + 2x_1 + 363x_2 + 614x_3 - 363, x_1^2 + 6x_1 x_3 - 10639x_3^2 - 726x_1 - 2178x_3 - 43923, x_3^3 - x_2 + 2x_3 + 1, x_1 x_3^2 + 363x_3^2 + 33x_1 + 3x_2 + 93x_3 + 1328 \rangle$. Then we calculate $I :_P \mathfrak{N}$ and get the $\mathfrak{M}$-primary component $\mathfrak{Q} = \langle x_2 - 5x_3 + 1, x_1 - x_3 + 3, (x_3 - 1)^2 \rangle$ of $I$.

Analogously, we add $\langle x_1^2 - 242x_1 + 14740 \rangle$ to $I$ and get the maximal ideal $\mathfrak{M}' = \langle x_1 + 3x_3 - 121, x_2 - 9x_3 - 1, x_3^2 + 11 \rangle$ of $I$. Proceeding as before, we calculate its $\mathfrak{M}'$-primary component $\mathfrak{Q}' = \langle x_1 - x_3^4 + 3x_3, x_2 - x_3^3 - 2x_3 - 1, (x_3^2 + 11)^3 \rangle$.

Finally, we verify that $I = \mathfrak{Q} \cap \mathfrak{Q}'$. Hence this is the primary decomposition of $I$. Moreover, we have $\mathrm{Rad}(I) = \mathfrak{M} \cap \mathfrak{M}'$.

## 5.5 The Separable Subalgebra

> *Don't worry about the world coming to an end today.*
> *It is already tomorrow in Australia.*
>
> (Charles Schultz)

Likewise, do not worry about the book coming to an end with this section. It is already followed by another chapter. For the time being, we are trying to complement the material of this chapter in a natural way. Courage is the complement of fear, humor is the complement of anger, but what is the complement of the nilradical of a zero-dimensional affine algebra $R$ over a perfect field $K$? As we shall see, it is the separable subalgebra, i.e., the subalgebra consisting of all separable elements. In this case an element of $R$ is neither more nor less than the sum of its separable and nilpotent parts. It is exactly equal to that sum, and Algorithm 5.5.9 shows you how to find the summands.

Most of the material presented in the following, in particular Proposition 5.5.6 and Algorithm 5.5.9, is inspired by the paper [19] which is a pretty young addition to a seemingly mature subject.

> *"Were you funny as a child?"*
> *"Well, no, I was an accountant."*

In this section we always work over a perfect field $K$. Recall that this implies that a polynomial $f \in K[x]$ is squarefree if and only if $\gcd(f, f') = 1$ (see [15], Proposition 3.7.9.d). Furthermore, let $P = K[x_1, \ldots, x_n]$, let $I$ be a zero-dimensional ideal in $P$, and let $R$ be the zero-dimensional affine $K$-algebra $R = P/I$.

**Definition 5.5.1**  In the above setting, an element $a \in R$ is said to be **separable** if its minimal polynomial is squarefree. Equivalently, the element $a$ is separable if $\gcd(\mu_a(z), \mu'_a(z)) = 1$.

Next we describe a consequence of Seidenberg's Lemma (cf. [15], Proposition 3.7.15).

**Proposition 5.5.2**  *In the above setting, let $\bar{x}_i$ denote the residue class of $x_i$ in $R$. The following conditions are equivalent.*

(a) *The ring $R$ is reduced, i.e., the ideal $I$ is a radical ideal.*
(b) *The elements $\bar{x}_1, \ldots, \bar{x}_n$ are separable.*
(c) *All elements in $R$ are separable.*

*Proof*  The equivalence between (a) and (b) follows from Seidenberg's Lemma and its corollary (see [15], Proposition 3.7.15 and Corollary 3.7.16). Since the implication (c) $\Rightarrow$ (b) is obviously true, it remains to prove (a) $\Rightarrow$ (c). Let $a \in R$, let $\mu_a(z)$ be its minimal polynomial, and let $g(z) = \mathrm{sqfree}(\mu_a(z))$. From the relation $\mu_a(a) \in I$ it follows that $g(a) \in \mathrm{Rad}(I) = I$. Consequently, we have $g(z) = \mu_a(z)$, and the fact that $\mu_a(z)$ is squarefree yields the claim.                     $\square$

Based on this characterization, we can describe the set of separable elements of $R$ as follows.

**Proposition 5.5.3**  *Let $K$ be a perfect field and let $R$ be a zero-dimensional affine $K$-algebra.*

(a) *An element $a \in R$ is separable if and only if there exists a reduced $K$-subalgebra $S$ of $R$ such that $a \in S$.*
(b) *The set $S$ of all separable elements of $R$ is a $K$-subalgebra of $R$.*

*Proof* To prove the implication "$\Rightarrow$" in (a) we observe that $a \in K[a]$, and if $a$ is separable then $K[a] \cong K[z]/\langle \mu_a(z) \rangle$ is reduced. The other implication follows from the observation that the injectivity of the embedding $S \hookrightarrow R$ implies that the minimal polynomials of $a$ viewed as an element of $S$ and as an element of $R$ coincide (see Proposition 5.1.4). Thus it suffices to apply the implication (a) $\Rightarrow$ (c) of Proposition 5.5.2 to the ring $S$.

To prove (b), it is enough to show for $a, b \in S$ that $K[a, b] \subseteq S$. The elements $a$ and $b$ are separable in $R$. Their minimal polynomials over $R$ and over $K[a, b]$ are identical. Hence they are separable elements in $K[a, b]$, and an application of Proposition 5.5.2 finishes the proof.                                                             $\square$

The following example shows that this proposition fails if $K$ is not a perfect field.

**Example 5.5.4**   Let $a$ be an indeterminate, and let $K = \mathbb{F}_2(a)$. In the polynomial ring $P = K[x, y]$ we consider the ideal $I = \langle x^2 + a, y^2 + a \rangle$. The $K$-algebra $R = P/I$ is generated by $\{\bar{x}, \bar{y}\}$. Since the polynomial $\mu_{\bar{x}}(z) = \mu_{\bar{y}}(z) = z^2 + a$ is squarefree, the elements $\bar{x}$ and $\bar{y}$ are separable elements of $R$. However, for the element $r = \bar{x} + \bar{y}$, we get $\mu_r(z) = z^2$, since $(\bar{x} + \bar{y})^2 = \bar{x}^2 + \bar{y}^2 = 2a = 0$. Consequently, the element $r \in R$ is not separable.

Proposition 5.5.3 motivates the following definition.

**Definition 5.5.5**   Let $K$ be a perfect field, and let $R$ be a zero-dimensional affine algebra over $K$. The $K$-subalgebra of $R$ which consists of all separable elements is called the **separable subalgebra** of $R$ and is denoted by $R^{\mathrm{sep}}$.

In view of this definition, the equivalence of conditions (a) and (c) in Proposition 5.5.2 can be rephrased by saying that a zero-dimensional affine $K$-algebra $R$ is reduced if and only if $R = R^{\mathrm{sep}}$. Our next goal is to show that, if $R$ is not reduced, the separable subalgebra is a complement of the nilradical of $R$. For this purpose we decompose every element of $R$ as follows.

**Proposition 5.5.6**   *Let $K$ be a perfect field, let $R$ be a zero-dimensional affine algebra over $K$, let $a \in R$, and let $f(z) = \mathrm{sqfree}(\mu_a(z))$.*

(a) *The element $f'(a) \in R$ is invertible.*
(b) *There exist elements $b, r \in R$ such that we have $a = b + r$ with $r \in \mathrm{Rad}(0)$ and $\mu_b(z) = f(z)$. In particular, we have $b \in R^{\mathrm{sep}}$.*

*Proof* First we prove claim (a). By the definition of $f(z)$, there exists a number $s > 0$ such that $f(a)^s = 0$. Since $K$ is perfect and $f(z)$ is squarefree, we have $\gcd(f(z), f'(z)) = 1$. Therefore there exist polynomials $g(z), h(z) \in K[z]$ such that $g(z) \cdot f(z)^s + h(z) \cdot f'(z) = 1$. Consequently, we get $h(a) \cdot f'(a) = 1$, and the claim follows.

To prove (b), we let $J = \langle f(a) \rangle$. We can obviously write $a = b_0 + r_0$ with $b_0 = a$ and $r_0 = 0$. Notice that we have $f(b_0) \in J^{2^0}$. In the following we want to improve this decomposition inductively, such that after performing $i$ steps we have a decomposition $a = b_i + r_i$ with $r_i \in \mathrm{Rad}(0)$ and $f(b_i) \in J^{2^i}$.

By induction, we may assume that we have a decomposition $a = b_i + r_i$ with these properties. Since we have the equality $b_i = a - r_i$ and $r_i \in \mathrm{Rad}(0)$, we see that $f'(b_i) \in f'(a) + \mathrm{Rad}(0)$, and part (a) implies that $f'(b_i)$ is a unit of $R$. In the ring $K[x, y]$, it is easy to check that $f(x + y) - f(x) - f'(x)y \in \langle y^2 \rangle$. By substituting $x$ with $b_i$ and $y$ with $-\frac{f(b_i)}{f'(b_i)}$, we get

$$f\left(b_i - \frac{f(b_i)}{f'(b_i)}\right) - f(b_i) + f'(b_i)\frac{f(b_i)}{f'(b_i)} = f\left(b_i - \frac{f(b_i)}{f'(b_i)}\right) \in \langle f(b_i)^2 \rangle \subseteq J^{2^{i+1}}$$

Thus, if we let $b_{i+1} = (b_i - \frac{f(b_i)}{f'(b_i)})$ and $r_{i+1} = (r + \frac{f(b_i)}{f'(b_i)})$, we have a decomposition $a = b_{i+1} + r_{i+1}$ with $f(b_{i+1}) \in J^{2^{i+1}}$ and $r_{i+1} \in \mathrm{Rad}(0)$.

When $i$ is sufficiently large, we have $f(b_i) \in J^{2^i} = \langle 0 \rangle$, and hence $b_i \in R^{\mathrm{sep}}$. Then the decomposition $a = b_i + r_i$ satisfies the claim.                     □

The decomposition in part (b) of this proposition is uniquely determined, as the following corollary shows.

**Corollary 5.5.7**  *In the setting of the proposition, every element $a \in R$ has a unique decomposition $a = b + r$ such that $b \in R^{\mathrm{sep}}$ and $r \in \mathrm{Rad}(0)$. In particular, we have a decomposition $R = R^{\mathrm{sep}} \oplus \mathrm{Rad}(0)$ into a direct sum of $K$-vector subspaces.*

*Proof*  By part (b) of the proposition, we have $R = R^{\mathrm{sep}} + \mathrm{Rad}(0)$. Now we choose an element $f \in R^{\mathrm{sep}} \cap \mathrm{Rad}(0)$. Since $f$ is nilpotent, its minimal polynomial is of the form $\mu_f(z) = z^m$ with $m \geq 1$. Since $f$ is separable, its minimal polynomial is squarefree. Hence we get $\mu_f(z) = z$, and therefore $f = 0$.                     □

This corollary suggests the following terminology.

**Definition 5.5.8**  In the setting of the proposition, we write $a = b + r$ with uniquely determined elements $b \in R^{\mathrm{sep}}$ and $r \in \mathrm{Rad}(0)$. Then $b$ is called the **separable part** of $a$ and is denoted by $a^{\mathrm{sep}}$. The element $r$ is called the **nilpotent part** of $a$ and is denoted by $a^{\mathrm{nil}}$.

As an immediate consequence of the proof of part (b) of this proposition, we have the following algorithm to compute a decomposition of an element $a \in R$ into its separable and its nilpotent part.

**Algorithm 5.5.9 (Computing the Separable and the Nilpotent Part)**
*Let $K$ be a perfect field, let $R$ be a zero-dimensional affine $K$-algebra, and let $a \in R$. Consider the following sequence of instructions.*

(1)  *Compute the minimal polynomial $\mu_a(z)$ of $a$ and $f(z) = \mathrm{sqfree}(\mu_a(z))$.*

(2) *Let $i = 0$, $b_0 = a$, and $r_0 = 0$.*
(3) *Increase $i$ by one, let $b_i = b_{i-1} - \frac{f(b_{i-1})}{f'(b_{i-1})}$, and let $r_i = r_{i-1} + \frac{f(b_{i-1})}{f'(b_{i-1})}$.*
(4) *Repeat Step (3) until $f(b_i) = 0$.*
(5) *Return the pair $(b_i, r_i)$.*

*This is an algorithm which computes a pair $(b, r)$ such that $b$ is the separable part and $r$ is the nilpotent part of $a$.*

In the following example we apply this algorithm. It shows that even in a relatively simple setting, it may yield rather intricate and non-obvious results.

**Example 5.5.10** Let $K = \mathbb{Q}$, let $P = K[x]$, let $I = \langle (x^2 - 2)^4 \rangle$, and let $R = P/I$. We apply Algorithm 5.5.9 to the element $a = \bar{x} \in R$ and follow the steps.

(1) We have $\mu_a(z) = (z^2 - 2)^4$ and $f(z) = z^2 - 2$.
(2) Let $b_0 = \bar{x}$ and $r_0 = 0$.
(3) We get $b_1 = -\frac{1}{16}\bar{x}^7 + \frac{1}{2}\bar{x}^5 - \frac{3}{2}\bar{x}^3 + \frac{5}{2}\bar{x}$ and $r_1 = \frac{1}{16}\bar{x}^7 - \frac{1}{2}\bar{x}^5 + \frac{3}{2}\bar{x}^3 - \frac{3}{2}\bar{x}$.
(3) We get $b_2 = -\frac{5}{128}\bar{x}^7 + \frac{21}{64}\bar{x}^5 - \frac{35}{32}\bar{x}^3 + \frac{35}{16}\bar{x}$ and $r_2 = \frac{5}{128}\bar{x}^7 - \frac{21}{64}\bar{x}^5 + \frac{35}{32}\bar{x}^3 - \frac{19}{16}\bar{x}$.
(4) Since $f(b_2) = 0$, the algorithm stops and returns the pair $(b_2, r_2)$.

Altogether, we have $a = b_2 + r_2$, the separable part of $a$ is $b_2$, and the nilpotent part of $a$ is $r_2$. Indeed, since $\mu_{b_2}(z) = z^2 - 2$ is squarefree, it follows that $b_2$ is separable, and $\mu_{r_2}(z) = z^4$ implies that $r_2$ is nilpotent.

The next proposition tells us how to compute the separable subalgebra of $R$.

**Proposition 5.5.11** *Let $K$ be a perfect field, let $P = K[x_1, \ldots, x_n]$, let $I$ be a zero-dimensional ideal in $P$, and let $R = P/I$.*

(a) *Given elements $a_1, \ldots, a_m \in R$ and a polynomial $f \in K[y_1, \ldots, y_m]$, we have*
$$f(a_1, \ldots, a_m)^{\mathrm{sep}} = f(a_1^{\mathrm{sep}}, \ldots, a_m^{\mathrm{sep}}).$$
(b) *We have $R^{\mathrm{sep}} = K[\bar{x}_1^{\mathrm{sep}}, \ldots, \bar{x}_n^{\mathrm{sep}}]$.*

*Proof* Since claim (b) follows immediately from (a), it suffices to prove (a). By Corollary 5.5.7, we have $f(a_1, \ldots, a_m) = f(a_1^{\mathrm{sep}} + a_1^{\mathrm{nil}}, \ldots, a_m^{\mathrm{sep}} + a_m^{\mathrm{nil}})$. When we expand this expression and use the fact that $\mathrm{Rad}(0)$ is an ideal, we see that $f(a_1, \ldots, a_m) = f(a_1^{\mathrm{sep}}, \ldots, a_m^{\mathrm{sep}}) + g$ with $g \in \mathrm{Rad}(0)$. Therefore the claim follows from the uniqueness of the decomposition in Corollary 5.5.7. $\qquad\square$

For instance, we can now compute the separable subalgebra of the following zero-dimensional affine algebra.

**Example 5.5.12** Let $K = \mathbb{F}_7$, let $P = K[x, y]$, let $I = \langle (x^2 + y^3)^2, x^2 - y^2 + 1 \rangle$, and let $R = P/I$. Using Algorithm 5.5.9, we compute the two decompositions

$\bar{x} = \bar{b}_1 + \bar{r}_1$ and $\bar{y} = \bar{b}_2 + \bar{r}_2$. We get

$$b_1 = -2xy^4 + 2xy^3 - 3xy^2 + 2xy - 3x \quad r_1 = 2xy^4 - 2xy^3 + 3xy^2 - 2xy - 3x$$
$$b_2 = y^5 + y^4 + 2y^3 + y^2 + y - 2 \qquad\qquad r_2 = -y^5 - y^4 - 2y^3 - y^2 + 2$$

Then $R^{\mathrm{sep}} = K[-2\bar{x}\bar{y}^4 + 2\bar{x}\bar{y}^3 - 3\bar{x}\bar{y}^2 + 2\bar{x}\bar{y} - 3\bar{x}, \bar{y}^5 + \bar{y}^4 + 2\bar{y}^3 + \bar{y}^2 + \bar{y} - 2]$.

The following corollary allows us to compute a $K$-basis of the separable subalgebra of $R$.

**Corollary 5.5.13** *Let $K$ be a perfect field, and let $R$ be a zero-dimensional affine $K$-algebra.*

(a) *The map $\sigma : R \longrightarrow R^{\mathrm{sep}}$ defined by $a \mapsto a^{\mathrm{sep}}$ is a surjective $K$-algebra homomorphism whose kernel is $\mathrm{Rad}(0)$.*

(b) *The map $\sigma$ induces a canonical $K$-algebra isomorphism $R/\mathrm{Rad}(0) \cong R^{\mathrm{sep}}$.*

(c) *Let $b_1, \dots, b_m \in R$ be elements whose residue classes form a $K$-basis of $R/\mathrm{Rad}(0)$. Then $(b_1^{\mathrm{sep}}, \dots, b_m^{\mathrm{sep}})$ is a $K$-basis of $R^{\mathrm{sep}}$.*

*Proof* Since claim (b) follows immediately from (a) and (c) follows from (b), it suffices to prove (a). The map $\sigma$ is a $K$-algebra homomorphism by part (a) of the proposition. The fact that $\mathrm{Rad}(0)$ is the kernel of $\sigma$ follows from Corollary 5.5.7. $\square$

Let us compute a $K$-basis of $R^{\mathrm{sep}}$ in the following case.

**Example 5.5.14** Let $K = \mathbb{Q}$, let $P = K[x, y]$, let $I = \langle (x^2 + y)^2, x^2 - y^2 + 1 \rangle$, and let $R = P/I$. In order to compute a $K$-basis of $R^{\mathrm{sep}}$, we start by applying Algorithm 5.9.9 to find the decompositions of $\bar{x}$ and $\bar{y}$ into their separable and nilpotent parts. We get

$$\bar{x}^{\mathrm{sep}} = \tfrac{2}{5}\bar{x}\bar{y}^3 + \tfrac{3}{5}\bar{x}\bar{y}^2 - \tfrac{1}{5}\bar{x}\bar{y} + \tfrac{4}{5}\bar{x} \qquad \bar{x}^{\mathrm{nil}} = -\tfrac{2}{5}\bar{x}\bar{y}^3 - \tfrac{3}{5}\bar{x}\bar{y}^2 + \tfrac{1}{5}\bar{x}\bar{y} + \tfrac{1}{5}\bar{x}$$
$$\bar{y}^{\mathrm{sep}} = -\tfrac{2}{5}\bar{y}^3 - \tfrac{3}{5}\bar{y}^2 + \tfrac{6}{5}\bar{y} + \tfrac{1}{5} \qquad \bar{y}^{\mathrm{nil}} = \tfrac{2}{5}\bar{y}^3 + \tfrac{3}{5}\bar{y}^2 - \tfrac{1}{5}\bar{y} - \tfrac{1}{5}$$

Next we compute $\mathrm{sqfree}(\mu_{\bar{x}}(z)) = z^4 - z^2 - 1$ and $\mathrm{sqfree}(\mu_{\bar{y}}(z)) = z^2 + z - 1$. We get $\mathrm{Rad}(I) = I + \langle x^4 - x^2 - 1, y^2 + y - 1 \rangle = \langle x^2 + y, y^2 + y - 1 \rangle$.

This implies that the nilradical of $R$ is $\mathrm{Rad}(0) = \langle \bar{x}^2 + \bar{y}, \bar{y}^2 + \bar{y} - 1 \rangle$. Since a $K$-basis of $R/\mathrm{Rad}(0)$ is given by the residue classes of the elements in the tuple $(1, x, y, xy)$, the corollary shows that $(1, \bar{x}^{\mathrm{sep}}, \bar{y}^{\mathrm{sep}}, \bar{x}^{\mathrm{sep}}\bar{y}^{\mathrm{sep}})$ is a $K$-basis of $R^{\mathrm{sep}}$.

*Life has so many chapters.*
*One funny chapter doesn't mean*
*it's the end of the book.*

# Chapter 6
# Solving Zero-Dimensional Polynomial Systems

> *When working towards the solution of a polynomial system,*
> *it helps if you know the answer.*

In this final chapter we converge towards the answer to the problem which started our work on this book: solving polynomial systems. Before attempting to solve it, let us discuss what the problem really is.

Given a field $K$, let $P = K[x_1, \ldots, x_n]$ be a polynomial ring over $K$, and let $f_1, \ldots, f_m \in P$. We consider the polynomial system

$$\begin{cases} f_1(x_1, \ldots, x_n) = 0 \\ \quad\vdots \\ f_m(x_1, \ldots, x_n) = 0 \end{cases}$$

What does it mean to solve this system?

The most natural answer is to ask for all tuples $(a_1, \ldots, a_n)$ in $K^n$ such that $f_i(a_1, \ldots, a_n) = 0$ for $i = 1, \ldots, m$. Unfortunately, it is known that, with complete generality, this task is algorithmically unsolvable, as Y. Matiyasevich's negative solution of Hilbert's Tenth Problem shows (see [20]).

Should we give up at this point? If at first we don't succeed, then skydiving may not be for us. But in Computational Commutative Algebra, we try, try, try again. Our first observation is that we are really looking for the zeros of the ideal $I = \langle f_1, \ldots, f_m \rangle$, because every common zero of $f_1, \ldots, f_m$ is a common zero of all polynomials in $I$.

Now recall from [15], Proposition 3.7.1 that the ideal $I$ has finitely many zeros with coordinates in the algebraic closure of $K$ if and only if $I$ is a zero-dimensional

**Electronic supplementary material** The online version of this chapter (DOI:10.1007/978-3-319-43601-2_6) contains supplementary material, which is available to authorized users.

M. Kreuzer and L. Robbiano, *Computational Linear and Commutative Algebra*,
DOI 10.1007/978-3-319-43601-2_6

ideal. Lo and behold, under this natural hypothesis the linear algebra methods developed in the first chapters apply and yield nice and efficient algorithms for computing the $K$-rational solutions of the system. This will be the topic of the first two sections of this chapter. Let us just mention the key idea here: a $K$-rational solution corresponds to a linear maximal ideal containing $I$, which, in turn, corresponds to a 1-dimensional joint eigenspace of the multiplication family.

Given the hypothesis that the ideal $I$ is zero-dimensional, we can also be more ambitious and ask for an extension field $L$ of $K$ such that all solutions of $I$ in $\overline{K}^n$ are actually in $L^n$. For this task it will turn out that the algorithms developed in Chap. 5 are just what we need. Given an extension field $L$ of $K$, a maximal component $\mathfrak{M}$ of $IL[x_1, \ldots, x_n]$ corresponds to a subset of the zeros of $I$ which we may view as a set of "conjugate points" over $L$. So, in some sense, the process of enlarging the base field until the extension ideal $IL[x_1, \ldots, x_n]$ splits completely into linear maximal components is the process of solving the system.

This is the theory, where we know everything, but nothing works. Practice is when everything works, but no one knows why. To combine theory and practice (i.e., nothing works and no one knows why), we split the problem into several cases. Depending on the base field $K$, we have very different tools available.

In Sect. 6.3 we study polynomial systems defined over finite fields. This is the world which is ruled by the *power of Frobenius*. Given an extension field $L$ of a field $K$ with $q$ elements, the $q$-Frobenius endomorphism $\phi_q$ of $L$ is the map defined by $x \mapsto x^q$. This map permutes the zeros of the ideal $I$ in $L^n$. If $I$ is maximal and $L^n$ contains one zero of $I$, repeated application of $\phi_q$ generates all zeros of $I$ from this single one. But how do we get a single solution? By cloning! For the details of this nice technique and various ways to apply it, we suggest that you look at the introduction of Sect. 6.3.

At this point, since we believe in free will, we have no choice but to mention Galois theory. Given a field extension $K \subseteq L$, the Galois group $\text{Gal}(L/K)$ is the group of all $K$-automorphims of $L$, i.e., the group of all automorphisms of $L$ which keep $K$ pointwise fixed. It is easy to see that the elements of $\text{Gal}(L/K)$ permute the solutions in $L^n$ of a polynomial system defined over $K$. The way we harness the power of Frobenius for extensions of finite fields can be motivated by the fact that in this case the Galois group is cyclic and the $q$-Frobenius endomorphism generates it. To keep this book as self-contained as possible, we do not require you to know Galois theory in order to understand the material presented here. But if you do, our occasional remarks in this direction will hopefully make some sense to you.

In the last section we tackle a very difficult problem, namely that of solving polynomial systems defined over the rational numbers in an exact way. Here the solutions have to be found in a number field, i.e., a finite extension $L$ of $\mathbb{Q}$. Even the case of one polynomial equation in one indeterminate is a classical, hard task. Except in some small or special cases, the Galois group, the number of solutions of the system, the defining equations of the number fields where they live, etc., will all be very large. Although we present a correct and complete solution to the task, it will tend to produce big outputs even on small and innocent-looking inputs.

Just look at our Burj al Arab Example 6.4.13, and you see what we mean. Unsurprisingly, attempting to solve big polynomial systems symbolically is, in general, a failure.

*Who is General Failure?*
*And why is he reading drive C?*

The topic of solving polynomial systems has a long history and has attracted attention from researchers working in a variety of branches of mathematics. For instance, for solving systems defined over $\mathbb{Q}$, there are highly developed methods to find approximate solutions. These approaches are not covered in this book, and we advise the interested reader to consult pertinent specialized books, for instance [25, 27] and [28]. To end these remarks on a positive note, we offer a final piece of advice.

*If Plan A fails, you still have 25 letters left.*

## 6.1 Rational Zeros via Commuting Families

*I don't have any solution,*
*but I certainly admire the problem.*
(Ashleigh Brilliant)

As explained above, the difficulty of solving a polynomial system depends a lot on the type of answer that we expect. The most natural requirement is to find all solutions defined over the given field $K$, i.e., the $K$-rational solutions. In this section and the next, we shall not be content with admiring this problem, but we shall solve it using the methods developed in the previous chapters.

Let $K$ be a field, let $P = K[x_1, \ldots, x_n]$, and let $f_1, \ldots, f_m \in P$ be polynomials defining a system

$$\begin{cases} f_1(x_1, \ldots, x_n) = 0 \\ \quad \vdots \\ f_m(x_1, \ldots, x_n) = 0 \end{cases}$$

We assume that the ideal $I = \langle f_1, \ldots, f_m \rangle$ is zero-dimensional, i.e., that the system has only finitely many solutions in $\overline{K}^n$. Our goal is to find those solutions which are actually contained in $K^n$.

In the present section we pursue this goal by using commuting families. More specifically, we form the multiplication family $\mathscr{F}$ of the zero-dimensional affine $K$-algebra $R = P/I$ and ask what a $K$-rational solution of the above system means for $\mathscr{F}$. The answer is straightforward: a $K$-rational solution $(a_1, \ldots, a_n)$ corresponds to a linear maximal ideal $\langle \bar{x}_1 - a_1, \ldots, \bar{x}_n - a_n \rangle$ of the ring $R$, and, via the isomorphism $\iota : R \longrightarrow \mathscr{F}$ defined by $f \mapsto \vartheta_f$, to a linear maximal ideal $\langle \varphi_1 - a_1, \ldots, \varphi_n - a_n \rangle$ of $\mathscr{F}$, where $\varphi_i = \vartheta_{\bar{x}_i}$.

Thus we have transformed the problem of finding all $K$-rational solutions of a polynomial system to the problem of computing the linear maximal ideals of a commuting family. This is the problem that we shall not only admire, but solve to the best of our abilities.

## 6.1.A  Computing One-Dimensional Joint Eigenspaces

*Always try to do things in chronological order;*
*it's less confusing that way.*

In the following we let $K$ be a field, let $V$ be a finite dimensional $K$-vector space, let $\mathscr{F}$ be a commuting family of endomorphisms of $V$, let $\Phi = (\varphi_1, \ldots, \varphi_n)$ be a system of $K$-algebra generators of $\mathscr{F}$, and let $\mathfrak{m}_1, \ldots, \mathfrak{m}_s$ be the maximal ideals of $\mathscr{F}$. Every joint eigenspace of the family $\mathscr{F}$ is of the form $\mathrm{Ker}(\mathfrak{m}_i)$ for some $i \in \{1, \ldots, s\}$. By Proposition 2.2.13.c, every 1-dimensional joint eigenspace corresponds to a linear maximal ideal $\mathfrak{m}_i = \langle \varphi_1 - \lambda_{i1}, \ldots, \varphi_n - \lambda_{in} \rangle$ with $\lambda_{i1}, \ldots, \lambda_{in} \in K$.

Unfortunately, not every linear maximal ideal of $\mathscr{F}$ comes from a 1-dimensional joint eigenspace, as we shall see later. But let us not jump ahead. Let's do things in chronological order and develop an algorithm for computing the 1-dimensional joint eigenspaces of $\mathscr{F}$. It is based on linear algebra techniques only and avoids the factorization of polynomials over extension fields of $K$ which was necessary in the algorithms of Sect. 5.3. It generalizes the approach described in [22], Sect. 7.

---

**Algorithm 6.1.1 (Computing 1-Dimensional Joint Eigenspaces)**
*In the above setting, consider the following sequence of instructions.*

(1) *Let $S = \emptyset$ and $T = (B)$ be the tuple whose entry is a basis $B$ of $V$.*

(2) *For $i = 1, \ldots, n$, perform the following steps. Then return $S$ and stop.*

(3) *Let $(B_1, \ldots, B_m)$ be the current tuple $T$. For $j = 1, \ldots, m$, remove $B_j$ from $T$ and perform the following steps.*

(4) *Compute a matrix representing the restriction $\varphi_{ij}$ of $\varphi_i$ to the vector subspace $\langle B_j \rangle$ of $V$.*

(5) *Compute the linear eigenfactors $p_{ij1}(z), \ldots, p_{ijk}(z)$ of $\varphi_{ij}$ and $K$-bases $B_1', \ldots, B_k'$ of the corresponding eigenspaces $\mathrm{Eig}(\varphi_{ij}, p_{ij\kappa}(z)) = \langle B_\kappa' \rangle$ of the map $\varphi_{ij}$.*

(6) *Append the 1-element tuples $B_\kappa'$ to $S$. Append the tuples in $\{B_1', \ldots, B_k'\}$ which have more than one entry to the tuple $T$.*

> This is an algorithm which computes a tuple $S$. For every 1-dimensional joint eigenspace of $\mathscr{F}$, there is an entry of $S$ which is a $K$-basis of it, and the correspondence is one-to-one.

*Proof* First of all, let us check that the steps of this algorithm can be executed. The critical point here is Step (4) where we need that the restriction of $\varphi_i$ to $\langle B_j \rangle$ is an endomorphism of this vector subspace of $V$. This follows from Proposition 1.1.6.a, since $\varphi_i$ commutes with $\varphi_1, \ldots, \varphi_{i-1}$.

It remains to prove correctness. Observe that in Step (5) only linear eigenfactors have to be detected, because a 1-dimensional joint eigenspace is by definition the kernel of a maximal ideal $\mathfrak{m}$ such that $\dim_K(\mathrm{Ker}(\mathfrak{m})) = 1$. This condition and Proposition 2.2.13.c imply that $\mathfrak{m}$ is linear, and therefore the corresponding eigenfactors of the maps $\varphi_{ij}$ are linear by Proposition 2.3.8.

Now we show by induction on $i$ that the union of $S$ and $T$ contains basis tuples of the joint eigenspaces of $\varphi_1, \ldots, \varphi_i$ corresponding to linear eigenfactors. In the case $i = 1$, the claim follows immediately from Step (5). For $i > 1$, the vectors in $\langle B'_\kappa \rangle$ are eigenvectors of $\varphi_1, \ldots, \varphi_{i-1}$ by the induction hypothesis, and eigenvectors of $\varphi_i$ by Step (5).

Conversely, let $W$ be a joint eigenspace corresponding to linear eigenfactors of $\varphi_1, \ldots, \varphi_i$. By the induction hypothesis, there exists an index $j \in \{1, \ldots, m\}$ such that $W \subseteq \langle B_j \rangle$. Since $W$ is contained in an eigenspace of $\varphi_i$, it is contained in one of the eigenspaces $\langle B'_\kappa \rangle$ of $\varphi_{ij}$. Hence the vector subspaces $\langle B'_\kappa \rangle$ with $\kappa \in \{1, \ldots, k\}$ are exactly the joint eigenspaces of $\varphi_1, \ldots, \varphi_i$ contained in $\langle B_j \rangle$ and corresponding to linear eigenfactors. Now, if one of these eigenspaces is 1-dimensional, it is automatically a joint eigenspace of $\varphi_1, \ldots, \varphi_n$ since it is an invariant space for every element of $\mathscr{F}$ by Proposition 1.1.6.a. Therefore the induction hypothesis shows that the algorithm computes bases of the joint eigenspaces of $\varphi_1, \ldots, \varphi_n$ corresponding to linear eigenfactors. $\qquad\square$

The following remark provides some ways to pass between 1-dimensional joint eigenspaces, their corresponding linear maximal ideals and their joint generalized eigenspaces.

**Remark 6.1.2** Let $\mathscr{F}$ be a commuting family of endomorphisms of $V$.

(a) Let $v \in V$ be the basis vector of a 1-dimensional joint eigenspace of $\mathscr{F}$. For $i = 1, \ldots, n$, compute the element $\lambda_i \in K$ such that $\varphi_i(v) = \lambda_i v$. Then the linear maximal ideal corresponding to $\langle v \rangle$ is $\mathfrak{m} = \langle \varphi_1 - \lambda_1 \, \mathrm{id}_V, \ldots, \varphi_n - \lambda_n \, \mathrm{id}_V \rangle$.

(b) Given a linear maximal ideal $\mathfrak{m}$ of $\mathscr{F}$, let $\lambda_1, \ldots, \lambda_n \in K$ be the elements such that $\mathfrak{m} = \langle \varphi_1 - \lambda_1 \, \mathrm{id}_V, \ldots, \varphi_n - \lambda_n \, \mathrm{id}_V \rangle$. Then one can compute the corresponding joint eigenspace by calculating $U_i = \mathrm{Ker}(\varphi_i - \lambda_i \, \mathrm{id}_V)$ for $i = 1, \ldots, n$ and then using Proposition 2.2.10.c to conclude that $\mathrm{Ker}(\mathfrak{m}) = \bigcap_{i=1}^n U_i$.

Furthermore, one can compute the corresponding joint generalized eigenspace by calculating $W_i = \mathrm{BigKer}(\varphi_i - \lambda_i\,\mathrm{id}_V)$ for $i = 1, \ldots, n$ and then using Proposition 2.2.10.c to conclude that $\mathrm{BigKer}(\mathfrak{m}) = \bigcap_{i=1}^{n} W_i$.

Let us try out Algorithm 6.1.1 in some examples.

**Example 6.1.3** Let $K = \mathbb{Q}$, let $V = K^3$ and let $\mathscr{F} = K[\varphi_1, \varphi_2]$ be the commuting family generated by the endomorphisms of $V$ given by

$$A_1 = \begin{pmatrix} 0 & 0 & 0 \\ 1 & 0 & 0 \\ 0 & 0 & 0 \end{pmatrix} \quad \text{and} \quad A_2 = \begin{pmatrix} 0 & 0 & 0 \\ 0 & 0 & 0 \\ 1 & 0 & 1 \end{pmatrix}$$

respectively. We apply Algorithm 6.1.1 and follow the steps.

(1) Let $S = \emptyset$ and $T = (B)$ where $B = (e_1, e_2, e_3)$.
(4) The restriction $\varphi_{11}$ of $\varphi_1$ to $V = \langle B \rangle$ is represented by the matrix $A_1$.
(5) Since we have $\mu_{\varphi_{11}}(z) = z^2$, the only linear eigenfactor of $\varphi_{11}$ is $p_{111}(z) = z$. We compute $\mathrm{Eig}(\varphi_{11}, p_{111}(z)) = \langle e_2, e_3 \rangle$.
(6) We let $S = \emptyset$ and $T = ((e_2, e_3))$.
(4) The restriction $\varphi_{21}$ of $\varphi_2$ to $\langle e_2, e_3 \rangle$ satisfies $\varphi_{21}(e_2) = 0$ as well as $\varphi_{21}(e_3) = e_3$. The corresponding matrix is $\begin{pmatrix} 0 & 0 \\ 0 & 1 \end{pmatrix}$.
(5) We calculate $\mu_{\varphi_{21}}(z) = z(z - 1)$ and the eigenspaces $\mathrm{Eig}(\varphi_{21}, z) = \langle e_2 \rangle$ and $\mathrm{Eig}(\varphi_{21}, z - 1) = \langle e_3 \rangle$.
(6) We let $S = ((e_2), (e_3))$ and $T = \emptyset$.
(2) Return the tuple $S$.

Thus the commuting family $\mathscr{F}$ has two 1-dimensional joint eigenspaces, namely $\mathrm{Ker}(\mathfrak{m}_1) = \langle e_2 \rangle$ and $\mathrm{Ker}(\mathfrak{m}_2) = \langle e_3 \rangle$. Using Remark 6.1.2.a, we see that we have $\mathfrak{m}_1 = \langle \varphi_1, \varphi_2 \rangle$ and $\mathfrak{m}_2 = \langle \varphi_1, \varphi_2 - \mathrm{id}_V \rangle$. Moreover, using Remark 6.1.2.c, we obtain the joint generalized eigenspaces $\mathrm{BigKer}(\mathfrak{m}_1) = \langle e_2, e_1 - e_3 \rangle$ and $\mathrm{BigKer}(\mathfrak{m}_2) = \mathrm{Ker}(\mathfrak{m}_2) = \langle e_3 \rangle$.

Next we apply Algorithm 6.1.1 to Example 2.3.10.

**Example 6.1.4** As in Example 2.3.10, let $K = \mathbb{Q}$, let $V = K^4$, and let $\mathscr{F}$ be the commuting family $\mathscr{F} = K[\varphi_1, \varphi_2]$ generated by the endomorphisms $\varphi_1, \varphi_2$ of $V$ given by

$$A_1 = \begin{pmatrix} 0 & 2 & 0 & 0 \\ 1 & 0 & 0 & 0 \\ 0 & 0 & 0 & 2 \\ 0 & 0 & 1 & 0 \end{pmatrix} \quad \text{and} \quad A_2 = \begin{pmatrix} 0 & 0 & 8 & 0 \\ 0 & 0 & 0 & 8 \\ 1 & 0 & 0 & 0 \\ 0 & 1 & 0 & 0 \end{pmatrix}$$

respectively. If we follow the steps of Algorithm 6.1.1, it immediately stops for $i = 1$ in Step (5) and returns $S = \emptyset$, because $\mu_{\varphi_1}(z) = z^2 - 2$ has no linear factors in $K[z]$.

The situation changes when we work over the larger base field $L = \mathbb{Q}(\sqrt{2})$ and consider the family $\mathscr{F}_L = L[\varphi_1, \varphi_2]$ of commuting endomorphisms of $V_L = L^4$. In this case the algorithm proceeds as follows.

(5) The minimal polynomial $\mu_{\varphi_1}(z) = (z - \sqrt{2})(z + \sqrt{2})$ yields two linear eigen-factors $p_{111}(z) = z - \sqrt{2}$ and $p_{112}(z) = z + \sqrt{2}$. The two corresponding eigenspaces turn out to be $\mathrm{Eig}(\varphi_1, p_{111}(z)) = \langle e_2 + \sqrt{2}e_1, e_4 + \sqrt{2}e_3 \rangle$ and $\mathrm{Eig}(\varphi_1, p_{112}(z)) = \langle e_2 - \sqrt{2}e_1, e_4 - \sqrt{2}e_3 \rangle$.

(6) We let $S = \emptyset$ and $T = (B_1, B_2)$ with tuples $B_1 = (e_2 + \sqrt{2}e_1, e_4 + \sqrt{2}e_3)$ and $B_2 = (e_2 - \sqrt{2}e_1, e_4 - \sqrt{2}e_3)$.

(3) Let $j = 1$ and $T = (B_2)$.

(4) The matrix of $\varphi_2$ restricted to the vector subspace $\langle B_1 \rangle$ is $\left( \begin{smallmatrix} 0 & 8 \\ 1 & 0 \end{smallmatrix} \right)$.

(5) The minimal polynomial $\mu_{\varphi_{21}}(z) = (z - 2\sqrt{2})(z + 2\sqrt{2})$ yields the two 1-dimensional eigenspaces $\langle (4, 2\sqrt{2}, \sqrt{2}, 1) \rangle$ and $\langle (4, 2\sqrt{2}, -\sqrt{2}, -1) \rangle$.

(6) The two 1-element bases computed in Step (5) are put into $S$.

(3) Let $j = 2$ and $T = \emptyset$.

(4) The matrix of $\varphi_2$ restricted to $\langle B_2 \rangle$ is $\left( \begin{smallmatrix} 0 & 8 \\ 1 & 0 \end{smallmatrix} \right)$.

(5) The minimal polynomial $\mu_{\varphi_{22}}(z) = (z - 2\sqrt{2})(z + 2\sqrt{2})$ yields the two 1-dimensional eigenspaces $\langle (-4, 2\sqrt{2}, -\sqrt{2}, 1) \rangle$ and $\langle (4, -2\sqrt{2}, -\sqrt{2}, 1) \rangle$.

(6) The 1-element bases computed in Step (5) are put into $S$.

(2) The algorithm returns the tuple $S = (((4, 2\sqrt{2}, \sqrt{2}, 1)), ((4, 2\sqrt{2}, -\sqrt{2}, -1)), ((-4, 2\sqrt{2}, -\sqrt{2}, 1)), ((4, -2\sqrt{2}, -\sqrt{2}, 1)))$ and stops.

Thus, over the base field $L$, the family $\mathscr{F}_L$ has four 1-dimensional joint eigenspaces.

After we have computed the 1-dimensional joint eigenspaces of a commuting family, a natural question arises: are these all joint eigenspaces? In general, this will not be the case, but the following remark can help us detect this special situation.

**Remark 6.1.5**   Given a commuting family $\mathscr{F}$, perform the following steps.

(1) Using Algorithm 6.1.1, compute all 1-dimensional joint eigenspaces $U_1, \ldots, U_\ell$ of $\mathscr{F}$.

(2) Compute the corresponding linear maximal ideals $\mathfrak{m}_1, \ldots, \mathfrak{m}_\ell$ of $\mathscr{F}$ using Remark 6.1.2.a.

(3) Compute the corresponding joint generalized eigenspaces $G_1, \ldots, G_\ell$ of $\mathscr{F}$ using Remark 6.1.2.b.

(4) Check whether we have $\dim_K(G_1) + \cdots + \dim_K(G_\ell) = \dim_K(V)$.

If the last equality holds, then $U_1, \ldots, U_\ell$ are all joint eigenspaces of $\mathscr{F}$ and $G_1, \ldots, G_\ell$ are all joint generalized eigenspaces by Corollary 2.4.4.

The following example illustrates this remark.

**Example 6.1.6**   Let $K = \mathbb{Q}$, let $V = K^4$, and let $\mathcal{F} = K[\varphi_1, \varphi_2]$ be the commuting family generated by the two endomorphisms of $V$ given by the matrices

$$A_1 = \begin{pmatrix} 0 & 1 & 0 & 0 \\ 0 & 0 & 0 & 1 \\ -6 & 5 & 2 & -1 \\ 4 & -8 & 0 & 5 \end{pmatrix} \qquad A_2 = \begin{pmatrix} 0 & 0 & 1 & 0 \\ -6 & 5 & 2 & -1 \\ 3 & -4 & 2 & 1 \\ -16 & 12 & 4 & -2 \end{pmatrix}$$

respectively. Let us apply the method of Remark 6.1.5.

(1) Using Algorithm 6.1.1, we find two 1-dimensional joint eigenspaces. They are $\mathrm{Ker}(\mathfrak{m}_1) = \langle (1, 1, 2, 1) \rangle$ and $\mathrm{Ker}(\mathfrak{m}_2) = \langle (1, 2, 1, 4) \rangle$.

(2) Using Remark 6.1.2.a, we determine the corresponding linear maximal ideals $\mathfrak{m}_1 = \langle \varphi_1 - \mathrm{id}_V, \varphi_2 - 2\,\mathrm{id}_V \rangle$ and $\mathfrak{m}_2 = \langle \varphi_1 - 2\,\mathrm{id}_V, \varphi_2 - \mathrm{id}_V \rangle$.

(3) Next we use Remark 6.1.2.b to compute the corresponding joint generalized eigenspaces. We get $\mathrm{BigKer}(\mathfrak{m}_1) = \mathrm{Ker}(\mathfrak{m}_1)$, while $\mathrm{BigKer}(\mathfrak{m}_2)$ is the 3-dimensional vector subspace of $V$ with basis $\{(1, 2, 1, 4), (1, 1, 0, 0), (1, 0, 0, -4)\}$.

(4) Since $\dim_K(\mathrm{BigKer}(\mathfrak{m}_1)) + \dim_K(\mathrm{BigKer}(\mathfrak{m}_2)) = 4 = \dim_K(V)$, we conclude that $\mathrm{Ker}(\mathfrak{m}_1)$ and $\mathrm{Ker}(\mathfrak{m}_2)$ are the only joint eigenspaces of $\mathcal{F}$.

Sometimes, even if Algorithm 6.1.5 does not produce the maximal number of 1-dimensional joint eigenspaces, but also some 2-dimensional candidates for joint eigenspaces, we can deduce that we have found all joint eigenspaces of $\mathcal{F}$ as in the following example.

**Example 6.1.7**   Let $K = \mathbb{Q}$, let $V = K^5$, and let the endomorphisms $\varphi_1$ and $\varphi_2$ of $V$ be given by the matrices

$$A_1 = \begin{pmatrix} 0 & 0 & 0 & 0 & 0 \\ 1 & 0 & 0 & 0 & 0 \\ 0 & 0 & 0 & 0 & 0 \\ 0 & 1 & 0 & -2 & 0 \\ 0 & 0 & 0 & 0 & 0 \end{pmatrix} \quad \text{and} \quad A_2 = \begin{pmatrix} 0 & 0 & 0 & 0 & 0 \\ 0 & 0 & 0 & 0 & 0 \\ 1 & 0 & 0 & 0 & 0 \\ 0 & 0 & 0 & 0 & 0 \\ 0 & 0 & 1 & 0 & 1 \end{pmatrix}$$

respectively. It is easy to check that $\mathcal{F} = K[\varphi_1, \varphi_2]$ is a commuting family of endomorphisms of $V$.

In this setting we run Algorithm 6.1.1. First we calculate $\mu_{\varphi_1}(z) = z^2(z + 2)$ and $\mathrm{Eig}(\varphi_1, 0) = \langle e_3, 2e_2 + e_4, e_5 \rangle$ as well as $\mathrm{Eig}(\varphi_1, -2) = \langle e_4 \rangle$. The latter is a 1-dimensional joint eigenspace of $\mathcal{F}$.

Then we calculate the restriction $\varphi_{21}$ of $\varphi_2$ to $U = \mathrm{Eig}(\varphi_1, 0)$. We get $\mu_{\varphi_{21}}(z) = z^2(z - 1)$ and $\mathrm{Eig}(\varphi_{21}, 0) = \langle e_3 - e_5, e_4 \rangle$ as well as $\mathrm{Eig}(\varphi_{21}, 1) = \langle e_5 \rangle$. The latter is again a 1-dimensional joint eigenspace of $\mathcal{F}$.

Now we know that $U = \langle e_3 - e_5, e_4 \rangle$ is a 2-dimensional $\mathscr{F}$-invariant subspace of $V$. It cannot contain a 1-dimensional joint eigenspace of $\mathscr{F}$, because this would have been detected by our algorithm. Hence we know that $U$ is a 2-dimensional joint eigenspace and we have found all joint eigenspaces.

Recall from Proposition 2.2.13.c that the dimension of a joint eigenspace $\mathrm{Ker}(\mathfrak{m})$ is a multiple of the dimension of the corresponding residue class field $\dim_K(\mathscr{F}/\mathfrak{m})$. This implies that Algorithm 6.1.1 may not compute the kernels of all linear maximal ideals, but only the ones corresponding to 1-dimensional joint eigenspaces. The following example, based on Example 3.3.4, shows that this actually happens in some cases.

**Example 6.1.8** Let $K = \mathbb{Q}$, let $V = K^3$, and let $\mathscr{F} = K[\varphi_1, \varphi_2]$ be the commuting family generated by the two endomorphisms of $V$ given by

$$A_1 = \begin{pmatrix} 0 & 0 & 0 \\ 1 & 0 & 0 \\ 0 & 0 & 0 \end{pmatrix} \quad \text{and} \quad A_2 = \begin{pmatrix} 0 & 0 & 0 \\ 0 & 0 & 0 \\ 1 & 0 & 0 \end{pmatrix}$$

respectively. We apply Algorithm 6.1.1 and follow the steps.

(5) Since we have $\mu_{\varphi_1}(z) = z^2$, the only linear eigenfactor of $\varphi_1$ is $p_{111}(z) = z$. We compute $\mathrm{Eig}(\varphi_1, p_{111}(z)) = \langle e_2, e_3 \rangle$.
(6) We let $S = \emptyset$ and $T = ((e_2, e_3))$.
(4) The restriction $\varphi_{21}$ of $\varphi_2$ to $\langle e_2, e_3 \rangle$ satisfies $\varphi_{21}(e_2) = 0$ as well as $\varphi_{21}(e_3) = 0$. The corresponding matrix is $\left(\begin{smallmatrix} 0 & 0 \\ 0 & 0 \end{smallmatrix}\right)$.
(5) We calculate $\mu_{\varphi_{21}}(z) = z^2$ and the eigenspace $\mathrm{Eig}(\varphi_{21}, z) = \langle e_2, e_3 \rangle$.
(2) We conclude that $S = \emptyset$.

If we compute the ideal of algebraic relations of $(\varphi_1, \varphi_2)$, we get $\langle x^2, xy, y^2 \rangle$. Therefore a $K$-basis of $\mathscr{F}$ is given by $(\mathrm{id}_V, \varphi_1, \varphi_2)$, and hence every endomorphism $\psi \in \mathscr{F}$ can be written as $\psi = a\, \mathrm{id}_V + b\varphi_1 + c\varphi_2$ with $a, b, c \in K$. The matrix which represents $\psi$ is

$$\begin{pmatrix} a & 0 & 0 \\ b & a & 0 \\ c & 0 & a \end{pmatrix}$$

The unique eigenfactor of $\psi$ is $z - a$. The corresponding eigenspace $\mathrm{Ker}(\psi - a\, \mathrm{id}_V)$ contains $\langle e_2, e_3 \rangle$. Hence there is a 2-dimensional joint eigenspace.

On the other hand, the family $\mathscr{F}$ has a unique maximal ideal $\mathfrak{m} = \langle \varphi_1, \varphi_2 \rangle$ which is linear. So, while Algorithm 6.1.1 correctly computed all 1-dimensional joint eigenspaces, it did not notice the existence of a further linear maximal ideal in $\mathscr{F}$. The reason for this failure is that $\mathscr{F}$ is a derogatory family (see Example 3.3.4). In the next subsection we describe how to remove this obstacle.

### 6.1.B  Computing Linear Maximal Ideals

> *People say nothing is impossible,*
> *but I do nothing every day.*
> (Winnie the Pooh)

Let us continue to use the setting introduced in the previous subsection. In particular, let $K$ be a field, let $V$ be a finite dimensional $K$-vector space, let $\mathscr{F}$ be a family of commuting endomorphisms of $V$, and let $\Phi = (\varphi_1, \ldots, \varphi_n)$ be a system of $K$-algebra generators of $\mathscr{F}$. In the following we consider the task of computing all linear maximal ideals of $\mathscr{F}$. As explained before, for a family $\mathscr{F}$ which is the multiplication family of a zero-dimensional affine $K$-algebra $R = P/I$, this amounts to finding all $K$-rational zeros of $I$.

Algorithm 6.1.1 computes the 1-dimensional joint eigenspaces of $\mathscr{F}$. Associated to each of them there is a linear maximal ideal of $\mathscr{F}$ which can be easily described (see Remark 6.1.2.a). However, not all linear maximal ideals of $\mathscr{F}$ can be computed in this way, as Example 6.1.8 shows. The deeper reason for this failure was that the family $\mathscr{F}$ was derogatory, and apparently nothing can be done to remedy this. But wait! Nothing is impossible! At least, if we assume that $V$ is $\mathscr{F}$-cyclic we know that the dual family is commendable by the Duality Theorem 3.5.12. Using this, we can extend Algorithm 6.1.1 suitably. The following method was first studied by H.-M. Möller and R. Tenberg in [22].

---

**Algorithm 6.1.9 (Computing Linear Maximal Ideals in the Cyclic Case)**
*Let $\mathscr{F}$ be a family of commuting endomorphisms of $V$ and $\Phi = (\varphi_1, \ldots, \varphi_n)$ a system of $K$-algebra generators of $\mathscr{F}$. Assuming that $V$ is $\mathscr{F}$-cyclic, consider the following sequence of instructions.*

(1) *Apply Algorithm 6.1.1 to the endomorphisms $\varphi_1^{\vee}, \ldots, \varphi_n^{\vee}$ and compute the set $S$ of bases of all 1-dimensional joint eigenspaces of $\mathscr{F}^{\vee}$. Then let $S = \{(\ell_1), \ldots, (\ell_s)\}$ with $\ell_i \in V^*$.*

(2) *Compute $\varphi_i^{\vee}(\ell_j) = \lambda_{ij} \ell_j$ with $\lambda_{ij} \in K$ for $i = 1, \ldots, n$ and $j = 1, \ldots, s$.*

(3) *Return $\mathfrak{m}_1, \ldots, \mathfrak{m}_s$, where $\mathfrak{m}_j = \langle \varphi_1 - \lambda_{1j} \operatorname{id}_V, \ldots, \varphi_n - \lambda_{nj} \operatorname{id}_V \rangle$.*

*This is an algorithm which computes all linear maximal ideals of $\mathscr{F}$.*

---

*Proof* Since $V$ is $\mathscr{F}$-cyclic, part (a) of the Duality Theorem 3.5.12 shows that the dual family $\mathscr{F}^{\vee}$ is commendable. Moreover, Proposition 3.5.2.a implies that it is generated by $\Phi^{\vee} = (\varphi_1^{\vee}, \ldots, \varphi_n^{\vee})$. By Remark 6.1.2.a, we know that if $(\ell_j)$ is a basis of a joint eigenspace of $\mathscr{F}^{\vee}$ and if we have $\varphi_i^{\vee}(\ell_j) = \lambda_{ij} \ell_j$, then the corresponding linear maximal ideal $\mathfrak{m}_j^{\vee}$ of $\mathscr{F}^{\vee}$ is given by $\mathfrak{m}_j^{\vee} = \langle \varphi_1^{\vee} - \lambda_{1j} \operatorname{id}_{V^*}, \ldots, \varphi_n^{\vee} - \lambda_{nj} \operatorname{id}_{V^*} \rangle$. Now we apply the inverse of the isomorphism $\delta : \mathscr{F} \longrightarrow \mathscr{F}^{\vee}$ of Proposition 3.5.2.a and conclude that the ideals $\mathfrak{m}_j = \langle \varphi_1 - \lambda_{1j} \operatorname{id}_V, \ldots, \varphi_n - \lambda_{nj} \operatorname{id}_V \rangle$ are linear maximal ideals of $\mathscr{F}$ for $J = 1, \ldots, s$.

Conversely, let $\mathfrak{m}$ be a linear maximal ideal of $\mathscr{F}$. By applying the isomorphism $\delta$ above, we get a linear maximal ideal $\mathfrak{m}^\vee$ of $\mathscr{F}^\vee$. Since we have $\dim_K(\mathscr{F}^\vee/\mathfrak{m}^\vee) = 1$ and since $\mathscr{F}^\vee$ is commendable, it follows that $\dim_K(\mathrm{Ker}(\mathfrak{m}^\vee)) = 1$. Therefore $\mathrm{Ker}(\mathfrak{m}^\vee)$ is one of the 1-dimensional joint eigenspaces computed by Algorithm 6.1.1. $\qquad\square$

Let us see this algorithm at work.

**Example 6.1.10** Let $K = \mathbb{Q}$, let $V = K^3$, and let the endomorphisms $\varphi_1$ and $\varphi_2$ of $V$ be given by the matrices

$$A_1 = \begin{pmatrix} 0 & -\frac{2}{3} & \frac{4}{3} \\ 1 & -\frac{5}{3} & \frac{4}{3} \\ 0 & \frac{2}{3} & -\frac{4}{3} \end{pmatrix} \quad \text{and} \quad A_2 = \begin{pmatrix} 0 & \frac{4}{3} & -\frac{2}{3} \\ 0 & \frac{4}{3} & -\frac{2}{3} \\ 1 & -\frac{4}{3} & \frac{5}{3} \end{pmatrix}$$

respectively. It is easy to check that $\mathscr{F} = K[\varphi_1, \varphi_2]$ is a commuting family, and using the Cyclicity Test 3.1.4 we can verify that $V$ is a cyclic $\mathscr{F}$-module. Let us follow the steps of Algorithm 6.1.9.

(1) As indicated in Algorithm 6.1.9, we apply Algorithm 6.1.1 to the endomorphisms $\varphi_1^\vee$ and $\varphi_2^\vee$ given by the matrices $A_1^{\mathrm{tr}}$ and $A_2^{\mathrm{tr}}$ and get $S = \{(\ell_1), (\ell_2), (\ell_3)\}$ with $\ell_1 = -e_1^* + 2e_2^* - 2e_3^*$, $\ell_2 = e_1^* - e_2^*$, and $\ell_3 = e_1^* + e_3^*$.

(2) We compute $\varphi_1^\vee(\ell_1) = -2\ell_1$, $\varphi_2^\vee(\ell_1) = 2\ell_1$, $\varphi_1^\vee(\ell_2) = -\ell_2$, $\varphi_2^\vee(\ell_2) = 0$, and $\varphi_1^\vee(\ell_3) = 0$, $\varphi_2^\vee(\ell_3) = \ell_3$.

(3) The algorithm returns the linear maximal ideals $\mathfrak{m}_1 = (\varphi_1 + 2\,\mathrm{id}_V, \varphi_2 - 2\,\mathrm{id}_V)$, $\mathfrak{m}_2 = (\varphi_1 + \mathrm{id}_V, \varphi_2)$, $\mathfrak{m}_3 = (\varphi_1, \varphi_2 - \mathrm{id}_V)$ of $\mathscr{F}$.

In our second example not all maximal ideals of $\mathscr{F}$ are linear.

**Example 6.1.11** Let $K = \mathbb{Q}$, let $V = K^4$, and let the endomorphisms $\varphi_1$ and $\varphi_2$ of $V$ be given by the matrices

$$A_1 = \begin{pmatrix} 0 & 0 & 0 & 0 \\ 1 & 0 & 0 & 0 \\ 0 & 0 & 0 & 0 \\ 0 & 0 & 0 & 0 \end{pmatrix} \quad \text{and} \quad A_2 = \begin{pmatrix} 0 & 0 & 0 & 0 \\ 0 & 0 & 0 & 0 \\ 1 & 0 & 0 & 0 \\ 0 & 0 & 1 & 2 \end{pmatrix}$$

respectively. It is easy to check that $\mathscr{F} = K[\varphi_1, \varphi_2]$ is a commuting family. Using the Cyclicity Test 3.1.4, we see that $V$ is a cyclic $\mathscr{F}$-module. Let us follow the steps of Algorithm 6.1.9 to compute the linear maximal ideals of $\mathscr{F}$.

(1) As indicated in Algorithm 6.1.9, we apply Algorithm 6.1.1 to the maps $\varphi_1^\vee$ and $\varphi_2^\vee$ represented by the matrices $A_1^{\mathrm{tr}}$ and $A_2^{\mathrm{tr}}$ and get the 1-dimensional joint eigenspace bases $S = \{(\ell_1), (\ell_2)\}$ where $\ell_1 = e_1^*$ and $\ell_2 = e_1^* + 2e_3^* + 4e_4^*$.

(2) We compute $\varphi_1^\vee(\ell_1) = 0$, $\varphi_2^\vee(\ell_1) = 0$, $\varphi_1^\vee(\ell_2) = 0$, and $\varphi_2^\vee(\ell_2) = 2\ell_2$.

(3) Finally, the algorithm returns the two linear maximal ideals $\mathfrak{m}_1 = \langle \varphi_1, \varphi_2 \rangle$ and $\mathfrak{m}_2 = \langle \varphi_1, \varphi_2 - 2\,\mathrm{id}_V \rangle$ of $\mathscr{F}$.

The Cyclicity Test 3.1.4 shows that $V^*$ is not a cyclic $\mathscr{F}^\vee$-module in this case. In particular, the family $\mathscr{F}$ is not commendable and a direct application of Algorithm 6.1.1 to $\mathscr{F}$ misses the linear maximal ideal $\mathfrak{m}_1$.

Notice that if we know that the family $\mathscr{F}$ is commendable, we can apply Algorithm 6.1.9 directly to $\mathscr{F}$ instead of $\mathscr{F}^\vee$, and we do not need the assumption that $V$ is $\mathscr{F}$-cyclic. However, this is an information which is normally not available. On the other hand, the situation when $V$ is $\mathscr{F}$-cyclic occurs quite frequently. For instance, it holds for the families of multiplication endomorphisms studied in Chap. 4.

Is this the end of the story? Or can we get rid of the assumption that $V$ has to be a cyclic $\mathscr{F}$-module? Nothing is impossible! If $V$ is not a cyclic $\mathscr{F}$-module, we replace it by a cyclic $\mathscr{F}$-module, namely by $\mathscr{F}$ itself. More precisely, we use Remark 4.1.3.c as follows.

---

**Algorithm 6.1.12 (Computing All Linear Maximal Ideals)**

*Let $\mathscr{F}$ be a family of commuting endomorphisms of $V$ and $\Phi = (\varphi_1, \dots, \varphi_n)$ a system of $K$-algebra generators of $\mathscr{F}$. Consider the following sequence of instructions.*

(1) *Using Algorithm 2.1.4, compute a Gröbner basis $G$ of $\mathrm{Rel}_P(\Phi)$ and a tuple $\Psi = (\psi_1, \dots, \psi_\delta)$ of elements of $\mathscr{F}$ which forms a $K$-basis of $\mathscr{F}$.*

(2) *Using the normal form with respect to $G$, compute the multiplication matrices representing $\vartheta_{\varphi_1}, \dots, \vartheta_{\varphi_n}$ with respect to the basis $\Psi$ (see also Remark 4.1.7).*

(3) *Apply Algorithm 6.1.9 to the multiplication family $\widetilde{\mathscr{F}} = K[\vartheta_{\varphi_1}, \dots, \vartheta_{\varphi_n}]$ of $\mathscr{F}$. Let $\widetilde{\mathfrak{m}}_1, \dots, \widetilde{\mathfrak{m}}_s$ be the linear maximal ideals it computes.*

(4) *For $i = 1, \dots, s$, let $\mathfrak{m}_i$ be the image of $\widetilde{\mathfrak{m}}_i$ in $\mathscr{F}$ via the canonical isomorphism $\iota : \widetilde{\mathscr{F}} \longrightarrow \mathscr{F}$ given by $\vartheta_f \mapsto f$. Return $\mathfrak{m}_1, \dots, \mathfrak{m}_s$ and stop.*

*This is an algorithm which computes all linear maximal ideals $\mathfrak{m}_1, \dots, \mathfrak{m}_s$ of $\mathscr{F}$.*

---

*Proof* Since the algorithm is clearly finite, let us check correctness. In Steps (1) and (2) we calculate the multiplication family $\widetilde{\mathscr{F}}$ of $\mathscr{F}$. Since $\mathscr{F}$ is clearly a cyclic $\widetilde{\mathscr{F}}$-module, we can apply Algorithm 6.1.9 and get all linear maximal ideals of $\widetilde{\mathscr{F}}$. Then it suffices to apply the isomorphism between $\widetilde{\mathscr{F}}$ and $\mathscr{F}$ given in Remark 4.1.3.c. $\qquad\square$

Let us illustrate this method by an example.

**Example 6.1.13**  Let $K = \mathbb{Q}$, let $V = K^4$, and let the endomorphisms $\varphi_1$ and $\varphi_2$ of $V$ be given by the matrices

$$A_1 = \begin{pmatrix} 0 & 0 & 1 & 0 \\ 0 & 0 & 0 & 0 \\ 0 & 0 & 0 & 0 \\ 0 & 0 & 0 & 0 \end{pmatrix} \quad \text{and} \quad A_2 = \begin{pmatrix} 0 & 0 & 0 & 1 \\ 0 & 0 & 0 & 0 \\ 0 & 0 & 0 & 0 \\ 0 & 0 & 0 & 0 \end{pmatrix}$$

respectively. Then it is easy to check that $\mathscr{F} = K[\varphi_1, \varphi_2]$ is a commuting family. Let us follow the steps of Algorithm 6.1.12.

(1) Using Algorithm 2.1.4 we find that $\Psi = (\mathrm{id}_V, \varphi_1, \varphi_2)$ is a $K$-basis of $\mathscr{F}$.
(2) The matrices representing the endomorphisms $\vartheta_{\varphi_1}$ and $\vartheta_{\varphi_2}$ with respect to the basis $\Psi$ are

$$B_1 = \begin{pmatrix} 0 & 0 & 0 \\ 1 & 0 & 0 \\ 0 & 0 & 0 \end{pmatrix} \quad \text{and} \quad B_2 = \begin{pmatrix} 0 & 0 & 0 \\ 0 & 0 & 0 \\ 1 & 0 & 0 \end{pmatrix}$$

respectively.
(3) The result of Algorithm 6.1.9 is $\tilde{\mathfrak{m}} = \langle \vartheta_{\varphi_1}, \vartheta_{\varphi_2} \rangle$.
(4) The output is the linear maximal ideal $\mathfrak{m} = \langle \varphi_1, \varphi_2 \rangle$ of $\mathscr{F}$.

Notice that we have $\dim_K(\mathscr{F}) = 3 < 4 = \dim_K(V)$ here. Hence Proposition 3.1.1.c shows that $V$ is not a cyclic $\mathscr{F}$-module. Therefore we could not have done this computation with the help of Algorithm 6.1.9 alone. Moreover, by Corollary 3.5.17, the family $\mathscr{F}$ is not commendable, so that a direct application of Algorithm 6.1.1 would not have done the job either. Thus we conclude that Algorithm 6.1.12 did it!

*A conclusion is the place where you got tired of thinking.*

## 6.2  Rational Zeros via Eigenvalues and Eigenvectors

*The stone age was marked*
*by man's clever use of crude tools;*
*the information age, to date, has been marked*
*by man's crude use of clever tools.*

In the preceding section we met some methods for computing the $K$-rational solutions of a polynomial system which were based on computing joint eigenspaces of a family of commuting matrices. In this section we explore some variants of this approach based on eigenvalues and eigenvectors. Although the algorithms presented below are still crude applications of these clever tools, they turn out to be surprisingly powerful.

In the following, we let $K$ be a field, let $P = K[x_1, \ldots, x_n]$, let $f_1, \ldots, f_m \in P$, let $I = \langle f_1, \ldots, f_m \rangle$, and assume that $I$ is a zero-dimensional ideal in $P$. Then

$R = P/I$ is a zero-dimensional affine $K$-algebra. Hence $R$ is a finite dimensional $K$-vector space. Its dimension is denoted by $d = \dim_K(R)$. Our goal is to find the set of $K$-rational zeros $\mathcal{Z}_K(I) = \mathcal{Z}(I) \cap K^n$ of $I$.

## 6.2.A  The Eigenvalue Method

> *To steal ideas from one person is plagiarism;*
> *to steal from many is research.*
> (Steven Wright)

The next proposition has a long history and has been rediscovered many times. Sometimes it is named Stickelberger's Eigenvalue Theorem after L. Stickelberger (1850–1936). One version of it was published in 1981 by D. Lazard (see [18]), another by W. Auzinger and H.J. Stetter in 1998 (see [1]). Since then, the result has been recounted and rephrased in literally dozens of papers and books. In our context, Proposition 2.3.8 provides a first connection between linear maximal ideals and linear eigenfactors, i.e., between rational solutions and eigenvalues. It is formulated using the multiplication family of $R$. Let us spell out the precise link between solutions of polynomial systems and eigenvalues of multiplication maps.

**Proposition 6.2.1** *Let $f \in P$, and let $\vartheta_f : R \longrightarrow R$ be the corresponding multiplication map.*

(a) *If there exists a point $p \in \mathcal{Z}_K(I)$ then $f(p) \in K$ is an eigenvalue of $\vartheta_f$.*

(b) *If $\lambda \in K$ is an eigenvalue of $\vartheta_f$ then there exist an algebraic field extension $L \supseteq K$ and a point $p \in \mathcal{Z}_L(I)$ such that $\lambda = f(p)$.*

*Proof* Let us first prove (a). We let $p = (a_1, \ldots, a_n)$ be a point of $\mathcal{Z}_K(I)$. Then we let $\langle x_1 - a_1, \ldots, x_n - a_n \rangle$ be the corresponding maximal ideal in $P$ which contains $I$, let $\mathfrak{m}$ be its image in $R$, and let $\iota(\mathfrak{m})$ be its image in $\mathcal{F}$ (see Proposition 4.1.2.b). Then Corollary 2.3.9 shows that $f(p)$ is an eigenvalue of $\vartheta_f$.

Now we prove (b). Let $\lambda \in K$ be an eigenvalue of $\vartheta_f$. Then $z - \lambda$ is an eigenfactor of $\vartheta_f$. By Proposition 2.3.2.f, there is a maximal ideal $\mathfrak{m}$ of $R$ with $f - \lambda \in \mathfrak{m}$. Let $\mathfrak{M}$ be the preimage of $\mathfrak{m}$ in $P$. By Hilbert's Nullstellensatz (cf. [15], Theorem 2.6.16), there exists a tuple $p = (a_1, \ldots, a_n) \in \overline{K}^n$ such that $\mathfrak{M}\overline{K}[x_1, \ldots, x_n]$ is contained in $\langle x_1 - a_1, \ldots, x_n - a_n \rangle$. Then the field $L = K(a_1, \ldots, a_n)$ is an algebraic extension of $K$ for which we have the inclusion $\mathfrak{m} \otimes_K L \subseteq \langle \bar{x}_1 - a_1, \ldots, \bar{x}_n - a_n \rangle$, and therefore $p = (a_1, \ldots, a_n) \in \mathcal{Z}_L(I)$. Finally, we note that $f - \lambda \in \mathfrak{m}$ implies $\lambda = f(p)$. $\qquad \square$

Even when the eigenvalue $\lambda$ is contained in $K$, the passage to an algebraic extension field of $K$ in part (b) is in general necessary, since our next example shows that the point $p$ need not be contained in $K^n$.

**Example 6.2.2**   Let $K = \mathbb{Q}$, let $P = K[x, y]$, let $I = \langle x^2, y^2 + 1 \rangle$, and let $R = P/I$. For the polynomial $x$, the multiplication map $\vartheta_x : R \longrightarrow R$ is nilpotent and has the eigenvalue $\lambda = 0$. However, there is no point $p = (a_1, a_2) \in \mathcal{Z}_K(I)$ such that we have $a_1 = x(p) = 0$. In fact, the zeros of $I$ in $\mathbb{Q}(i)^2$ are $(0, i)$ and $(0, -i)$.

If we apply the preceding proposition to the multiplication endomorphisms corresponding to the indeterminates $x_i$, we obtain the following corollary.

**Corollary 6.2.3**   *For $i \in \{1, \ldots, n\}$, the $i$-th coordinates of the points of $\mathcal{Z}_K(I)$ are the eigenvalues of the multiplication map $\vartheta_{x_i}$.*

This corollary suggests the following simple method for computing $\mathcal{Z}_K(I)$.

**Remark 6.2.4**   For $i = 1, \ldots, n$, compute the eigenvalues $\lambda_{i1}, \ldots, \lambda_{i\nu_i} \in K$ of the multiplication map $\vartheta_{x_i} : R \longrightarrow R$. Then check for every tuple $(\lambda_{1j_1}, \ldots, \lambda_{nj_n})$ with $1 \le j_i \le \nu_i$ whether it solves the system. Those which do form the set $\mathcal{Z}_K(I)$ of $K$-rational solutions of $I$.

This method has a major drawback, namely that we have to check for each of $\prod_{i=1}^{n} \nu_i$ tuples whether it solves the system. This is, in general, a computationally expensive task.

The following example contains a case where the above naive method does have some advantages.

**Example 6.2.5**   Let $K = \mathbb{Q}$, let $P = K[x_1, x_2, x_3]$, and consider the polynomial system defined by

$$f_1 = x_2 x_3 + \tfrac{2}{3} x_3^2 - \tfrac{5}{9} x_1 - \tfrac{58}{9} x_2 + \tfrac{68}{9} x_3 - \tfrac{10}{3},$$

$$f_2 = x_1 x_3 + \tfrac{17}{3} x_3^2 - \tfrac{23}{9} x_1 - \tfrac{202}{9} x_2 + \tfrac{275}{9} x_3 - \tfrac{46}{3},$$

$$f_3 = x_2^2 + \tfrac{7}{3} x_3^2 - \tfrac{10}{9} x_1 - \tfrac{125}{9} x_2 + \tfrac{145}{9} x_3 - \tfrac{20}{3},$$

$$f_4 = x_1 x_2 + 7 x_3^2 - 3 x_1 - 28 x_2 + 37 x_3 - 18,$$

$$f_5 = x_1^2 + \tfrac{14}{3} x_3^2 + \tfrac{55}{9} x_1 - \tfrac{316}{9} x_2 + \tfrac{314}{9} x_3 + \tfrac{2}{3},$$

$$f_6 = x_3^3 + 10 x_3^2 - \tfrac{10}{3} x_1 - \tfrac{98}{3} x_2 + \tfrac{115}{3} x_3 - 20$$

Then the ideal $I = \langle f_1, f_2, f_3, f_4, f_5, f_6 \rangle$ is zero-dimensional and the reduced Gröbner basis of $I$ with respect to `DegRevLex` is $\{f_1, f_2, f_3, f_4, f_5, f_6\}$. Since a $K$-basis of $R = P/I$ is given by the residue classes of $\{1, x_1, x_2, x_3, x_3^2\}$, there are at most five zeros in $\mathcal{Z}_K(I)$.

To find them using Remark 6.2.4, we first compute $\mu_{x_1}(z)$ and get the polynomial $\mu_{x_1}(z) = z^5 + 12z^4 + 37z^3 - 18z^2 - 164z - 120$. Factoring this polynomial yields $\mu_{x_1}(z) = (z + 6)(z + 5)(z + 2)(z + 1)(z - 2)$. Hence the eigenvalues of $\vartheta_{x_1}$ are $\{-6, -5, -2, -1, 2\}$, and these are the candidates for the $x_1$-coordinates of the

$K$-rational solutions of the system. Analogously, we get $\mu_{x_2}(z) = z^5 - z^4 - 7z^3 + z^2 + 6z = (z+2)(z+1)z(z-1)(z-3)$ and $\mu_{x_3}(z) = z^5 + z^4 - 7z^3 - z^2 + 6z = (z+3)(z+1)z(z-1)(z-2)$.

Now we should check which of the resulting $5^3 = 125$ triples are zeros of $I$. We would find that five triples yield a solution point. However, suppose that we are interested only in positive solutions whose coordinates are greater than 1. Then only one triple is a good candidate, namely $(2, 3, 2)$, and a direct check shows that it is indeed a zero of $I$.

Another merit of the method in Remark 6.2.4 shows up when we work over $K = \mathbb{R}$ and the minimal polynomials $\mu_{x_i}(z)$ cannot be factored exactly, i.e., their zeros can be determined only approximately. In this case substituting the candidate eigenvalues of $\vartheta_{x_1}$ into the minimal polynomial of $\vartheta_{x_2}$ and continuing in this way may lead to an undesirable accumulation of rounding errors. A direct substitution of all candidate points as in Remark 6.2.4 may avoid this problem.

The algorithms in Sect. 6.1.A have one drawback in common: they require the computation of the multiplication family and some eigenspace calculations before coming back to linear maximal ideals in $R$. In the following we construct an algorithm which leads more directly to the rational zeros of $I$. The next example points us in the right direction.

**Example 6.2.6** In Example 3.3.4 we saw a derogatory commuting family of endomorphisms $\mathscr{F} = \mathbb{Q}[\varphi_1, \varphi_2]$ such that $\mathrm{Rel}_P(\varphi_1, \varphi_2) = \langle x^2, xy, y^2 \rangle$. In Example 6.1.8 we noticed that the computation of the 1-dimensional joint eigenspaces of $\mathscr{F}$ misses the linear maximal ideal $\mathfrak{m} = \langle \varphi_1, \varphi_2 \rangle$. Besides the fact that the answer is straightforward, we could get it from the observation that $\mu_x(z) = z^2$ has only one linear factor, namely $z$. Then, by adding $x$ to the ideal of algebraic relations, we get the ideal $J = \langle x, y^2 \rangle$. Modulo this ideal we have $\mu_y(z) = z^2$ and this polynomial has again only one linear factor, namely $z$. Hence, if we add $y$ to the ideal $J$, we see that there is only one linear maximal ideal in $\mathscr{F}$, namely $\mathfrak{m} = \langle x, y \rangle$.

The argument given in this example to compute the linear maximal ideal can be generalized. Notice that the following algorithm proceeds exactly like Algorithm 5.3.12, but at every step only the linear factors of the minimal polynomial are kept and processed further. And since we adjoin linear polynomials, no extensions of the base field are required.

---

**Algorithm 6.2.7 (The Eigenvalue Method)**
Let $K$ be a field, let $P = K[x_1, \ldots, x_n]$, let $I$ be a zero-dimensional ideal in $P$, and let $R = P/I$. Consider the following sequence of instructions.

(1) Let $S_0$ be the tuple consisting of the zero ideal, and let $S_1, \ldots, S_n$ be empty tuples.

(2) For $i = 1, \ldots, n$, perform the following steps. Then return $S_n$ and stop.

(3) Let $S_{i-1} = (L_1, \ldots, L_m)$. For $j = 1, \ldots, m$, remove $L_j$ from $S_{i-1}$ and perform the following steps.

(4) Write $L_j = \langle x_1 - a_{1j_1}, \ldots, x_{i-1} - a_{i-1j_{i-1}} \rangle$ with $a_{ik} \in K$.

(5) By substituting $x_k \mapsto a_{kj_k}$ for $k = 1, \ldots, i-1$, split the ideal $L_j + I$ in the form $L_j + I = L_j + L'_j$ with $L'_j \subseteq K[x_i, \ldots, x_n]$.

(6) Compute the minimal polynomial $\mu_{\bar{x}_i}(z)$ of $\bar{x}_i \in K[x_i, \ldots, x_n]/L'_j$ and its distinct linear factors $z - a_{i1}, \ldots, z - a_{i\ell_i}$ with $a_{ik} \in K$.

(7) For each $z - a_{ik}$, append the ideal $\langle x_1 - a_{1j_1}, \ldots, x_{i-1} - a_{i-1j_{i-1}}, x_i - a_{ik} \rangle$ to $S_i$.

This is an algorithm which computes a tuple $S_n$ consisting of all linear maximal ideals in $P$ which contain $I$.

*Proof* The algorithm is clearly finite. When the iteration of Steps (3)–(7) has finished for a certain value $i$, the tuple $S_i$ contains only ideals $L_j$ of the form $L_j = \langle x_1 - a_{1j_1}, \ldots, x_i - a_{ij_i} \rangle$. Hence, after the iteration is finished for $i = n$, the tuple $S_n$ contains linear maximal ideals of $P$. By construction, we have

$$I \subseteq \langle x_1 - a_{1j_1} \rangle + I = \langle x_1 - a_{1j_1} \rangle + L'_{j_1} \subseteq \cdots \subseteq \langle x_1 - a_{1j_1}, \ldots, x_n - a_{nj_n} \rangle + \langle 0 \rangle$$

It remains to show that the algorithm constructs all linear maximal ideals of $P$ containing $I$. Let $\mathfrak{M} = \langle x_1 - a_1, \ldots, x_n - a_n \rangle$ be a linear maximal ideal in $P$ containing $I$, where $a_1, \ldots, a_n \in K$. Since we have $x_1 - a_1 \in \mathfrak{M}$, the polynomial $z - a_1$ is the $\mathfrak{m}$-eigenfactor of the multiplication map $\vartheta_{\bar{x}_1}$ and therefore divides $\mu_{\bar{x}_1}(z)$. When the iteration of Steps (3)–(7) is in effect for a certain value $i$ and we are working on the ideal $L_j = \langle x_1 - a_1, \ldots, x_{i-1} - a_{i-1} \rangle$, the ideal $L'_j$ is the image of $I$ in the ring $K[x_i, \ldots, x_n]$ under the epimorphism corresponding to $L_j$. The ideal $\mathfrak{m}_i = \langle \bar{x}_i - a_i, \ldots, \bar{x}_n - a_n \rangle$ is the image of $\mathfrak{M}$ under the same epimorphism. Since $\mathfrak{m}_i$ is a linear maximal ideal containing $L'_j$, the polynomial $z - a_i$ has to be one of the linear factors of the minimal polynomial $\mu_{\bar{x}_i}(z)$ of the image $\bar{x}_i$ of $x_i$ in $K[x_i, \ldots, x_n]/L'_j$. Therefore is one of the factors computed in Step (6). Altogether, it follows inductively that the algorithm constructs the ideal $\langle x_1 - a_1, \ldots, x_n - a_n \rangle$ in $S_n$. □

Let us follow the steps of the Eigenvalue Method and compute the rational solutions of some non-trivial polynomial systems.

**Example 6.2.8** Let $K = \mathbb{F}_{101}$, let $P = K[x_1, x_2, x_3]$, and let

$$f_1 = x_3^2 + x_1,$$

$$f_2 = x_1 x_3 - x_2 - x_3 - 1,$$

$$f_3 = x_1^2 + x_2 x_3 - x_1 + x_3,$$

$$f_4 = x_2^3 + 8x_2^2 x_3 - 4x_1 x_2 - 4x_2^2 + 16x_2 x_3 - 8x_1 - 19x_2 - 14,$$

$$f_5 = x_1 x_2^2 + 3x_2^2 x_3 + x_1 x_2 - 2x_2^2 + 7x_2 x_3 - 2x_1 - 6x_2 + 2x_3 - 4$$

To find all $K$-rational solutions of the system defined by these polynomials, we form the ideal $I = \langle f_1, f_2, f_3, f_4, f_5 \rangle$ in $P$ and check that it is a zero-dimensional ideal. A $K$-basis of $R = P/I$ is given by $B = (1, x_1, x_2, x_3, x_1 x_2, x_2^2, x_2 x_3, x_2^2 x_3)$. We follow the steps of Algorithm 6.2.7.

(1) We let $S_0 = (\langle 0 \rangle)$ and $S_1 = S_2 = S_3 = \emptyset$.
(2) Let $i = 1$.
(5) Let $L_1' = I$.
(6) We compute $\mu_{\bar{x}_1}(z) = z^7 + z^6 - 2z^5 - z^4 + 4z^3 + 4z^2 + z$ for $\bar{x}_1 \in P/I$ and factor it as $\mu_{\bar{x}_1}(z) = z(z+1)(z^3 - 2z^2 + z + 1)$.
(7) We form $S_1 = (\langle x_1 \rangle, \langle x_1 + 1 \rangle)$.
(2) Now we consider $i = 2$.
(3) Choose $L_1 = \langle x_1 \rangle$ and delete it from $S_1$.
(5) We have $L_1 + I = \langle x_1 \rangle + I = L_1 + L_1'$ with $L_1' = \langle x_2 + x_3 + 1, x_3^2 \rangle$.
(6) Now we compute the minimal polynomial of $\bar{x}_2$ in $K[x_2, x_3]/L_1'$. It is given by
$$\mu_{\bar{x}_2}(z) = (z+1)^2$$ and has only one linear factor, namely $z + 1$.
(7) Hence we append $\langle x_1, x_2 + 1 \rangle$ to $S_2$.
(3) Choose $L_2 = \langle x_1 + 1 \rangle$ and delete it from $S_1$.
(5) We have $L_2 + I = \langle x_1 + 1 \rangle + I = L_2 + L_2'$ with $L_2' = \langle x_2 + 3, x_3 - 1 \rangle$.
(6) Now we compute the minimal polynomial of $\bar{x}_2$ in $K[x_2, x_3]/L_2'$ and get the linear polynomial $\mu_{\bar{x}_2}(z) = z + 3$.
(7) Hence we append $\langle x_1 + 1, x_2 + 3 \rangle$ to $S_2$.
(3) Next we consider $i = 3$.
(4) Choose $L_1 = \langle x_1, x_2 + 1 \rangle$ and delete it from $S_2$.
(5) We have $L_1 + I = \langle x_1, x_2 + 1, x_3 \rangle$ and $L_1' = \langle x_3 \rangle$.
(6) The minimal polynomial of $\bar{x}_3$ is $\mu_{\bar{x}_3}(z) = z$.
(7) We append $\langle x_1, x_2 + 1, x_3 \rangle$ to $S_3$.
(4) Choose $L_2 = \langle x_1 + 1, x_2 + 3 \rangle$ and delete it from $S_2$.
(5) We have $L_2 + I = \langle x_1 + 1, x_2 + 3, x_3 - 1 \rangle$ and $L_2' = \langle x_3 - 1 \rangle$.
(6) The minimal polynomial of $\bar{x}_3$ $\mu_{\bar{x}_3}(z) = z - 1$.
(7) We append $\langle x_1 + 1, x_2 + 3, x_3 - 1 \rangle$ to $S_3$.

Altogether, we find that the ideal $I$ has two linear maximal components, namely $\mathfrak{M}_1 = \langle x_1, x_2 + 1, x_3 \rangle$ and $\mathfrak{M}_2 = \langle x_1 + 1, x_2 + 3, x_3 - 1 \rangle$, and that the given polynomial system has two $K$-rational solutions, namely $(0, -1, 0)$ and $(-1, -3, 1)$.

Next we redo Example 6.1.11 using Algorithm 6.2.7.

**Example 6.2.9** Let us apply Algorithm 6.2.7 in the setting of Example 6.1.11. In particular, let $K = \mathbb{Q}$, let $V = K^3$, and let $\mathscr{F} = K[\varphi_1, \varphi_2]$ be the commuting family defined there. Using the Buchberger-Möller Algorithm for Matrices 2.1.4, we get $R = P/I$, where $P = K[x_1, x_2]$ and $I = \text{Rel}_P(\varphi_1, \varphi_2) = \langle x_1 x_2, x_1^2, x_2^3 - 2x_2^2 \rangle$. Our goal is to compute the linear maximal ideals of $P$ containing $I$. To simplify the presentation, we only mention the key steps of Algorithm 6.2.7.

(6) We calculate $\mu_{\bar{x}_1}(z) = z^2$.

(5) Let $L_1 = \langle x_1 \rangle$ and form $L_1 + I = \langle x_1, x_2^3 - 2x_2^2 \rangle$. This yields $L_1' = \langle x_2^3 - 2x_2^2 \rangle$.

(6) Modulo $L_1'$ we have $\mu_{\bar{x}_2}(z) = z^2(z - 2)$.

(2) The algorithm returns the list $S_2 = (\langle x_1, x_2 \rangle, \langle x_1, x_2 - 2 \rangle)$.

Hence we conclude that the linear maximal ideals of $P$ containing the ideal $I$ are $\mathfrak{M}_1 = \langle x_1, x_2 \rangle$ and $\mathfrak{M}_2 = \langle x_1, x_2 - 2 \rangle$. This agrees with the result in Example 6.1.11.

Our next example revisits the system of equations given in Example 6.2.5 and solves it more efficiently.

**Example 6.2.10**   In the setting of Example 6.2.5, we compute the linear maximal ideals of $P = K[x_1, x_2, x_3]$ containing the ideal $I = \langle f_1, f_2, f_3, f_4, f_5 \rangle$ and follow the main steps of Algorithm 6.2.7.

(6) We calculate $\mu_{\bar{x}_1}(z) = (z + 6)(z + 5)(z + 2)(z + 1)(z - 2)$.

(5) Let $L_1 = \langle x_1 + 6 \rangle$ and $L_1 + I = \langle x_1 + 6, x_2, x_3 \rangle$. This yields $L_1' = \langle x_2, x_3 \rangle$.

(5) Let $L_2 = \langle x_1 + 5 \rangle$ and $L_2 + I = \langle x_1 + 5, x_2 + 1, x_3 + 1 \rangle$. This yields the ideal $L_2' = \langle x_2 + 1, x_3 + 1 \rangle$.

(5) Let $L_3 = \langle x_1 + 2 \rangle$ and $L_3 + I = \langle x_1 + 2, x_2 + 2, x_3 + 3 \rangle$. This yields the ideal $L_3' = \langle x_2 + 2, x_3 + 3 \rangle$.

(5) Let $L_4 = \langle x_1 + 1 \rangle$ and $L_4 + I = \langle x_1 + 1, x_2 - 1, x_3 - 1 \rangle$. This yields the ideal $L_4' = \langle x_2 - 1, x_3 - 1 \rangle$.

(5) Let $L_5 = \langle x_1 - 2 \rangle$ and $L_5 + I = \langle x_1 - 2, x_2 - 3, x_3 - 2 \rangle$. This yields the ideal $L_5' = \langle x_2 - 3, x_3 - 2 \rangle$.

(2) The algorithm returns the list $S_3 = (\langle x_1 + 6, x_2, x_3 \rangle, \langle x_1 + 5, x_2 + 1, x_3 + 1 \rangle,$ $\langle x_1 + 2, x_2 + 2, x_3 + 3 \rangle, \langle x_1 + 1, x_2 - 1, x_3 - 1 \rangle, \langle x_1 - 2, x_2 - 3, x_3 - 2 \rangle)$.

Altogether, we see that in $P$ there are five linear maximal ideals $\mathfrak{M}_1, \ldots, \mathfrak{M}_5$ containing $I$, namely the entries of $S_3$. Notice that we did not have to try 125 candidates here. Instead, for each factor of $\mu_{\bar{x}_1}(z)$, we found the corresponding solution right away.

In the last part of this subsection we apply Algorithm 6.2.7 to compute rational zeros of not necessarily zero-dimensional ideals. Here we assume that either the base field is finite or we are looking for rational zeros in a predefined finite set.

Let $K$ be a field, let $P = K[x_1, \ldots, x_n]$, and let $I$ be a not necessarily zero-dimensional ideal in $P$. Furthermore, let $\mathbb{X} = \{p_1, \ldots, p_s\}$ be a finite set of points in $K^n$, and let $\mathcal{I}(\mathbb{X})$ be the vanishing ideal of $\mathbb{X}$. Then the ideals $\mathcal{I}(p_1), \ldots, \mathcal{I}(p_s)$ are linear maximal ideals in $P$, and $\mathcal{I}(\mathbb{X})$ is their intersection (see [16], Proposition 6.3.3.a). Our goal is to compute $\mathcal{Z}_K(I) \cap \mathbb{X}$, i.e., the rational zeros of $I$ contained in $\mathbb{X}$. We need the following auxiliary result.

**Lemma 6.2.11**   *Let $R$ be a ring, let $\mathfrak{m}_1, \ldots, \mathfrak{m}_t$ be maximal ideals of $R$, and let $J$ be a proper ideal of $R$ containing $\mathfrak{m}_1 \cap \cdots \cap \mathfrak{m}_t$. Then there exist indices $i_1, \ldots, i_r$ in the set $\{1, \ldots, t\}$ such that $J = \mathfrak{m}_{i_1} \cap \cdots \cap \mathfrak{m}_{i_r}$.*

*Proof* Clearly, we may assume that $t \geq 2$. Since the ideals $\mathfrak{m}_1, \ldots, \mathfrak{m}_t$ are comaximal, we can apply the Chinese Remainder Theorem 2.2.1 and get an isomorphism

$$R/(\mathfrak{m}_1 \cap \cdots \cap \mathfrak{m}_t) \cong R/\mathfrak{m}_1 \times \cdots \times R/\mathfrak{m}_t$$

Now it suffices to consider the image of $J$ in this ring and to note that each factor $R/\mathfrak{m}_i$ has only two ideals: the zero ideal and the unit ideal. $\qquad\square$

The following proposition solves the task we set ourselves above.

**Proposition 6.2.12**  *In the above setting, let* $J = I + \mathscr{I}(\mathbb{X})$.

(a) *The ideal $J$ is a zero-dimensional ideal in $P$ which is an intersection of linear maximal ideals. In particular, it is a radical ideal.*
(b) *We have* $\mathscr{Z}_K(I) \cap \mathbb{X} = \mathscr{Z}_K(J)$.
(c) *We have* $\#(\mathscr{Z}_K(I) \cap \mathbb{X}) = \dim_K(P/J)$

*Proof* Claim (a) follows immediately from the lemma. Claim (b) is a consequence of the fact that the zero set of a sum of ideals is the intersection of the zero sets of the individual ideals (see for instance [15], Sect. 2.6.B). Finally, claim (c) follows from (b) and the Chinese Remainder Theorem 2.2.1. $\qquad\square$

A common situation in which we can apply this proposition is the case where we look for the $K$-rational zeros of an ideal $I$ defined over a finite field $K$. In this case the proposition can be rephrased as follows.

**Corollary 6.2.13**  *Let $K$ be a finite field having $q$ elements.*

(a) *The vanishing ideal of the finite set $\mathbb{X} = K^n$ is* $\mathscr{I}(\mathbb{X}) = \langle x_1^q - x_1, \ldots, x_n^q - x_n \rangle$.
(b) *For every ideal $I$ in $P$, we have* $\mathscr{Z}_K(I) = \mathscr{Z}_K(I + \langle x_1^q - x_1, \ldots, x_n^q - x_n \rangle)$. *The latter ideal is the intersection of the linear maximal ideals containing it.*
(c) *We have* $\#(\mathscr{Z}_K(I)) = \dim_K(P/(I + \langle x_1^q - x_1, \ldots, x_n^q - x_n \rangle))$.

*Proof* Claim (a) follows from the fact that $x_i^q - x_i = \prod_{c \in K}(x_i - c)$ (see [10], Sect. 4.13) and from [16], Proposition 6.3.8.a. Claims (b) and (c) follow from (a) and the proposition. $\qquad\square$

The ideal $\mathscr{I}(K^n) = \langle x_1^q - x_1, \ldots, x_n^q - x_n \rangle$ is also called the **field ideal** in $P$. Let us see two examples which illustrate this corollary.

**Example 6.2.14**  Let $K = \mathbb{F}_7$, let $P = K[x, y, z]$, let $f_1 = x^2 - x - y - z$, and let $f_2 = z^2 + xy + xz$. Our goal is to compute the $K$-rational zeros of $I = \langle f_1, f_2 \rangle$. According to the corollary, we have to compute the linear maximal ideals in $P$ containing $I + \langle x^7 - x, y^7 - y, z^7 - z \rangle$. We apply Algorithm 6.2.7 and get the six linear maximal ideals $\mathfrak{M}_1 = \langle x, y, z \rangle$, $\mathfrak{M}_2 = \langle x + 1, y + 1, z - 3 \rangle$, $\mathfrak{M}_3 = \langle x + 3, y + 1, z + 1 \rangle$, $\mathfrak{M}_4 = \langle x + 1, y + 2, z + 3 \rangle$, $\mathfrak{M}_5 = \langle x + 3, y + 3, z - 1 \rangle$, as well as $\mathfrak{M}_6 = \langle x - 1, y, z \rangle$. Therefore the ideal $I$ has six $K$-rational zeros, namely $(0, 0, 0)$, $(-1, -1, 3)$, $(-3, -1, -1)$, $(-1, -2, -3)$, $(-3, -3, 1)$, and $(1, 0, 0)$.

A similar, but slightly more involved case is given by the following example.

**Example 6.2.15**   Let $K = \mathbb{F}_5$, let $P = K[x_1, x_2, x_3]$, let $f_1 = x_1^3 - x_2^2 - x_2 - x_3$, let $f_2 = x_3^2 + x_1 x_2 + x_1 x_3$, and let $I = \langle f_1, f_2 \rangle$. Our goal is to compute the $K$-rational zeros of $I$. By the corollary, it is sufficient to compute the $K$-rational zeros of the ideal $J = I + \langle x_1^5 - x_1, x_2^5 - x_2, x_3^5 - x_3 \rangle$. We use Algorithm 6.2.7 and follow the main steps.

(6)   We calculate $\mu_{\bar{x}_1}(z) = z^3 + z = z(z - 2)(z + 2)$.

(5)   For $L_1 = \langle x_1 \rangle$, we have $L_1 + J = \langle x_1 \rangle + L_1'$ with $L_1' = \langle x_2^2 + x_2, x_3 \rangle$.

(6)   Modulo $L_1'$, we calculate $\mu_{\bar{x}_2}(z) = z(z + 1)$.

(5)   For $L_2 = \langle x_1 - 2 \rangle$, we have $L_2 + J = \langle x_1 - 2 \rangle + L_2'$ where $L_2' = \langle x_2 - 2x_3 + 1, x_3^2 + x_3 - 2 \rangle$.

(6)   Modulo $L_2'$, we calculate $\mu_{\bar{x}_2}(z) = z(z - 1)$.

(5)   For $L_3 = \langle x_1 + 2 \rangle$, we have $L_3 + J = \langle x_1 + 2 \rangle + L_3'$ where $L_3' = \langle x_2 + 2x_3 + 1, x_3^2 + 2x_3 + 2 \rangle$.

(6)   Modulo $L_3'$, we calculate $\mu_{\bar{x}_2}(z) = z(z - 2)$.

(3)   For the next iteration, we start with the list $S_2 = (\tilde{L}_1, \ldots, \tilde{L}_6)$, where $\tilde{L}_1 = \langle x_1, x_2 \rangle$, $\tilde{L}_2 = \langle x_1, x_2 + 1 \rangle$, $\tilde{L}_3 = \langle x_1 - 2, x_2 \rangle$, $\tilde{L}_4 = \langle x_1 - 2, x_2 - 1 \rangle$, $\tilde{L}_5 = \langle x_1 + 2, x_2 \rangle$, and $\tilde{L}_6 = \langle x_1 + 2, x_2 - 2 \rangle$.

(3)   The algorithm returns the list $S_3 = (G_1, \ldots, G_6)$, where we have $G_1 = \langle x_1, x_2, x_3 \rangle$, $G_2 = \langle x_1, x_2 + 1, x_3 \rangle$, $G_3 = \langle x_1 - 2, x_2, x_3 + 2 \rangle$, $G_4 = \langle x_1 - 2, x_2 - 1, x_3 - 1 \rangle$, $G_5 = \langle x_1 + 2, x_2, x_3 \rangle$, and $G_6 = \langle x_1 + 2, x_2 - 2, x_3 - 1 \rangle$.

Corresponding to the linear maximal ideals in $S_3$, we have six $K$-rational zeros of $I$, namely $(0, 0, 0)$, $(0, -1, 0)$, $(2, 0, -2)$, $(2, 1, 1)$, $(-2, 0, 2)$, and $(-2, 2, 1)$.

Finally, we point out that if we are looking for the $K$-rational zeros of an ideal defined over a finite field for which we already know that it is zero-dimensional, it is still beneficial to add the field equations before starting the computation of the zeros. The following lite example sheds some light on this approach.

**Example 6.2.16**   Let $K = \mathbb{F}_2$, let $P = K[x_1, x_2]$, and let $I = \langle x_1^7 + x_1, x_2^3 + x_1^5 + x_1 \rangle$. To compute the $K$-rational zeros of $I$ using Algorithm 6.2.7, we have to calculate and factor the minimal polynomial $\mu_{\bar{x}_1}(z) = z^7 + z = z(z + 1)^2(z^2 + z + 1)^2$. On the other hand, if we first add the field ideal and then compute the $K$-rational zeros of $J = I + \langle x_1^2 - x_1, x_2^2 - x_2 \rangle = \langle x_1^2 + x_1, x_2 \rangle$, we have to calculate and factor a simpler minimal polynomial, namely $\mu_{\bar{x}_1}(z) = z(z + 1)$.

*Introducing "Lite"—*
*The new way to spell "Light",*
*but with twenty percent fewer letters.*
(Jerry Seinfeld)

## 6.2.B  The Eigenvector Method

*"Why are you late?"*
*"There was a man who lost a hundred dollar bill."*
*"Oh! Were you helping him find it?"*
*"No, I was standing on it."*

In Sect. 6.1, we calculated all 1-dimensional joint eigenspaces of a commuting family and pointed out that this allows us to find all $K$-rational zeros of a polynomial system. Although we had this application right under our shoes, it is now time to pull it out, make good use of it, and present some valuable improvements.

Let us continue to use the setting of the preceding subsection. In particular, we let $K$ be a field, let $P = K[x_1, \ldots, x_n]$, let $f_1, \ldots, f_m \in P$, and let $I = \langle f_1, \ldots, f_m \rangle$. We assume that $I$ is a zero-dimensional ideal and want to compute the $K$-rational zeros of $I$, i.e., the linear maximal ideals of $P$ containing $I$.

The residue class ring $R = P/I$ is a zero-dimensional $K$-algebra, and its multiplication family is given by $\mathscr{F} = K[\vartheta_{x_1}, \ldots, \vartheta_{x_n}]$. If we translate Algorithm 6.1.9 to the present context, we obtain the following result.

---

**Algorithm 6.2.17 (The First Eigenvector Method)**
*In the above setting, let $\mathscr{F}$ be the multiplication family of $R = P/I$, and let $\mathscr{F}^{\vee}$ be its dual family. Consider the following sequence of instructions.*

(1) *Apply Algorithm 6.1.1 to the dual endomorphisms $\vartheta_{x_1}^{\vee}, \ldots, \vartheta_{x_n}^{\vee}$ and compute a set $S$ whose elements are bases of all 1-dimensional joint eigenspaces of $\mathscr{F}^{\vee}$. Let $S = \{(\ell_1), \ldots, (\ell_s)\}$ with $\ell_i \in \omega_R$.*
(2) *Compute $\vartheta_{x_i}^{\vee}(\ell_j) = \lambda_{ij}\ell_j$ with $\lambda_{ij} \in K$ for $i = 1, \ldots, n$ and $j = 1, \ldots, s$.*
(3) *Return the set $\{\wp_1, \ldots, \wp_s\}$, where $\wp_j = (\lambda_{1j}, \ldots, \lambda_{nj})$*

*This is an algorithm which computes the set $\mathcal{Z}_K(I)$ of all $K$-rational zeros of $I$.*

---

*Proof* Since $R$ is a cyclic $\mathscr{F}$-module, we can apply Algorithm 6.1.9 and get this algorithm.                                                                 $\square$

To show the workings of this algorithm, we apply it to the system of equations given in Example 6.2.8.

**Example 6.2.18** Let $K = \mathbb{F}_{101}$, let $K[x_1, x_2, x_3]$, and let $I = \langle f_1, \ldots, f_5 \rangle$ be the zero-dimensional ideal of $P$ defined in Example 6.2.8. A $K$-basis of $R = P/I$ is given by $B = (1, x_1, x_2, x_3, x_1x_2, x_2x_3, x_2^2, x_2^2x_3)$. The multiplication matrices $A_{x_1}$,

$A_{x_2}$, and $A_{x_3}$ are given by

$$
\begin{pmatrix}
0 & 0 & 0 & 1 & 0 & 4 & 0 & 14 \\
1 & 1 & 0 & 0 & 0 & 2 & 0 & 8 \\
0 & 0 & 0 & 1 & 0 & 6 & 1 & 19 \\
0 & -1 & 0 & 1 & 0 & -2 & 0 & 0 \\
0 & 0 & 1 & 0 & 1 & -1 & 0 & 4 \\
0 & 0 & 0 & 0 & 0 & 2 & 1 & 5 \\
0 & -1 & 0 & 0 & -1 & -7 & 1 & -16 \\
0 & 0 & 0 & 0 & -1 & -3 & 0 & -7
\end{pmatrix}
\begin{pmatrix}
0 & 0 & 0 & 0 & 4 & 14 & 0 & 40 \\
0 & 0 & 0 & 0 & 2 & 8 & 0 & 16 \\
1 & 0 & 0 & 0 & 6 & 19 & 0 & -41 \\
0 & 0 & 0 & 0 & -2 & 0 & 0 & 6 \\
0 & 1 & 0 & 0 & -1 & 4 & 0 & 8 \\
0 & 0 & 1 & 0 & 2 & 4 & 0 & 20 \\
0 & 0 & 0 & 1 & -7 & -16 & 0 & -33 \\
0 & 0 & 0 & 0 & -3 & -8 & 1 & -20
\end{pmatrix}
\begin{pmatrix}
0 & 1 & 0 & 0 & 0 & 0 & 0 & -4 \\
0 & 0 & 0 & -1 & 0 & 0 & 0 & -2 \\
0 & 1 & 0 & 0 & 1 & 0 & 0 & -6 \\
1 & 1 & 0 & 0 & 0 & 0 & 0 & 2 \\
0 & 0 & 0 & 0 & 0 & 0 & -1 & 1 \\
0 & 0 & 0 & 0 & 1 & 0 & 0 & -2 \\
0 & 0 & 1 & 0 & 1 & 0 & 0 & 7 \\
0 & 0 & 0 & 0 & 0 & 1 & 0 & 3
\end{pmatrix}
$$

respectively. They represent the generators $\vartheta_{x_1}, \vartheta_{x_2}, \vartheta_{x_3}$ of the multiplication family $\mathscr{F}$. Now let us follow the steps of the algorithm.

(1) We consider the dual family $\mathscr{F}^\vee$ whose generators are represented by $A_{x_1}^{\mathrm{tr}}$, $A_{x_2}^{\mathrm{tr}}$, and $A_{x_3}^{\mathrm{tr}}$. Using Algorithm 6.1.1, we calculate the 1-dimensional joint eigenspaces of this family and get $S = \{(\ell_1), (\ell_2)\}$ where the coordinate column of $\ell_1$ in the basis $B^*$ is $w_1 = (1, 0, -1, 0, 0, 1, 0, 0)^{\mathrm{tr}}$ and the coordinate column of $\ell_2$ in the basis $B^*$ is $w_2 = (1, -1, -3, 1, 3, 9, -3, 9)^{\mathrm{tr}}$.

(2) We calculate $A_{x_1}^{\mathrm{tr}} \cdot w_1 = 0$, $A_{x_2}^{\mathrm{tr}} \cdot w_1 = -w_1$, and $A_{x_3}^{\mathrm{tr}} \cdot w_1 = 0$. Furthermore, we calculate $A_{x_1}^{\mathrm{tr}} \cdot w_2 = -w_2$, $A_{x_2}^{\mathrm{tr}} \cdot w_2 = -3w_2$, and $A_{x_3}^{\mathrm{tr}} \cdot w_2 = w_2$.

(3) The algorithm returns the points $(0, -1, 0)$ and $(-1, -3, 1)$.

Hence the $K$-rational zeros of $I$ are $(0, -1, 0)$ and $(-1, -3, 1)$.

In the following we introduce a refinement of the first eigenvector method. It was initially proposed by W. Auzinger and H.J. Stetter in [1] and further analyzed in [22]. The basis of this refinement is the following theorem which describes evaluation forms as the joint eigenvectors of the dual multiplication family. Given a zero-dimensional ideal $I$ in $P = K[x_1, \ldots, x_n]$ and a point $p \in \mathscr{Z}_K(I)$, the map $\mathrm{ev}_p : P/I \longrightarrow K$ given by $f \mapsto f(p)$ is well-defined. It is called the **evaluation form** associated to $p$.

**Theorem 6.2.19 (Characterization of Evaluation Forms)**
*Let $K$ be a field, let $P = K[x_1, \ldots, x_n]$, let $I$ be a zero-dimensional ideal in $P$, let $R = P/I$, let $\mathscr{F}$ be the multiplication family of $R$, and let $p \in \mathscr{Z}_K(I)$.*

(a) *The evaluation form $\mathrm{ev}_p \in \omega_R$ is a joint eigenvector of the dual family $\mathscr{F}^\vee$ and generates a 1-dimensional joint eigenspace.*
(b) *Every 1-dimensional joint eigenspace of $\mathscr{F}^\vee$ is generated by an evaluation form.*
(c) *For every element $f \in R$, the eigenvalue of $\vartheta_f^\vee$ corresponding to the eigenvector $\mathrm{ev}_p$ is $f(p)$.*
(d) *Given a $K$-basis $B = (t_1, \ldots, t_d)$ of $R$, we have $\mathrm{ev}_p = t_1(p)t_1^* + \cdots + t_d(p)t_d^*$ in $\omega_R$.*

*Proof* First we prove (a). We write $p = (\lambda_1, \ldots, \lambda_n)$ with $\lambda_i \in K$. Then the corresponding linear maximal ideal of $\mathscr{F}$ is $\tilde{\mathfrak{m}} = \langle \vartheta_{x_1} - \lambda_1 \mathrm{id}_R, \ldots, \vartheta_{x_n} - \lambda_n \mathrm{id}_R \rangle$ and the corresponding ideal of $\mathscr{F}^\vee$ is $\tilde{\mathfrak{m}}^\vee = \langle \vartheta_{x_1}^\vee - \lambda_1 \mathrm{id}_R^\vee, \ldots, \vartheta_{x_n}^\vee - \lambda_n \mathrm{id}_R^\vee \rangle$. The ideal $\tilde{\mathfrak{m}}^\vee$, in turn, corresponds to the joint eigenspace

$$
\mathrm{Ker}(\tilde{\mathfrak{m}}^\vee) = \big\{ \ell \in \omega_R \mid \varphi^\vee(\ell) = \ell \circ \varphi = 0 \text{ for all } \varphi \in \tilde{\mathfrak{m}} \big\}
$$

of $\mathcal{F}^{\vee}$ which is 1-dimensional since $\mathcal{F}^{\vee}$ is commendable. In other words, the joint eigenspace $\mathrm{Ker}(\tilde{\mathfrak{m}}^{\vee})$ is the set of elements of $\omega_R$ which vanish at all elements of $\tilde{\mathfrak{m}}$. Therefore it suffices to show that $\mathrm{ev}_p$ is a non-zero linear form which vanishes at $\tilde{\mathfrak{m}}$. Under the isomorphism $\mathcal{F} \cong R$ given by $\vartheta_f \mapsto f$, the ideal $\tilde{\mathfrak{m}}$ corresponds to $\mathfrak{m} = \langle x_1 - \lambda_1, \ldots, x_n - \lambda_n \rangle$. The claim that $(\mathrm{ev}_p \circ \vartheta_f)(g) = \mathrm{ev}_p(fg) = 0$ for all $\vartheta_f \in \tilde{\mathfrak{m}}$ and all $g \in R$ then follows from the fact that the elements $f \in \mathfrak{m}$ vanish at $p$. Since $\mathrm{ev}_p(1) = 1$, we have $\mathrm{ev}_p \neq 0$, and hence $\mathrm{ev}_p$ generates the 1-dimensional joint eigenspace $\mathrm{Ker}(\tilde{\mathfrak{m}}^{\vee})$.

Next we show (b). A 1-dimensional joint eigenspace of $\mathcal{F}^{\vee}$ is of the form $\mathrm{Ker}(\tilde{\mathfrak{m}}^{\vee})$ with a linear maximal ideal $\tilde{\mathfrak{m}}^{\vee} = \langle \vartheta_{x_1}^{\vee} - \lambda_1 \mathrm{id}_R^{\vee}, \ldots, \vartheta_{x_n}^{\vee} - \lambda_n \mathrm{id}_R^{\vee} \rangle$ of $\mathcal{F}^{\vee}$, where $\lambda_1, \ldots, \lambda_n \in K$. By Algorithm 6.2.17, the point $p = (\lambda_1, \ldots, \lambda_n)$ is a $K$-rational zero of $I$. Hence the proof of (a) shows that the evaluation form $\mathrm{ev}_p$ generates the eigenspace $\mathrm{Ker}(\tilde{\mathfrak{m}}^{\vee})$.

Since claim (c) follows from Corollary 2.3.9, only claim (d) remains to be proved. Every element of $\omega_R$ can be written as $\sum_{i=1}^d c_i t_i^*$ with $c_i \in K$. So, we let $c_i \in K$ be such that $\mathrm{ev}_p = \sum_{i=1}^d c_i t_i^*$. By definition, we have $\mathrm{ev}_p(t_i) = t_i(p)$, while $(\sum_{i=1}^d c_i t_i^*)(t_i) = c_i$. From this we deduce that $t_i(p) = c_i$ for $i = 1, \ldots, d$, and the proof is complete.                                                                                     $\square$

Using this theorem, we can improve the first eigenvector method as follows: if the residue classes of $x_1, \ldots, x_n$ are part of the $K$-basis of $R$ that we use, the coordinates $x_i(p)$ of a $K$-rational zero $p$ of $I$ can be read off the coefficients of the linear form $\mathrm{ev}_p$, and hence off a generator of the corresponding 1-dimensional joint eigenspace of $\mathcal{F}^{\vee}$. The precise formulation of this method reads as follows.

---

**Algorithm 6.2.20 (The Second Eigenvector Method)**
*In the setting of the theorem, let $B = (t_1, \ldots, t_d)$ be a $K$-basis of $R$ such that $t_1 = 1$ and $t_{i+1} = x_i$ for $i = 1, \ldots, n$. Consider the following instructions.*

(1) *Using Algorithm 6.1.1, compute a set $S$ of bases of all 1-dimensional joint eigenspaces of $\mathcal{F}^{\vee}$. Let $S = \{(\ell_1), \ldots, (\ell_s)\}$ with $\ell_i \in \omega_R$.*
(2) *For $i = 1, \ldots, s$, write $\ell_i = a_{i1} t_1^* + \cdots + a_{id} t_d^*$ with $a_{ij} \in K$. Return the set of tuples $\{(a_{i2}/a_{i1}, \ldots, a_{in+1}/a_{i1}) \mid i = 1, \ldots, s\}$.*

*This is an algorithm which computes the set $\mathcal{Z}_K(I)$ of all $K$-rational zeros of $I$.*

---

*Proof* The algorithm is clearly finite. Therefore it suffices to prove correctness. Let $i \in \{1, \ldots, s\}$. By the theorem, the generator $\ell_i$ of the 1-dimensional joint eigenspace $K \cdot \ell_i$ of $\mathcal{F}^{\vee}$ is of the form $\ell_i = c_i \cdot \mathrm{ev}_{p_i}$, where $c_i \in K$ and $p_i \in \mathcal{Z}_K(I)$ is the corresponding zero of $I$. We write $\ell_i = a_{i1} t_1^* + \cdots + a_{id} t_d^*$ with $a_{ij} \in K$ and $p_i = (\lambda_{i1}, \ldots, \lambda_{in})$ with $\lambda_{ij} \in K$. According to part (d) of the theorem, we have $\mathrm{ev}_{p_i} = t_1^* + \lambda_{i1} t_2^* + \cdots + \lambda_{in} t_{n+1}^* + \sum_{j=n+2}^d e_j t_j^*$ with $e_j \in K$. By comparing coefficients in $\ell_i = c_i \cdot \text{ev}_{p_i}$, we see that we have $c_i = a_{i1}$, and for $j = 1, \ldots, n$, we get $\lambda_{ij} = a_{ij+1}/a_{i1}$. □

Let us apply this algorithm in some concrete cases.

**Example 6.2.21** Over the field $K = \mathbb{Q}$, consider the system defined by the polynomials

$$f_1 = x^2 + \tfrac{4}{3}xy + \tfrac{1}{3}y^2 - \tfrac{7}{3}x - \tfrac{5}{3}y + \tfrac{4}{3}$$

$$f_2 = y^3 + \tfrac{10}{3}xy + \tfrac{7}{3}y^2 - \tfrac{4}{3}x - \tfrac{20}{3}y + \tfrac{4}{3}$$

$$f_3 = xy^2 - \tfrac{7}{3}xy - \tfrac{7}{3}y^2 - \tfrac{2}{3}x + \tfrac{11}{3}y + \tfrac{2}{3}$$

Let $I$ be the ideal in $P = K[x, y]$ generated by the set $\{f_1, f_2, f_3\}$. It is easy to check that this set is a Gröbner basis of $I$ with respect to the term ordering $\sigma = \text{DegRevLex}$. Hence $\mathcal{O} = \mathbb{T}^2 \setminus \text{LT}_\sigma(I) = \{1, x, y, xy, y^2\}$ is an order ideal of terms whose residue classes form a $K$-basis of $P/I$. We note that $\mathcal{O}$ satisfies the hypotheses of the algorithm. Let us apply the eigenvector method in this case.

(1) First we compute the multiplication matrices $A_x$ and $A_y$ with respect to $\mathcal{O}$ and get

$$A_x = \begin{pmatrix} 0 & -\tfrac{4}{3} & 0 & \tfrac{4}{3} & -\tfrac{2}{3} \\ 1 & \tfrac{7}{3} & 0 & -\tfrac{4}{3} & \tfrac{2}{3} \\ 0 & \tfrac{5}{3} & 0 & \tfrac{4}{3} & -\tfrac{11}{3} \\ 0 & -\tfrac{4}{3} & 1 & \tfrac{1}{3} & \tfrac{7}{3} \\ 0 & -\tfrac{1}{3} & 0 & -\tfrac{2}{3} & \tfrac{7}{3} \end{pmatrix} \quad A_y = \begin{pmatrix} 0 & 0 & 0 & -\tfrac{2}{3} & -\tfrac{4}{3} \\ 0 & 0 & 0 & \tfrac{2}{3} & \tfrac{4}{3} \\ 1 & 0 & 0 & -\tfrac{11}{3} & \tfrac{20}{3} \\ 0 & 1 & 0 & \tfrac{7}{3} & -\tfrac{10}{3} \\ 0 & 0 & 1 & \tfrac{7}{3} & -\tfrac{7}{3} \end{pmatrix}$$

The minimal polynomial of $A_x^{\text{tr}}$ is $\mu_{\vartheta_x^{\vee}} = (z+1)(z-1)(z-2)$. The map $\vartheta_x^{\vee}$ has one 1-dimensional and two 2-dimensional eigenspaces. If we restrict $\vartheta_y^{\vee}$ to the 2-dimensional eigenspaces, the fact that $\mu_{\vartheta_y^{\vee}}(z) = z(z-1)(z+1)(z-2)(z+2)$ implies that these eigenspaces split further and there are five 1-dimensional joint eigenspaces of the family $\mathcal{F}^{\vee}$. Algorithm 6.1.1 then returns the five basis vectors whose coordinate tuples with respect to $\mathcal{O}^*$ are $(1, 1, 0, 0, 0)$, $(1, 1, 1, 1, 1)$, $(1, 2, -1, -2, 1)$, $(1, -1, 2, -2, 4)$, and $(1, 2, -2, -4, 4)$.

(2) From Step (1) we get $\mathcal{Z}_K(I) = \{(1, 0), (1, 1), (2, -1), (-1, 2), (2, -2)\}$.

In the following example, we apply Algorithm 6.2.20 to a case where every single endomorphism in $\mathcal{F}^{\vee}$ is derogatory (see Example 6.1.6).

**Example 6.2.22** Over the field $K = \mathbb{Q}$, consider the system defined by the polynomials

$$g_1 = x^3 - 5x^2 + 8x - 4$$
$$g_2 = x^2y + 2x^2 - 12x - 4y + 16$$
$$g_3 = xy + x^2 - 5x - 2y + 6$$
$$g_4 = y^2 - x^2 + 4x - 2y - 3$$

Let $P = K[x, y]$, and let $I \subset P$ be the ideal generated by $\{g_1, g_2, g_3, g_4\}$. Our goal is to determine $\mathscr{Z}_K(I)$. It is clear that $G = (g_1, g_2, g_3, g_4)$ is an $\mathscr{O}$-border prebasis of $I$ with respect to $\mathscr{O} = (1, x, y, x^2)$. The corresponding formal multiplication matrices are

$$A_x = \begin{pmatrix} 0 & 0 & -6 & 4 \\ 1 & 0 & 5 & -8 \\ 0 & 0 & 2 & 0 \\ 0 & 1 & -1 & 5 \end{pmatrix} \quad \text{and} \quad A_y = \begin{pmatrix} 0 & -6 & 3 & -16 \\ 0 & 5 & -4 & 12 \\ 1 & 2 & 2 & 4 \\ 0 & -1 & 1 & -2 \end{pmatrix}$$

Since $A_x A_y = A_y A_x$, the set $G$ is actually an $\mathscr{O}$-border basis of $I$, i.e., the residue classes of the elements of $\mathscr{O}$ form a $K$-basis of $R = P/I$.

The transposed matrices $A_x^{\text{tr}}$ and $A_y^{\text{tr}}$ agree with the matrices $A_1$ and $A_2$ in Example 6.1.6. Therefore this example shows that every endomorphism in the family $\mathscr{F}^{\vee} = K[\vartheta_x{}^{\vee}, \vartheta_y{}^{\vee}]$ is derogatory. Let us now find the $K$-rational solutions of the given system using Algorithm 6.2.20.

(1) The minimal polynomial of $A_x^{\text{tr}}$ is $\mu_{\vartheta_x{}^{\vee}}(z) = (z-1)(z-2)^2$. The corresponding eigenspaces of $\vartheta_x{}^{\vee}$ are $\text{Eig}(\vartheta_x{}^{\vee}, z-1) = \langle (1, 1, 2, 1)\rangle$ and $\text{Eig}(\vartheta_x{}^{\vee}, z-2) = \langle (1, 2, 0, 4), (0, 0, 1, 0)\rangle$ where the coordinate tuples are given w.r.t. $\mathscr{O}^*$. The matrix of the restriction of $\vartheta_y{}^{\vee}$ to the second eigenspace of $\vartheta_x{}^{\vee}$ w.r.t. the basis vectors $\ell_1 = 1^* + 2x^* + 4(xy)^*$ and $\ell_2 = y^*$ is $\begin{pmatrix} 0 & 1 \\ -1 & 2 \end{pmatrix}$, and the minimal polynomial is $(z-1)^2$. The corresponding eigenspace is $\langle \ell_1 + \ell_2\rangle$. Thus Algorithm 6.1.1 returns the two basis vectors whose coordinates with respect to $\mathscr{O}^*$ are $(1, 1, 2, 1)$ and $(1, 2, 1, 4)$.

(2) Consequently, we get $\mathscr{Z}_K(I) = \{(1, 2), (2, 1)\}$.

What happens if the $K$-basis $B$ of $R$ does not satisfy the hypothesis of Algorithm 6.2.20? The following remark shows us a way out.

### Remark 6.2.23 (Missing Indeterminates)
Let $I$ be a zero-dimensional ideal in $P = K[x_1, \ldots, x_n]$, and let $R = P/I$.

(a) Clearly, we can assume that the $K$-basis $B$ of $R$ required by Algorithm 6.2.20 contains the element 1. But it may happen that a polynomial $a_0 + a_1 x_1 + \cdots + a_n x_n$ with $a_i \in K$ is contained in $I$. In this case we cannot find a $K$-basis $B$ of $R$ containing $(1, x_1, \ldots, x_n)$. Conversely, if the ideal $I$ does not contain a polynomial of degree 1 then the residue classes of $1, x_1, \ldots, x_n$ are $K$-linearly independent and we can extend them to a $K$-basis $B$ of $R$ satisfying the hypothesis of Algorithm 6.2.20.

(b) Now assume that we have a $K$-basis $B = (t_1, \ldots, t_d)$ of $R$ consisting of residue classes of polynomials $T_1, \ldots, T_d \in P$ which do not involve some of the indeterminates. Without loss of generality, assume that $T_1, \ldots, T_d \in K[x_1, \ldots, x_m]$ for some $m < n$. Then the residue class in $R$ of any indeterminate $x_i$ with

$i \in \{m + 1, \ldots, n\}$ can be written as a $K$-linear combination of the elements of $B$. Thus it is given by a polynomial expression of the residue classes of $x_1, \ldots, x_m$.

Now we consider the polynomial ring $P' = K[x_1, \ldots, x_m]$ and the ideal $I'$ in $P'$ obtained by substituting $x_{m+1}, \ldots, x_n$ by their expressions in $x_1, \ldots, x_m$. We have $P'/I' \cong P/I$ and the residue classes of $(T_1, \ldots, T_d)$ are a $K$-basis of $P'/I'$. Hence we can run Algorithm 6.2.20 for this setting and compute the first $m$ coordinates of every point in $\mathcal{Z}_K(I)$. Finally, for $i = m + 1, \ldots, n$, we use the expressions of $x_i$ in terms of $x_1, \ldots, x_m$ to compute the remaining coordinates of the points in $\mathcal{Z}_K(I)$.

In the next example we show how one can apply this remark in practice.

**Example 6.2.24** Let $K = \mathbb{Q}$, let $P = K[x_1, x_2]$, and let $I$ be the ideal of $P$ defined as $I = \langle x_1 - x_2^2 + x_2 - 1, x_2^3 - 8 \rangle$. If we calculate the reduced `DegRevLex`-Gröbner basis of $I$, we get $G = (x_1^2 - 3x_1 - 9x_2 + 18, x_1 x_2 + x_1 - 9, x_2^2 - x_1 - x_2 + 1)$. Hence $B = (1, x_1, x_2)$ is a $K$-basis of $R = P/I$ which satisfies the hypothesis of Algorithm 6.2.20.

On the other hand, the reduced `Lex`-Gröbner basis of $I$ is $G' = (x_1 - x_2^2 + x_2 - 1, x_2^3 - 8)$. Therefore $B' = (1, x_2, x_2^2)$ is also a $K$-basis of $R = P/I$. The indeterminate $x_1$ is missing, but we have the equality $x_1 + I = x_2^2 - x_2 + 1 + I$ in $R$. We run Algorithm 6.2.20 with $P' = K[x_2]$ and $I' = \langle x_2^3 - 8 \rangle$ and get the $K$-rational zero $(a_2)$ of $I'$, where $a_2 = 2$. Now we calculate the corresponding $x_1$-value from $a_1 = a_2^2 - a_2 + 1 = 3$ and conclude that $\mathcal{Z}_K(I)$ consists of one point $(a_1, a_2)$, namely the point $(a_1, a_2) = (2, 3)$.

## 6.3 Solving Polynomial Systems over Finite Fields

> *I didn't say that I didn't say it.*
> *I said that I didn't say that I said it.*
> *I want to make that very clear.*
>
> (George Romney)

In this section we start to look for solutions of polynomial systems which aren't there. Are we looking for a black cat in a dark cellar which isn't there? Not if we are working over a finite field. In this case we can easily extend the field to a larger finite field and find the solutions there. In order to make this process very clear, we have to tread carefully.

As a first step, let us reconsider the meaning of the symbol $\mathbb{F}_q$ for a finite field with $q$ elements. Here $q = p^e$ is a power of a prime number $p$ and $e > 0$. But how exactly do the elements of $\mathbb{F}_q$ look like when $e > 1$? Didn't we say at the beginning of Sect. 5.2 that $\mathbb{F}_q$ is an isomorphism class? So, in order to construct a field in which we can find our desired solutions, we have to construct a *representative* of $\mathbb{F}_q$, i.e., a presentation using generators and relations. Any two representatives are isomorphic, because every field with $q$ elements is a splitting field of the polynomial $z^q - z$ in $\mathbb{F}_p[z]$ (see Definition 6.3.14), and it is known that splitting fields are unique up to isomorphism (see for instance [21], Proposition 2.7). If we are actually given two representatives of $\mathbb{F}_q$, how do we find such an isomorphism? Can we calculate all isomorphisms?

The first subsection deals with this question. Since we are assuming that the polynomial system is given over a finite field $K$ having $q$ elements, we are looking for finite extension fields of $K$. Thus, in Theorem 6.3.4, we compute all $K$-algebra isomorphisms between finite extension fields of $K$. An immediate application is then the possibility to determine all $K$-automorphisms of a finite extension field $L$ of $K$. Recall that the set of these $K$-automorphisms is the Galois group of $L/K$ and that it is known that this is a cyclic group of order $\dim_K(L)$ which is generated by the $q$-Frobenius automorphism of $L$.

How can we apply these algorithms to solving polynomial systems? First of all, for a polynomial system defined by polynomials $f_1, \ldots, f_m \in K[x_1, \ldots, x_n]$, we can compute the maximal components of the ideal $I = \langle f_1, \ldots, f_m \rangle$ using the methods presented in Chap. 5. Hence the problem of solving the given system is reduced to the problem of solving a polynomial system whose associated ideal is a maximal ideal $\mathfrak{M}$ of $P = K[x_1, \ldots, x_n]$.

The next step is the central idea: cloning! Since $L = P/\mathfrak{M}$ is an extension field of $K$, should we try to find the solutions there? Not exactly. It is better to produce a *clone* $L'$ of this field and find the coordinates of the zeros of the ideal $\mathfrak{M}$ in the clone $L' = P'/\mathfrak{M}'$, where $P' = K[y_1, \ldots, y_n]$. Then we know already one solution, namely $(\bar{y}_1, \ldots, \bar{y}_n)$. And now the *power of the Frobenius* automorphism kicks in: all other solutions are nothing but coordinatewise powers of this given solution. There is one disadvantage that could spoil the fun, namely that we may have to bring high powers to their normal form. For this task we offer a good approach via Algorithm 6.3.16.

A second way of using an isomorphism $P/\mathfrak{M} \cong L'$ is studied in the third subsection, where we assume that we know a univariate representation of $L'$, for instance, because $L'$ is represented in our computer algebra system via a built-in polynomial $f(y)$ such that $L' = K[y]/\langle f(y) \rangle$. Finally, the last subsection suggests the iterative approach to split one generator of $\mathfrak{M}$ at a time. This may be a viable alternative if the $K$-vector space dimension of $P/\mathfrak{M}$ is very large. The iterative approach will also turn out to be useful for solving polynomial systems over $\mathbb{Q}$ in the last section.

As we said before, and we never repeat ourselves, the algorithms in this section are rather tricky. We want to make it very clear that we tried to make them very clear. Don't say that we didn't say it!

> *Drawing on my fine command of language,*
> *I said nothing.*

## 6.3.A Computing Isomorphisms of Finite Fields

> *I have just changed my diet.*
> *The cookies are now to the left of the laptop.*

In the following we approach the problem of solving polynomial systems over a finite field from a rather general point of view. As we mentioned above, to find the solutions we may have to extend the base field and to perform certain calculations over the extended field. How should we represent this field extension?

Nowadays many computer algebra systems have efficient implementations of finite fields available. Can we make our algorithms somehow feed on their speed? Clearly, the representative of $\mathbb{F}_q$ in your favourite computer algebra system may differ from the representative obtained from the polynomial system. Hence we have to change the representative. But how do you do that? Is it as easy as putting the cookie jar on the other side of the laptop? Or do we have to keep some cookies inside the computer?

Let us do a hop, step and jump. The first part of this subsection provides a quick hop through some material about isomorphisms of affine algebras which extends [15], Sect. 3.6. Then we step into the calculation of all $\mathbb{F}_q$-isomorphisms between two representatives of $\mathbb{F}_{q^d}$. This includes the calculation of all $\mathbb{F}_p$-automorphisms of $\mathbb{F}_q$. Finally, a third jump brings us to a small cookie reward, namely an algorithm for computing the primitive elements of an extension of a finite field which have the same minimal polynomial.

> *Come to the dark side, we have cookies!*

In the following we let $K$ be a field, let $P = K[x_1, \ldots, x_n]$, let $I$ be an ideal in $P$, let $P' = K[y_1, \ldots, y_m]$ be another polynomial ring, let $I'$ be an ideal in $P'$, and let $\varphi : P/I \longrightarrow P'/I'$ be a $K$-algebra homomorphism. Our goal is to extend some results of [15], Sect. 3.6. In particular, we shall characterize when $\varphi$ is an isomorphism. The next lemma provides some background for maps between polynomial rings. In the following arguments we use [15], Sect. 3.6. We note that $J$ has a different meaning here.

**Lemma 6.3.1** *In the above setting, assume that $\Phi : P \longrightarrow P'$ is the $K$-algebra homomorphism which is given by $\Phi(x_i) = f_i$ with $f_i \in P'$ for $i = 1, \ldots, n$. Let $Q$ be the ring $Q = K[x_1, \ldots, x_n, y_1, \ldots, y_m]$ and $J$ the ideal $J = \langle x_1 - f_1, \ldots, x_n - f_n \rangle$ in $Q$.*

(a) *The map $\Phi$ induces a $K$-algebra homomorphism $\varphi : P/I \longrightarrow P'/I'$ if and only if $IQ \subseteq I'Q + J$.*

*For the following claims, assume that the map $\Phi$ induces a $K$-algebra homomorphism $\varphi : P/I \longrightarrow P'/I'$.*

(b) *The map $\varphi : P/I \longrightarrow P'/I'$ is injective if and only if we have $I = (I'Q + J) \cap P$.*

(c) *The map* $\varphi : P/I \longrightarrow P'/I'$ *is surjective if and only if there exist polynomials* $g_1, \ldots, g_m \in P$ *such that for the ideal* $J' = \langle y_1 - g_1, \ldots, y_m - g_m \rangle$ *in* $Q$ *we have* $J' \subseteq I'Q + J$. *In this case we have* $\varphi(\bar{g}_i) = \bar{y}_i$ *for* $i = 1, \ldots, m$.

(d) *Assume that the map* $\varphi : P/I \longrightarrow P'/I'$ *is surjective. Let* $g_1, \ldots, g_m \in P$ *be polynomials such that* $\varphi(\bar{g}_i) = \bar{y}_i$ *for* $i = 1, \ldots, m$, *and let* $\Psi : P' \longrightarrow P$ *be the* $K$-*algebra homomorphism defined by* $y_i \mapsto g_i$. *Then* $\Psi$ *induces a* $K$-*algebra homomorphism* $\psi : P'/I' \longrightarrow P/I$ *such that* $\varphi \circ \psi = \mathrm{id}_{P'/I'}$.

*Proof* First we prove (a). Notice that [15], Proposition 3.6.1 yields the relation $g \in g(f_1, \ldots, f_n) + J$ for every $g \in P$. If we have $IQ \subseteq I'Q + J$, then this shows $g(f_1, \ldots, f_n) \in I'Q + J$ for every $g \in I$. Now we substitute $x_i \mapsto f_i$ for $i = 1, \ldots, n$ here and get $g(f_1, \ldots, f_n) \in I'$. Hence the map $\varphi$ is well-defined. Conversely, assume that $\varphi$ is well-defined. This means that every $g \in I$ satisfies $g(f_1, \ldots, f_n) \in I'$. Therefore we get $IQ \subseteq I'Q + J$.

Claim (b) follows immediately from (a) and [15], Proposition 3.6.2. In part (c), if the map $\varphi$ is surjective, then [15], Proposition 3.6.6.d shows that polynomials $g_1, \ldots, g_m \in P$ with $y_i - g_i \in I'Q + J$ exist. Conversely, suppose that such polynomials $g_1, \ldots, g_m \in P$ exist, and let $\sigma$ be an elimination ordering for $(y_1, \ldots, y_m)$ on $\mathbb{T}(x_1, \ldots, x_n, y_1, \ldots, y_m)$. Then the normal form $\mathrm{NF}_{\sigma, I'Q+J}(y_i)$ equals $\mathrm{NF}_{\sigma, I'Q+J}(g_i)$ and $g_i \in P$ implies that also the latter normal form is in $P$. Consequently, we have $y_i + I' \in \mathrm{Im}(\varphi)$ by [15], Proposition 3.6.6.a.

Finally, claim (d) follows from (c), since $(\varphi \circ \psi)(\bar{y}_i) = \varphi(\bar{g}_i) = \bar{y}_i$ for all $i = 1, \ldots, m$.                                                                                                    $\square$

Based on this lemma, we can characterize isomorphisms of finitely generated $K$-algebras as follows.

**Proposition 6.3.2 (Isomorphisms of Affine $K$-Algebras)**
*Let* $K$ *be a field, let* $P$ *be the polynomial ring* $P = K[x_1, \ldots, x_n]$, *let* $I \subseteq P$ *be an ideal, let* $P' = K[y_1, \ldots, y_m]$, *and let* $I' \subseteq P'$ *be an ideal. Moreover, given* $f_1, \ldots, f_n \in P'$, *let* $\Phi : P \longrightarrow P'$ *be the* $K$-*algebra homomorphism defined by* $\Phi(x_i) = f_i$ *for* $i = 1, \ldots, n$, *and let* $J$ *be the ideal* $J = \langle x_1 - f_1, \ldots, x_n - f_n \rangle$ *in* $Q = K[x_1, \ldots, x_n, y_1, \ldots, y_m]$.

*Furthermore, let* $g_1, \ldots, g_m$ *be polynomials in* $P$, *let* $\Psi : P' \longrightarrow P$ *be the* $K$-*algebra homomorphism defined by* $\Psi(y_j) = g_j$ *for* $j = 1, \ldots, m$, *and let* $J' = \langle y_1 - g_1, \ldots, y_m - g_m \rangle \subseteq Q$. *Then the following conditions are equivalent.*

(a) *The homomorphisms* $\Phi$ *and* $\Psi$ *defined above induce* $K$-*algebra isomorphisms* $\varphi : P/I \longrightarrow P'/I'$ *and* $\psi : P'/I' \longrightarrow P/I$ *which are inverse to each other.*

(b) *We have the equality* $IQ + J' = I'Q + J$.

*Proof* First we show (a) $\Rightarrow$ (b). By parts (a) and (c) of the lemma, we have the inclusion $IQ + J' \subseteq I'Q + J$. The converse inclusion follows by interchanging the roles of $\Phi$ and $\Psi$.

Now we prove (b) $\Rightarrow$ (a). By part (a) of the lemma, the maps $\varphi$ and $\psi$ are well-defined. Since the inclusion $I \subseteq (IQ + J') \cap P$ clearly holds, and since the inclusion

$(IQ + J') \cap P \subseteq I$ can be shown using the same arguments as in the proof of part (a) of the lemma, we have $I = (IQ + J') \cap P$. Then the hypothesis implies the equality $I = (I'Q + J) \cap P$, and part (b) of the lemma shows that $\varphi$ is injective. From part (c) of the lemma we get that $\varphi$ is surjective. Hence $\varphi$ is an isomorphism. By interchanging the roles of $\Phi$ and $\Psi$, it follows that also $\psi$ is an isomorphism. Finally, we note that part (d) of the lemma yields that $\psi$ is inverse to $\varphi$. $\qquad\square$

In the following we apply this proposition to compute all isomorphisms between two representatives of a finite field $\mathbb{F}_q$ with $q = p^e$ for some prime number $p$ and $e > 0$. In field theory it is shown that the group of all automorphisms of $\mathbb{F}_q$ is a cyclic group of order $e$ generated by the Frobenius homomorphism. Since we are interested in $\mathbb{F}_q$-isomorphisms between fields having $q^d$ elements, we generalize this result as follows.

**Lemma 6.3.3** *Let $K$ be a field having $q$ elements, and let $L$ and $L'$ be two representatives of $\mathbb{F}_{q^d}$ for some $d > 0$. Then there exist exactly $d$ distinct isomorphisms of $K$-algebras $\varphi_i : L \longrightarrow L'$ with $i \in \{1, \dots, d\}$.*

*Proof* Let $a \in L$ be a generator of the cyclic group $L^\times = L \setminus \{0\}$. Consequently, the $K$-algebra homomorphism $K[x] \longrightarrow L$ defined by $x \mapsto a$ is surjective. Hence there exists an irreducible polynomial $f \in K[x]$ such that $\varepsilon : K[x]/\langle f \rangle \longrightarrow L$ is an isomorphism. Similarly, we find an isomorphism $\eta : K[y]/\langle g \rangle \longrightarrow L'$ for some irreducible polynomial $g \in K[y]$.

Now we consider the $L'$-algebra $R = L'[x]/f L'[x]$. Since $f$ is irreducible and $L'$ is a perfect field, the polynomial $f$ is squarefree in $L'[x]$ (see [15], Proposition 3.9.7.d). Hence $R$ is a reduced ring. From $a^{q^d} = a$ we get the relation $x^{q^d} - x \in f K[x] \subseteq f L'[x]$, and hence $\bar{x}^{q^d} = \bar{x}$ in $R$. Consequently, every $\bar{h} \in R$ satisfies $\bar{h}^{q^d} = \bar{h}$. By Theorem 5.2.4.c, it follows that the zero ideal of $R$ is the intersection of $d$ linear maximal ideals. Hence $f L'[x]$ is the intersection of $d$ linear maximal ideals in $L'[x]$. They correspond to $d$ distinct zeros $b_1, \dots, b_d$ of $f$ in $L'$. For $i \in \{1, \dots, d\}$, the $K$-algebra homomorphism $\varphi_i : K[x]/\langle f \rangle \longrightarrow K[y]/\langle g \rangle$ given by $\bar{x} \mapsto \eta^{-1}(b_i)$ is well-defined. Since both fields have the same number of elements, the map $\varphi_i$ is bijective. Thus we have found $d$ distinct $K$-algebra isomorphisms between $L$ and $L'$. The fact that these are all $K$-algebra isomorphisms follows from the observation that a well-defined map $\varphi_i$ has to map $\bar{x}$ to a zero of $f$ in $K[y]/\langle g \rangle \cong L'$. $\qquad\square$

Now we are ready to state and prove the main theorem of this subsection.

**Theorem 6.3.4 (Isomorphisms between Representatives of $\mathbb{F}_{q^d}$)**
*Let $p$ be a prime number, let $q = p^e$, let $K$ be a field with $q$ elements, and let $d > 0$. Assume that we are given two representatives of $\mathbb{F}_{q^d}$, namely $L = P/\mathfrak{M}$ with a maximal ideal $\mathfrak{M}$ of $P = K[x_1, \dots, x_n]$ and $L' = P'/\mathfrak{M}'$ with a maximal ideal $\mathfrak{M}'$ of $P' = K[y_1, \dots, y_m]$.*

(a) *In the ring $Q = K[x_1, \ldots, x_n, y_1, \ldots, y_m]$, the ideal $\mathfrak{M}Q + \mathfrak{M}'Q$ is a radical ideal. It has a primary decomposition of the form $\mathfrak{M}Q + \mathfrak{M}'Q = \mathfrak{M}_1 \cap \cdots \cap \mathfrak{M}_d$ with maximal ideals $\mathfrak{M}_i$.*

(b) *There exists polynomials $f_{ij} \in P'$ and $g_{ik} \in P$ with $i \in \{1, \ldots, d\}$, with $j \in \{1, \ldots, n\}$, and with $k \in \{1, \ldots, m\}$ such that, for $i = 1, \ldots, d$, the ideal $\mathfrak{M}_i$ in (a) is given by*

$$\mathfrak{M}_i = \mathfrak{M}'Q + \langle x_1 - f_{i1}, \ldots, x_{in} - f_{in} \rangle = \mathfrak{M}Q + \langle y_1 - g_{i1}, \ldots, y_m - g_{im} \rangle$$

*In particular, the primary decomposition of the ideal $\mathfrak{M}L'[x_1, \ldots, x_n]$ is given by $\mathfrak{M}L'[x_1, \ldots, x_n] = \bigcap_{i=1}^{d} \langle x_1 - \bar{f}_{i1}, \ldots, x_n - \bar{f}_{in} \rangle$, and an analogous formula holds for $\mathfrak{M}'L[y_1, \ldots, y_m]$.*

(c) *For $i = 1, \ldots, d$, let $\Phi_i : P \longrightarrow P'$ be the $K$-algebra homomorphism defined by $\Phi_i(x_j) = f_{ij}$ for $j = 1, \ldots, n$, and let $\Psi_i : P' \longrightarrow P$ be the $K$-algebra homomorphism defined by $\Psi_i(y_k) = g_{ik}$ for $k = 1, \ldots, m$. Then, for $i = 1, \ldots, d$, the maps $\Phi_i$ and $\Psi_i$ induce $K$-algebra isomorphisms $\varphi_i : P/\mathfrak{M} \longrightarrow P'/\mathfrak{M}'$ and $\psi_i : P'/\mathfrak{M}' \longrightarrow P/\mathfrak{M}$ which are inverse to each other.*

(d) *The isomorphisms $\varphi_1, \ldots, \varphi_d$ constructed in (c) are all isomorphisms between $P/\mathfrak{M}$ and $P'/\mathfrak{M}'$.*

*Proof* To prove (a), we first show that $\mathfrak{M}Q + \mathfrak{M}'Q$ is a radical ideal. This is equivalent to the claim that $\mathfrak{M}L'[x_1, \ldots, x_n]$ is a radical ideal. The minimal polynomials of the elements $\bar{x}_1, \ldots, \bar{x}_n$ in $P/\mathfrak{M}$ are squarefree. Thus we have $\gcd(\mu_{\bar{x}_i}(z), \mu_{\bar{x}_i}(z)') = 1$ for $i = 1, \ldots, n$. By Proposition 1.2.3, these polynomials coincide with the minimal polynomials of the residue classes $\bar{x}_1, \ldots, \bar{x}_n$ in $L'[x_1, \ldots, x_n]/\mathfrak{M}L'[x_1, \ldots, x_n]$. Since they are coprime to their derivatives and $L'$ is a perfect field, they are squarefree polynomials by [15], Proposition 3.7.9.d. Hence Seidenberg's Lemma (see [15], Proposition 3.7.15) implies the claim.

Next we prove that $\mathfrak{M}Q + \mathfrak{M}'Q$ has exactly $d$ maximal components of the form described in (b). By the lemma, we know that there are exactly $d$ $K$-algebra isomorphisms $\varphi_i : P/\mathfrak{M} \longrightarrow P'/\mathfrak{M}'$. For $i = 1, \ldots, d$ and $j = 1, \ldots, n$, let $\varphi_i(\bar{x}_j) = \bar{f}_{ij}$ with $f_{ij} \in P'$. Then Proposition 6.3.2 shows that $\mathfrak{M}Q + \mathfrak{M}'Q$ is contained in $\mathfrak{M}_i = \mathfrak{M}'Q + \langle x_1 - f_{i1}, \ldots, x_n - f_{in} \rangle$ for $i = 1, \ldots, d$. The ideals $\mathfrak{M}_i$ are maximal ideals of $Q$, because $Q/\mathfrak{M}_i \cong P'/\mathfrak{M}'$ is a field.

Translating this to $L'[x_1, \ldots, x_n]$, we see that $\mathfrak{M}L'[x_1, \ldots, x_n]$ is contained in $d$ distinct linear maximal ideals. Now we use the Chinese Remainder Theorem 2.2.1 and $\dim_{L'}(L'[x_1, \ldots, x_n]/\mathfrak{M}L'[x_1, \ldots, x_n]) = \dim_K(P/\mathfrak{M}) = d$ to conclude that these are all maximal components of $\mathfrak{M}L'[x_1, \ldots, x_n]$, and hence that the ideals $\mathfrak{M}_1, \ldots, \mathfrak{M}_d$ are all maximal components of $\mathfrak{M}Q + \mathfrak{M}'Q$.

To prove (c), let $J_i = \langle x_1 - f_{i1}, \ldots, x_n - f_{in} \rangle$ and $J_i' = \langle y_1 - g_{i1}, \ldots, y_m - g_{im} \rangle$ in $Q$ for $i = 1, \ldots, d$. Then the equalities in (b) yield $J_i + \mathfrak{M}'Q = J_i' + \mathfrak{M}Q$, and the claim follows from Proposition 6.3.2. Finally, we note that claim (d) follows from (c) and the lemma.  $\square$

When we translate this theorem to a result about a tuple of endomorphisms of a finite dimensional vector space, we obtain the following corollary about simultaneous diagonalizability after a suitable base field extension.

**Corollary 6.3.5**  *Let $K$ be a finite field, let $V$ be a finite-dimensional $K$-vector space, let $\Phi = (\varphi_1, \ldots, \varphi_n)$ be a tuple of pairwise commuting endomorphisms of $V$, and let $\mathscr{F} = K[\varphi_1, \ldots, \varphi_n]$ be the commuting family they generate. We assume that $\mathrm{Rel}_P(\Phi)$ is a maximal ideal in $P = K[x_1, \ldots, x_n]$ and let $d = \dim_K(P/\mathrm{Rel}_P(\Phi))$.*

*Then we introduce new indeterminates $y_1, \ldots, y_m$, we let $P' = K[y_1, \ldots, y_m]$, and let $\mathfrak{M}'$ be a maximal ideal in $P'$ such that $\dim_K(L') = d$ for $L' = P'/\mathfrak{M}'$. Finally, we let $\Phi_{L'} = ((\varphi_1)_{L'}, \ldots, (\varphi_n)_{L'})$, where $(\varphi_i)_{L'}$ is the extension of $\varphi_i$ to $V_{L'} = V \otimes_K L'$ for $i = 1, \ldots, n$, and we let $\mathscr{F}_{L'}$ be the commuting family generated by $\Phi_{L'}$. Then the family $\mathscr{F}_{L'}$ is simultaneously diagonalizable.*

*Proof*  Using $\mathfrak{M} = \mathrm{Rel}_P(\Phi)$, the proof follows by combining claim (b) of the theorem and Proposition 2.6.1.g.                                                                              $\square$

Next we turn the theorem into an algorithm for computing all $\mathbb{F}_q$-isomorphisms between two finite fields having $q^d$ elements.

---

**Algorithm 6.3.6 (Computing All Isomorphisms of Finite Fields)**
*In the setting of Theorem 6.3.4, consider the following sequence of instructions.*

(1)  *Form the ring $Q = K[x_1, \ldots, x_n, y_1, \ldots, y_m]$ and compute the primary decomposition of the ideal $\mathfrak{M}Q + \mathfrak{M}'Q$. Let $\mathfrak{M}_1, \ldots, \mathfrak{M}_d$ be the resulting maximal ideals.*

(2)  *Choose a term ordering $\sigma$ on $\mathbb{T}(x_1, \ldots, x_n, y_1, \ldots, y_m)$ which is an elimination ordering for $(y_1, \ldots, y_m)$. For $i \in \{1, \ldots, d\}$, compute the reduced $\sigma$-Gröbner basis $G_i$ of $\mathfrak{M}_i$ and write $G_i = H_i \cup \{y_1 - g_{i1}, \ldots, y_m - g_{im}\}$ with $H_i \subset P$ and $g_{ik} \in P$.*

(3)  *Choose a term ordering $\sigma'$ on $\mathbb{T}(x_1, \ldots, x_n, y_1, \ldots, y_m)$ which is an elimination ordering for $(x_1, \ldots, x_n)$. For every index $i \in \{1, \ldots, d\}$, compute the reduced $\sigma'$-Gröbner basis $G'_i$ of the ideal $\mathfrak{M}_i$ and write $G'_i = H'_i \cup \{x_1 - f_{i1}, \ldots, x_n - f_{in}\}$ with $H'_i \subset P'$ and $f_{ij} \in P'$.*

(4)  *Return the pairs $((f_{i1}, \ldots, f_{in}), (g_{i1}, \ldots, g_{im}))$, where $i = 1, \ldots, d$.*

*This is an algorithm which computes $d$ pairs of tuples of polynomials. Each pair defines two isomorphisms $\varphi_i : P/\mathfrak{M} \longrightarrow P'/\mathfrak{M}'$ such that $\bar{x}_j \mapsto \bar{f}_{ij}$ and $\psi_i : P'/\mathfrak{M}' \longrightarrow P/\mathfrak{M}$ such that $\bar{y}_k \mapsto \bar{g}_{ik}$ which are inverse to each other.*

---

Let us see some examples which illustrate this algorithm.

**Example 6.3.7**  Let $K = \mathbb{F}_{101}$, and let $L = K[x]/\langle f(x) \rangle$ be the field with $q = 101^4$ elements defined by $f(x) = x^4 + 41x^3 - 36x^2 + 39x - 12$. In order to compute the four $K$-algebra automorphisms of the field $L$, we create an isomorphic copy $L' = K[y]/\langle f(y) \rangle$ of $L$ and use Algorithm 6.3.6 to compute the distinct isomorphisms $\varphi_i : L \longrightarrow L'$. (In Sect. 6.3.B the field $L'$ will be called a *clone* of $L$.) Let us follow the steps of the algorithm.

(1) First we form $Q = K[x, y]$ and compute the primary decomposition of the ideal $I = \langle f(x), f(y) \rangle$ in $Q$. The result is $I = \mathfrak{M}_1 \cap \mathfrak{M}_2 \cap \mathfrak{M}_3 \cap \mathfrak{M}_4$, where

$$\mathfrak{M}_1 = \langle f(y), x - y \rangle$$
$$\mathfrak{M}_2 = \langle f(y), x - 34y^3 - 20y^2 + 3y + 41 \rangle$$
$$\mathfrak{M}_3 = \langle f(y), x - 4y^3 - 47y^2 + 29y + 35 \rangle$$
$$\mathfrak{M}_4 = \langle f(y), x + 38y^3 - 34y^2 - 31y - 35 \rangle$$

(2) We choose an elimination ordering $\sigma$ for $y$ and compute the reduced $\sigma$-Gröbner basis of the ideals $\mathfrak{M}_i$. We get

$$\mathfrak{M}_1 = \langle f(x), y - x \rangle$$
$$\mathfrak{M}_2 = \langle f(x), y + 38x^3 - 34x^2 - 31x - 35 \rangle$$
$$\mathfrak{M}_3 = \langle f(x), y - 4x^3 - 47x^2 + 29x + 35 \rangle$$
$$\mathfrak{M}_4 = \langle f(x), y - 34x^3 - 20x^2 + 3x + 41 \rangle$$

(3) Next we choose an elimination ordering $\sigma'$ for $x$ and compute the reduced $\sigma'$-Gröbner bases of the ideals $\mathfrak{M}_i$. We get the generators given in Step (1).

(4) The algorithm returns the four pairs $(y, x)$, $(34y^3 + 20y^2 - 3y - 41, -38x^3 + 34x^2 + 31x + 35)$, $(4y^3 + 47y^2 - 29y - 35, 4x^3 + 47x^2 - 29x - 35)$, and $(-38y^3 + 34y^2 + 31y + 35, 34x^3 + 20x^2 - 3x - 41)$.

Altogether, it follows that there are four isomorphisms between $P/\mathfrak{M}$ and $P'/\mathfrak{M}'$, the second one of which is given by $\bar{x} \mapsto 34\bar{y}^3 + 20\bar{y}^2 - 3\bar{y} - 41$ and $\bar{y} \mapsto -38\bar{x}^3 + 34\bar{x}^2 + 31\bar{x} + 35$, etc. For instance, the second of four automorphisms of $L$ is given by $\bar{x} \mapsto 34\bar{x}^3 + 20\bar{x}^2 - 3\bar{x} - 41$, and its inverse is $\bar{x} \mapsto -38\bar{x}^3 + 34\bar{x}^2 + 31\bar{x} + 35$.

In our second example we find five isomorphisms.

**Example 6.3.8** We let $K = \mathbb{F}_{71}$ and let $L = P/\langle f(x) \rangle$, where $P = K[x]$ and $f(x) = x^5 + 2x^4 + 7x^3 - 9x^2 - x + 16 \in K[x]$. Furthermore, we use the ring $L' = P'/\langle g(y) \rangle$, where $P' = K[y]$ and $g(y) = y^5 - 23y^4 + 17y - 21$. Since both $f(x)$ and $g(y)$ are irreducible of degree five, the fields $L$ and $L'$ are isomorphic. They both represent $\mathbb{F}_{71^5}$. Let us compute all $K$-isomorphisms between $L$ and $L'$ using the algorithm.

(1) In the ring $Q = K[x, y]$ we consider the ideal $I = \langle f(x), g(y) \rangle$ and compute its primary decomposition. We get $I = \mathfrak{M}_1 \cap \mathfrak{M}_2 \cap \mathfrak{M}_3 \cap \mathfrak{M}_4 \cap \mathfrak{M}_5$ where

$$\mathfrak{M}_1 = \langle g(y), x + 22y^4 - 21y^3 + 19y^2 + 23y + 21 \rangle$$

$$\mathfrak{M}_2 = \langle g(y), x + 13y^4 - 4y^3 - 17y^2 - 29y - 7 \rangle$$

$$\mathfrak{M}_3 = \langle g(y), x - 17y^3 - 10y^2 + 35y + 6 \rangle$$

$$\mathfrak{M}_4 = \langle g(y), x - 8y^4 + 26y^3 + 19y^2 - 30y - 23 \rangle$$

$$\mathfrak{M}_5 = \langle g(y), x - 27y^4 + 16y^3 - 11y^2 + y + 5 \rangle$$

(2) Next we compute the reduced Gröbner basis of the ideals $\mathfrak{M}_1, \ldots, \mathfrak{M}_5$ with respect to an elimination ordering for $y$ and get

$$\mathfrak{M}_1 = \langle f(x), y - 20x^4 - 8x^3 + 14x^2 + 6x - 27 \rangle$$

$$\mathfrak{M}_2 = \langle f(x), y - 35x^4 - 20x^3 - 8x^2 + 32x - 27 \rangle$$

$$\mathfrak{M}_3 = \langle f(x), y - x^4 + 3x^3 + 29x^2 - x - 11 \rangle$$

$$\mathfrak{M}_4 = \langle f(x), y - 29x^4 + 25x^3 - 16x^2 - 4x - 16 \rangle$$

$$\mathfrak{M}_5 = \langle f(x), y + 14x^4 - 19x^2 - 33x - 13 \rangle$$

(3) The reduced Gröbner bases of the ideals $\mathfrak{M}_1, \ldots, \mathfrak{M}_5$ with respect to an elimination ordering for $x$ are the systems of generators given in Step (1).

(4) Altogether, the algorithm returns the five tuples

$$\left( -22y^4 + 21y^3 - 19y^2 - 23y - 21, 20x^4 + 8x^3 - 14x^2 - 6x + 27 \right)$$

$$\left( -13y^4 + 4y^3 + 17y^2 + 29y + 7, 35x^4 + 20x^3 + 8x^2 - 32x + 27 \right)$$

$$\left( 17y^3 + 10y^2 - 35y - 6, x^4 - 3x^3 - 29x^2 + x + 11 \right)$$

$$\left( 8y^4 - 26y^3 - 19y^2 + 30y + 23, 29x^4 - 25x^3 + 16x^2 + 4x + 16 \right)$$

$$\left( 27y^4 - 16y^3 + 11y^2 - y - 5, -14x^4 + 19x^2 + 33x + 13 \right)$$

which define the five isomorphisms $\varphi_i : L \longrightarrow L'$ together with their inverses $\psi_i : L' \longrightarrow L$.

In the last part of this subsection we apply the preceding algorithm to the following problem. Suppose we are given a finite field $K$ and a finite field extension $K \subseteq L$. What are the primitive elements for $K \subseteq L$ (see Definition 3.4.3) which share the same minimal polynomial? The next proposition provides the basic link between primitive elements and irreducible polynomials.

**Proposition 6.3.9** *Let $K$ be a finite field with $q$ elements, let $K \subseteq L$ be a finite field extension, and let $\dim_K(L) = d$. We denote the set of primitive elements for $K \subseteq L$*

by $\mathrm{Prim}_K(L)$ *and the set of monic irreducible polynomials of degree d in* $K[z]$ *by* $\mathrm{Irr}_K(d)$. *Then the map* $\alpha : \mathrm{Prim}_K(L) \longrightarrow \mathrm{Irr}_K(d)$ *given by* $a \mapsto \mu_a(z)$ *is surjective and d-to-1.*

*Proof* To show that $\alpha$ is surjective, we first note that, for $f \in \mathrm{Irr}_K(d)$, the field $K[x]/\langle f(x) \rangle$ is a $d$-dimensional $K$-vector space, and therefore has $q^d$ elements. Hence this field is isomorphic to $L$ and $f(z)$ is the minimal polynomial of the element corresponding to $\bar{x}$.

Now we fix a polynomial $f(z) \in \mathrm{Irr}_K(d)$ and let $a \in L$ be a primitive element for $K \subseteq L$ such that $\mu_a(z) = f(z)$. Then we have $L = K[a] \cong K[x]/\langle \mu_a(x) \rangle$. Therefore the map $\varphi_a : L \longrightarrow K[z]/\langle f(z) \rangle$ defined by $a \mapsto \bar{z}$ is an isomorphism of $K$-algebras. By Theorem 6.3.4, there are exactly $d$ such isomorphisms and the proof is complete. □

The following algorithm is an application of Algorithm 6.3.6 to the setting of this proposition. Notice that Proposition 5.3.10 provides a similar algorithm. However, since there we do not assume that $K$ is finite, we may get fewer than $d$ elements sharing a minimal polynomial (see for instance Example 5.3.11).

---

**Algorithm 6.3.10 (Primitive Elements Sharing a Minimal Polynomial)**
*In the setting of the proposition, let* $L = P/\mathfrak{M}$ *with* $P = K[x_1, \ldots, x_n]$ *and a maximal ideal* $\mathfrak{M}$ *in* $P$, *and let* $f(z) \in \mathrm{Irr}_K(d)$. *Consider the following sequence of instructions.*

(1) *Compute the primary decomposition* $\mathfrak{M}_1 \cap \cdots \cap \mathfrak{M}_d$ *of the ideal* $\mathfrak{M} + \langle f(z) \rangle$ *in* $K[x_1, \ldots, x_n, z]$.
(2) *Choose a term ordering* $\sigma$ *on* $\mathbb{T}(x_1, \ldots, x_n, z)$ *which is an elimination ordering for* $z$ *and compute the reduced* $\sigma$-*Gröbner bases* $G_1, \ldots, G_d$ *of the maximal ideals* $\mathfrak{M}_1, \ldots, \mathfrak{M}_d$, *respectively.*
(3) *For* $i = 1, \ldots, d$, *let* $g_1, \ldots, g_d \in P$ *be such that* $z - g_i$ *is contained in* $G_i$.
(4) *Return the polynomials* $g_1, \ldots, g_d$.

*This is an algorithm which computes polynomials* $g_1, \ldots, g_d$ *in* $P$ *such that the residue classes* $\bar{g}_1, \ldots, \bar{g}_d$ *in* $L = P/\mathfrak{M}$ *are precisely those primitive elements of* $L$ *whose minimal polynomial is* $f(z)$.

---

The final example in this subsection provides an explicit application of the preceding algorithm.

**Example 6.3.11** Let $K = \mathbb{F}_{71}$, let $P = K[x_1, x_2]$, and let $\mathfrak{M}$ be the ideal given by $\mathfrak{M} = \langle x_1^2 + x_1 + 1, x_2^3 + 11x_2 + 1 \rangle$. Then $\mathfrak{M}$ is a maximal ideal of $P$ and $L = P/\mathfrak{M}$ is a field with $\dim_K(L) = 6$. The minimal polynomial of the element $\bar{x}_1 + \bar{x}_2 \in L$ is $f(z) = z^6 + 3z^5 + 28z^4 - 18z^3 + 21z^2 + 12z - 21$, and this polynomial is irreducible. Hence the field $K[z]/\langle f(z) \rangle$ is isomorphic to $L$.

Our goal is to find all primitive elements of the field $L$ whose minimal polynomial is $f(z)$. To this end, we apply the above algorithm. The primary decomposition of the ideal $\mathfrak{M}Q + \langle f(z) \rangle$ in $Q = K[x_1, x_2, z]$ is $\mathfrak{M}_1 \cap \cdots \cap \mathfrak{M}_6$, where

$$\mathfrak{M}_1 = \mathfrak{M} + \langle z - x_1 - x_2 \rangle$$

$$\mathfrak{M}_2 = \mathfrak{M} + \langle z + x_1 - x_2 + 1 \rangle$$

$$\mathfrak{M}_3 = \mathfrak{M} + \left\langle z + x_1 + 9x_2^2 - 33x_2 - 4 \right\rangle$$

$$\mathfrak{M}_4 = \mathfrak{M} + \left\langle z - x_1 + 9x_2^2 - 33x_2 - 5 \right\rangle$$

$$\mathfrak{M}_5 = \mathfrak{M} + \left\langle z + x_1 - 9x_2^2 + 34x_2 + 6 \right\rangle$$

$$\mathfrak{M}_6 = \mathfrak{M} + \left\langle z - x_1 - 9x_2^2 + 34x_2 + 5 \right\rangle$$

Hence there are six elements in $L$ whose minimal polynomial is $f(z)$, namely $\bar{x}_1 + \bar{x}_2$, $-\bar{x}_1 + \bar{x}_2 - 1$, $-\bar{x}_1 - 9\bar{x}_2^2 + 33\bar{x}_2 + 4$, $\bar{x}_1 - 9\bar{x}_2^2 + 33\bar{x}_2 + 5$, $-\bar{x}_1 + 9\bar{x}_2^2 - 34\bar{x}_2 - 6$, and $\bar{x}_1 + 9\bar{x}_2^2 - 34\bar{x}_2 - 5$.

## 6.3.B  Solving over Finite Fields via Cloning

*The following is an example of a clone.*
*The following is an example of a clone.*

In this subsection we introduce a powerful technique: cloning. In science, cloning refers to the development of offspring that are genetically identical to their parent. Although it is sometimes viewed controversially, according to Wikipedia, cloning is a natural form of reproduction that has allowed life forms to spread for more than 50 thousand years. It is the reproduction method used by many bacteria, plants, and fungi. Can it also be used in computational algebra? Will it help this book to spread for more than 50 thousand years?

Let us continue to use the notation introduced before. In particular, let $K$ be a finite field with $q$ elements, where $q = p^e$ for some prime number $p$ and $e > 0$, let $P = K[x_1, \ldots, x_n]$, let $\mathfrak{M}$ be a maximal ideal of $P$, and let $L = P/\mathfrak{M}$. In order to find the solutions of the polynomial system given by a set of generators of $\mathfrak{M}$, we have to enlarge $K$ suitably. In the following we investigate a straightforward method to do this. It is based on a construction that is our way of introducing the technique of cloning into this book.

**Definition 6.3.12**  Let $y_1, \ldots, y_n$ be new indeterminates, let $P' = K[y_1, \ldots, y_n]$, and let $\mathfrak{M}'$ be the maximal ideal of $P'$ which is obtained by applying the substitution $x_1 \mapsto y_1, \ldots, x_n \mapsto y_n$ to the polynomials in $\mathfrak{M}$.

(a) The ideal $\mathfrak{M}'$ is called the **clone** of the ideal $\mathfrak{M}$ in $P'$.
(b) The field $L' = P'/\mathfrak{M}'$ is called a **clone** of the field $L = P/\mathfrak{M}$.

Recall that in Theorem 6.3.4 we showed that a zero of $\mathfrak{M}$ in an extension field $L' = P'/\mathfrak{M}'$ of $L = P/\mathfrak{M}$, where $P' = K[y_1, \ldots, y_m]$ and $\mathfrak{M}'$ is a maximal ideal of $P'$, corresponds uniquely to a linear maximal component of $\mathfrak{M}L'[x_1, \ldots, x_n]$. The following proposition extends this result as follows: if we know one such zero, we know them all, since the other ones can be obtained by applying powers of the $q$-Frobenius automorphism of $L'$.

**Proposition 6.3.13** *Let $K$ be field with $q$ elements, let $P = K[x_1, \ldots, x_n]$, let $\mathfrak{M}$ be a maximal ideal of $P$, and let $L = P/\mathfrak{M}$. Furthermore, let $P' = K[y_1, \ldots, y_m]$, let $\mathfrak{M}'$ be a maximal ideal of $P'$, let $L' = P'/\mathfrak{M}'$, and let $f_1, \ldots, f_n \in P'$ be polynomials such that $\langle x_1 - \bar{f}_1, \ldots, x_n - \bar{f}_n \rangle$ is one of the maximal components of $\mathfrak{M}L'[x_1, \ldots, x_n]$. Then the set of all zeros of $\mathfrak{M}$ in $(L')^n$ is given by*

$$\mathcal{Z}_{L'}(\mathfrak{M}) = \left\{ \left( \bar{f}_1^{q^i}, \ldots, \bar{f}_n^{q^i} \right) \mid i = 0, \ldots, d-1 \right\}$$

*Proof* First we introduce further indeterminates $t_1, \ldots, t_n$ and consider the polynomial ring $P'' = K[t_1, \ldots, t_n]$. Let $\mathfrak{M}''$ be the clone of $\mathfrak{M}$ in $P''$, and let $L'' = P''/\mathfrak{M}''$. By Theorem 6.3.4, there exists an isomorphism of $K$-algebras $\varphi : L' \longrightarrow L''$. It extends to an isomorphism $\tilde{\varphi} : L'[x_1, \ldots, x_n] \longrightarrow L''[x_1, \ldots, x_n]$ of $K$-algebras which maps $\mathfrak{M}L'[x_1, \ldots, x_n]$ to $\mathfrak{M}L''[x_1, \ldots, x_n]$. A tuple $(\bar{g}_1, \ldots, \bar{g}_n)$ is a zero of $\mathfrak{M}$ in $(L')^n$ if and only if the tuple $(\varphi(\bar{g}_1), \ldots, \varphi(\bar{g}_n))$ is a zero of $\mathfrak{M}$ in $(L'')^n$, and the map $\varphi$ commutes with the $q$-Frobenius automorphism $\phi_q$. Therefore it suffices to prove the claim for the zeros of $\mathfrak{M}$ in $(L'')^n$.

Clearly, the tuple $(\bar{t}_1, \ldots, \bar{t}_n)$ is a zero of $\mathfrak{M}$ in $(L'')^n$. By applying $\phi_q$ repeatedly, it follows that all tuples $(\bar{t}_1^{q^i}, \ldots, \bar{t}_n^{q^i})$ with $0 \leq i \leq d-1$ are zeros of $\mathfrak{M}$ in $(L'')^n$. In view of Theorem 6.3.4, it remains to show that these tuples are pairwise distinct. Assume that $0 \leq i < j \leq d-1$ satisfy $(\phi_q^i(\bar{t}_1), \ldots, \phi_q^i(\bar{t}_n)) = (\phi_q^j(\bar{t}_1), \ldots, \phi_q^j(\bar{t}_n))$. Since $\phi_q$ is bijective, we get $\phi_q^{j-i}(\bar{t}_k) = \bar{t}_k$ for $k = 1, \ldots, n$. Now the fact that the residue classes $\bar{t}_1, \ldots, \bar{t}_n$ form a system of $K$-algebra generators of $L''$ implies that $\phi_q^{j-i}(\bar{g}) = \bar{g}$ for all $\bar{g} \in L''$. If we use a generator $\bar{g}$ of the multiplicative cyclic group $L'' \setminus \{0\}$ here, the inequality $j - i < d$ yields a contradiction to the fact that the smallest power $k$ such that $\bar{g}^{q^k} = \bar{g}$ is $k = d$.  $\square$

Clearly, this proposition can be applied to the setting where $L' = P'/\mathfrak{M}'$ is a clone of $L = P/\mathfrak{M}$. We start by looking at the univariate case.

**Definition 6.3.14** Let $K$ be a field, and let $K \subseteq L$ be a field extension.

(a) Given a polynomial $f \in K[x] \setminus K$, we say that $f$ **splits** in $L$ if there exist elements $c_1, \ldots, c_s \in L$ such that we have $f = \mathrm{LC}(f)(x - c_1) \cdots (x - c_s)$ in $L[x]$.
(b) If a polynomial $f \in K[x] \setminus K$ splits in $L$ as in (a), and if $L = K(c_1, \ldots, c_s)$, we say that $L$ is a **splitting field** of $K$.

It is a standard result in Algebra that the splitting field of a polynomial exists and is uniquely determined up to an isomorphism of $K$-algebras (see [21], Chap. 2).

It is also known that, if $K$ is a perfect field, the $K$-vector space dimension of the splitting field $L$ is the number of elements of $\mathrm{Gal}(L/K)$. The first observation is that, by cloning it, we can split an irreducible polynomial over a finite field.

**Proposition 6.3.15 (Splitting a Polynomial over a Finite Field)**
*Let $K$ be a finite field with $q$ elements, let $f(x) \in K[x]$ be a monic irreducible polynomial of degree $d \geq 1$, and let $L' = K[y]/\mathfrak{M}'$ be a clone of the field $K[x]/\mathfrak{M}$, where $\mathfrak{M} = \langle f(x) \rangle$ and $\mathfrak{M}' = \langle f(y) \rangle$.*

(a) *The field $L'$ has $q^d$ elements.*

(b) *In $L'[x]$ we have $f(x) = (x - \bar{y})(x - \bar{y}^q) \cdots (x - \bar{y}^{q^{d-1}})$, where $\bar{y}$ denotes the residue class of $y$ in $L'$. In particular, the field $L'$ is a splitting field of $f(x)$ and the elements $\bar{y}, \bar{y}^q, \ldots, \bar{y}^{q^{d-1}}$ are the $d$ roots of $f(x)$ in $L'$.*

(c) *Let $I = \langle f(x), f(y) \rangle \subset K[x, y]$. Then the ideal $\mathfrak{M}_i = \langle f(y), x - y^{q^i} \rangle$ is a maximal ideal of $K[x, y]$ for every $i \in \{0, \ldots, d - 1\}$, and the primary decomposition of $I$ is given by $I = \mathfrak{M}_0 \cap \cdots \cap \mathfrak{M}_{d-1}$.*

(d) *Let $i \in \{0, \ldots, d - 1\}$, and let $g_i(y) = \mathrm{NF}_{\mathfrak{M}'}(y^{q^i})$ be the normal form of $y^{q^i}$ with respect to $\mathfrak{M}'$. Then we can substitute $\bar{y}^{q^i}$ with $g_i(\bar{y})$ in (b) and $y^{q^i}$ with $g_i(y)$ in (c).*

*Proof* Claim (a) follows from the fact that $\dim_K(L') = \deg(f(y)) = d$. Claim (b) is a special case of Proposition 6.3.13. Then claim (c) follows from Theorem 6.3.4, and claim (d) is a consequence of the observation that $y^{q^i}$ and $g_i(y)$ have the same residue class in $L'$.                                                                       $\square$

This proposition yields a direct method for computing the splitting field of an irreducible polynomial over a finite field. Apparently, one drawback in part (d) is that we have to compute the normal forms of big powers of polynomials. But there is a remedy which is described in the following algorithm.

**Algorithm 6.3.16 (Computing Big Powers over Finite Fields)**
*Let $K$ be a finite field with $q$ elements, let $f(y)$ be a monic irreducible polynomial of degree $d \geq 1$ in $K[y]$, let $\mathfrak{M}' = \langle f(y) \rangle$, and let $L' = K[y]/\mathfrak{M}'$. Given $g(y) \in K[y]$, we write $\mathrm{NF}_{\mathfrak{M}'}(g(y)) = c_0 y^0 + \cdots + c_{d-1} y^{d-1}$ with $c_i \in K$. Furthermore, let $\alpha \in \mathbb{N}$.*

(1) *Let $B$ be the $K$-basis $B = (1, \bar{y}, \ldots, \bar{y}^{d-1})$ of $L'$. Compute the matrix $M_B(\phi_q)$ representing the $q$-Frobenius endomorphism $\phi_q$ of $L'$ in this basis.*

(2) *Let $v \in K^d$ be the column vector $v = (c_0, c_1, \ldots, c_{d-1})^{\mathrm{tr}}$. Calculate the column vector $w = M_B(\phi_q)^\alpha \cdot v$.*

(3) *Return the polynomial $(1, y, \ldots, y^{d-1}) \cdot w$.*

*This is an algorithm which computes $\mathrm{NF}_{\mathfrak{M}'}(g(y)^{q^\alpha})$.*

*Proof* Since the vector $v$ contains the coordinates of the residue class of $\mathrm{NF}_{\mathfrak{M}'}(g(y))$ in $L'$ with respect to the basis $B$, the column vector $w$ contains the coordinates of the residue class of $\mathrm{NF}_{\mathfrak{M}'}(g(y)^{q^\alpha})$. From this the claim follows. $\qquad\square$

Let us add a few remarks about the efficiency of this algorithm.

**Remark 6.3.17** The calculation of $M_B(\phi_q)^\alpha \cdot v = M_B(\phi_q)^{\alpha-1} \cdot (M_B(\phi_q) \cdot v)$ requires at most $\alpha$ matrix-vector products which are computationally less costly than matrix products. Similarly, we can improve the algorithm by noting that it suffices to compute $\log_2(\alpha)$ powers $M_B(\phi_q)^{2^i}$ and at most $\log_2(\alpha)$ matrix-vector products of some of these powers with $v$.

It is time to see Proposition 6.3.15 and Algorithm 6.3.16 in action.

**Example 6.3.18** Let $K = \mathbb{F}_{101}$, let the monic irreducible polynomial $f(x) \in K[x]$ be given by $f(x) = x^4 + 41x^3 - 36x^2 + 39x - 12$, and let $L' = K[y]/\langle f(y)\rangle$ be a clone of $L = K[x]/\langle f(x)\rangle$.

(a) By Proposition 6.3.15.b, we get $f(x) = (x - \bar{y})(x - \bar{y}^{101})(x - \bar{y}^{101^2})(x - \bar{y}^{101^3})$ in $L'[x]$.

(b) Another representation of the zeros of $f$ can be found by combining Proposition 6.3.15.d with Algorithm 6.3.16. We get

$$f(x) = (x - \bar{y}) \cdot (x - 34\bar{y}^3 - 20\bar{y}^2 + 3\bar{y} + 41) \cdot (x - 4\bar{y}^3 - 47\bar{y}^2 + 29\bar{y} + 35)$$
$$\cdot (x + 38\bar{y}^3 - 34\bar{y}^2 - 31\bar{y} - 35)$$

(c) A third way to compute the roots of $f(x)$ in $L$ is to use the primary decomposition of the ideal $I = \langle f(x), f(y)\rangle \subset K[x, y]$ and Proposition 6.3.15.c. We obtain the decomposition $I = \mathfrak{M}_1 \cap \mathfrak{M}_2 \cap \mathfrak{M}_3 \cap \mathfrak{M}_4$ where

$$\mathfrak{M}_1 = \langle f(y), x - y\rangle$$
$$\mathfrak{M}_2 = \langle f(y), x - 34y^3 - 20y^2 + 3y + 41\rangle$$
$$\mathfrak{M}_3 = \langle f(y), x - 4y^3 - 47y^2 + 29y + 35\rangle$$
$$\mathfrak{M}_4 = \langle f(y), x + 38y^3 - 34y^2 - 31y - 35\rangle$$

In other words, the roots of $f(x)$ in $L'$ are the residue classes of the polynomials $y$, $34y^3 + 20y^2 - 3y - 41$, $4y^3 + 47y^2 - 29y - 35$, and $-38y^3 + 34y^2 + 31y + 35$.

Now we treat the case where the maximal ideal $\mathfrak{M}$ is not necessarily principal. The next proposition follows by combining Theorem 6.3.4 and Proposition 6.3.13

in the case where the field $L'$ is a clone of $L$. Notice that we always know one zero of $\mathfrak{M}$ in $(L')^n$, namely the clone $(\bar{y}_1, \ldots, \bar{y}_n)$ of the tuple $(\bar{x}_1, \ldots, \bar{x}_n)$.

**Proposition 6.3.19** *Let $K$ be a finite field with $q$ elements, let $P = K[x_1, \ldots, x_n]$, let $\mathfrak{M}$ be a maximal ideal in $P$, let $L = P/\mathfrak{M}$, and let $d = \dim_K(L)$. We introduce new indeterminates $y_1, \ldots, y_n$, let $P' = K[y_1, \ldots, y_n]$, denote the clone of $\mathfrak{M}$ in $P'$ by $\mathfrak{M}'$, and let $L' = P'/\mathfrak{M}'$ be the clone of $L$. Then we have*

$$\mathcal{Z}_{L'}(\mathfrak{M}) = \left\{ \left( \bar{y}_1^{q^i}, \ldots, \bar{y}_n^{q^i} \right) \mid i = 0, \ldots, d-1 \right\}$$

The formula given in this proposition contains some polynomials of potentially very high degrees. When we combine this formula with Algorithm 6.3.16, we get the following algorithm.

---

**Algorithm 6.3.20 (Solving over Finite Fields via Cloning)**
*In the setting of the proposition, consider the following instructions.*

(1) *Introduce new indeterminates $y_1, \ldots, y_n$, let $P' = K[y_1, \ldots, y_n]$, let $\mathfrak{M}'$ be the clone of $\mathfrak{M}$ in $P'$, and let $L' = P'/\mathfrak{M}'$.*

(2) *For every $i \in \{0, \ldots, d-1\}$, use Algorithm 6.3.16 to compute the tuple $(NF_{\mathfrak{M}'}(y_1^{q^i}), \ldots, NF_{\mathfrak{M}'}(y_n^{q^i}))$. Then return the set of all these tuples.*

*This is an algorithm which computes a set of n-tuples of polynomials in $K[y_1, \ldots, y_n]$ which are in normal form with respect to $\mathfrak{M}'$ and whose residue classes in $(L')^n$ are the distinct zeros of $\mathfrak{M}$.*

---

The following remark provides a variant of this algorithm which relies on the computation of a primary decomposition instead of Algorithm 6.3.16.

**Remark 6.3.21** Suppose we are in the setting of the algorithm. Another version of Step (2) is obtained by computing the primary decomposition of $\mathfrak{M}Q + \mathfrak{M}'Q$ in $Q = K[x_1, \ldots, x_n, y_1, \ldots, y_n]$ and using an elimination ordering $\sigma$ for $(x_1, \ldots, x_n)$ on $\mathbb{T}(x_1, \ldots, x_n, y_1, \ldots, y_n)$ in order to write the reduced $\sigma$-Gröbner bases $G_i$ of the primary components in the form $G_i = H_i \cup \{x_1 - f_{i1}, \ldots, x_n - f_{in}\}$ with $H_i \subset P'$ and $f_{ij} \in P'$. Then the residue classes of the tuples $(f_{i1}, \ldots, f_{in})$ in $(L')^n$ are the zeros of $\mathfrak{M}$ by Theorem 6.3.4 and Algorithm 6.3.6.

Notice that the output of this variant is the same set of tuples of polynomials as the output of the algorithm above, since the set $H_i$ contains a reduced Gröbner basis of $\mathfrak{M}'$.

The following example illustrates the preceding algorithm. Later we will see that, if the maximal ideal $\mathfrak{M}$ is given via a triangular set of generators, we can get an even simpler representation of its zeros (see Example 6.3.29).

**Example 6.3.22**  Let $K = \mathbb{F}_2$, let $P = K[x_1, x_2, x_3]$, and let $\mathfrak{M} = \langle f_1, f_2, f_3 \rangle$, where $f_1 = x_1^2 + x_1 + 1$, $f_2 = x_2^3 + x_1$, and $f_3 = x_3^2 + x_3 + x_1$. Using the methods of Chap. 5, we can check that $\mathfrak{M}$ is a maximal ideal in $P$. Hence $L = P/\mathfrak{M}$ is a finite field. Our goal is to enlarge the field $K$ such that we find the zeros of $\mathfrak{M}$ in the larger field.

(a) According to Proposition 6.3.19, we form the ring $P' = K[y_1, y_2, y_3]$, the clone $\mathfrak{M}'$ of $\mathfrak{M}$ and the clone $L' = P'/\mathfrak{M}'$ of the field $L = P/\mathfrak{M}$. Since the field $K$ has $q = 2$ elements and $d = \dim_K(L) = 2 \cdot 3 \cdot 2 = 12$, the zeros of $\mathfrak{M}$ in $(L')^3$ are $(\bar{y}_1^{2^i}, \bar{y}_2^{2^i}, \bar{y}_3^{2^i})$ for $i = 0, \ldots, 11$.

(b) Another way of writing these triples uses Algorithm 6.3.20. With the help of Algorithm 6.3.16, we compute the necessary normal forms and get the following representations:

$$
\begin{array}{lll}
(y_1, y_2, y_3), & (y_1{+}1, y_2^2, y_1{+}y_3), & (y_1, y_1 y_2, y_3{+}1), \\
(y_1{+}1, y_1 y_2^2{+}y_2^2, y_1{+}y_3{+}1), & (y_1, y_1 y_2{+}y_2, y_3), & (y_1{+}1, y_1 y_2^2, y_3{+}y_1), \\
(y_1, y_2, y_3{+}1), & (y_1{+}1, y_2^2, y_1{+}y_3{+}1), & (y_1, y_1 y_2, y_3), \\
(y_1{+}1, y_1 y_2^2{+}y_2^2, y_1{+}y_3), & (y_1, y_1 y_2{+}y_2, y_3{+}1), & (y_1{+}1, y_1 y_2^2, y_1{+}y_3{+}1)
\end{array}
$$

A further example in which the generators of $\mathfrak{M}$ are not in triangular form is given as follows.

**Example 6.3.23**  Over the field $K = \mathbb{F}_7$, let us consider the system of polynomial equations defined by

$$\tilde{f}_1 = x_1^4 - 2x_1 + 3,$$

$$\tilde{f}_2 = x_1^2 x_2^3 + 2x_1^2 - 2x_2 - 1$$

To solve it, we let $P = K[x_1, x_2]$ and $\mathfrak{M} = \langle \tilde{f}_1, \tilde{f}_2 \rangle$. To check that $\mathfrak{M}$ is a maximal ideal in $P$, we first check that $\mathfrak{M}$ is a zero-dimensional ideal. Then we apply Algorithm 5.3.12. The minimal polynomial of $\bar{x}_1$ in $P/\mathfrak{M}$ is $\mu_{\bar{x}_1}(z) = z^4 - 2z + 3$ which is irreducible in $K[z]$. Thus we let $K_1 = K[x_1]/\langle x_1^4 - 2x_1 + 3 \rangle$. Next we compute the minimal polynomial of $\bar{x}_2$ in $K_1[x_2]/\mathfrak{M}K_1[x_2]$. It is given by the polynomial $\mu_{\bar{x}_2, K_1}(z) = z^3 + (2\bar{x}_1^3 + 3\bar{x}_1^2 + 3)z + \bar{x}_1^3 - 2\bar{x}_1^2$ and is irreducible in $K_1[z]$.

Consequently, the preimage of the ideal $\langle \mu_{\bar{x}_2, K_1}(x_2) \rangle$ in $K[x_1, x_2]$, i.e., the ideal $\langle x_1^4 - 2x_1 + 3, x_2^3 + (2x_1^3 + 3x_1^2 + 3)x_2 + x_1^3 - 2x_1^2 \rangle$ is a maximal ideal contained in $\mathfrak{M}$, and hence equal to $\mathfrak{M}$. Thus we have a triangular set of generators of $\mathfrak{M}$, and we get $d = \dim_K(P/\mathfrak{M}) = 12$.

Now we want to solve $\mathfrak{M}$ via cloning. We introduce two new indeterminates $y_1$ and $y_2$, we let $P' = K[y_1, y_2]$, we let $\mathfrak{M}'$ be the clone of $\mathfrak{M}$ in $P'$, and we let $L' = P'/\mathfrak{M}'$ be the clone of $L = P/\mathfrak{M}$. Then the set of zeros of $\mathfrak{M}$ in $(L')^2$ is $\{(y_1^{q^i}, y_2^{q^i}) \mid i = 0, \ldots, 11\}$. When we compute the corresponding normal forms via

Algorithm 6.3.16, we get the following representations:

$(y_1,$  $\quad y_2)$

$(-3y_1^3 - 3y_1 + 1, \quad y_1^3 + 3y_1^2 y_2 + 3y_1 y_2^2 + 3y_2^3 + 3y_1^2 - 3y_1 y_2 + y_2^2 + 3y_2 - 3)$

$(-2y_1^3 - 2y_1^2 - 2y_1 + 3, \quad 3y_1^2 y_2^2 - 2y_1^3 - 2y_1 y_2^2 - 2y_2^3 + 3y_1^2 - 2y_1 y_2 + 3y_2^2 + 3)$

$(-2y_1^3 + 2y_1^2 - 3y_1 + 3, \quad -2y_1^2 y_2^2 + 2y_2^4 + y_1^3 - 3y_1^2 y_2 + 2y_1 y_2^2 - 2y_2^3 - y_1^2 - 3y_1 y_2 + 3y_1 - 2y_2)$

$(y_1, \quad -y_1^2 y_2^2 - y_2^4 - y_1^3 - y_1^2 y_2 - 3y_1 y_2^2 - 3y_1^2 + y_1 y_2 - y_2^2 + 2y_1 + 3y_2 - 3)$

$(-3y_1^3 - 3y_1 + 1, \quad y_1^2 y_2^2 + 2y_1^3 + 3y_1^2 y_2 + 2y_1 y_2^2 - 2y_2^3 - 3y_1^2 - y_1 y_2 + y_2)$

$(-2y_1^3 - 2y_1^2 - 2y_1 + 3, \quad y_1^2 y_2^2 - y_2^4 - y_1^3 + 3y_1^2 y_2 + 2y_1 y_2^2 + 2y_2^3 + 3y_1^2 - 3y_2^2 + 2y_1 + y_2 + 3)$

$(-2y_1^3 + 2y_1^2 - 3y_1 + 3, \quad 2y_1^2 y_2^2 - y_2^4 - 2y_1^2 y_2 + 3y_1 y_2^2 + 2y_2^3 - 2y_1^2 - 2y_2^2 + 2y_1 + 2y_2 + 2)$

$(y_1, \quad y_1^2 y_2^2 + y_2^4 + y_1^3 + y_1^2 y_2 + 3y_1 y_2^2 + 3y_1^2 - y_1 y_2 + y_2^2 - 2y_1 + 3y_2 + 3)$

$(-3y_1^3 - 3y_1 + 1, \quad 3y_1^2 y_2^2 + y_2^4 + 3y_1^3 - 3y_1^2 y_2 + y_1^2 + 2y_1 y_2 - 2y_1 - y_2 + 1)$

$(-2y_1^3 - 2y_1^2 - 2y_1 + 3, \quad 3y_1^2 y_2^2 + y_2^4 + 3y_1^3 - 3y_1^2 y_2 + y_1^2 + 2y_1 y_2 - 2y_1 - y_2 + 1)$

$(-2y_1^3 + 2y_1^2 - 3y_1 + 3, \quad -y_2^4 - y_1^3 - 2y_1^2 y_2 + 2y_1 y_2^2 + 3y_1^2 + 3y_1 y_2 + 2y_2^2 + 2y_1 - 2)$

## 6.3.C Solving over Finite Fields via Univariate Representations

*Patience is something you admire*
*in the driver behind you,*
*but not in the one ahead.*
(Bill McClashen)

As before, we let $K$ be a field with $q = p^e$ elements, where $p$ is a prime and $e > 0$, let $P = K[x_1, \ldots, x_n]$, and let $\mathfrak{M}$ be a maximal ideal in $P$ whose zeros we want to compute in a suitably chosen extension field of $K$.

Suppose we are using a method based on Theorem 6.3.4, and for every calculation, we are impatiently awaiting it to finish. Can we speed up and pass the obstacles before us? If the computer algebra system we are using has a fast implementation of finite fields, it knows an irreducible polynomial $f(y)$ of degree $d$ in $K[y]$, namely the defining polynomial of the built-in representative of $\mathbb{F}_{q^d}$. Using the univariate representation $L' = K[y]/\langle f(y) \rangle$, we can solve the system quickly, as the following algorithm shows. An added advantage is that, after we finish the calculation, we have the solutions represented via the built-in field representative and can use them for further computations without delay.

**Algorithm 6.3.24 (Solving via a Univariate Representation)**
*Let $K$ be a finite field with $q$ elements, let $P = K[x_1, \ldots, x_n]$, let $\mathfrak{M}$ be a maximal ideal of $P$, let $L = P/\mathfrak{M}$, let $d = \dim_K(L)$, let $f(y) \in K[y]$ be a monic irreducible polynomial of degree $d$, and let $L' = K[y]/\langle f(y) \rangle$. Consider the following sequence of instructions.*

(1) *Compute a triangular set* $\{g_1(x_1), g_2(x_1, x_2), \ldots, g_n(x_1, \ldots, x_n)\}$ *of generators of* $\mathfrak{M}$ *(e.g., use Algorithm 5.3.12). Then for* $i = 1, \ldots, n$, *let* $d_i = \deg(g_i)$.

(2) *Let* $S_0$ *be the set consisting of the empty tuple.*

(3) *For* $i = 1, \ldots, n$, *perform the following steps. Then return* $S_n$ *and stop.*

(4) *Let* $S_i = \emptyset$. *For every tuple* $(h_1(y), \ldots, h_{i-1}(y))$ *in* $S_{i-1}$, *perform the following steps.*

(5) *Substitute* $x_1 \mapsto h_1(y), \ldots, x_{i-1} \mapsto h_{i-1}(y)$ *in* $g_1, \ldots, g_i$ *and get polynomials* $\tilde{g}_1(y), \ldots, \tilde{g}_{i-1}(y) \in K[y]$ *as well as* $\tilde{g}_i(x_i, y) \in K[x_i, y]$.

(6) *In the ring* $Q = K[x_i, y]$, *calculate the primary decomposition of the ideal* $I_i = \langle f(y), \tilde{g}_1(y), \ldots, \tilde{g}_{i-1}(y), \tilde{g}_i(x_i, y) \rangle$ *and get* $p_j(y) \in K[y]$ *such that*

$$I_i = \langle f(y), x_i - p_1(y) \rangle \cap \cdots \cap \langle f(y), x_i - p_{d_i}(y) \rangle$$

(7) *For* $j = 1, \ldots, d_i$, *append the tuple* $(h_1(y), \ldots, h_{i-1}(y), p_j(y))$ *to* $S_i$.

*This is an algorithm which computes a set of d tuples* $S_n$ *whose residue classes in* $(L')^n$ *are precisely the zeros of* $\mathfrak{M}$.

*Proof* By Theorem 6.3.4 and Algorithm 6.3.6, the primary decomposition of the ideal $I_i$ has the indicated form. Moreover, if we let $\mathfrak{M}_i = \langle g_1, \ldots, g_i \rangle$, then the homomorphism $K[x_1, \ldots, x_i]/\mathfrak{M}_i \longrightarrow L'$ given by $\bar{x}_k \mapsto h_k(\bar{y})$ for $k = 1, \ldots, i-1$ and $\bar{x}_i \mapsto p_j(\bar{y})$ is an injective $K$-algebra homomorphism. Thus the tuples in $S_i$ represent zeros of the ideal $\mathfrak{M}_i$ in $(L')^i$. In particular, the final tuple $S_n$ represents zeros of $\mathfrak{M}$ in $(L')^n$. The construction shows that the tuples in $S_n$ represent $d_1 \cdots d_n$ pairwise isomorphisms $P/\mathfrak{M} \longrightarrow L'$, and hence pairwise distinct zeros of $\mathfrak{M}$.  □

Notice that this algorithm requires chiefly factorizations of univariate polynomials over the field $L' = K[y]/\langle f(y) \rangle$. This special type of primary decompositions may be faster to determine than the general one (see Algorithm 5.3.8). Let us see a case in point.

**Example 6.3.25** Let $K = \mathbb{F}_{101}$, let $P = K[x_1, x_2]$, let $\mathfrak{M}$ be the maximal ideal of $P$ generated by the triangular set $\{x_1^2 - 2x_1 - 1, x_2^3 - x_2^2 - x_1x_2 - 12\}$, and let $L = P/\mathfrak{M}$. We calculate $\dim_K(L) = 6$. Now assume that we know the monic irreducible polynomial $f(y) = y^6 - y - 1$ in $K[y]$. Then the field $L' = K[y]/\mathfrak{M}'$, where we let $\mathfrak{M}' = \langle f(y) \rangle$, satisfies $\dim_K(L') = 6$, and thus $(L')^2$ contains the six zeros of $\mathfrak{M}$.

(a) First we compute these six zeros by applying Theorem 6.3.4 directly. In the ring $Q = K[x_1, x_2, y]$, we calculate the primary decomposition of the ideal $\mathfrak{M}Q + \mathfrak{M}'Q$ and get six maximal components $\mathfrak{M}_1, \ldots, \mathfrak{M}_6$. We represent these ideas $\mathfrak{M}_i$ by their reduced Gröbner bases with respect to an elimination ordering for $(x_1, x_2)$. The result is

$$\mathfrak{M}_1 = \mathfrak{M}'Q + \langle x_1 + 19y^5 - 45y^4 - 35y^3 + 9y^2 + 21y,$$
$$x_2 + 3y^5 + 2y^4 - 12y^3 - 19y^2 + 22y + 14 \rangle$$
$$\mathfrak{M}_2 = \mathfrak{M}'Q + \langle x_1 + 19y^5 - 45y^4 - 35y^3 + 9y^2 + 21y,$$
$$x_2 + 29y^5 + 42y^4 + 38y^3 + 44y^2 + 31 \rangle$$
$$\mathfrak{M}_3 = \mathfrak{M}'Q + \langle x_1 + 19y^5 - 45y^4 - 35y^3 + 9y^2 + 21y,$$
$$x_2 - 32y^5 - 44y^4 - 26y^3 - 25y^2 + 48y - 41 \rangle$$
$$\mathfrak{M}_4 = \mathfrak{M}'Q + \langle x_1 - 19y^5 + 45y^4 + 35y^3 - 9y^2 - 21y - 2,$$
$$x_2 - 46y^5 - 23y^4 - 22y^3 - 31y^2 + 14y + 38 \rangle$$
$$\mathfrak{M}_5 = \mathfrak{M}'Q + \langle x_1 - 19y^5 + 45y^4 + 35y^3 - 9y^2 - 21y - 2,$$
$$x_2 + 40y^5 + 6y^4 - 41y^2 + 6y \rangle$$
$$\mathfrak{M}_6 = \mathfrak{M}'Q + \langle x_1 - 19y^5 + 45y^4 + 35y^3 - 9y^2 - 21y - 2,$$
$$x_2 + 6y^5 + 17y^4 + 22y^3 - 29y^2 - 20y - 39 \rangle$$

(b) Now we solve the polynomial system defined by the generators of $\mathfrak{M}$ using Algorithm 6.3.24. Let us follows its steps.

(1) The generators $g_1(x_1) = x_1^2 - 2x_1 - 1$ and $g_2(x_1, x_2) = x_2^3 - x_2^2 - x_1 x_2 - 12$ of $\mathfrak{M}$ form already a triangular set.

(6) The primary decomposition of the ideal $I_1 = \langle f(y), g_1(x_1) \rangle$ is of the form $I_1 = \langle f(y), x_1 - p_1(y) \rangle \cap \langle f(y), x_1 - p_2(y) \rangle$ where $p_1(y) = -19y^5 + 45y^4 + 35y^3 - 9y^2 - 21y$ and $p_2(y) = 19y^5 - 45y^4 - 35y^3 + 9y^2 + 21y + 2$.

(7) Hence we let $S_1 = \{(p_1(y)), (p_2(y))\}$.

(5) For $i = 2$ and $h_1(y) = p_1(y)$, the substitution $x_1 \mapsto p_1(y)$ yields the polynomial $\tilde{g}_2(x_2, y) = x_2^3 - x_2^2 + 19y^5 x_2 - 45y^4 x_2 - 35y^3 x_2 + 9y^2 x_2 + 21y x_2 - 12$.

(6) The primary decomposition of the ideal $\langle f(y), \tilde{g}_2(x_2, y) \rangle$ is of the form $I_{21} \cap I_{22} \cap I_{23}$ where

$$I_{21} = \langle f(y), x_2 + 3y^5 + 2y^4 - 12y^3 - 19y^2 + 22y + 14 \rangle$$
$$I_{22} = \langle f(y), x_2 + 29y^5 + 42y^4 + 38y^3 + 44y^2 + 31y + 26 \rangle$$
$$I_{23} = \langle f(y), x_2 - 32y^5 - 44y^4 - 26y^3 - 25y^2 + 48y - 41 \rangle$$

(7) Hence we let $S_2 = \{(p_1(y), q_1(y)), (p_1(y), q_2(y)), (p_1(y), q_3(y))\}$, where
$q_1(y) = -3y^5 - 2y^4 + 12y^3 + 19y^2 - 22y - 14$, $q_2(y) = -29y^5 - 42y^4 - 38y^3 - 44y^2 - 31y - 26$, and $q_3(y) = 32y^5 + 44y^4 + 26y^3 + 25y^2 - 48y + 41$.

(5) For $i = 2$ and $h_1(y) = p_2(y)$, the substitution $x_1 \mapsto p_2(y)$ yields the polynomial $\tilde{g}_2'(x_2, y) = x_2^3 - x_2^2 - 19y^5 x_2 + 45y^4 x_2 + 35y^3 x_2 - 9y^2 x_2 - 21yx_2 - 2x_2 - 12$.

(6) The primary decomposition of $\langle f(y), \tilde{g}_2'(x_2, y) \rangle$ is $I_{31} \cap I_{32} \cap I_{33}$ where

$$I_{31} = \langle f(y), x_2 - 46y^5 - 23y^4 - 22y^3 - 31y^2 + 14y + 38 \rangle$$

$$I_{32} = \langle f(y), x_2 + 40y^5 + 6y^4 - 41y^2 + 6y \rangle$$

$$I_{33} = \langle f(y), x_2 + 6y^5 + 17y^4 + 22y^3 - 29y^2 - 20y - 3 \rangle$$

(7) Now we append the three pairs $(p_2(y), q_1'(y))$, $(p_2(y), q_2'(y))$, and $(p_2(y), q_3'(y))$ to $S_2$, where $q_1'(y) = 46y^5 + 23y^4 + 22y^3 + 31y^2 - 14y + 38$, $q_2'(y) = -40y^5 - 6y^4 + 41y^2 - 6y$, and $q_3'(y) = -6y^5 - 17y^4 - 22y^3 + 29y^2 + 20y + 3$.

(3) The algorithm returns the set $S_2$ consisting of six pairs.

Altogether, we find the six zeros of $\mathfrak{M}$ in $L'$ which are put into $S_2$ in the two iterations of Step (7) for $i = 2$, and these six zeros correspond to the six linear maximal ideals found in (a).

Now we use Algorithm 6.3.24 to solve a system of polynomial equations over the field $\mathbb{F}_2$.

**Example 6.3.26** Let $K = \mathbb{F}_2$, and let us consider the system of polynomial equations defined by

$$f_1 = x_1^2 + x_1 + 1,$$

$$f_2 = x_2^3 + x_2 + 1$$

Let $P = K[x_1, x_2]$, and let $\mathfrak{M} = \langle f_1, f_2 \rangle$. Since both $f_1$ and $f_2$ are irreducible over $K$, it follows that $\mathfrak{M}$ is a zero-dimensional radical ideal. By calculating the primary decomposition of $\mathfrak{M}$, we see that $\mathfrak{M}$ is a maximal ideal, and that the field $L = P/\mathfrak{M}$ satisfies $\dim_K(L) = 6$.

Furthermore, we assume that we are given the monic irreducible polynomial $f(y) = y^6 + y + 1$ in $K[y]$. Then also the ideal $\mathfrak{M}' = \langle f(y) \rangle$ is maximal in the ring $P' = K[y]$, and the field $L' = P'/\mathfrak{M}'$ is isomorphic to $L$. Again we use this information to find the zeros of $\mathfrak{M}$ in $(L')^2$ in two ways.

(a) First we follow the method suggested by Theorem 6.3.4. We compute the primary decomposition of the ideal $\mathfrak{M}Q + \mathfrak{M}'Q$ in the ring $Q = K[x_1, x_2, y]$ and get the result $\mathfrak{M}Q + \mathfrak{M}'Q = \mathfrak{M}_1 \cap \cdots \cap \mathfrak{M}_6$, where

$$\mathfrak{M}_1 = \langle y^6 + y + 1,\ x_1 + y^5 + y^4 + y^3 + y + 1,\ x_2 + y^3 + y^2 + y \rangle$$
$$\mathfrak{M}_2 = \langle y^6 + y + 1,\ x_1 + y^5 + y^4 + y^3 + y,\ \quad x_2 + y^4 + y^3 + 1 \rangle$$
$$\mathfrak{M}_3 = \langle y^6 + y + 1,\ x_1 + y^5 + y^4 + y^3 + y + 1,\ x_2 + y^4 + y^2 + y + 1 \rangle$$
$$\mathfrak{M}_4 = \langle y^6 + y + 1,\ x_1 + y^5 + y^4 + y^3 + y + 1,\ x_2 + y^4 + y^3 + 1 \rangle$$
$$\mathfrak{M}_5 = \langle y^6 + y + 1,\ x_1 + y^5 + y^4 + y^3 + y,\ \quad x_2 + y^4 + y^2 + y + 1 \rangle$$
$$\mathfrak{M}_6 = \langle y^6 + y + 1,\ x_1 + y^5 + y^4 + y^3 + y,\ \quad x_2 + y^3 + y^2 + y \rangle$$

For $i = 1, \ldots, 6$, the given generators form a $\sigma$-Gröbner basis of the maximal ideal $\mathfrak{M}_i$ for every elimination ordering $\sigma$ for $(x_1, x_2)$ on $\mathbb{T}(x_1, x_2, y)$. Hence the zeros of $\mathfrak{M}$ in the field $L' = K[y]/\langle f(y) \rangle$ are given by the six pairs

$$(\bar{y}^5 + \bar{y}^4 + \bar{y}^3 + \bar{y} + 1,\ \bar{y}^3 + \bar{y}^2 + \bar{y}),\qquad (\bar{y}^5 + \bar{y}^4 + \bar{y}^3 + \bar{y},\ \bar{y}^4 + \bar{y}^3 + 1)$$
$$(\bar{y}^5 + \bar{y}^4 + \bar{y}^3 + \bar{y} + 1,\ \bar{y}^4 + \bar{y}^2 + \bar{y} + 1),\ (\bar{y}^5 + \bar{y}^4 + \bar{y}^3 + \bar{y} + 1,\ \bar{y}^4 + \bar{y}^3 + 1)$$
$$(\bar{y}^5 + \bar{y}^4 + \bar{y}^3 + \bar{y},\ \bar{y}^4 + \bar{y}^2 + \bar{y} + 1),\qquad (\bar{y}^5 + \bar{y}^4 + \bar{y}^3 + \bar{y},\ \bar{y}^3 + \bar{y}^2 + \bar{y})$$

(b) Next we solve the given polynomial system using Algorithm 6.3.24. The primary decomposition of the ideal $I_1 = \langle f(y), x_1^2 + x_1 + 1 \rangle$ is

$$I_1 = \langle y^6 + y + 1,\ x_1 + y^5 + y^4 + y^3 + y \rangle \cap \langle y^6 + y + 1,\ x_1 + y^5 + y^4 + y^3 + y + 1 \rangle$$

Thus the set $S_1$ consists of $(y^5 + y^4 + y^3 + y)$ and $(y^5 + y^4 + y^3 + y + 1)$.

For each of these elements, the substitution into $f_2$ yields the same result, because $f_2$ does not involve $x_1$. A more thorough treatment of this case will be given in Proposition 6.3.32. So, in both iterations of the loop in Steps (4)–(7), we calculate the primary decomposition of $I_2 = \langle f(y), x_2^3 + x_2 + 1 \rangle$ and get

$$I_2 = \langle y^6 + y + 1,\ x_2 + y^3 + y^2 + y \rangle \cap \langle y^6 + y + 1,\ x_2 + y^4 + y^2 + y + 1 \rangle$$
$$\cap \langle y^6 + y + 1,\ x_2 + y^4 + y^3 + 1 \rangle$$

Therefore the set $S_2$ is obtained by appending each of the three polynomials $y^3 + y^2 + y$, $y^4 + y^2 + y + 1$, and $y^4 + y^3 + 1$ to each of the two tuples in $S_1$. Altogether, we find the same six pairs as in (a).

> *Doing 170 km/h, I passed that slowcoach.*
> *Then I had to wait for half an hour*
> *until he administered me first aid.*
> (Harald Grill)

### 6.3.D  Solving over Finite Fields via Recursion

*To define recursion, we must first define recursion.*

In the setting of the preceding subsection, suppose that we do not know an irreducible univariate polynomial of degree $d$ over $K$. Can we still imitate Algorithm 6.3.24 and recursively split the task of computing a potentially huge primary decomposition into smaller ones? To answer this question, we first have to define which kind of recursion we look for.

Let $K$ be a finite field having $q$ elements, let $P = K[x_1, \ldots, x_n]$, and suppose that the polynomials defining the given system form a triangular set of a maximal ideal $\mathfrak{M}$ in $P$. Then we can split the first polynomial in a suitable extension field of $K$ and work recursively over this field with the maximal ideal defined by substituting the various values of $x_1$ in the other generators of $\mathfrak{M}$. In the next section, when we try to solve polynomial systems over the rational numbers, this kind of recursive approach will become important.

Our first observation is that some of the powers needed to write down the zeros in Proposition 6.3.19 can be reduced substantially if we know a triangular set of generators of $\mathfrak{M}$, as the next proposition shows.

**Proposition 6.3.27**  *Let $K$ be a finite field with $q$ elements, let $P = K[x_1, \ldots, x_n]$, let $\mathfrak{M}$ be a maximal ideal of $P$, and let $L = P/\mathfrak{M}$. Furthermore, assume that we have a triangular set $\{f_1(x_1), f_2(x_1, x_2), \ldots, f_n(x_1, \ldots, x_n)\}$ of generators of $\mathfrak{M}$, and let $d_i = \deg_{x_i}(f_i)$ for $i = 1, \ldots, n$. After introducing new indeterminates $y_1, \ldots, y_n$, let $P' = K[y_1, \ldots, y_n]$, let $\mathfrak{M}'$ be the clone of $\mathfrak{M}$ in $P'$, and let $L' = P'/\mathfrak{M}'$ be the clone of $L$. Then we have*

$$\mathscr{Z}_{L'}(\mathfrak{M}) = \left\{ \left(\bar{y}_1^{q^{\beta_1}}, \ldots, \bar{y}_n^{q^{\beta_n}}\right) \in \left(L'\right)^n \,\middle|\, 0 \le \alpha_i \le d_i - 1 \text{ and}\right.$$
$$\left. \beta_i = \alpha_1 + d_1\alpha_2 + \cdots + (d_1 \cdots d_{i-1})\alpha_i \text{ for } i = 1, \ldots, n \right\}$$

*Proof* For $i = 0, \ldots, n$, we let $P'_i = K[y_1, \ldots, y_i]$, let $\mathfrak{M}'_i$ be the clone of the ideal $\mathfrak{M}_i = \langle f_1(x_1), \ldots, f_i(x_1, \ldots, x_i) \rangle$ in $P'_i$, and let $L'_i = P'_i/\mathfrak{M}'_i$. Here $\mathfrak{M}_i = \mathfrak{M} \cap P_i$ is a maximal ideal, and hence $L'_i$ is a subfield of $L'$. Notice that the number of elements of $L'_i$ is $q^{d_1 \cdots d_i}$.

Now we prove the proposition by induction on $i$. For $i = 1$, it follows from Proposition 6.3.15.b that the zeros of the ideal $\mathfrak{M}_1$ in $L'_1$ are given by $\bar{y}_1^{q^j}$ with $j = 0, \ldots, d_1 - 1$. To prove the induction step, we let $(\bar{y}_1^{q^{\beta_1}}, \ldots, \bar{y}_{i-1}^{q^{\beta_{i-1}}})$ be a zero of $\mathfrak{M}_{i-1}$ in $(L'_{i-1})^{i-1}$. We substitute this zero for $(x_1, \ldots, x_{i-1})$ in the polynomial $f_i(x_1, \ldots, x_i)$ and get an irreducible polynomial $g_i(x_i)$ in $L'_{i-1}[x_i]$.

To find the zeros of this polynomial, we use Proposition 6.3.13. Notice that, in view of the orders of the fields $L'_1, \ldots, L'_{i-1}$, the given zero of $\mathfrak{M}_{i-1}$ is equal to $(\bar{y}_1^{q^{\beta_{i-1}}}, \ldots, \bar{y}_{i-1}^{q^{\beta_{i-1}}})$. Now the fact that $(\bar{y}_1, \ldots, \bar{y}_i)$ is a zero of $\mathfrak{M}_i$ in the field $L'_i$ and a $\beta_{i-1}$-fold application of the Frobenius homomorphism show that

$(\bar{y}_1^{q^{\beta_i-1}},\ldots,\bar{y}_i^{q^{\beta_i-1}})$ is a zero of $\mathfrak{M}_i$. Hence one zero of $g_i(x_i)$ is $\bar{y}_i^{q^{\beta_i-1}}$. Since the field $L'_{i-1}$ has $d_1\cdots d_{i-1}$ elements, Proposition 6.3.13 implies that the other zeros of $g_i(x_i)$ are $\bar{y}_i^{q^{\beta_i-1+d_1\cdots d_{i-1}\alpha_i}}$ with $1 \le \alpha_i \le d_i - 1$. Since $\beta_i = \beta_{i-1} + d_1\cdots d_{i-1}\alpha_i$, the induction step follows. When we reach $i = n$, the proof is complete. $\qquad\square$

For practical computations, it is usually preferable that the output of an algorithm is in normal form, if it consists of residue classes. Thus the preceding proposition yields the following algorithm.

---

**Algorithm 6.3.28 (Solving over $\mathbb{F}_q$ via Recursion)**
*Let $K$ be a finite field with $q$ elements, let $P = K[x_1,\ldots,x_n]$, let $\mathfrak{M}$ be a maximal ideal of $P$, let $L = P/\mathfrak{M}$, and let $d = \dim_K(L)$. Consider the following sequence of instructions.*

(1) *Compute a triangular set $\{f_1(x_1), f_2(x_1,x_2),\ldots, f_n(x_1,\ldots,x_n)\}$ of generators of $\mathfrak{M}$ (e.g., use Algorithm 5.3.12), and let $d_i = \deg_{x_i}(f_i)$ for $i = 1,\ldots,n$.*
(2) *Let $S_0 = \{(\emptyset, 0)\}$. For $i = 1,\ldots,n$, perform the following steps.*
(3) *Let $S_i = \emptyset$. Introduce a new indeterminate $y_i$, and in $K[y_1,\ldots,y_i]$ define the ideal $\mathfrak{M}'_i = \langle f_1(y_1),\ldots, f_i(y_1,\ldots,y_i)\rangle$.*
(4) *For every pair $(G_{i-1}, \beta_{i-1})$ in $S_{i-1}$, perform the following step.*
(5) *For every $j \in \{0,\ldots,d_i - 1\}$, let $\beta_i = \beta_{i-1} + d_1\cdots d_{i-1}j$, compute the normal form $h(y_1,\ldots,y_i) = \mathrm{NF}_{\mathfrak{M}'_i}(y_i^{q^{\beta_i}})$, append this polynomial to $G_{i-1}$ to get a tuple $G_i$, and put the pair $(G_i, \beta_i)$ into $S_i$.*

*This is an algorithm which computes a set $S_n$ of tuples of polynomials in $K[y_1,\ldots, y_n]$ whose residue classes in $(L')^n$ are precisely the zeros of $\mathfrak{M}$, where $L' = K[y_1,\ldots, y_n]/\mathfrak{M}'$ is the clone of $L$.*

---

*Proof* In view of the proposition, it suffices to note that $y_i^{q^j}$ and $\mathrm{NF}_{\mathfrak{M}'_i}(y_i^{q^j})$ have the same residue class in the field $L'_i = K[y_1,\ldots, y_i]/\mathfrak{M}'_i$, and that $L'_i$ is a subfield of $L'$. $\qquad\square$

The following example shows that Proposition 6.3.27 allows us to reduce the powers necessary in Proposition 6.3.19.

**Example 6.3.29** In the setting of Example 6.3.22, notice that the given set of generators $f_1 = x_1^2 + x_1 + 1$, $f_2 = x_2^3 + x_1$, and $f_3 = x_3^2 + x_3 + x_1$ is a triangular set of generators of the maximal ideal $\mathfrak{M}$ in $P = K[x_1, x_2, x_3]$. Thus we can apply Proposition 6.3.27 and get the following twelve zeros of $\mathfrak{M} = \langle f_1, f_2, f_3\rangle$ in the clone

$L' = K[y_1, y_2, y_3]/\mathfrak{M}'$ of $L = P/\mathfrak{M}$.

$$(y_1, y_2, y_3), \quad (y_1, y_2, y_3^{2^6}), \quad (y_1, y_2^{2^2}, y_3^{2^2}), \quad (y_1, y_2^{2^2}, y_3^{2^8}),$$
$$(y_1, y_2^{2^4}, y_3^{2^4}), \quad (y_1, y_2^{2^4}, y_3^{2^{10}}), \quad (y_1^2, y_2^2, y_3^2), \quad (y_1^2, y_2^2, y_3^{2^7})$$
$$(y_1^2, y_2^{2^3}, y_3^{2^3}), \quad (y_1^2, y_2^{2^3}, y_3^{2^9}), \quad (y_1^2, y_2^{2^5}, y_3^{2^5}), \quad (y_1^2, y_2^{2^5}, y_3^{2^{11}})$$

When we apply Algorithm 6.3.28 to this system, we get the representations of the zeros of $\mathfrak{M}$ given in Example 6.3.22.b.

Another way to perform the last step of Algorithm 6.3.28 is given in the following version of the algorithm.

**Algorithm 6.3.30 (Solving over $\mathbb{F}_q$ via Recursive Factorizations)**
*In the preceding algorithm, replace Steps (4) and (5) by the following steps.*

(4') *For every tuple G in $S_{i-1}$, perform the following step.*
(5') *Let $L_i' = K[y_1, \ldots, y_i]/\mathfrak{M}_i'$. Substitute the entries of G for $x_1, \ldots, x_{i-1}$ in $f_i$ and get a polynomial $\bar{f}_i(x_i)$ in $L_{i-1}'[x_i]$. Then compute the factorization $\bar{f}_i(x_i) = (x_i - \bar{h}_{i1}) \cdots (x_i - \bar{h}_{id_i})$ with $\bar{h}_{ij} \in L_i'$ and represent $\bar{h}_{ij}$ by a polynomial $h_{ij}$ in $K[y_1, \ldots, y_i]$. For $j = 1, \ldots, d_i$, append the polynomial $h_{ij}$ to G and put the resulting tuple into $S_i$.*

*The result is an algorithm which computes a set $S_n$ of tuples of polynomials in $K[y_1, \ldots, y_n]$ whose residue classes in $(L')^n$ are precisely the zeros of $\mathfrak{M}$.*

*Proof* Clearly, a zero $(\bar{p}_1, \ldots, \bar{p}_n)$ of $\mathfrak{M}$ in $(L')^n$ yields a zero $G = (\bar{p}_1, \ldots, \bar{p}_{i-1})$ of $\mathfrak{M}_{i-1}$ in $(L_{i-1}')^{i-1}$ and a zero $\bar{p}_i$ of the polynomial $\bar{f}_i(x_i)$ in $L_i'$, where $\bar{f}_i(x_i)$ is obtained by substituting G for $(x_1, \ldots, x_{i-1})$. Thus the zero $(\bar{p}_1, \ldots, \bar{p}_n)$ is computed during the course of the algorithm. Conversely, it is clear that every tuple $(\bar{h}_1, \ldots, \bar{h}_i)$ in $S_i$ yields a zero $(\bar{h}_1, \ldots, \bar{h}_i)$ of $\mathfrak{M}_i$ in $(L')^n$. This proves the correctness of the algorithm. $\square$

Comparing the last algorithm to Theorem 6.3.4, we see that it reduces the task of finding the primary decomposition of a big ideal $\mathfrak{M}Q + \mathfrak{M}'Q$ in a big polynomial ring $Q = K[x_1, \ldots, x_n, y_1, \ldots, y_m]$ to a number of univariate factorizations over extension fields of $K$.

In the next example we apply several solving methods to the polynomial system given in Example 6.3.23.

**Example 6.3.31** Let $K = \mathbb{F}_7$, and let us consider the system of polynomial equations defined by

$$\tilde{f}_1 = x_1^4 - 2x_1 + 3,$$
$$\tilde{f}_2 = x_1^2 x_2^3 + 2x_1^2 - 2x_2 - 1$$

Let $P = K[x_1, x_2]$, let $\mathfrak{M} = \langle \tilde{f}_1, \tilde{f}_2 \rangle$, and let $L = P/\mathfrak{M}$. In Example 6.3.23 we checked that $\mathfrak{M}$ is a maximal ideal in the ring $P$, noted that the field $L$ satisfies $d = \dim_K(L) = 12$, and found the twelve zeros of $\mathfrak{M}$ in a suitable extension of $L$. Here our goal is to find these zeros using different strategies.

(a) First we solve the task via Theorem 6.3.4. We have to form the polynomial ring $Q = K[x_1, x_2, y_1, y_2]$ and to compute the primary decomposition of the ideal $\mathfrak{M}Q + \mathfrak{M}'Q$, where $\mathfrak{M}'$ is the clone of $\mathfrak{M}$ in $P' = K[y_1, y_2]$. In this way we find again the 12 solutions listed in Example 6.3.23.

(b) Next we use Algorithm 6.3.28. We calculate a triangular set of generators of the ideal $\mathfrak{M}$ and get $\mathfrak{M} = \langle f_1(x_1), f_2(x_1, x_2) \rangle$, where $f_1(x_1) = \tilde{f}_1(x_1)$ and $f_2(x_1, x_2) = x_2^3 + 2x_1^3 x_2 + 3x_1^2 x_2 + 3x_2 + x_1^3 - 2x_1^2$. Thus we have $d_1 = 4$ and $d_2 = 3$, and the zeros of $\mathfrak{M}$ in $(L')^2$ are given by $(\bar{y}_1^{q^\alpha}, \bar{y}_2^{q^{4\beta}})$ with $0 \le \alpha \le 3$ and $0 \le \beta \le 2$.

(c) Finally, we use Algorithm 6.3.30. For $i = 1$, we have to factorize the polynomial $f_1(x_1) = x_1^4 - 2x_1 + 3$ over the field $L_1' = K[y_1]/\langle y_1^4 - 2y_1 + 3 \rangle$. We get

$$f_1(x_1) = (x_1 - \bar{y}_1)(x_1 + 3\bar{y}_1^3 + 3\bar{y}_1 - 1)(x_1 + 2\bar{y}_1^3 + 2\bar{y}_1^2 + 2\bar{y}_1 - 3)$$
$$\cdot (x_1 + 2\bar{y}_1^3 - 2\bar{y}_1^2 + 3\bar{y}_1 - 3)$$

Therefore, when we start with $i = 2$, we have $S_1 = \{(y_1), (-3y_1^3 - 3y_1 + 1), (-2y_1^3 - 2y_1^2 - 2y_1 + 3), (-2y_1^3 + 2y_1^2 - 3y_1 + 3)\}$ and $L_1' = K[y_1]/\langle f_1(y_1) \rangle$. In the first iteration of Step (5'), we have to factorize the polynomial $f_2(\bar{y}_1, x_2) = x_2^3 + 2\bar{y}_1^3 x_2 + 3\bar{y}_1^2 x_2 + 3x_2 + \bar{y}_1^3 - 2\bar{y}_1^2$, and we get three irreducible factors, namely

$x_2 - \bar{y}_2$

$x_2 - \bar{y}_1^2 \bar{y}_2^2 - \bar{y}_2^4 - \bar{y}_2^3 - \bar{y}_1^2 \bar{y}_2 - 3\bar{y}_1 \bar{y}_2^2 - 3\bar{y}_2^2 + \bar{y}_1 \bar{y}_2 - \bar{y}_2^2 + 2\bar{y}_1 - 3\bar{y}_2 - 3$

$x_2 + \bar{y}_1^2 \bar{y}_2^2 + \bar{y}_2^4 + \bar{y}_2^3 + \bar{y}_1^2 \bar{y}_2 + 3\bar{y}_1 \bar{y}_2^2 + 3\bar{y}_2^2 - \bar{y}_1 \bar{y}_2 + \bar{y}_2^2 - 2\bar{y}_1 - 3\bar{y}_2 + 3$

They correspond to the three zeros $(\bar{y}_1, \bar{y}_2)$, $(\bar{y}_1, \bar{y}_1^2 \bar{y}_2^2 + \bar{y}_2^4 + \bar{y}_2^3 + \bar{y}_1^2 \bar{y}_2 + 3\bar{y}_1 \bar{y}_2^2 + 3\bar{y}_1^2 - \bar{y}_1 \bar{y}_2 + \bar{y}_2^2 - 2\bar{y}_1 + 3\bar{y}_2 + 3)$, and $(\bar{y}_1, -\bar{y}_1^2 \bar{y}_2^2 - \bar{y}_2^4 - \bar{y}_2^3 - \bar{y}_1^2 \bar{y}_2 - 3\bar{y}_1 \bar{y}_2^2 - 3\bar{y}_1^2 + \bar{y}_1 \bar{y}_2 - \bar{y}_2^2 + 2\bar{y}_1 + 3\bar{y}_2 - 3)$ of $\mathfrak{M}$ in $(L')^2$.

In the next iteration of Step (5'), we factorize $f_2(-3\bar{y}_1^3 - 3\bar{y}_1 + 1, x_2)$, and we get another three zeros of $\mathfrak{M}$, and so on. Altogether, we find again the 12 solutions listed in Example 6.3.23.

There is still one very nice situation in which we do not have to make any recursive calls at all. It is described in the following proposition which takes care of

maximal ideals generated by special triangular sets, namely triangular sets consisting of univariate polynomials.

**Proposition 6.3.32 (Solving Sequences of Univariate Polynomials)**
*Let $K$ be a finite field with $q$ elements, let $P = K[x_1, \ldots, x_n]$, and let $\mathfrak{M}$ be a maxi-
mal ideal in $P$ of the form $\mathfrak{M} = \langle f_1(x_1), \ldots, f_n(x_n) \rangle$, where $f_i(x_i) \in P$. Moreover,
for $i = 1, \ldots, n$, let $d_i = \deg(f_i)$, and let $d = \prod_{i=1}^{n} d_i$.*

(a) *For every subset $\{i_1, \ldots, i_k\} \subseteq \{1, \ldots, n\}$, the ideal $\langle f_{i_1}(x_{i_1}), \ldots, f_{i_k}(x_{i_k}) \rangle$ is a
    maximal ideal in the polynomial ring $K[x_{i_1}, \ldots, x_{i_k}]$.*
(b) *Let $\{i_1, \ldots, i_k\} \subset \{1, \ldots, n\}$, and let $j \in \{1, \ldots, n\} \setminus \{i_1, \ldots, i_k\}$. Then the image
    of $f_j(x_j)$ in $(K[x_{i_1}, \ldots, x_{i_k}]/\langle f_{i_1}(x_{i_1}), \ldots, f_{i_k}(x_{i_k}) \rangle)[x_j]$ is irreducible.*
(c) *Let $P' = K[y_1, \ldots, y_n]$, let $\mathfrak{M}' = \langle f_1(y_1), f_2(y_2), \ldots, f_n(y_n) \rangle$ be the clone
    of $\mathfrak{M}$ in $P'$, and let $L' = P'/\mathfrak{M}'$ be a clone of $P/\mathfrak{M}$. Then the ideal $\mathfrak{M}$ has $d$
    zeros in $(L')^n$, namely*

$$\mathcal{Z}_{L'}(\mathfrak{M}) = \left\{ \left( \bar{y}_1^{q^{i_1}}, \ldots, \bar{y}_n^{q^{i_n}} \right) \mid 0 \le i_j \le d_j - 1 \text{ for } j = 1, \ldots, n \right\}$$

*Proof* Clearly, the ring $K[x_{i_1}, \ldots, x_{i_k}]/\langle f_{i_1}(x_{i_1}), \ldots, f_{i_k}(x_{i_k}) \rangle$ is a zero-dimensional
affine $K$-algebra. It is an integral domain, since it is contained in the field $P/\mathfrak{M}$.
Consequently, it is a field by Corollary 2.2.6, and this proves (a).

Since claim (b) follows from (a), it remains to prove (c). By Proposition 6.3.13,
for every $j \in \{1, \ldots, n\}$, the residue classes $\bar{y}_j^{q^0}, \bar{y}_j^{q^1}, \ldots, \bar{y}_j^{q^{d_j-1}}$ are pairwise distinct
zeros of $f_j(x_j)$ in $K[y_j]/\langle f_j(y_j) \rangle$. Therefore they are distinct zeros of $f_j(x_j)$ in the
larger field $L'$. Consequently, we have found $d$ distinct zeros of $\mathfrak{M}$ in $(L')^n$, and the
proof is complete.                                                                    $\square$

Let us reconsider Example 6.3.26.

**Example 6.3.33** As in Example 6.3.26, we let $K = \mathbb{F}_2$, let $P = K[x_1, x_2]$, and
let $\mathfrak{M} = \langle f_1(x_1), f_2(x_2) \rangle$, where $f_1(x_1) = x_1^2 + x_1 + 1$ and $f_2(x_2) = x_2^3 + x_2 + 1$.
Then $\mathfrak{M}$ is a maximal ideal of $P$ which satisfies the hypothesis of the proposi-
tion. Therefore, to find the zeros of $\mathfrak{M}$, we let $P' = K[y_1, y_2]$, let $\mathfrak{M}'$ be the clone
of $\mathfrak{M}$ in $P'$, and let $L' = P'/\mathfrak{M}'$. By the proposition, the set of zeros of $\mathfrak{M}$ in $(L')^2$
is

$$\mathcal{Z}_{L'}(\mathfrak{M}) = \left\{ (\bar{y}_1, \bar{y}_2), (\bar{y}_1, \bar{y}_2^2), (\bar{y}_1, \bar{y}_2^{2^2}), (\bar{y}_1^2, \bar{y}_2), (\bar{y}_1^2, \bar{y}_2^2), (\bar{y}_1^2, \bar{y}_2^{2^2}) \right\}$$

Using normal forms with respect to $\mathfrak{M}'$, we can also write this set of zeros as
$\mathcal{Z}_{L'}(\mathfrak{M}) = \{(\bar{y}_1, \bar{y}_2), (\bar{y}_1, \bar{y}_2^2), (\bar{y}_1, \bar{y}_2^2 + \bar{y}_2), (\bar{y}_1+1, \bar{y}_2), (\bar{y}_1+1, \bar{y}_2^2), (\bar{y}_1+1, \bar{y}_2^2+\bar{y}_2)\}$.

# 6.4 Solving Polynomial Systems over the Rationals

> *What is rational is real;*
> *And what is real is rational.*
> (Georg Wilhelm Friedrich Hegel)

In this, the final section of the chapter and the book, we endeavour to solve the following problem: given a finite system of polynomial equations over the field $K$ of rational numbers, find a suitable algebraic number field $L$ and write down the solutions of the system in $L^n$.

This question has intrigued legions of researchers throughout the ages. The point of view from which we approach it is diametrically opposed to the words of the famous philosopher. What is real is not necessarily rational. In particular, we are not interested in describing the solutions of the problem as tuples of real numbers via their floating point representations. Instead, we strive to find explicit symbolic representations via polynomials with rational coefficients. And we reinterpret German idealism to mean that we should really study the polynomial ideals generated by the equations in a rational manner.

It seems impossible to trace the origins and developments of all approaches to solving over the rationals which have been taken in the past. Instead we shall focus on pointing out a few possible applications of the material presented earlier in this book to the problem at hand.

This section is subdivided into two subsections. The first one is devoted to the study of the splitting field of a univariate polynomial in $\mathbb{Q}[x]$. The main idea we use is the cloning technique introduced in the previous section, and more specifically the recursive way of using this cloning technique presented in Sect. 6.3.D.

In the following sense, computing the splitting field of a univariate polynomial seems to be "all we have to do". Via the computation of the maximal components of the ideal $I$ in $P = \mathbb{Q}[x_1, \ldots, x_n]$ generated by the polynomials defining the given system, we reduce the problem to the case of a maximal ideal $\mathfrak{M}$. Since the base field is infinite, we can bring $\mathfrak{M}$ into $x_n$-normal position and then use the Shape Lemma to show $P/\mathfrak{M} \cong K[x_n]/\langle f(x_n)\rangle$ for some polynomial $f(x_n) \in \mathbb{Q}[x_n]$ (see [15], Sect. 3.7). Unfortunately, in all but the smallest cases this will lead to very hard computations.

Another approach, which is in general infeasible, is to use the splitting algebra of a univariate polynomial which we also mention briefly. Its main disadvantage is its size: for a polynomial of degree $d$, the splitting algebra is a $d!$-dimensional vector space.

In the second subsection we consider the general task of solving polynomial systems over $\mathbb{Q}$. After reducing the task to the case of polynomial systems which correspond to a single maximal ideal of $P = \mathbb{Q}[x_1, \ldots, x_n]$, we can calculate a triangular set of generators and get a better handle on the problem. Moreover, if the degrees of the generators are not too high, we can iteratively compute their splitting fields after substituting the "partial" zeros of the previous polynomials and thus divide the problem into manageable pieces.

*It is so much easier to suggest solutions*
*when you don't know too much about the problem.*
(Malcolm Forbes)

If the degrees of the generators or the size of the polynomial system get too large, there is no feasible way to find the exact solutions of the system, and all that is left is to resort to approximations. Since we do not know much about this technique, and since this would carry us far away from the topic of the book, we content ourselves with an example and a few hints in this direction.

**Some Closing Remarks About Strategies**

*However beautiful the strategy,*
*you should occasionally look at the results.*
(Sir Winston Churchill)

Throughout this section we follow the strategy of first computing the maximal components of the given ideal and then determining their zeros separately. This simplifies the presentation, but is it the best strategy?

Let us look at the results. In certain cases, it may not produce the simplest possible description of the zeros of the ideal $I$. For instance, consider the ideal $I = \langle x_1^2 - 2, x_2^2 - x_2 \rangle$ in $\mathbb{Q}[x_1, x_2]$. Clearly, its solutions are defined over the field $\mathbb{Q}[\sqrt{2}] \cong \mathbb{Q}[y]/\langle y^2 - 2 \rangle$. According to the strategy we adhere to in this section, the zeros of $I$ are described by first computing the primary decomposition $I = \mathfrak{M}_1 \cap \mathfrak{M}_2$ with $\mathfrak{M}_1 = \langle x_1^2 - 2, x_2 \rangle$ and $\mathfrak{M}_2 = \langle x_1^2 - 2, x_2 - 1 \rangle$, and then finding the zeros of $\mathfrak{M}_i$ in $L'_{1i} = \mathbb{Q}[y_{1i}]/\langle y_{1i}^2 - 2 \rangle$ without noticing that $L'_{11} \cong L'_{12}$.

On the other hand, in general we should not expect that the fields corresponding to different maximal components of a radical ideal are strongly related. For instance, trying to split the ideal $\mathfrak{M}_3$ in Example 6.4.13 over $\mathbb{Q}[\sqrt{-2}] \cong \mathbb{Q}[y]/\langle y^2 + 2 \rangle$ which splits $\mathfrak{M}_1$ and $\mathfrak{M}_2$ would probably lead to unsurmountable computational difficulties.

Following our *divide and conquer* strategy, shouldn't we also split the computation of the zeros of a single maximal ideal into smaller parts and fork the computation every time we have to factor several univariate polynomials obtained by substituting different "partial" zeros? A look at the results in Example 6.4.11 shows that this may not be optimal either. The reason is that the zeros of a single irreducible polynomial or a single maximal ideal are indeed related to each other. More precisely, the elementary symmetric polynomials in the zeros of a polynomial are (up to sign) its coefficients. Therefore a polynomial which splits after substituting some partial zeros also splits after substituting the remaining ones, i.e., that we do not have to enlarge the field any further.

*"When you come to a fork in the road, take it."*
*"I will, if it is a silver one."*

Summing up, we can say that, in the end, it is left to the reader to decide in every example which path to follow and if he wants to take the fork.

## 6.4.A  Splitting Fields in Characteristic Zero

> *Let the gentle bush dig its root deep*
> *and spread upward to split one boulder.*
> (Carl Sandburg)

In the following we let $K = \mathbb{Q}$, let $P = K[x]$, and let $f(x) \in P$ be a monic irreducible polynomial of degree $d \geq 2$. Recall that the Galois group of $f(x)$ is the group of $K$-automorphisms of the field $L = P/\langle f(x) \rangle$. In contrast to the previous section, the Galois group need not be cyclic, and there is no Frobenius automorphism which generates it. This makes it much harder to compute the Galois group and the splitting field of $f(x)$. Nevertheless, Proposition 6.3.15 indicates one way we may go: clone the field $L$, try to dig the roots of $f(x)$ deep in $L' = K[y]/\langle f(y) \rangle$, and spread them upward to split each irreducible factor. This idea leads to the construction of the following algorithm.

---

**Algorithm 6.4.1 (Splitting a Polynomial over the Rationals)**
*Let $K = \mathbb{Q}$, let $P = K[x]$, and let $f(x) \in P$ be a monic irreducible polynomial of degree $d \geq 2$. Consider the following sequence of instructions.*

(1) *Let $i = 0$, let $L'_0 = K$, let $g_1(x) = f(x)$, and let $S_0 = \emptyset$. Repeat the following steps until $S_i$ contains $d$ elements. Then return the tuple of polynomials $G = (g_1(y_1), g_2(y_1, y_2), \ldots, g_i(y_1, \ldots, y_i))$ and the set $S_i$ and stop.*

(2) *Increase $i$ by one. Then introduce a new indeterminate $y_i$ and form the field $L'_i = L'_{i-1}[y_i]/\langle g_i(\bar{y}_1, \ldots, \bar{y}_{i-1}, y_i) \rangle$.*

(3) *Factorize $f(x)$ in $L'_i[x]$. Append all linear factors to $S_i$.*

(4) *If not all factors are linear, let $g_{i+1}(y_1, \ldots, y_i, x) \in K[y_1, \ldots, y_i, x]$ be a representative of one of the irreducible factors.*

*This is an algorithm which computes both a tuple of polynomials $G$ such that the splitting field $L'_i$ of $f(x)$ is given by $L'_i = K[y_1, \ldots, y_i]/\langle G \rangle$ and the set $S_i$ of the linear factors of $f(x)$ in $L'_i[x]$.*

---

*Proof* For every $i \geq 1$, the field $L'_i$ contains at least one zero of $g_i(\bar{y}_1, \ldots, \bar{y}_{i-1}, x)$, namely $\bar{y}_i$. Hence at least one of the irreducible factors of $f(x)$ splits off a linear factor when we pass from $L'_{i-1}$ to $L'_i$. Therefore the algorithm terminates after finitely many steps. Clearly, it produces a complete factorization of $f(x)$ in $L'_i[x]$ into linear factors. Moreover, the field $L'_i$ is generated by $\bar{y}_1, \ldots, \bar{y}_i$ over $K$. Since these are zeros of $f(x)$ in $L'_i$, it follows that $L'_i$ is the splitting field of $f(x)$. $\qquad \square$

Notice that Step (3) of this algorithm can be implemented more efficiently by remembering the factorization of $f(x)$ in $L'_{i-1}[x]$ and refining this factorization in the next iteration. Let us try out this algorithm for some polynomials in $\mathbb{Q}[x]$ whose degree is not too high.

**Example 6.4.2** Let $K = \mathbb{Q}$, let $P = K[x]$, and let $f(x) = x^4 + 14x^3 + 57x^2 + 56x - 19$. Then $f(x)$ is an irreducible polynomial in $P$, and we can apply the algorithm. Let us follow its main steps.

(1) Let $g_1(x) = f(x)$.
(2) Let $L'_1 = K[y_1]/\langle g_1(y_1)\rangle$.
(3) To factorize $f(x)$ in $L'_1[x]$, we use Algorithm 5.3.6. We define $\mathfrak{M} = \langle f(x)\rangle$ and $\mathfrak{M}' = \langle g_1(y_1)\rangle = \langle f(y_1)\rangle$. We calculate the primary decomposition of the ideal $\mathfrak{M}Q + \mathfrak{M}'Q$ in $Q = K[x_1, y_1]$ and get $\mathfrak{M}Q + \mathfrak{M}'Q = \mathfrak{N}_1 \cap \mathfrak{N}_2 \cap \mathfrak{N}_3 \cap \mathfrak{N}_4$, where

$$\mathfrak{N}_1 = \mathfrak{M}'Q + \langle x - y_1 \rangle$$

$$\mathfrak{N}_2 = \mathfrak{M}'Q + \langle x + y_1 + 7 \rangle$$

$$\mathfrak{N}_3 = \mathfrak{M}'Q + \langle x - \tfrac{4}{23}y_1^3 - \tfrac{42}{23}y_1^2 - \tfrac{81}{23}y_1 + \tfrac{140}{23}\rangle$$

$$\mathfrak{N}_4 = \mathfrak{M}'Q + \langle x + \tfrac{4}{23}y_1^3 + \tfrac{42}{23}y_1^2 + \tfrac{81}{23}y_1 + \tfrac{21}{23}\rangle$$

It follows that $f(x)$ is the product of the above four linear polynomials in $L'_1[x]$.

The result of the algorithm is the tuple $G = (g_1(y_1)) = (f(y_1))$ such that the field $L'_1 = K[y_1]/\langle g_1(y_1)\rangle$ is the splitting field of $f(x)$. Notice that we have $\dim_K(L'_1) = 4$.

Our next example shows that already for polynomials of degree four the task can be much harder than the preceding example suggests. In Example 6.4.10 we examine the similar polynomial $x^4 - 10x^2 + 1$ and discover that its behaviour is very different.

**Example 6.4.3** Let $K = \mathbb{Q}$, let $P = K[x]$, and let $f(x) = x^4 - 10x + 1$. Then $f(x)$ is a monic irreducible polynomial in $P$. To compute its splitting field, we apply Algorithm 6.4.1 again and follows its main steps.

(1) Let $g_1(x) = f(x)$.
(2) Let $L'_1 = K[y_1]/\langle g_1(y_1)\rangle$.
(3) Using Algorithm 5.3.6, factorize $f(x) = (x - \bar{y}_1)(x^3 + \bar{y}_1 x^2 + \bar{y}_1^2 x + \bar{y}_1^3 - 10)$ in $L'_1[x]$.
(4) Let $g_2(y_1, x) = x^3 + y_1 x^2 + y_1^2 x + y_1^3 - 10$.
(2) Let $i = 2$ and $L'_2 = L'_1[y_2]/\langle g_2(\bar{y}_1, y_2)\rangle$.
(3) To factor $f(x)$ in $L'_2[x]$, we use Algorithm 5.3.6 again. We let $\mathfrak{M} = \langle f(x)\rangle$ and $\mathfrak{M}'_2 = \langle g_1(y_1), g_2(y_1, y_2)\rangle$. In the ring $Q = K[y_1, y_2, x]$, we compute the

primary decomposition $\mathfrak{M}Q + \mathfrak{M}'_2Q = \mathfrak{N}_1 \cap \mathfrak{N}_2$, where

$$\mathfrak{N}_1 = \mathfrak{M}'_2Q + \langle x - y_2 \rangle$$

$$\mathfrak{N}_2 = \mathfrak{M}'_2Q + \langle x^2 + y_1x + y_2x + y_1^2 + y_1y_2 + y_2^2 \rangle$$

(4) We let $g_3(y_1, y_2, x) = x^2 + y_1x + y_2x + y_1^2 + y_1y_2 + y_2^2$.
(2) Let $i = 3$ and $L'_3 = L'_2[y_3]/\langle g_3(\bar{y}_1, \bar{y}_2, y_3) \rangle$.
(3) In $L'_3[x]$ we get the factorization $f(x) = (x - \bar{y}_1)(x - \bar{y}_2)(x - \bar{y}_3)(x + \bar{y}_1 + \bar{y}_2 + \bar{y}_3)$.
(1) The algorithm returns the tuple $G = (g_1(y_1), g_2(y_1, y_2), g_3(y_1, y_2, y_3))$ and the set of four linear factors given above.

Altogether, it follows that $L'_3 = K[y_1, y_2, y_3]/\langle G \rangle$ is the splitting field of $f(x)$. Notice that we have $\dim_K(L'_3) = 24$ and that the Galois group of $K[x]/\langle f(x) \rangle$ over $K$ is the full symmetric group $S_4$.

The following example is inspired by [3], Example 2.5.6.

**Example 6.4.4**  Let $K = \mathbb{Q}$, and let $P = K[x]$. We consider the monic irreducible polynomials $f(x) = x^3 - 3x + 1$ and $g(x) = x^3 - 4x + 1$ in $P$ and determine their splitting fields.

(a) To compute the splitting field of $f(x)$, we form $L'_1 = K[y_1]/\langle f(y_1) \rangle$ and factor $f(x)$ in $L'_1[x]$. The result is $f(x) = (x - \bar{y}_1)(x - \bar{y}_1^2 + 2)(x + \bar{y}_1^2 + \bar{y}_1 - 2)$. Hence $L'_1$ is the splitting field of $f(x)$. Notice that we have $\dim_K(L'_1) = 3$.
(b) Next we compute the splitting field of $g(x)$. Let $L'_1 = K[y_1]/\langle g(y_1) \rangle$. We factor $g(x)$ in $L'_1[x]$ and get $g(x) = (x - \bar{y}_1)(x^2 + \bar{y}_1x + \bar{y}_1^2 - 4)$. Therefore we let $g_2(y_1, x) = x^2 + y_1x + y_1^2 - 4$ and form the field $L'_2 = L'_1[y_2]/\langle g_2(\bar{y}_1, y_2) \rangle$. The factorization of $g(x)$ in $L'_2[x]$ is $g(x) = (x - \bar{y}_1)(x - \bar{y}_2)(x + \bar{y}_1 + \bar{y}_2)$. Consequently, the field $L'_2$ is the splitting field of $g(x)$, and we have $\dim_K(L'_2) = 6$.

Why do these two similar looking polynomials behave so differently? The discriminant of a cubic polynomial $x^3 + ax + b$ is $\Delta = -4a^3 - 27b^2$. Thus the discriminant of $f(x)$ is 81, a square number. As explained in [3], Example 2.5.6, this implies that the Galois group of $K[x]/\langle f(x) \rangle$ is $\mathbb{Z}/3\mathbb{Z}$. On the other hand, the discriminant of $g(x)$ is 229, and thus the Galois group of $K[x]/\langle g(x) \rangle$ is the full symmetric group $S_3$. If fact, the sizes of these Galois groups are reflected in the different dimensions of the corresponding splitting fields.

*Have you ever tried to split sawdust?*
(Eugene McCarthy)

Another approach to split a polynomial is to use the splitting algebra. In the following we introduce only some very basic facts about this algebra, because it can be used to split merely the tiniest of polynomials and hence will be applied sparingly.

Recall that the **elementary symmetric polynomials** $s_1, \ldots, s_d$ in the indeterminates $y_1, \ldots, y_d$ are defined by the identity

$$(x - y_1)(x - y_2) \cdots (x - y_d) = x^d - s_1 x^{d-1} + \cdots + (-1)^d s_d$$

in $K[x, y_1, \ldots, y_d]$. Consequently, we have $s_i = \sum_{1 \le j_1 < \cdots < j_i \le d} y_{j_1} \cdots y_{j_i}$ for every $1 \le i \le d$. Now we introduce the following algebra.

**Definition 6.4.5**  Let $K$ be a field, let $c_1, \ldots, c_d \in K$, and assume that the polynomial $f(x) = x^d - c_1 x^{d-1} + \cdots + (-1)^d c_d$ is irreducible in $K[x]$. We let $y_1, \ldots, y_d$ be new indeterminates, let $P = K[y_1, \ldots, y_d]$, let $s_1, \ldots, s_d \in P$ be the elementary symmetric polynomials in the indeterminates $y_1, \ldots, y_d$, and finally let $I = \langle s_1 - c_1, s_2 - c_2, \ldots, s_d - c_d \rangle \subseteq P$. Then the $K$-algebra $S_f = P/I$ is called the **splitting algebra** of $f(x)$.

The splitting algebra has many interesting properties. For instance, it is a zero-dimensional affine $K$-algebra and satisfies $\dim_K(S_f) = d!$. A more detailed description of the splitting algebra with proofs of these claims is contained in [5]. For our purposes, the following property is the most useful one.

**Proposition 6.4.6**  *Let $K$ be a field, let $c_1, \ldots, c_d \in K$, and assume that the polynomial $f(x) = x^d - c_1 x^{d-1} + \cdots + (-1)^d c_d$ is irreducible in $K[x]$. If the splitting algebra $S_f$ is a field then it is the splitting field of $f(x)$.*

*Proof*  From the presentation $S_f = K[y_1, \ldots, y_d]/\langle s_1 - c_1, \ldots, s_d - c_d \rangle$ we get

$$S_f[x]/\langle f(x) \rangle \cong K[y_1, \ldots, y_d, x]/\langle s_1 - c_1, \ldots, s_d - c_d, f(x) \rangle$$

Let $g(x) = x^d - s_1 x^{d-1} + \cdots + (-1)^d s_d$. Then we have $\langle s_1 - c_1, \ldots, s_d - c_d, f(x) \rangle = \langle s_1 - c_1, \ldots, s_d - c_d, g(x) \rangle$. On the other hand, by definition, we have the factorization $g(x) = \prod_{i=1}^{d}(x - y_i)$. Therefore we obtain $f(x) = \prod_{i=1}^{d}(x - \bar{y}_i)$ in $S_f[x]$, and the proof is complete.  $\square$

Let us see the splitting algebra at work and split the polynomial given in Example 6.4.3.

**Example 6.4.7**  Let $K = \mathbb{Q}$, and let $f(x) = x^4 - 10x + 1$ be the irreducible polynomial in $K[x]$ already studied in Example 6.4.3. In the setting of the proposition, we have $c_1 = 0$, $c_2 = 0$, $c_3 = 10$, and $c_4 = 1$. Hence the splitting algebra of $f(x)$ is given by

$$S_f = K[y_1, y_2, y_3, y_4]/\langle s_1, s_2, s_3 - 10, s_4 - 1 \rangle$$

By the observation above, it satisfies $\dim_K(S_f) = 4! = 24$.

In order to show that $S_f$ is a field, we randomly choose a linear form, for instance $\ell = y_1 - y_2 + 2y_3 - 4y_4$ and compute the minimal polynomial of its residue class $\bar{\ell}$

in $S_f$. We get the polynomial

$$\mu_{\bar{\ell}}(z) = z^{24} - 880z^{20} - 22000z^{18} + 417120z^{16} + 134184000z^{14}$$
$$+ 47866930400z^{12} - 57240480000z^{10} - 1312347332060z^8$$
$$+ 1333453318560000z^6 + 5581211448086016z^4$$
$$+ 674721947418496000z^2 + 56779112448710560000$$

which is irreducible. Using Corollary 5.4.10.b, we conclude that $S_f$ is a field. By the proposition, it follows that $S_f$ is the splitting field of $f(x)$. Hence it is isomorphic to the field $L_3'$ in Example 6.4.3.

Finally, we generalize Algorithm 6.4.1 in two ways: we do not assume that $f(x)$ is irreducible, and we allow the case where the base field $K$ is a number field. Recall that a **number field** is a finite field extension of $\mathbb{Q}$. The following algorithm will be a key ingredient in the recursive solution of polynomial systems over the field of rational numbers presented in the next subsection.

---

**Algorithm 6.4.8 (Splitting a Polynomial over a Number Field)**

*Let $\mathbf{t} = (t_1, \ldots, t_m)$ be a tuple of indeterminates, let $K = \mathbb{Q}[\mathbf{t}]/\mathfrak{m}$ be a number field, where $\mathfrak{m}$ is a maximal ideal in $\mathbb{Q}[\mathbf{t}]$, and denote the image of $\mathbf{t}$ in $K$ by $\bar{\mathbf{t}}$. Furthermore, let $P = K[x]$, and let $f(\mathbf{t}, x)$ be a polynomial in $\mathbb{Q}[\mathbf{t}, x]$ whose image $\bar{f}(x) = f(\bar{\mathbf{t}}, x)$ in $P$ is monic and of degree $d \geq 1$ in $x$. Consider the following sequence of instructions.*

(1) *Let $i = 0$, let $L_0' = K$, and let $S_0 = \emptyset$.*
(2) *Repeat the following steps until $S_i$ contains $d$ elements. Then return the tuple of polynomials $F = (f_1(\mathbf{t}, y_1), f_2(\mathbf{t}, y_1, y_2), \ldots, f_i(\mathbf{t}, y_1, \ldots, y_i))$ and the set $S_i$ and stop.*
(3) *In the ring $Q = \mathbb{Q}[\mathbf{t}, y_1, \ldots, y_i, x]$, compute the maximal components of the ideal $\mathfrak{m}Q + \langle f_1(\mathbf{t}, y_1), \ldots, f_i(\mathbf{t}, y_1, \ldots, y_i), f(\mathbf{t}, x)\rangle$ and get maximal ideals $\mathfrak{N}_1, \ldots, \mathfrak{N}_k$ of $Q$.*
(4) *For every $j \in \{1, \ldots, k\}$, use an elimination ordering for $x$ to write $\mathfrak{N}_j$ as $\mathfrak{N}_j = \mathfrak{m}Q + \langle f_1(\mathbf{t}, y_1), \ldots, f_i(\mathbf{t}, y_1, \ldots, y_i)\rangle + \langle g_j(\mathbf{t}, y_1, \ldots, y_i, x)\rangle$, where the image $\bar{g}_j(x) = g_j(\bar{\mathbf{t}}, \bar{y}_1, \ldots, \bar{y}_i, x)$ is an irreducible polynomial in $L_i'[x]$. Append the representatives of all linear factors $\bar{g}_j(x)$ to $S_i$.*
(5) *If not all factors are linear, increase $i$ by one. Let $f_i(\mathbf{t}, y_1, \ldots, y_i, x)$ be a representative in $\mathbb{Q}[\mathbf{t}, y_1, \ldots, y_i, x]$ of one of the non-linear irreducible factors, introduce a new indeterminate $y_i$, and form the field $L_i' = L_{i-1}'[y_i]/\langle f_i(\bar{\mathbf{t}}, \bar{y}_1, \ldots, \bar{y}_{i-1}, y_i)\rangle$.*

*This is an algorithm which computes both a tuple of polynomials $F$ such that the splitting field $L_i'$ of $\bar{f}(x)$ is given by $L_i' = K[y_1, \ldots, y_i]/\langle F\rangle$ and a set $S_i$ of polynomials representing the linear factors of $\bar{f}(x)$ in $L_i'[x]$.*

---

*Proof* The correctness follows in the same way as for Algorithm 6.4.1 if we show that Steps (3) and (4) compute the factorization of $\bar{f}(x)$ in $L'_i[x]$. This follows from Algorithm 5.3.6. □

In the next subsection we will see many applications of this algorithm.

### 6.4.B  Solving over the Rational Numbers via Cloning

> *This algorithm has been proved to work,*
> *but has never been observed to do so.*
> (Alexander Barvinok)

Finally, we attempt to solve actual polynomial systems over the rational numbers. In other words, we construct a suitable extension field $L$ of $\mathbb{Q}$ such that the solution of a given polynomial system can be written down as a set of tuples in $L^n$. Although the algorithms for performing this task have been proven to work, they can be observed to do so only in relatively simple cases. As a first step, we assume that the methods of Chap. 5 have been used to compute the maximal components of the ideal generated by the polynomials defining the system. In other words, we assume that the ideal whose zeros we want to find is maximal.

Secondly, we note that the Lex-method explained in [15], Sect. 3.7, can be used to solve the system. However, this method requires us to perform a coordinate transformation which puts the maximal ideal into $x_n$-normal position, and then to calculate a Lex-Gröbner basis. This technique tends to be computationally very heavy.

> *Why do they call it rush hour*
> *when nothing moves?*
> (Robin Williams)

Based on the cloning technique underlying Algorithm 6.3.28, we propose the following alternative strategy which works particularly well if we are given a triangular system of generators having low degrees.

---

**Algorithm 6.4.9 (Solving over the Rationals via Recursive Cloning)**
Let $K = \mathbb{Q}$, let $P = K[x_1, \ldots, x_n]$, let $\mathfrak{M}$ be a maximal ideal of $P$, let $L = P/\mathfrak{M}$, and let $d = \dim_K(L)$. Consider the following sequence of instructions.

(1)  *Compute a triangular set* $\{f_1(x_1), f_2(x_1, x_2), \ldots, f_n(x_1, \ldots, x_n)\}$ *of generators of* $\mathfrak{M}$ *(e.g., use Algorithm 5.3.12), and let* $d_i = \deg(f_i)$ *for* $i = 1, \ldots, n$.

(2) *Let $S_0 = \emptyset$. For $i = 1, \ldots, n$, perform the following steps. Then return the set $S_n$ and the presentation of the field $L'_n$.*

(3) *Let $S_i = \emptyset$. For every tuple $G$ in $S_{i-1}$, perform the following steps.*

(4) *Substitute the entries of $G$ for $x_1, \ldots, x_{i-1}$ in $f_i$ and get a polynomial $\bar{f}_i(x_i)$ in $L'_{i-1}[x_i]$.*

(5) *Using Algorithm 6.4.8, compute the splitting field $L'_i = L'_{i-1}[\mathbf{y}_i]/\langle F_i \rangle$ of $\bar{f}_i(x_i)$, where $\mathbf{y}_i = (y_{ik_i}, \ldots, y_{i\ell_i})$ is a tuple of new indeterminates and $F_i$ is a tuple of polynomials in $K[\mathbf{y}_1, \ldots, \mathbf{y}_i]$ whose residue classes generate a maximal ideal in $L'_{i-1}[\mathbf{y}_i]$. Replace $L'_{i-1}$ by $L'_i$.*

(6) *Algorithm 6.4.8 also returns a set of polynomials $x_i - h_{ij}$ such that we have the factorization $\bar{f}_i(x_i) = (x_i - \bar{h}_{i1}) \cdots (x_i - \bar{h}_{id_i})$ in $L'_i[x_i]$. For $j = 1, \ldots, d_i$, append the polynomial $h_{ij}$ to $G$ and put the resulting tuple into $S_i$.*

*This is an algorithm which computes a set $S_n$ of tuples of polynomials in $K[\mathbf{y}_1, \ldots, \mathbf{y}_n]$ whose residue classes in $(L'_n)^n$ are precisely the zeros of $\mathfrak{M}$, where $L'_n = K[\mathbf{y}_1, \ldots, \mathbf{y}_n]/\mathfrak{M}'_n$ and $\mathfrak{M}'_n = \langle F_1 \cup \cdots \cup F_n \rangle$.*

*Proof* For every $i \in \{1, \ldots, n\}$, Step (5) computes the splitting field $L'_i$ of $\bar{f}_i(x_i)$ over the previous field $L'_{i-1}$. Here $\bar{f}_i(x_i)$ denotes the polynomial in $L'_{i-1}[x_i]$ which is obtained from $f_i$ by replacing $x_1, \ldots, x_{i-1}$ with a zero of $\langle f_1(x_1), \ldots, f_{i-1}(x_1, \ldots, x_{i-1}) \rangle$. Moreover, Step (5) computes polynomials $h_{ij}$ representing all possible elements of $L'_i$ which extend this zero to a zero of $\langle f_1(x_1), \ldots, f_i(x_1, \ldots, x_i) \rangle$. So, in the end we construct all zeros of $\mathfrak{M}$ in $L'_n$. $\square$

Notice that in Step (5) of this algorithm we may have to extend the original field $L'_{i-1}$ several times by new tuples of indeterminates $\mathbf{y}_i$. This is achieved by replacing it with $L'_i$ at the end of Step (5) and incrementing the range $\{k_i, \ldots, \ell_i\}$ for the new indeterminates $y_{ij}$ in the next iteration.

Let us see this algorithm at work in some examples. We start with an easy one.

**Example 6.4.10** Let $K = \mathbb{Q}$, let $P = K[x_1, x_2]$, and let $\mathfrak{M} = \langle f_1(x_1), f_2(x_2) \rangle$, where $f_1(x_1) = x_1^2 - 2$ and $f_2(x_2) = x_2^2 - 3$. Using the methods described in Chap. 5 it is easy to check that $\mathfrak{M}$ is a maximal ideal of $P$.

(a) To begin with, we apply the algorithm to find the zeros of $\mathfrak{M}$ in a suitable number field. First we have to determine the splitting field of $f_1(x_1)$. The result of Algorithm 6.4.1 is $L'_1 = K[y_1]/\langle y_1^2 - 2 \rangle$, and the set of zeros is given by $S_1 = \{(y_1), (-y_1)\}$.

Next we have to compute the splitting field of $\bar{f}_2(x_2) = x_2^2 - 3$ in the ring $L'_1[x_2]$. We get $L'_2 = L'_1[y_2]/\langle y_2^2 - 3 \rangle$, and the factorization is given by $\bar{f}_2(x_2) = (x_2 - \bar{y}_2)(x_2 + \bar{y}_2)$. Notice that this happens both for $x_1 \mapsto \bar{y}_1$ and for $x_1 \mapsto -\bar{y}_1$, since $f_2(x_2)$ does not depend on $x_1$.

   Thus we obtain $S_2 = \{(y_1, y_2), (y_1, -y_2), (-y_1, y_2), (-y_1, -y_2)\}$. This is the
set of zeros of $\mathfrak{M}$ in $(L_2')^2$, where $L_2' = K[y_1, y_2]/\langle y_1^2 - 2, y_2^2 - 3\rangle$.

(b) Now let us compare this method to the Lex-method described in [15], Sect. 3.7.
Since we know that $\mathfrak{M}$ is a maximal ideal, the next step is to check whether $\mathfrak{M}$
is in $x_2$-normal position. Clearly, this is not the case. Therefore we perform
the linear change of coordinates $x_1 \mapsto x_1$ and $x_2 \mapsto x_1 + x_2$. Then the reduced
Lex-Gröbner basis of the resulting ideal $J$ is $\{x_1 + \frac{1}{2}x_2^3 - \frac{9}{2}x_2, x_2^4 - 10x_2^2 + 1\}$.

   The second polynomial has the splitting field $L' = K[y]/\langle y^4 - 10y^2 + 1\rangle$. It
factorizes in $L'[x_2]$ as $x_2^4 - 10x_2^2 + 1 = (x_2 - y)(x_2 + y)(x_2 - y^3 + 10y)(x_2 + y^3 - 10y)$. Notice that $\dim_K(L') = 4$ and the four solutions of the transformed
system are given by $(\frac{1}{2}y^3 - \frac{9}{2}y, y)$, $(-\frac{1}{2}y^3 + \frac{9}{2}y, -y)$, $(\frac{1}{2}y^3 - \frac{9}{2}y, y^3 - 10y)$,
and $(-\frac{1}{2}y^3 + \frac{9}{2}y, -y^3 + 10y)$.

(c) Recall that the polynomial $x_2^4 - 10x_2^2 + 1$ which appeared in the second solution was already mentioned in Example 5.1.14. We noted that it is irreducible
over $\mathbb{Q}$, but reducible over every finite field $\mathbb{F}_p$. The ideal $\mathfrak{M} = \langle x_1^2 - 2, x_2^2 - 3\rangle$
is a maximal ideal in $\mathbb{Q}[x_1, x_2]$. Since the ideal $J$ resulted from $\mathfrak{M}$ by a linear
change of coordinates, it follows that, for every prime $p$, the residue class ideal
$\overline{\mathfrak{M}}$ in $\mathbb{F}_p[x_1, x_2]$ is not maximal and not even primary.

   For $p = 2$ and $p = 3$, this is clearly true. Let us check some other primes.

$$
\begin{aligned}
p = 5: \quad & \overline{\mathfrak{M}} = \langle x_1 + 2x_2, x_2^2 + 2\rangle \ \cap\ \langle x_1 - 2x_2, x_2^2 + 2\rangle\\
p = 11: \quad & \overline{\mathfrak{M}} = \langle x_1^2 - 2, x_2 - 5\rangle \quad\ \cap\ \langle x_1^2 - 2, x_2 + 5\rangle\\
p = 31: \quad & \overline{\mathfrak{M}} = \langle x_1 - 8, x_2^2 - 3\rangle \ \ \cap\ \langle x_1 + 8, x_2^2 - 3\rangle\\
p = 71: \quad & \overline{\mathfrak{M}} = \langle x_1^2 - 2, x_2 - 28\rangle \cap \langle x_1^2 - 2, x_2 + 28\rangle\\
p = 101: \quad & \overline{\mathfrak{M}} = \langle x_1 + 13x_2, x_2^2 - 3\rangle \cap \langle x_1 - 13x_2, x_2^2 - 3\rangle\\
p = 32003: \quad & \overline{\mathfrak{M}} = \langle x_1^2 - 2, x_2 + 253\rangle \cap \langle x_1^2 - 2, x_2 - 253\rangle
\end{aligned}
$$

It is interesting to observe how different the various decompositions look. For
$p = 71$, the ideal $\overline{\mathfrak{M}}$ splits even further into four linear maximal ideals.


   The next example illustrates an important point we made in the introduction to
this section, namely that the coordinates of the zeros of a maximal ideal may be
related to each other in unexpected ways.


**Example 6.4.11** Let $K = \mathbb{Q}$, let $P = K[x_1, x_2]$, let $f_1 = x_1^3 - 2$, let $f_2 = x_2^2 + 3$,
and let $\mathfrak{M} = \langle f_1, f_2\rangle$. It is easy to check that $\mathfrak{M}$ is a maximal ideal of $P$. Let us use
the algorithm to solve the polynomial system given by $f_1 = 0$ and $f_2 = 0$.

(1) The system of generators $\{f_1, f_2\}$ of $\mathfrak{M}$ is already triangular.
(5) The splitting field of $f_1(x_1)$ is $L_1' = K[y_{11}, y_{12}]/\langle y_{11}^3 - 2, y_{12}^2 + y_{11}y_{12} + y_{11}^2\rangle$.
(6) In $L_1'[x_1]$, we factor $f_1(x_1) = (x_1 - \bar{y}_{11})(x_1 - \bar{y}_{12})(x_1 + \bar{y}_{11} + \bar{y}_{12})$. Thus we
   have $S_1 = \{(y_{11}), (y_{12}), (-y_{11} - y_{12})\}$.
(5) For $i = 1$ and $G = (y_{11})$, the splitting field of $f_2(x_2)$ is $L_1'$.
(6) Factoring $f_2(x_2)$ in $L_1'[x_2]$ yields $f_2(x_2) = (x_2 - \bar{y}_{11}\bar{y}_{12}^2 - 1)(x_2 + \bar{y}_{11}\bar{y}_{12}^2 + 1)$.

(5) For $i = 2$ and $G = (y_{12})$, and for $i = 2$ and $G = (-y_{11} - y_{12})$, the same factorization results.

(6) Thus we get the set $S_2 = \{(\bar{y}_{11}, \bar{y}_{11}\bar{y}_{12}^2 + 1), (\bar{y}_{11}, -\bar{y}_{11}\bar{y}_{12}^2 - 1), (\bar{y}_{12}, \bar{y}_{11}\bar{y}_{12}^2 + 1),$
$(\bar{y}_{12}, -\bar{y}_{11}\bar{y}_{12}^2 - 1), (-\bar{y}_{11} - \bar{y}_{12}, \bar{y}_{11}\bar{y}_{12}^2 + 1), (-\bar{y}_{11} - \bar{y}_{12}, -\bar{y}_{11}\bar{y}_{12}^2 - 1)\}$.

Altogether, the system has the six solutions in $(L_1')^2$ given by the set $S_2$. Notice that it was not necessary to enlarge the field $L_1'$ for the iterations involving $i = 2$.

Although the polynomial system appears to be very similar to the preceding one, the following example is significantly more involved.

**Example 6.4.12**  Let $K = \mathbb{Q}$, let $P = K[x_1, x_2]$, and let $\mathfrak{M} = \langle x_1^2 + x_1 - 1,$
$x_2^3 - x_1 - 1 \rangle$. Using the lexicographic term ordering with $x_2 > x_1$, we get that the reduced Gröbner basis of $\mathfrak{M}$ is $(x_2^6 - x_2^3 - 1, x_1 - x_2^3 + 1)$. From the fact that the polynomial $x_2^6 - x_2^3 - 1$ is irreducible in $K[x_2]$ we then deduce that $\mathfrak{M}$ is a maximal ideal of $P$. Even better, we can use Corollary 5.4.10 and obtain the same result. Now we use Algorithm 6.4.9 to solve $\mathfrak{M}$, i.e., to find a number field in which we can write down all zeros of $\mathfrak{M}$. We follow the main steps of the algorithm.

(1) The polynomials $f_1(x_1) = x_1^2 + x_1 - 1$ and $f_2(x_2) = x_2^3 - x_1 - 1$ form a set of generators $\{f_1, f_2\}$ of $\mathfrak{M}$ which is already triangular.

(5) We compute the splitting field of $f_1(x_1)$ and get $L_1' = K[y_1]/\langle y_1^2 + y_1 - 1 \rangle$. The polynomial $f_1(x_1)$ splits in $L_1'[x_1]$ in the form $f_1(x_1) = (x_1 - \bar{y}_1)(x_1 + \bar{y}_1 + 1)$.

(6) Let $S_1 = \{(y_1), (-y_1 - 1)\}$.

(4) For $i = 2$ and $G = (y_1)$, we let $\bar{f}_2(x_2) = x_2^3 - \bar{y}_1 - 1$ in $L_1'[x_2]$.

(5) Using Algorithm 6.4.8, we calculate the splitting field of $\bar{f}_2(x_2)$ and obtain $L_2' = L_1'[y_{21}, y_{22}]/\langle y_{21}^3 - y_1 - 1, y_{22}^2 + y_{21}y_{22} + y_{21}^2 - 1 \rangle$ and the factorization $\bar{f}_2(x_2) = (x_2 - \bar{y}_2)(x_2^2 + \bar{y}_{21}x_2 + \bar{y}_{21}^2) = (x_2 - \bar{y}_{21})(x_2 - \bar{y}_{22})(x_2 + \bar{y}_{21} + \bar{y}_{22})$. (Here we have $k_2 = 1$ and $\ell_2 = 2$ for the range of the new indeterminates $y_{2j}$.) We replace $L_1'$ by $L_2'$.

(6) Let $S_2 = \{(y_1, y_{21}), (y_1, y_{22}), (y_1, -y_{21} - y_{22})\}$.

(4) For $i = 2$ and $G = (-y_1 - 1)$, we let $\bar{f}_2(x_2) = x_2^3 + \bar{y}_1 - 1$ in $L_1'[x_2]$, where $L_1' = K[y_1, y_{21}, y_{22}]/\langle y_1^2 + y_1 - 1, y_{21}^3 - y_1 - 1, y_{22}^2 + y_{21}y_{22} + y_{21}^2 - 1 \rangle$.

(5) The polynomial $\bar{f}_2(x_2)$ splits in the ring $L_1'[x_2]$ into three linear factors: $\bar{f}_2(x_2) = (x_2 - \bar{y}_1\bar{y}_{21}\bar{y}_{22})(x_2 - \bar{y}_1\bar{y}_{21}\bar{y}_{22} - \bar{y}_1\bar{y}_{21}^2)(x_2 + \bar{y}_1\bar{y}_{21}^2)$. Thus we use this field as $L_2'$.

(6) Appending the new tuples yields $S_2 = \{(y_1, y_{21}), (y_1, y_{22}), (y_1, -y_{21} - y_{22}),$
$(-y_1 - 1, y_1y_{21}y_{22}), (-y_1 - 1, y_1y_{21}y_{22} + y_1y_{21}^2), (-y_1 - 1, y_1y_{21}^2)\}$.

(2) Finally, the algorithm returns the current set $S_2$ and the field $L_2'$ with its presentation $L_2' = K[y_1, y_{21}, y_{22}]/\langle y_1^2 + y_1 - 1, y_{21}^3 - y_1 - 1, y_{22}^2 + y_{21}y_{22} + y_{21}^2 - 1 \rangle$.

Thus the polynomial system has 6 solutions in $(L_2')^2$, where $\dim_K(L_2') = 12$.

*Well, if I called the wrong number,*
*why did you answer the phone?*
(James Thurber)

Now we present the one and only seven star example in the world. It goes to the limit of what is possible to provide you with the utmost luxury: exact solutions of a polynomial system defined over the rational numbers.

**Example 6.4.13 (The Burj al Arab Example)**
Over the field $K = \mathbb{Q}$, consider the system of polynomial equations associated to

$$f_1 = x_2^4 + \tfrac{3}{5}x_1x_2^2 - x_2^2 - \tfrac{9}{5}x_1 - 6$$

$$f_2 = x_1x_2^3 - x_1x_2^2 - 3x_1x_2 + 3x_1$$

$$f_3 = x_1^2x_2^2 + 5x_1x_2^2 - 3x_1^2 - 15x_1$$

$$f_4 = x_1^5 + \tfrac{7803}{25}x_1x_2^2 - \tfrac{2}{5}x_2^2 - \tfrac{23434}{25}x_1 - \tfrac{4}{5}$$

Let $P = K[x_1, x_2]$, and let $I = \langle f_1, f_2, f_3, f_4 \rangle$. Our goal is to find the zeros of $I$ in a suitable number field. Since $I$ is not a maximal ideal, we have to compute its maximal components first.

For this purpose we use the method of Algorithm 5.1.15. Thus we calculate $\mu_{x_1+1}(z) = z^7 + 5z^6 - z^3 - 7z^2 - 10z = (z+5)z(z^5 - z - 2)$ in $K[z]$. The ideals

$$\mathfrak{M}_1 = I + \langle x_1 + 5 \rangle = \langle x_1 + 5, x_2 - 1 \rangle$$

$$\mathfrak{M}_2 = I + \langle x_1 \rangle = \langle x_1, x_2^2 + 2 \rangle$$

$$\mathfrak{M}_3 = I + \langle x_1^5 - x_1 - 2 \rangle = \langle x_1^5 - x_1 - 2, x_2^2 - 3 \rangle$$

turn out to be maximal, and we have $I = \mathfrak{M}_1 \cap \mathfrak{M}_2 \cap \mathfrak{M}_3$. The ideal $\mathfrak{M}_1$ corresponds to the $K$-rational solution $(-5, 1)$ of the system. The ideal $\mathfrak{M}_2$ can be split easily in the field $L' = K[y]/\langle y^2 + 2 \rangle$. It corresponds to the two solutions $(0, \bar{y})$ and $(0, -\bar{y})$ in $(L')^2$.

Consequently, it remains to find the zeros of $\mathfrak{M}_3$. Let us verify that this is indeed a maximal ideal. We calculate the minimal polynomial of $f = \bar{x}_1 + \bar{x}_2$ in $P/\mathfrak{M}_3$ and get $\mu_f(z) = z^{10} - 15z^8 + 88z^6 + 4z^5 - 300z^4 + 120z^3 + 496z^2 + 176z - 188$. We check that $\mu_f(z)$ is irreducible and conclude from Corollary 5.4.10 that $\mathfrak{M}_3$ is a maximal ideal. Moreover, since $\dim_K(P/\mathfrak{M}_3) = 10$, we see that $f$ is a primitive element of the field $L = P/\mathfrak{M}_3$. At this point we could be tempted to proceed by finding the zeros of $\mu_f(z)$. However, this is not an easy task. For instance, the dimension of the splitting algebra $S_f$ is $\dim_K(S_f) = 10! = 3,628,800$, a prohibitive number.

Instead, we can use the fact that $\mathfrak{M}_3 = \langle x_1^5 - x_1 - 2, x_2^2 - 3 \rangle$ is given by a triangular set of generators and apply Algorithm 6.4.9. The main problem that we encounter during the course of this algorithm is to compute the splitting field of the polynomial $f_1(x_1) = x_1^5 - x_1 - 2$. For this purpose we use the splitting algebra of $f_1(x_1)$ (see Definition 6.4.5). Let $y_1, \ldots, y_5$ be new indeterminates, let $s_1, \ldots, s_5$ be the elementary symmetric polynomials in $y_1, \ldots, y_5$, and let $P = K[y_1, \ldots, y_5]$. Then the splitting algebra of $f_1(x_1)$ is given by $S_{f_1} = P/\langle s_1, s_2, s_3, s_4 + 1, s_5 - 2 \rangle$ and its dimension is $\dim_K(S_{f_1}) = 5! = 120$.

In view of Proposition 6.4.6, we want to show that $S_{f_1}$ is a field. To this end, we choose the linear form $\ell = y_1 + 2y_2 + 3y_3 + 4y_4 + 5y_5$ in $P$ and compute the minimal polynomial $\mu_{\bar{\ell}}(z)$ of its residue class $\bar{\ell}$ in $S_{f_1}$. We get the world's first seven star polynomial

$$
\begin{aligned}
&z^{120} \\
&+900\,z^{116} \\
&+508450\,z^{112} \\
&+12375000\,z^{110} \\
&+233445700\,z^{108} \\
&+24978500000\,z^{106} \\
&+86356219375\,z^{104} \\
&+24591849375000\,z^{102} \\
&-643754458910680\,z^{100} \\
&+13397352506250000\,z^{98} \\
&-193920628655076100\,z^{96} \\
&+5239487723543750000\,z^{94} \\
&+15197805079748967200\,z^{92} \\
&+160191401739550880000000\,z^{90} \\
&+163905271872491333822575\,z^{88} \\
&+18571115131411094249750000\,z^{86} \\
&+89120445389616729179232500\,z^{84} \\
&+10283912347310917452507500000\,z^{82} \\
&+198669965254843100047656058210\,z^{80} \\
&+407016930923397347921319687500 0\,z^{78} \\
&+13221410515800598227911779540290 0\,z^{76} \\
&+1087516052780449382583495637500000\,z^{74} \\
&+6050884960239353039220093814444582 5\,z^{72} \\
&+347660911413224431990937140564875000\,z^{70} \\
&+165882002991963937429717712065707512 00\,z^{68} \\
&+339815786915571165896688517485792250000\,z^{66} \\
&+17912716377537485011344833574489313000 00\,z^{64} \\
&+1033998920928855617282714375810270750000 00\,z^{62} \\
&+42781400263648014224518386947751638108944 0\,z^{60} \\
&+791500704500872447665068096938863965000000 00\,z^{58} \\
&+8106330824665600447801185509758795013200640 0\,z^{56} \\
&-58166475752540506274387604979739567990000000 00\,z^{54} \\
&+25602517447858531609031664885636569764001139200\,z^{52} \\
&-336337098220208207279010809728251098368912000000\,z^{50} \\
&+187770153248498834970727485376767128636388187520 0\,z^{48} \\
&+73307905946928174406128646864582738317564096000000\,z^{46} \\
&-2698210899109936993573223447490346830465277440000000\,z^{44} \\
&+38479638527556330568997409673280988610752236800000 00\,z^{42} \\
&+29901468960195230389988192738702179827930958042308608 0\,z^{40} \\
&+425589988881037317207821575923201869421791616000000000\,z^{38} \\
&-347214071244366277347046508843786041565632974054281216 00\,z^{36} \\
&-285013531165903449792106996342027484217520412416000000 00\,z^{34} \\
&-50490001735908587203119862921024550428531818527437946880 0\,z^{32} \\
&+20153310691018337975446390185898297311743029640726528000000\,z^{30} \\
&+21088321914096711690237205513699115109847237886097503027200\,z^{28} \\
&+91973334995806935195413729241109562880392423979305369600000 0\,z^{26} \\
&+20811017573464598985212561659428623245292245157485824819200 00\,z^{24} \\
&+2492091129111886515045158969432948941690459083580665856000000 0\,z^{22} \\
&+1078963492292720272970178457320826033486205060293360539297382 4\,z^{20} \\
&+1071798900103034067638174981362775976075695132766720000000000000\,z^{18} \\
&-44382053095550524848343631937377293229538844407267104522240000000\,z^{16} \\
&+6178375123253158526487182611516019681755559868893741056000000000\,z^{14} \\
&-196511735034978223557675583812991755356031074199102668800000000000\,z^{12} \\
&+17057058181130156999312934476928723862270895272267776000000000000 0\,z^{10} \\
&+24365639973684917858679310359271336781135285690368000000000000000\,z^{8} \\
&+40615589227675119914100324987801198834835078208000000000000000000\,z^{6} \\
&+21508468021796121272673650799457544328788048878080000000000000000000\,z^{4} \\
&+12368003557410966199224287239940288183979630080000000000000000000000\,z^{2} \\
&+53899832988697131573013337804587950520590400000000000000000000000000
\end{aligned}
$$

which is irreducible. Therefore Corollary 5.4.10 shows that $S_{f_1}$ is a field, and hence the splitting field of $f_1(x_1)$ by Proposition 6.4.6. More precisely, the zeros of $f_1(x_1)$ are the residue classes $\bar{y}_1, \ldots, \bar{y}_5$ in $S_{f_1} = K[y_1, \ldots, y_5]/\langle s_1, s_2, s_3, s_4 + 1, s_5 - 2\rangle$.

Now let us return to the task of finding the zeros of $\mathfrak{M}_3$. We may check that the polynomial $x_2^2 - 3$ is irreducible in $S_{f_1}[x_2]$. Thus Algorithm 6.4.9 yields that the zeros of $\mathfrak{M}_3$ are contained in $(L_2')^2$, where $L_2'$ is the field

$$L_2' = K[y_1, \ldots, y_5, z]/\langle s_1, s_2, s_3, s_4 + 1, s_5 - 2, z^2 - 3\rangle$$

There are 10 solutions $\{(\bar{y}_i, \bar{z})\ i = 1, \ldots, 5\} \cup \{(\bar{y}_i, -\bar{z}) \mid i = 1, \ldots, 5\}$. Notice that the field $L_2'$ satisfies $\dim_K(L_2') = 240$.

Looking at the examples above, we see that they contain polynomial systems which correspond to *small ideals*. So, what about *big ideals*? Clearly, our symbolic approaches are going to fail when the dimensions of the necessary splitting fields become too large. In other words, we will not be able to write down the *exact* zeros.

The obvious way out is to try to compute *approximate* zeros. Given a univariate polynomial over $\mathbb{Q}$, there are, of course, classical numerical methods to determine its real and complex zeros to arbitrary given precision. However, many of our algorithms require us to perform further computations with these approximate zeros. Hence they lead to questions such as understanding error propagation and the numerical stability of algorithms. Rather than delving into this new world of research, let us end our discussions here with an example which illustrates a couple of easy initial suggestions:

(1) Split the problem into smaller subproblems as well as you can.
(2) Proceed symbolically as long as it is somehow viable.

**Example 6.4.14**  Over the field $K = \mathbb{Q}$, consider the polynomial system associated to

$$f_1 = 3x_1^2 - x_2^2 + 2x_2x_3 - x_3^2 - 8x_1 - 8x_2 + 5x_3 - 5$$
$$f_2 = x_1^3 - 6x_1^2 - 6x_1x_2 - 4x_2^2 + x_3^2 + 3x_1 + 7x_2 - 7x_3 + 15$$
$$f_3 = x_3^3 + 4x_1^2 + 2x_1x_2 - 3x_3^2 - 13x_1 - 5x_2 + 6x_3 + 5$$

Let $P = K[x_1, x_2, x_3]$, and let $I = \langle f_1, f_2, f_3\rangle$. Our goal is to find the zeros of $I$.

The first step is to find the maximal components of $I$. We compute the minimal polynomial of $\bar{x}_1$ in $P/I$ and get $\mu_{\bar{x}_1}(z) = f(z)(z - 2)^3$, where

$$f(z) = z^{15} - 102z^{14} + 4254z^{13} - 91420z^{12} + 1108030z^{11} - 7617460z^{10}$$
$$+ 26368737z^9 - 16709678z^8 - 1634014027^7 + 408143116z^6$$
$$- 102334132z^5 - 561462068z^4 + 119765981z^3 + 1225188788z^2$$
$$- 1326547350z + 524273780$$

The ideal $\mathfrak{Q}_1 = I + \langle(x_1 - 2)^3\rangle$ is a primary ideal. We calculate its radical and get $\mathfrak{M}_1 = \langle x_1 - 2, x_2 + 1, x_3 - 1\rangle$. This maximal ideal corresponds to the $K$-rational zero $(2, -1, 1)$ of $I$.

Next we form the ideal $\mathfrak{M}_2 = I + \langle f(x_1) \rangle$. Let us check that $\mathfrak{M}_2$ is a maximal ideal. For this purpose we compute the reduced Lex-Gröbner basis of $\mathfrak{M}_2$ and obtain $(x_1 - g_1(x_3), x_2 - g_2(x_3), g_3(x_3))$, where

$$
\begin{aligned}
g_3(x_3) = x_3^{15} &- 15x_3^{14} + 453x_3^{13} - 1723x_3^{12} + 1653x_3^{11} + 39878x_3^{10} - 143949x_3^{9} \\
&+ 44358x_3^{8} + 1493331x_3^{7} - 3545819x_3^{6} + 1822878x_3^{5} + 12465182x_3^{4} \\
&- 23275915x_3^{3} + 35305982x_3^{2} - 40170300x_3 + 15978680
\end{aligned}
$$

and where $g_1(x_3)$ and $g_2(x_3)$ are polynomials of degree 14 whose coefficients are fractions involving 40-digit numerators and denominators. We verify that $g_3(x_3)$ is irreducible and conclude that $\mathfrak{M}_2$ is indeed a maximal ideal.

Now what? There seems to be no way to find the zeros of $\mathfrak{M}_3$ symbolically. The reason is that to split an irreducible polynomial with rational coefficients and of high degree is apparently a prohibitively difficult computational task. One way to continue would be to find good approximations to the zeros of $g_3(x_3)$ and to plug the resulting floating point numbers into the polynomials $g_1(x_3)$ and $g_2(x_3)$. This brings us to the topics of floating point calculations and error estimates which are well outside the scope of this volume.

> *A man only becomes wise when he begins to calculate*
> *the approximate depth of his ignorance.*
> (Gian Carlo Menotti)

Thus we have reached the end of this book. Or should we say that we have *approximately* reached the end of the book?

# Notation

*Chapter 1*

| | |
|---|---|
| $\mathrm{End}_K(V)$ | set of endomorphisms of a vector space |
| $\mathrm{BigKer}(\varphi)$ | big kernel of an endomorphism |
| $\mathrm{SmIm}(\varphi)$ | small image of an endomorphism |
| $\mu_\varphi(z)$ | minimal polynomial of an endomorphism |
| $\mathrm{Eig}(\varphi, p(z))$ | eigenspace associated to the eigenfactor $p(z)$ |
| $\mathrm{Eig}(\varphi, \lambda)$ | eigenspace associated to the eigenvalue $\lambda$ |
| $\mathrm{Gen}(\varphi, p(z))$ | generalized eigenspace associated to the eigenfactor $p(z)$ |
| $\mathrm{Gen}(\varphi, \lambda)$ | generalized eigenspace associated to the eigenvalue $\lambda$ |
| $I_d$ | identity matrix of size $d$ |
| $M_B(\varphi)$ | matrix representing $\varphi$ in the basis $B$ |
| $\chi_\varphi(z)$ | characteristic polynomial of an endomorphism |
| $\mathrm{adj}(A)$ | adjugate matrix of $A$ |
| $\mathrm{mmult}(\varphi, p(z))$ | minimal multiplicity of the eigenfactor $p(z)$ |
| $\mathrm{amult}(\varphi, p(z))$ | algebraic multiplicity of the eigenfactor $p(z)$ |
| $\mathrm{gmult}(\varphi, p(z))$ | geometric multiplicity of the eigenfactor $p(z)$ |
| $\mathrm{Ann}_{K[z]}(v)$ | annihilator of a vector $v$ in $K[z]$ |
| $\mathfrak{C}(v, W)$ | conductor ideal of a vector $v$ into a subspace $W$ |
| $\mathfrak{c}_{v,W}(z)$ | conductor of a vector $v$ into a subspace $W$ |
| $C_{p(z)}$ | companion matrix of a monic polynomial $p(z)$ |

*Chapter 2*

| | |
|---|---|
| $\mathrm{Rel}_P(\Phi)$ | ideal of algebraic relations of a tuple of endomorphisms |
| $\mathscr{F}_U$ | restriction of a family to a subspace |
| $\mathrm{Ann}_{\mathscr{F}}(U)$ | annihilator of a subspace in a family |
| $\mathrm{Rad}_{\mathscr{F}}(0)$ | nilradical of a family |
| $\mathrm{Ker}(I)$ | kernel of an ideal |
| $\mathrm{BigKer}(I)$ | big kernel of an ideal |
| $\mathrm{Col}(A_1, \ldots, A_s)$ | block column matrix of $(A_1, \ldots, A_s)$ |
| $p_{\mathfrak{m},\varphi}(z)$ | $\mathfrak{m}$-eigenfactor of the endomorphism $\varphi$ |
| $\mathrm{card}(K)$ | cardinality of a field $K$ |

© Springer International Publishing Switzerland 2016
M. Kreuzer and L. Robbiano, *Computational Linear and Commutative Algebra*,
DOI 10.1007/978-3-319-43601-2

| | |
|---|---|
| $\mathrm{Spl}(\mathscr{F})$ | splitting locus of a family |
| $\mathrm{NSpl}(\mathscr{F})$ | non-splitting locus of a family |
| $\Theta_{y,\Psi}$ | generically extended endomorphism for $\Psi = (\psi_1, \ldots, \psi_\delta)$ using a tuple of indeterminates $y = (y_1, \ldots, y_\delta)$ |

*Chapter 3*

| | |
|---|---|
| $\mathrm{gr}_{\mathfrak{m}}(\mathscr{F})$ | associated graded ring of a local family |
| $V^*$ | dual vector space of $V$ |
| $\varphi^{\vee}$ | dual endomorphism of $\varphi$ |
| $\mathscr{F}^{\vee}$ | dual family of endomorphisms |
| $\ell^*$ | dual linear form of $\ell$ |

*Chapter 4*

| | |
|---|---|
| $\vartheta_f$ | multiplication endomorphism by $f$ |
| $\iota$ | isomorphism between a ring and its multiplication family |
| $\mathscr{O}_\sigma(I)$ | order ideal of $I$ w.r.t. the term ordering $\sigma$ |
| $\mathrm{BigAnn}_M(\mathfrak{a})$ | big annihilator of $\mathfrak{a}$ in $M$ |
| $\tilde{\mathfrak{a}}$ | ideal in the multiplication family corresponding to $\mathfrak{a}$ |
| $\mathrm{Soc}(R)$ | socle of a local ring |
| $D_f(K)$ | set of non-zeros of a polynomial $f$ |
| $\mathrm{nix}(f)$ | nilpotency index of an element $f$ |
| $\mathrm{nix}(\mathfrak{m})$ | nilpotency index of the maximal ideal $\mathfrak{m}$ |
| $p_f(z)$ | unique eigenfactor of $\vartheta_f$ in the local case |
| $\omega_R$ | canonical module of $R$ |
| $\mathrm{Eval}(L, B)$ | evaluation matrix of $L$ at $B$ |
| $F_i R$ | $i$-th component of the degree filtration of $R$ |
| $\mathrm{HF}_R^a$ | affine Hilbert function of $R$ |
| $\Delta\,\mathrm{HF}_R^a$ | Hilbert difference function of $R$ |
| $\mathrm{ri}(R)$ | regularity index of $R$ |
| $\#S$ | number of elements of a finite set $S$ |
| $\mathrm{ord}(f)$ | order of an element w.r.t. the degree filtration |
| $\mathrm{sepdeg}(\mathfrak{m})$ | separator degree of a maximal ideal |
| $G_i \omega_R$ | $i$-th component of the degree filtration of $\omega_R$ |
| $\mathrm{HF}_{\omega_R}^a$ | affine Hilbert function of the canonical module |

*Chapter 5*

| | |
|---|---|
| $\phi_q$ | $q$-Frobenius endomorphism of a ring |
| $\mathrm{Frob}_q(R)$ | $q$-Frobenius space of a ring |
| $R^{\mathrm{sep}}$ | separable subalgebra of $R$ |
| $a^{\mathrm{sep}}$ | separable part of an element |
| $a^{\mathrm{nil}}$ | nilpotent part of an element |

*Chapter 6*

| | |
|---|---|
| $\mathrm{Gal}(L/K)$ | Galois group of a field extension |
| $\mathscr{Z}_K(I)$ | set of $K$-rational zeros of an ideal |
| $\mathbb{X}$ | finite set of $K$-rational points |
| $\mathscr{I}(\mathbb{X})$ | vanishing ideal of $\mathbb{X}$ |

| | |
|---|---|
| $\mathrm{ev}_p$ | evaluation form associated to a point $p$ |
| $\mathrm{Prim}_K(L)$ | set of primitive elements of an extension field |
| $\mathrm{Irr}_K(d)$ | set of monic irreducible polynomials of degree $d$ |
| $S_f$ | splitting field of the polynomial $f$ |

# References

1. Auzinger, W., Stetter, H.J.: An elimination algorithm for the computation of all zeros of a system of multivariate polynomial equations. In: Agarwal, R.G., Chow, Y.M., Wilson, S.J. (eds.) Int. Conf. on Numerical Mathematics (Singapore 1988), vol. 86, pp. 11–30. Birkhäuser, Basel (1988)
2. The CoCoA Team: CoCoA: a system for doing computations in commutative algebra. Available at http://cocoa.dima.unige.it
3. Cox, D.A.: Solving equations via algebras. In: Dickenstein, A., Emiris, I. (eds.) Solving Polynomial Equations. Algorithms and Comp. in Math., vol. 14, pp. 63–124. Springer, Berlin (2005)
4. Dickenstein, A., Emiris, I.Z. (eds.): Solving Polynomial Equations. Alg. and Comp. in Math., vol. 14. Springer, Berlin (2005)
5. Ekedahl, T., Laksov, D.: Splitting algebras, symmetric functions, and Galois theory. J. Algebra Appl. **4**, 59–75 (2005)
6. Gao, S., Wan, D., Wang, M.: Primary decomposition of zero-dimensional ideals over finite fields. Math. Comput. **78**, 509–521 (2008)
7. Geramita, A.V., Kreuzer, M., Robbiano, L.: Cayley-Bacharach schemes and their canonical modules. Trans. Am. Math. Soc. **339**, 163–189 (1993)
8. Gerstenhaber, M.: On dominance and varieties of commuting matrices. Ann. Math. **73**, 324–348 (1961)
9. Hoffman, K., Kunze, R.: Linear Algebra, 2nd edn. Prentice-Hall, Englewood Cliffs (1971)
10. Jacobson, N.: Basic Algebra I, 2nd edn. Dover Publications, Mineola (2009)
11. Jacobson, N.: Basic Algebra II, 2nd edn. Dover Publications, Mineola (2009)
12. Jacobson, N.: Schur's theorems on commutative matrices. Bull. Am. Math. Soc. **50**, 431–436 (1944)
13. Kemper, G.: The calculation of radical ideals in positive characteristic. J. Symb. Comput. **34**, 229–238 (2002)
14. Kehrein, A., Kreuzer, M., Robbiano, L.: An algebraist's view on border bases. In: Dickenstein, A., Emiris, I. (eds.) Solving Polynomial Equations. Alg. and Comp. in Math., vol. 14, pp. 169–202. Springer, Heidelberg (2005)
15. Kreuzer, M., Robbiano, L.: Computational Commutative Algebra 1. Springer, Heidelberg (2000)
16. Kreuzer, M., Robbiano, L.: Computational Commutative Algebra 2. Springer, Heidelberg (2005)
17. Lasker, E.: Zur Theorie der Moduln und Ideale. Math. Ann. **60**, 19–116 (1905)
18. Lazard, D.: Résolution des systèmes d'equations algébriques. Theor. Comput. Sci. **15**, 77–110 (1981)

© Springer International Publishing Switzerland 2016       315
M. Kreuzer and L. Robbiano, *Computational Linear and Commutative Algebra*,
DOI 10.1007/978-3-319-43601-2

19. Lenstra, H.W., Silvenberg, A.: Algorithms for commutative algebras over the rational numbers. Preprint (2015). Available at http://arxiv.org/1509.08843v1
20. Matiyasevich, Y.: Hilbert's Tenth Problem. MIT Press, Cambridge (1993)
21. Milne, J.S.: Fields and Galois theory. On-line lecture notes, Ver. 4.50 (2014). Available at http://www.jmilne.org/math/CourseNotes/FT.pdf
22. Möller, H.-M., Tenberg, R.: Multivariate polynomial system solving using intersections of eigenspaces. J. Symb. Comput. **32**, 513–531 (2001)
23. Monico, C.: Computing the primary decomposition of zero-dimensional ideals. J. Symb. Comput. **34**, 451–459 (2002)
24. Mourrain, B.: A new criterion for normal form algorithms. In: Fossorier, M., Imai, H., Lin, S., Poli, A. (eds.) Proc. Conf. AAECC-13, Honolulu, 1999. LNCS, vol. 1719, pp. 440–443. Springer, Heidelberg (1999)
25. Robbiano, L., Abbott, J. (eds.): Approximate Commutative Algebra. Springer, Vienna (2009)
26. Schur, J.: Zur Theorie der vertauschbaren Matrizen. J. Reine Angew. Math. **130**, 66–76 (1905)
27. Sommese, A., Wampler, C.W. II: The Numerical Solution of Systems of Polynomial Equations Arising in Engineering and Science. World Scientific Publ., Singapore (2005)
28. Stetter, H.: Numerical Polynomial Algebra. SIAM, Philadelphia (2004)
29. Sun, Y., Wang, D.: An efficient algorithm for factoring polynomials over algebraic extension fields. Preprint (2010). Available at http://arxiv.org/abs/0907.2300
30. Zariski, O., Samuel, P.: Commutative Algebra, vol. 1. Van Nostrand, Princeton (1958)
31. Zariski, O., Samuel, P.: Commutative Algebra, vol. 2. Van Nostrand, Princeton (1960)

# Index

## A
Adjoint matrix, 14
Adjugate matrix, 14
Affine
  algebra, 131
  algebra isomorphisms, 272
  Hilbert function, 177, 180
Algebraic multiplicity, 20
Algorithm
  heuristic, 195
  Las Vegas, 195
  Monte Carlo, 195
Annihilator
  big, 137
  of a joint eigenspace, 72
  of a subspace, 52
Associated graded ring, 115
Associated point, 147

## B
Base field extension, 214
Bidual family, 120
Big annihilator, 137
Big kernel, 4
  of an ideal, 60
Big powers, 281
Block column matrix, 63
Buchberger-Möller algorithm for matrices, 50
Burj al Arab
  example, 306
  polynomial, 307

## C
Canonical module, 162
  degree filtration, 180
  explicit membership, 168
  Hilbert function, 180

  minimal generators, 164, 166
  of a Gorenstein ring, 174
  of a residue class ring, 163
Cardinality, 79
Castelnuovo function, 177
Cayley-Bacharach property, 179
  characterization, 182
  checking, 180, 183
Cayley-Hamilton theorem, 15
Characteristic polynomial, 13
  prime factors, 16
Chinese remainder theorem, 56
Clone
  of a field, 279
  of an ideal, 279
Coherent family
  of eigenfactors, 70
  of eigenvalues, 70
Comaximal ideals, 56
  kernels and big kernels, 61
Commendable endomorphism, xvi, 35
  characterization, 37, 39, 104, 105, 112, 122
  direct sum, 40
  existence, 107
  in a field, 114
  in a multiplication family, 146
  locally, 105
Commendable family, 110
  checking, 126
  dimension, 126
  local nature, 112
Commuting family, 49
  reduced, 73, 80
Companion matrix, 38
Component
  maximal, 141
  primary, 12, 73, 141, 142

© Springer International Publishing Switzerland 2016
M. Kreuzer and L. Robbiano, *Computational Linear and Commutative Algebra*,
DOI 10.1007/978-3-319-43601-2